Middle Proterozoic to Cambria
Central North America

Edited by

Richard W. Ojakangas
Department of Geology
University of Minnesota
Duluth, Minnesota 55812

Albert B. Dickas
Department of Geosciences
University of Wisconsin
Superior, Wisconsin 54880

and

John C. Green
Department of Geology
University of Minnesota
Duluth, Minnesota 55812

SPECIAL PAPER
312
1997

Published by The Geological Society of America, Inc.
3300 Penrose Place, P.O. Box 9140, Boulder, Colorado 80301

Printed in U.S.A.

GSA Books Science Editor Abhijit Basu

Library of Congress Cataloging-in-Publication Data
Middle Proterozoic to Cambrian rifting, central North America / edited
    by Richard W. Ojakangas, Albert B. Dickas, and John C. Green
        p.   cm. -- (Special paper ; 312)
    Includes bibliographical references (p. -) and index.
    ISBN 0-8137-2312-4
    1. Rifts (Geology)--Middle West. 2. Geology, Stratigraphic-
-Proterozoic. 3. Geology, Stratigraphic--Cambrian. 4. Geology,
Structural--Middle West. I. Ojakangas, Richard W. II. Dickas,
Albert B. III. Green, John C., 1932- . IV. Series: Special
papers (Geological Society of America) ; 312.
QE606.5.U6M53   1997
551.8'72'0977--dc21   96-48160
                                                         CIP

Cover: View to the northeast of thin-bedded pahoehoe basalt flow units, North Shore Volcanic Group, dipping gently into Lake Superior. Sugarloaf Point, Cook County, Minnesota. Prominent hill left of center on horizon is Carlton Peak, made of huge anorthosite blocks rafted up into the Midcontinent Rift plateau lavas in a diabase intrusion. Photograph by J. C. Green.

10 9 8 7 6 5 4 3 2 1

Contents

Preface

This volume of 19 papers—13 on the Middle Proterozoic Midcontinent Rift and 5 on other rift basins—is an outgrowth of the 10th International Basement Tectonics Conference held at the University of Minnesota–Duluth, in August 1992. It was decided before the conference that papers dealing with rifting in central North America, with an emphasis on the Midcontinent Rift System, should be published together in an accessible medium.

Abstracts and extended abstracts of most of these papers were included in a volume entitled, *Basement Tectonics 10*, published in 1995 by Kluwer Academic Publishers, Dordrecht, Boston, and London (442 p.). That volume also included papers and abstracts about basement control on overlying structures and shear zones.

The following people dutifully served as reviewers of the manuscripts, and we appreciate their efforts: Raymond Anderson, Pieter Berendsen, William Cambray, William Cannon, Ken Card, Val Chandler, Paul Daniels, R. E. Denison, Albert B. Dickas, Douglas Elmore, John C. Green, William Gregg, Dennis Kolata, Steve Marshak, John McGinnis, James Mosler, G. B. Morey, Richard W. Ojakangas, Jill Pasteris, James Robertson, Bernard Saini-Eidukat, Ron Sage, Douglas Shrake, Richard Tollo, George Viele, Laurel Woodruff, The Kentucky Geological Survey, and some anonymous persons.

Sponsors of the 1992 conference included the Petroleum Research Fund of the American Chemical Society; U.S. Geological Survey; ARCO Oil and Gas Company; Phelps Dodge Mining Company; Department of Geology, University of Minnesota–Duluth; the Graduate School, University of Minnesota; and Cooperative Extension, University of Wisconsin–Superior.

We also express our sincere appreciation to former GSA Books Science Editor Richard Hoppin for his continued encouragement since the International Basement Tectonics Conference was held in 1992.

Richard W. Ojakangas
Albert B. Dickas
John C. Green

Geological Society of America
Special Paper 312
1997

Introduction: Middle Proterozoic to Cambrian rifting, central North America

Richard W. Ojakangas
Department of Geology, University of Minnesota, Duluth, Minnesota 55812
Albert B. Dickas
Department of Geosciences, University of Wisconsin, Superior, Wisconsin 54880
John C. Green
Department of Geology, University of Minnesota, Duluth, Minnesota 55812

INTRODUCTION

This volume includes, in addition to this introductory paper, 18 papers on rifting. The Midcontinent Rift System (MRS) is one of the longest Precambrian intracontinental extensional features on earth, and 13 of the 18 papers deal with the MRS. Eight of these are on geophysics, tectonics, or structure (Hinze et al.; Dickas and Mudrey; Allen et al.; Miller and Chandler; Witthuhn; Craddock et al.; Carlson; Berendsen), three are on mineralization (Bornhorst; Hauck et al.; McCuaig and Kissen), and two are on sedimentary rocks (Suszek; Anderson).

The other five papers relate to other rifts in the central part of North America. The paper by Richard et al. deals with pre–Mount Simon basins of Ohio. Two papers by Stark are on the English Graben and the newly proposed East Continent Rift Complex. Kolata and Nelson's contribution is on the role of the Reelfoot Rift/Rough Creek Graben in the evolution of the Illinois Basin. The final paper by Hogan and Gilbert deals with A-type sheet granites in the Oklahoma Aulacogen of Cambrian age.

GENERAL GEOLOGY AND GEOPHYSICS

The study of Precambrian rifting in North America effectively began with the discovery by Woollard (1943) in northern Kansas of a strong positive gravity anomaly associated with a basement source. This anomaly was extended northeast into the Lake Superior region as the Midcontinent Gravity High (Coons et al., 1967), and then southeast through the Lower Peninsula of Michigan as the Mid-Michigan High (Oray et al., 1973). With the recognition that this feature also displayed a sharp linear magnetic signature (King and Zeitz, 1971), the interpretation of this geophysical trend as the Middle Proterozoic Midcontinent Rift System was generally accepted. Extensive field work has shown that the only exposures of this intracontinental rift are in the Lake Superior basin, where the Keweenawan Supergroup of dominantly mafic igneous and derivative sedimentary rocks crops out.

The geology, tectonics, and geophysics of the latter region as related to the rift interpretation were the subject in 1982 of The Geological Society of America Memoir 156 entitled *Geology and Tectonics of the Lake Superior Basin* (Wold and Hinze, 1982). This landmark volume of 21 papers served as a foundation for further work, including papers, for example, by Green (1983), Van Schmus and Hinze (1985), Dickas (1986), and Sims et al. (1993).

In 1983, following the public announcement that the MRS had been classified as a Precambrian hydrocarbon exploration frontier, a new phase of analysis of this structure began. During the remainder of the decade, approximately 4,000 km of contract and speculative reflection seismology were collected by the hydrocarbon industry, both on land and offshore in the Lake Superior region. Several thousand kilometers of additional profiles were collected in eastern Minnesota, central Iowa, and northeastern Kansas. On occasion, these surveys included the collection of gravity and magnetic information. This flurry of seismic reflection activity resulted in the drilling of four record-depth boreholes, one each in Kansas in 1984/1985 (Berendsen et al., 1988), Iowa in 1987 (Anderson, 1990, this volume), the Upper Peninsula of Michigan in 1987, and northern Wisconsin in 1992 (Dickas and Mudrey, 1996).

Prompted by this activity, the Great Lakes International Multidisciplinary Program on Crustal Evolution (GLIMPCE), a public consortium of American and Canadian scientists, was formed in 1985. In addition to the recording of a wide range of aeromagnetic, gravity, and high-resolution reflection/refraction surveys, 1,350 km of multichannel seismic reflection profiles

Ojakangas, R. W., Dickas, A. B., and Green, J. C., 1997, Introduction: Middle Proterozoic to Cambrian rifting, central North America, *in* Ojakangas, R. W., Dickas, A. B., and Green, J. C., eds., Middle Proterozoic to Cambrian Rifting, Central North America: Boulder, Colorado, Geological Society of America Special Paper 312.

were collected by GLIMPCE in Lake Huron, Lake Michigan, and Lake Superior. The GLIMPCE profiles were recorded to a depth of 20-s two-way travel time, imaging both the Moho and the upper mantle.

Through the combined resources of the GLIMPCE consortium and those industrial organizations exploring the hydrocarbon potential of the midwestern Middle Proterozoic rift terrane, an unprecedented modern period of evaluation of the southwestern arm of the MRS was inaugurated. Since the late 1980s, numerous papers and field guides have been published on the results of this second phase of evaluations. A few of the more informative include Behrendt et al. (1988), Berendsen et al. (1988), Cannon et al. (1989), Chandler and others (1989), Dickas (1989), Anderson (1990), Davis and Paces (1990), and Hinze et al. (1992). Additional papers on the MRS are contained within the following dedicated volumes: *Eos* (Transactions, American Geophysical Union), v. 70, September, 1989; *Tectonophysics*, v. 213, October 1992; *Canadian Journal of Earth Sciences*, v. 31, April 1994. The aforementioned eight papers of this volume that pertain to the geophysics, tectonics, and structure of the MRS are but the latest in this long series of informative papers.

In 1982 when *The Geological Society of America Memoir 156* was published, the MRS was being defined as a symmetrical feature with a central horst and flanking half grabens. Newer data show a partially asymmetrical and segmented rift (e.g., Dickas and Mudrey, this volume).

MIDCONTINENT RIFT SYSTEM MAGMATISM— A BRIEF REVIEW

Igneous rocks constitute one of the most important lines of evidence for the origin, nature, and evolution of the Midcontinent Rift System (MRS). In the creation of one of the world's largest igneous provinces, MRS/Keweenawan magmatism has been estimated to have produced more than 1×10^6 km^3 of mainly basaltic rocks and their intrusive equivalents (Hutchinson et al., 1990). These mafic rocks produce the major gravity and magnetic signatures of the MRS.

Accumulating radiometric ages for MRS igneous rocks show that they span a 23 m.y. range from 1,109 to 1,086 Ma, with magmatism concentrated in two intense bursts of activity at 1,109 to 1,107 and 1,099 to 1,094 Ma (Davis and Sutcliffe, 1985; Palmer and Davis, 1987; Davis and Paces, 1990; Paces and Miller, 1993; Davis et al., 1995). This double-pulse feature is a characteristic of many of the world's more recent very large continental and oceanic basalt provinces. Bercovici and Mahoney (1994) have attributed such double stages to the activity of a rising mantle plume that became detached from its stem, which subsequently also reached the crust.

Physically, this magmatism is expressed mainly as plateau lavas, which have been described by Green (1977, 1983, 1989) and BVSP (1981) and references therein. Several very large flood rhyolites, erupted mainly as high-temperature rheoignimbrites and fluid lavas, occur in Minnesota (Green and Fitz,

1993), Michigan, and Wisconsin. The MRS is remarkable among the world's continental rifts for the great thickness of basalts erupted. Plateau lava sequences as thick as 12 km are exposed around the Lake Superior basin, and as much as 20 km of lavas are imaged along the broad MRS axis beneath the lake by seismic reflection studies (Behrendt et al., 1988; Cannon et al., 1989). Several mafic dike swarms associated with the MRS testify to the large fissure eruptions that fed these flows (Green et al., 1987). Other subvolcanic intrusions, for example, the Beaver Bay Complex, Minnesota (Miller and Chandler, this volume) and Lake Nipigon plutons, Ontario (Sutcliffe, 1987) and larger mafic plutons, are comagmatic with the lavas. The largest are in Minnesota (the Duluth Complex: Weiblen and Morey, 1980; Miller and Weiblen, 1990; Miller, 1995) and Wisconsin (Mellen Complex: Olmsted, 1968, 1979; Klewin, 1990; Seifert et al., 1992). Many of the mafic intrusions are layered and some (especially the Duluth Complex) contain significant resources of Cu, Ni, and Ti (Nicholson et al., 1992; Hauck et al., this volume). Interpretation of deep seismic reflections and gravity modeling suggest the presence of large volumes of mafic rocks in the lower crust and as underplating along the rift axis (e.g., Behrendt et al., 1990).

Geochemically, the MRS igneous rocks constitute a fairly typical, bimodal tholeiitic suite that shows strong iron enrichment (BVSP, 1981). The Coldwell Complex on the north shore of Lake Superior in Ontario is an exception; composed principally of alkali gabbro and nepheline syenite, it is the largest alkaline complex in North America (Mitchell and Platt, 1978; Currie, 1980). The greatly predominating mafic rocks of the MRS are olivine tholeiites and slightly "enriched" transitional basalts (e.g., Brannon, 1984; Paces, 1988; BVSP, 1981; Klewin and Berg, 1991). The olivine tholeiites tend to be rich in Al$_2$O$_3$ and have fairly primitive geochemical characteristics, representing 15 to 20% melting of undepleted mantle source rocks (Paces, 1988; Nicholson and Shirey, 1990). Continuous trends in basalt compositions indicate lower- or midcrustal fractionation in both closed and open systems, and these have locally evolved into basaltic andesites, ferroandesites, icelandites, and rhyolites. Although most of the olivine tholeiites (and some basal transitional lavas) show trace-element and isotopic evidence of an enriched aesthenospheric plume source, many basalts have significant contributions from lithospheric mantle and Archean crust (e.g., Lightfoot et al., 1991; Nicholson and Shirey, 1990; Nicholson et al., 1991; Shirey et al., 1994). The rhyolites, which are particularly abundant in Minnesota, show isotopic evidence of moderate to large contributions from Archean crust (Nicholson and Shirey, 1990; Vervoort and Green, 1995).

SEDIMENTARY FILL OF THE MIDCONTINENT RIFT SYSTEM

In the Lake Superior region, four sequences of sedimentary rocks reflect the tectonic-sedimentary framework immediately before, during, and after the ca. 1,100-Ma magmatic event of the Midcontinent Rift System (MRS). The post-magmatic sedimentary rocks are overwhelmingly dominant in both outcrops on land and beneath Lake Superior. They may comprise 7,000 m or more of the 30-km-thick volcanic-sedimentary rift fill beneath Lake Superior (Cannon et al., 1989).

The oldest sequence includes four geographically separated thin (~100 m) pre-volcanic white to pink quartzose sandstone units (with minor basal conglomerates) that were largely deposited upon the eroded surface of Early Proterozoic turbidites (Ojakangas and Morey, 1982a, b) and immediately underlie the lava flows. The Bessemer Quartzite is present along the Gogebic Range in Michigan and Wisconsin, the Nopeming Sandstone is found near Duluth, Minnesota, and the western tip of Lake Superior, and the Puckwunge Quartzite occurs in northeasternmost Minnesota. The basal unit of the Osler Group in Ontario on the northern shore of Lake Superior is located at a similar stratigraphic position, beneath the lowest Osler volcanic rocks that have been dated at 1,109 Ma (Davis and Sutcliffe, 1985). Each of these four units is composed largely of quartzose fine-grained sand interpreted to have been deposited in distal braided alluvial plains and perhaps partly in lakes. The Bessemer and the Nopeming formations appear to have been unlithified when the first flows were extruded into the same bodies of water in which the sands were being deposited, thereby indicating an age equal to the lowermost lava flows (ca. 1,100 Ma). It is suggested that a gentle downwarping formed the basin in which the quartzose sands accumulated and were preserved, thus presaging the actual rifting and volcanism. This contrasts with conventional rift models that feature an initial phase of uplift (Green, 1983).

The second sedimentary sequence consists of red interflow sediments—muds, silts, sands, and gravels—that were deposited in alluvial fan, fluvial, and lacustrine environments between pulses of volcanism (Merk and Jirsa, 1982; Jirsa, 1984). They are numerous; more than 100 are found on the North Shore of Lake Superior in Minnesota, and many others are present in Ontario, Michigan, and Wisconsin. Most are thin units (<1 m), but some are more than 30 m thick. The provenance of these sediments is variable; most were locally derived from within the volcanically active rift, but some had extrabasinal cratonic contributions. The conglomerates are largely composed of volcanic clasts, whereas the sands are lithic and/or feldspathic. Quartz is generally only a minor constituent. In some of the interflow conglomerates in the Portage Lake Volcanics of Michigan, subvolcanic granophyre boulders are common.

Cessation of volcanism in the rift was followed by the deposition of the third sedimentary sequence, the Oronto Group, which includes the Copper Harbor Conglomerate, the Nonesuch Formation, and the Freda Sandstone (Daniels, 1982). On the Keweenaw Peninsula of Michigan, the Copper Harbor Conglomerate, which is as thick as 1,340 m, contains in its lower portion a few basaltic flows that represent some of the last volcanism within the basin; these Lakeshore Traps have been dated at 1,094 Ma (Paces and Davis, 1988). The Copper Harbor is

largely a coarse, clast-supported conglomerate; the clasts are dominantly volcanic and were derived from a variety of volcanic rocks. Sandstones that are largely composed of volcanic detritus are common, especially in the upper half of the formation. Siltstones and shales are less abundant, at least in the exposed portion of the formation, but may be more abundant at depth nearer the axis of the basin. Deposition was on alluvial fans and associated braided alluvial plains; rare stromatolite zones formed in standing bodies of water during pauses in sedimentation. A unit that is probably correlative is the subsurface Solor Church Formation of southeastern Minnesota (Morey, 1974). The Nonesuch Formation, which is as thick as 200 m, is noted for its copper content (mainly chalcocite) and the presence of petroleum seeps in the White Pine Copper Mine in the Upper Peninsula of Michigan. The Nonesuch is largely a fine-grained sequence of gray to black siltstone with minor shale, sandstone, and conglomerate. The fine sediment apparently was deposited in a restricted lacustrine basin in which organic material was accumulating as a result of a reducing chemical environment (Elmore et al., 1988; Suszek, 1991, this volume). The Freda Sandstone, which is as thick as 1,525 m, is composed dominantly of red to buff sandstone with intercalated siltstone and shale; the sandstones are more quartzose than those lower in the Oronto Group. The environment of deposition of the Freda Sandstone fits a braided-fluvial model.

The fourth and final sedimentary sequence is the 2,100-m-thick Bayfield Group, which is exposed on Wisconsin's Bayfield Peninsula, and equivalent units in Minnesota, Michigan, and Ontario (Morey and Ojakangas, 1982; Ojakangas and Morey, 1982b; Kalliokoski, 1982). The contact between the Bayfield Group and the underlying Oronto Group is not exposed; some field data can be interpreted as being indicative of an unconformable relationship, but a fault relationship is also possible. The basal unit, the Orienta Formation, and its equivalent to the west in Minnesota (the Fond du Lac Formation), are red lithic-feldspathic sandstone-siltstone-shale units with fining-upward sequences that are typical of deposition in meandering river systems. The relatively thin Devils Island Sandstone, its equivalent to the west in Minnesota (the Hinckley Sandstone), and part of the Jacobsville Sandstone to the east (Kalliokoski, 1982) are quartzose sandstone, which is probably the product of the maturation of lithic-feldspathic sand by eolian (and lacustrine?) processes on alluvial plains. It has long been assumed that the Devils Island is overlain by another submature sandstone, the Chequamegon Formation, but this may be a misinterpretation related to low dips of the Orienta Sandstone.

Paleocurrent data indicate a general transport of sediment toward the axis of the elongate rift basin during deposition of all four sequences (Ojakangas and Morey, 1982b). There is a general upward increase in textural and mineralogical maturity of the sandstones of the combined Oronto and Bayfield Groups, and whereas intrabasinal volcanogenic detritus is dominant in the Oronto Group, extrabasinal granitic detritus dominates in the Bayfield Group.

Sediments were obviously deposited in the MRS along its entire length. Off the southeastern end of Lake Superior in the Lower Peninsula of Michigan, a deep petroleum test hole showed sedimentary rocks similar to those described above (Fowler and Kuenzi, 1978; Ojakangas and Morey, 1982b). Still farther southeastward in northern Ohio, subsurface sedimentary rocks are petrographically similar to those in the Lake Superior region (Dickas et al., 1992). Richard et al. (this volume) have presented arguments for a post-Grenvillian (i.e., Cambrian?) age for several subsurface sedimentary rock units in Ohio.

Southwestward from Lake Superior, sedimentary rocks have been penetrated by deep drilling in Iowa (Anderson, this volume). Information on the subsurface sedimentary rocks of Nebraska and Kansas is presented in this volume by Carlson and Berendsen, respectively. Berendsen and Barczuk (1993) made general petrographic comparisons between the exposed sedimentary rocks of the Lake Superior region and subsurface rocks along the rift to the southwest.

REFERENCES CITED

Anderson, R. R., editor, 1990, The Amoco M. G. Eischeid #1 Deep Petroleum Test, Carroll County, Iowa: Iowa Department of Natural Resources Special Report Series No. 2, 185 p.

Behrendt, J. C., Green, A. G., Cannon, W. F., Hutchinson, D. R., Lee, M. W., Milkereit, B., Agena, W. F., and Spencer, C., 1988, Crustal structure of the Midcontinent Rift System: Results from GLIMPCE deep seismic reflection profiles: Geology, v. 16, p. 81–85.

Behrendt, J. C., Hutchinson, D. R., Lee, M., Thornber, C. R., Tréhu, A., Cannon, W., and Green, A., 1990, GLIMPCE seismic reflection evidence of deep-crustal and upper-mantle intrusions and magmatic underplating associated with the Midcontinent Rift System of North America: Tectonophysics, v. 173, p. 595–615.

Bercovici, D., and Mahoney, J., 1994, Double fluid basalts and plume head separation at the 600-kilometer discontinuity: Science, v. 266, p. 1367–1369.

Berendsen, P, and Barczuk, A., 1993, Petrography and correlation of Precambrian clastic sedimentary rocks associated with the Midcontinent Rift System: U.S. Geological Survey Bulletin 1989-E, 20 p.

Berendsen, P, Borcherding, R. M., Doveton, J., Gerhard, L., Newell, K. D., Steeples, D., and Watney, W. L., 1988, Texaco Poersch #1, Washington County, Kansas—Preliminary geologic report of the pre-Phanerozoic rocks: Kansas Geological Survey Open-File Report 88-22, 116 p.

Brannon, J. C., 1984, Geochemistry of successive lava flows of the Keweenawan North Shore Volcanic Group [Ph.D.thesis]: St. Louis, Missouri, Washington University, 312 p.

BVSP (Basaltic Volcanism Study Project), 1981, Basaltic volcanism on the terrestrial planets: New York, Pergamon Press, 1286 p.

Cannon, W. F., and 11 others, 1989, The North American Midcontinent Rift beneath Lake Superior from GLIMPCE seismic reflection profiling: Tectonics, v. 8, p. 305–332.

Chandler, V. W., McSwiggen, P. L., Morey, G. B., Hinze, W. J., and Anderson, R. R., 1989, Interpretation of seismic reflection, gravity and magnetic data across the Middle Proterozoic Midcontinent Rift System in western Wisconsin, eastern Minnesota and central Iowa: American Association of Petroleum Geologists, v. 73, p. 261–275.

Coons, R. L., Woollard, G. P., and Hershey, G. 1967, Structural significance and analysis of Mid-Continent Gravity High: The American Association of Petroleum Geologists Bulletin, v. 51, p. 2381–2399.

Currie, K. L., 1980, A contribution to the petrology of the Coldwell Alkaline Complex, northern Ontario: Geological Survey of Canada, Bulletin 287, 42 p.

Daniels, P. A., Jr., 1982, Upper Precambrian sedimentary rocks: Oronto Group, Michigan-Wisconsin, in Wold, R. J. and Hinze, W. F., eds., Geology and tectonics of the Lake Superior basin: Geological Society of America Memoir 156, p. 107–133.

Davis, D. W., and Paces, J. B., 1990, Time resolution of geologic events on the Keweenaw Peninsula and implications for development of the Midcontinent Rift System: Earth and Planetary Science Letters, v. 97, p. 54–64.

Davis, D. W., and Sutcliffe, R. H., 1985, U-Pb ages from the Nipigon plate and northern Lake Superior: Geological Society of America Bulletin, v. 96, p. 1572–1579.

Davis, D. W., Green, J. C., and Manson, M., 1995, Geochronology of the 1.1 Ga North American Midcontinent Rift [abs.]: Institute on Lake Superior Geology, 41st, Marathon, Ontario, Program and Abstracts, p. 9–10.

Dickas, A. B., 1986, Comparative Precambrian stratigraphy and structure along the Midcontinent Rift: The American Association of Petroleum Geologists Bulletin, v. 70, p. 225–238.

Dickas, A. B., 1989, Lake Superior basin segment of the Midcontinent Rift System, Lake Superior basin (Minnesota, Wisconsin, Michigan), in International Geological Congress, 28th, Field Trip Guidebook T344: Washington, D. C., American Geophysical Union, 62 p.

Dickas, A. B., and Mudrey M. G., Jr., editors, 1996, The Terra-Patrick No. 7-22 deep hydrocarbon test, Bayfield County, Wisconsin: Investigations and final report: Madison, Wisconsin, Wisconsin Geologic and Natural History Survey (in press).

Dickas, A. B., Mudrey, M. G., Jr., Ojakangas, R. W., and Shrala, D. L., 1992, A possible southeastern extension of the Midcontinent Rift System located in Ohio: Tectonics, v. 11, p. 1406–1414.

Elmore, R. D., Milavec, G., Imbus, S., and Engel, M., 1988, The Precambrian Nonesuch Formation of the North American Mid-Continent Rift: Sedimentology and organic geochemical aspects of lacustrine deposition: Precambrian Research 43, 191–213.

Fowler, J. H., and Kuenzi, W. D., 1978, Keweenawan turbidites in Michigan (deep borehole red beds): A foundered basin sequence developed during evolution of a protoceanic rift system: Journal of Geophysical Research, v. 83, p. 5833–5843.

Green, J. C., 1977, Keweenawan plateau volcanism in the Lake Superior region, in Baragar, W. R. A., Coleman, L. C., and Hall, J. M., eds., Volcanic regimes in Canada: Geological Association of Canada, Special Paper 16, p. 407–422.

Green, J C., 1983, Geologic and geochemical evidence for the nature and development of the Middle Proterozoic (Keweenawan) Midcontinent Rift of North America: Tectonophysics, v. 94, p. 413–437.

Green, J. C., 1989, Physical volcanology of mid-Proterozoic plateau lavas: The Keweenawan North Shore Volcanic Group, Minnesota: Geological Society of America Bulletin, v. 101, p. 486–500.

Green, J. C., and Fitz, T. J., III, 1993, Extensive felsic lavas and rheoignimbrites in the Keweenawan Midcontinent Rift plateau volcanics, Minnesota: petrographic and field recognition: Journal of Volcanology and Geothermal Research, v. 54, p. 177–196.

Green, J. C., Bornhorst, T. J., Chandler, V. W., Mudrey, M. G., Jr., Myers, P. E., Pesonen, L. J., and Wilband, J. T., 1987, Keweenawan dykes of the Lake Superior region: Evidence for evolution of the Middle Proterozoic Midcontinent Rift of North America, in Halls, H. C., and Fahrig, W. F., eds., Mafic dyke swarms: Geological Association of Canada Special Paper 34, p. 289–302.

Hinze, W. J., Allen, D. J., Fox, A. J., Sunwoo, D., Woelk, T., and Green, A. G., 1992, Geophysical investigations and crustal structure of the North American Midcontinent Rift System: Tectonophysics, v. 213, p. 17–32.

Hutchinson, D. R., White, R. S., Cannon, W. F., and Schulz, K. J., 1990, Keweenaw hotspot: Geophysical evidence for a 1.1-Ga mantle plume beneath the Midcontinent Rift System: Journal of Geophysical Research, v. 95, p. 10869-10884.

Jirsa, M. A., 1984 Interflow sedimentary rocks in the Keweenawan North

Shore Volcanic Group, Northeastern Minnesota: Minnesota Geological Survey Report of Investigations 30, 20 p.

Kalliokoski, J., 1982, Jacobsville Sandstone, *in* Wold, R. J., and Hinze, W. J., eds., Geology and tectonics of the Lake Superior basin: Geological Society of America Memoir 156, p. 147–156.

King, E. R., and Zietz, I., 1971, Aeromagnetic study of the Midcontinent Gravity High of central United States: Geological Society of America Bulletin, v. 82, p. 2187–2208.

Klewin, K. W., 1990, Petrology of the Proterozoic Potato River layered intrusion, northern Wisconsin, USA: Journal of Petrology, v. 31, p. 1115–1139.

Klewin, K.W., and Berg, J. H., 1991, Petrology of the Keweenawan Mamainse Point lavas, Ontario: Petrogenesis and continental rift evolution: Journal of Geophysical Research, v. 96, no. B1, p. 457–474.

Lightfoot, P. C., Sutcliffe, R. H., and Doherty, W., 1991, Crustal contamination identified in Keweenawan Osler Group tholeiites, Ontario: A trace element perspective: The Journal of Geology, v. 99, no. 5, p. 739–760.

Merk, G. P., and Jirsa, M. A., 1982, Provenance and tectonic significance of the Keweenawan interflow sedimentary rocks, *in* Wold, R. J., and Hinze, W. J., eds., Geology and tectonics of the Lake Superior basin: Geological Society of America Memoir 156, p. 97–106.

Miller, J. D., Jr., 1995, Prediction and discovery of PGE occurrences in the Duluth Complex at Duluth [abs.]: Institute on Lake Superior Geology, 41st, Program and Abstracts, Marathon, Ontario, p. 45–46.

Miller, J. D., Jr., and Weiblen, P. W., 1990, Anorthositic rocks of the Duluth Complex: Examples of rocks formed from plagioclase crystal mush: Journal of Petrology, v. 31, pp. 295–339.

Mitchell, R. H., and Platt, R. G., 1978, Mafic mineralogy of ferroaugite syenite from the Coldwell Alkaline Complex, Ontario, Canada: Journal of Petrology, v. 19, p. 627–651.

Morey, G. B., 1974, Cyclic sedimentation of the Solar Church Formation (upper Precambrian, Keweenawan), southeastern Minnesota: Journal of Sedimentary Petrology, v. 44, p. 872–884.

Morey, G. B., and Ojakangas, R. W., 1982, Keweenawan sedimentary rocks of eastern Minnesota and northwestern Wisconsin, *in* Wold, R.J., and Hinze, W. J., eds., Geology and tectonics of the Lake Superior basin: Geological Society of America Memoir 156, p. 135–146.

Nicholson, S. W., and Shirey, S. B., 1990, Midcontinent Rift volcanism in the Lake Superior region: Sr, Nd, and Pb isotopic evidence for a mantle plume origin: Journal of Geophysical Research, v. 95 (B7), p. 10851–10868.

Nicholson, S. W., Shirey, S. B., and Green, J. C., 1991, Regional Nd and Pb isotopic variations among the earliest Midcontinent Rift basalts in western Lake Superior [abs.]: Geological Association of Canada—Mineralogical Association of Canada Annual Meeting, Program with Abstracts, v. 16, p. A90.

Nicholson, S. W., Cannon, W. F., and Schulz, K. J., 1992, Metallogeny of the Midcontinent Rift System of North America: Precambrian Research, v. 58, p. 355–386.

Ojakangas, R. W. and Morey, G. B., 1982a, Keweenawan pre-volcanic quartz sandstones and related rocks of the Lake Superior region, *in* Wold, R. J., and Hinze, W. J., eds., Geology and tectonics of the Lake Superior basin: Geological Society of America Memoir 156, p. 85–96.

Ojakangas, R. W. and Morey, G. B., 1982b, Keweenawan sedimentary rocks of the Lake Superior region: A summary, *in* Wold, R.J., and Hinze, W. J., eds., Geology and tectonics of the Lake Superior basin: Geological Society of America Memoir 156, p. 157–164.

Olmsted, J. F., 1968, Petrology of the Mineral Lake intrusion, northwestern Wisconsin, *in* Isachsen, Y. W., ed., Origin of anorthosite and related rocks: New York State Museum and Science Service Memoir 18, p. 149–162.

Olmsted, J. F., 1979, Crystallization history and textures of the Rearing Pond gab-

bro, northwestern Wisconsin: American Mineralogist, v. 64, p. 844–855.

Oray, E., Hinze, W. J., and O'Hara, N. W., 1973, Gravity and magnetic evidence for the eastern termination of the Lake Superior syncline: Geological Society of America Bulletin, v. 84, p. 2763–2780.

Paces, J. B., 1988, Magmatic processes, evolution and mantle source characteristics contributing to the petrogenesis of Midcontinent Rift basalts: Portage Lake Volcanics, Keweenaw Peninsula, Michigan [Ph.D. thesis]: Houghton, Michigan, Michigan Technological University, 413 p.

Paces, J. B., and Davis, D. W., 1988, Implications of high precision U-Pb age dates on zircons from Portage Lake volcanic basalts on Mid-continent Rift subsidence rates, lava flow repose periods and magma production rates: Proceedings and abstracts, Annual Institute on Lake Superior Geology, 34th, p. 85–86.

Paces, J. B. and Miller, J. D., Jr., 1993, Precise U-Pb ages of Duluth Complex and related mafic intrusions, northeastern Minnesota: Geochronological insights to physical, petrogenetic, paleomagnetic, and tectonomagmatic processes associated with the 1.1 Ga Midcontinent Rift System: Journal of Geophysical Research, v. 98, no. B8, p. 13997–14013.

Palmer, H. C., and Davis, D. W., 1987, Paleomagnetism and U-Pb geochronology of Michipicoten Island: Precise calibration of the Keweenawan polar wander track: Precambrian Research, v. 37, p. 157–171.

Seifert, K. E., Peterman, Z. E., and Thieben, S. E., 1992, Possible contamination of Midcontinent Rift igneous rocks: examples from the Mineral Lake intrusions, Wisconsin: Canadian Journal of Earth Science, v. 29, no. 6, p. 1140–1153.

Shirey, S. B., Klewin, K. W., Berg, J. H., and Carlson, R. W., 1994, Temporal changes in the sources of flood basalts: isotopic and trace element evidence from the 1100 Ma old Keweenawan Mamainse Point Formation, Ontario, Canada: Geochimica et Cosmochimica Acta, v. 58, p. 4475–4490.

Sims, P. K., and 15 others, 1993, The Lake Superior region and trans-Hudson orogen, *in* Redd, J. C., Jr., Bickford, M. E., Houston, R. S., Link, P. K., Rankin, D. W., Sims, P. K., and Van Schmus, W. R., eds., Precambrian: Conterminous U.S.: Boulder, Colorado, Geological Society of America, The Geology of North America, v. C-2, p. 11–120.

Suszek, T. J., 1991, Petrography and sedimentation of the Middle Proterozoic (Keweenawan) Nonesuch Formation, western Lake Superior region, Midcontinent Rift System [M.S. thesis]: Duluth, University of Minnesota, 198 p.

Sutcliffe, R. H., 1987, Petrology of Middle Proterozoic diabases and picrites from Lake Nipigon, Canada: Contributions to Mineralogy and Petrology, v. 96, p. 201–11.

Van Schmus, W. R., and Hinze, W. J., 1985, The Midcontinent Rift System: Annual Reviews of Earth and Planetary Science, v. 13, p. 345–383.

Vervoort, J. D., and Green, J. C., 1995. Origin of evolved magmas in the Midcontinent Rift System, NE Minnesota: Nd isotope evidence for melting of Archean crust: International Geological Correlation Program Project 336, Petrology and Metallogeny of Volcanic and Intrusive Rocks of the Midcontinent Rift System, Duluth, Minnesota, Program with Abstracts, p. 197–198.

Weiblen, P. W., and Morey, G. B., 1980, A summary of the stratigraphy, petrology, and structure of the Duluth Complex: American Journal of Science, Jackson Volume 280-A, p. 88–133.

Wold, R. J., and Hinze, W. J., eds., 1982, Geology and tectonics of the Lake Superior basin: Geological Society of America Memoir 156, 280 p.

Woollard, G. P., 1943, Transcontinental gravitational and magnetic profile of North America and its relation to geologic structure: Geological Society of America Bulletin, v. 54, p. 747–790.

MANUSCRIPT ACCEPTED BY THE SOCIETY JANUARY 16, 1996

Geological Society of America
Special Paper 312
1997

The Midcontinent Rift System:
A major Proterozoic continental rift

W. J. Hinze, David J. Allen*, Lawrence W. Braile, and John Mariano*
Department of Earth and Atmospheric Sciences, Purdue University, West Lafayette, Indiana 47907

ABSTRACT

The Middle Proterozoic Midcontinent Rift System is one of the world's great continental rifts. This 1,100-Ma rift, which extends more than 2,000 km across the North American craton, is a major Precambrian magmatic province and tectonic disruption of the lithosphere. The volcanic and sedimentary rocks within the ~30-km-thick rift basin record an early igneous event, the evidence of which is preserved by an enormous volume of mantle-derived basaltic volcanic flows extruded onto a highly extended crust, followed by a sedimentary episode during which great volumes of clastic sediments were deposited into basins above the subsiding volcanic grabens. The rift was aborted just prior to continental separation, probably as the result of the same regional stresses that resulted in reverse faulting and other compressional structures within the rift. During the past decade, intensive geophysical, geochemical, and isotopic investigations of the rift have produced a much clearer understanding of the rift's magmatic and tectonic evolution and their manifestations throughout the lithosphere. This paper summarizes our current knowledge of the rift, with special emphasis on the great progress that has been made in the last ten years. It is intended to be a temporary benchmark in our understanding of the rift, for the progress of the last decade has raised a series of significant new questions and has opened the door for a new generation of investigations of the Midcontinent Rift.

INTRODUCTION

Early regional geophysical studies in the central stable interior of the United States (Woollard, 1943; Lyons, 1950; Woollard and Joesting, 1964) showed the presence of an intense, curvilinear gravity anomaly extending from Kansas to Lake Superior. Except in the Lake Superior region, the surface rocks are relatively flat-lying Phanerozoic sedimentary strata that provide no clue to the origin of the anomaly. This anomaly has come to be known as the "Midcontinent gravity high" or, more appropriately, the "Midcontinent geophysical anomaly," for it is both an

intense gravity and magnetic anomaly (King and Zietz, 1971). Investigations by Thiel (1956) showed that this feature extends into mafic volcanic and clastic sedimentary rocks of the Middle Proterozoic Keweenawan sequence in the Lake Superior region, thereby providing an explanation for the anomaly. Subsequently, geological and geophysical studies in the mid-1960s in the Lake Superior region (e.g., Smith et al., 1966; Ocola and Meyer, 1973) supported previous evidence that suggested a possible rift origin for the Keweenawan rock package. This led Wold and Hinze (1982) to apply the genetic/location name of the Midcontinent Rift System to the source of the anomaly. The Midcontinent Rift

---

*Present address: Exxon Exploration Company, 222 Benmar, Houston, Texas 77060.

Hinze, W. J., Allen, D. J., Braile, L. W., and Mariano, J., 1997, The Midcontinent Rift System: A major Proterozoic continental rift, *in* Ojakangas, R. W., Dickas, A. B., and Green, J. C., eds., Middle Proterozoic to Cambrian Rifting, Central North America: Boulder, Colorado, Geological Society of America Special Paper 312.

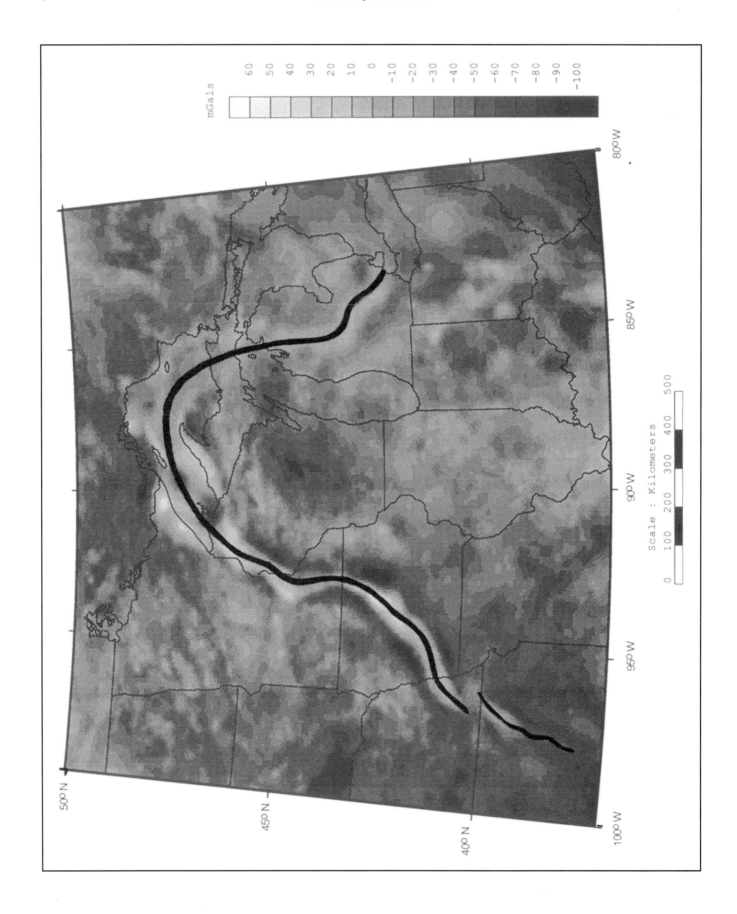

can be traced (Figs. 1 and 2) from Kansas into western Lake Superior, through the eastern portion of the lake (Hinze et al., 1966), and beneath the Michigan basin into southeastern Michigan (Hinze et al., 1975). The associated gravity and magnetic anomalies extend for more than 2,000 km and over most of their lengths are the most obvious and sharply defined features of regional geophysical anomaly maps of North America.

More than 80 years ago Van Hise and Leith (1911) related the Keweenawan rocks of the Lake Superior region to a Precambrian rifting event. Since the 1950s the rift origin has been supported by interpretations of geophysical anomalies. However, acceptance of a rift as the source of the midcontinent geophysical anomaly has come slowly due to the limited exposures of rift-related rocks and the lack of clear evidence for normal faults where the rocks do crop out. Nonetheless, acceptance of the rift origin has made a dramatic turn in the past decade because of the increasing availability of deeply penetrating seismic reflection profiles, extensive regional gravity and magnetic anomaly data, drilling through the overlying sedimentary rocks into the rift-rock package, and geochemical and isotopic investigations of the rift's igneous rocks.

Several lines of evidence have led to general acceptance of a rift origin for the geophysical anomaly. First, a great volume of mantle-derived magma was extruded and intruded within a relatively short time interval spanning a few tens of millions of years. The magmas were emplaced into and adjacent to a series of structural troughs related in part to normal faulting, although the presence of normal faults is equivocal because many of the major normal faults have been modified by late-stage reverse movement. Furthermore, seismic reflection data suggest that the normal faults are buried beneath several kilometers of volcanic flows and late-stage sedimentary strata. Hence the normal faults are not directly observable, but are instead inferred by zones of rapid thickening of the rift fill. Second, the Midcontinent Rift discordantly transects older basement geological provinces and their structural patterns. There are notable exceptions, however, where the rift pattern was deflected by preexisting zones of weakness. Third, although the present crust is generally thicker along the axis of the rift (up to 55 km), the pre-rift crust has been thinned to less than one-third of its original thickness. The present increased thickness of the crust is due to the great thickness (up to 30 km) of the rift-rock package and underplating of the crust. Fourth, mafic dikes of Keweenawan age strike roughly parallel to the axis of the rift, testifying to the presence of tensional stresses at the time of generation of the Midcontinent Rift.

Although numerous journal and monograph articles have been written on specific geographic segments or geologic aspects of the Midcontinent Rift, only a few have attempted to present a comprehensive view of the entire rift (Halls, 1978; Green, 1983; Van Schmus and Hinze, 1985; Dickas, 1986). Since the latest of these reviews our knowledge of the Midcontinent Rift has increased significantly. In this paper we present an overview of our current understanding of this "world-class" paleorift and the processes that controlled its evolution.

SIGNIFICANCE OF THE MIDCONTINENT RIFT

During the past decade, studies of the Midcontinent Rift have increased markedly in terms of both the number and range of expertise of investigators and the sophistication of the experiments and observations. This increase reflects the acknowledged scientific and pragmatic significance of the rift. The Midcontinent Rift is truly a dramatic continental lithospheric feature on any basis of comparison and rivals the East African Rift System (EARS) in length and impact upon the crust.

The Midcontinent Rift is significant because of the following:

1. It is unusual among paleorifts in its great length and enormous volume of igneous rocks. Although the modern EARS is approximately the same length, it has experienced significantly less igneous activity than the Midcontinent Rift. The extension of the crust beneath the Midcontinent Rift was arrested just prior to production of oceanic crust. Whatever the source of the tensional stresses, they were eliminated or drastically reduced. Igneous activity continued, however, leading to an unusually thick stack of volcanic rocks without the production of an oceanic crust. Thus, the Midcontinent Rift records ancient plate tectonic processes not preserved in oceans and largely obliterated or hidden beneath continental margins.

2. Rift-associated mafic igneous rocks serve as mantle probes because the Midcontinent Rift transects crustal segments ranging in age from 1.65 to 3.6 Ga (Fig. 3). Present evidence indicates that the magmatic activity occurred over a relatively short span of geologic time (~10 to 20 m.y.) near 1.1 Ga. Potentially, these mantle rocks may be used to investigate: (a) the heterogeneity of the mantle, (b) the relation between the heterogeneity and the age of crustal stabilization, (c) mantle dynamics associated with the rifting process, and (d) crust/mantle interaction.

3. Potentially economic mineral deposits may be associated with the rift. The Midcontinent Rift's sedimentary and igneous rocks have been important sources of copper in the past and numerous other metallic minerals have been discovered in near-ore grades, primarily within the mafic intrusive rocks (Nicholson et al., 1992). In addition, organic hydrocarbons occur in the late-stage rift sedimentary rocks that overlie the volcanic rocks. The potential of the Midcontinent Rift as a frontier hydrocarbon province has been the impetus for numerous seismic reflection profiles that are now available over the length of the Midcontinent Rift (Dickas, 1986). These profiles have been of great value in providing critical information concerning the nature of the rift.

Figure 1. Bouguer gravity anomaly map of the Midcontinent of North America. Anomaly values increase from dark to light shading. Wide dark line follows the approximate center of the Midcontinent Rift System along the generally higher anomaly values (lighter shading) (after Hinze et al., 1992).

W. J. Hinze and Others

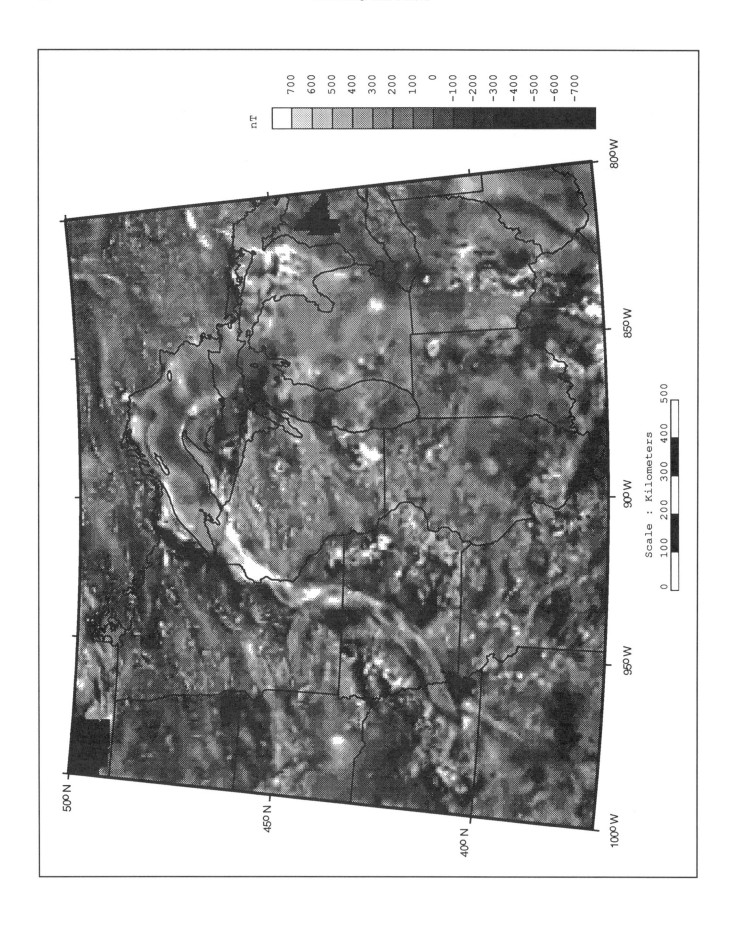

Figure 2. Total magnetic intensity anomaly map of the Midcontinent of North America. The approximate center of the Midcontinent Rift System is generally marked by higher anomaly values and flanked by smooth minima over the marginal clastic sedimentary basins (modified from Hinze et al., 1992).

← ─────────────────────────────────

GEOLOGIC FRAMEWORK

Introduction

Direct geologic information concerning the rift-related rocks of the Midcontinent Rift is rather limited despite over a century of study in the copper mining regions of Lake Superior.

The rift rocks crop out only in the Lake Superior region; moreover, the exposed rocks predominantly represent the margins of the structure or the upper parts of the stratigraphic section. Available drill holes are shallow and poorly distributed for scientific purposes, and generally provide only limited samples for analysis. Recently, deeper exploratory holes have been drilled into the Midcontinent Rift in Iowa, Kansas, Michigan, Minnesota, and Wisconsin. These holes, however, have been targeted at the upper sedimentary section of the rift-rock package.

The rocks of the Midcontinent Rift in the Lake Superior region constitute the 1.1-Ga Keweenawan Supergroup (Morey and Van Schmus, 1988). They are divided into two major suites: a lower, largely volcanic sequence, and an upper, overlying

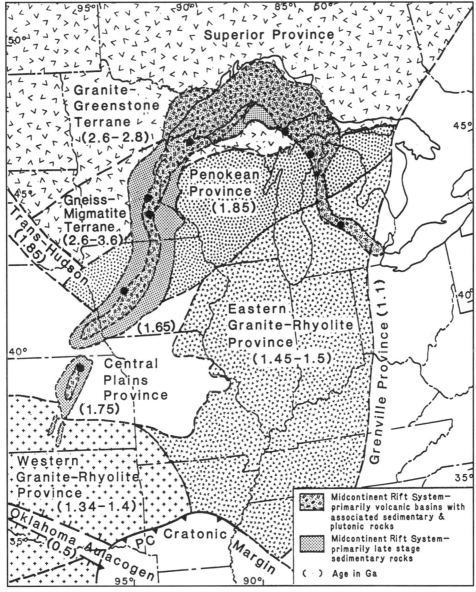

Figure 3. Basement provinces of the midcontinent of North America (modified from Van Schmus et al., 1987) showing the position of the Midcontinent Rift System (after Hinze and Kelly, 1988). Filled in dots are the locations of deep drill holes in the Midcontinent Rift that have penetrated through the late-stage clastic sedimentary strata of the Midcontinent Rift or into the igneous rocks of the rift.

clastic sedimentary package. The volcanic sequence, extruded contemporaneously with the rifting event, consists of a range of lithologies from picrite to rhyolite, although basalt is the predominant volcanic rock type. The volcanic sequence is locally intruded by predominantly gabbroic rocks such as the Duluth and Mellen Complexes. In addition, the volcanic flows are interbedded with a minor volume of interflow clastic sedimentary strata. The upper sedimentary package consists of sandstones and siltstones deposited in crustal depressions above the rift. These clastic rocks increase in compositional and textural maturity upsection. The present edges of both the volcanic and sedimentary units are erosional in nature, indicating that the original extent of these rocks were greater than they are today.

Drillhole data and geophysical interpretations indicate that the rock sequence observed in the Lake Superior region is in general duplicated along the limbs of the rift extending southwest into Kansas and southeast across Michigan (Dickas, 1986). However, some rock units may be absent along particular segments of the rift and, locally, units may be stratigraphically inverted by high-angle reverse faults. Furthermore, not all units occurring within the Lake Superior segment or the limbs of the rift may be represented in current correlation charts of the Keweenawan Supergroup. Units that do not crop out may not have been drilled. The total thickness of the Keweenawan Supergroup has been difficult to determine because of limited outcrops and problems in correlation. Recent seismic reflection studies (e.g., Behrendt et al., 1988) show total maximum thicknesses of the order of 30 km in the Lake Superior region whereas 15 km thicknesses are common in the limbs of the rift (e.g., Fox, 1988; Chandler et al., 1989; Woelk and Hinze, 1991). These thicknesses are much greater than those predicted on the basis of surface exposures in the Lake Superior region.

Volcanic unit

Introduction. Volcanism commenced at or shortly after the initiation of rifting. In the western Lake Superior region, however, a thin basal clastic sedimentary unit, which is exposed at several isolated localities, was deposited prior to the onset of volcanism. We might anticipate that these sedimentary rocks would thicken toward the center of the rift if early, nonvolcanic subsidence was significant. These thickened sedimentary rocks could potentially be mapped using seismic reflection data. However, there is no evidence of them in the present seismic reflection interpretations. Rather, volcanic rocks predominate in the earliest stages of rifting. The volcanic units include fissure-fed flood basalts with some ignimbrites that were derived from eruptive centers. Eight of these centers have been identified in Lake Superior by White (1972) and Green (1982) and, undoubtedly, many additional centers occur along the axis of the rift. Individual flows show great continuity both along strike in outcrop and also across the basins as interpreted from seismic reflection sections. Many flows can be traced for distances exceeding 100 km. Measured thicknesses vary from a few

meters to over 100 m. Seismic-reflection data show reflections extending from surface exposures of volcanic rock to depths of 15 to 20 km (Hinze et al., 1990), testifying to the reflective nature of the volcanic section.

Reflections within the volcanic sequence may be caused in part by seismic impedance contrasts between the volcanic and interflow sedimentary rocks, although the limited (<5%) volume of sedimentary rocks within the volcanic pile (Wallace, 1981) suggests that the reflections are caused by physical property variations, both velocity and density, within and between the volcanic flows (Halls, 1969; Lippus, 1988; Allen and Chandler, 1993). The internal reflectivity of the volcanic section is relatively intense, although both lateral and vertical reflectivity variations are observed (e.g., Cannon et al., 1989; Chandler et al., 1989; Hinze et al., 1990; Milkereit et al., 1990; Samson and West, 1992). Some of these variations in reflectivity may be due to data acquisition and processing parameters, but most are related to the reflectivity of the volcanic sequence. For example, on the southern margin of the Lake Superior segment of the Midcontinent Rift Hinze et al. (1990) observe an upper, more reflective volcanic zone underlain by a less reflective volcanic sequence (Fig. 4). An interpretation of the two-fold division in

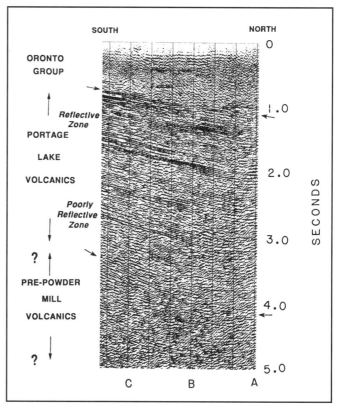

Figure 4. Portion of migrated seismic reflection record (courtesy of Petty-Ray Geophysical Group of Geosource, Inc.) on the southern margin of the Midcontinent Rift in western Lake Superior showing highly reflective (upper) and poorly reflective (lower) zones of Keweenawan volcanic rocks (after Hinze et al., 1990). Refer to Hinze et al. (1990) for interpretational details.

the seismic reflections, consistent with the modeling of Lippus (1988), is that the upper portion of the volcanic section consists mainly of thinner flows that are separated toward the top of the section by increasing amounts of interflow sedimentary rocks. This interpretation suggests that the volcanism was initiated abruptly with enormous volumes of magma extruded, and that the volcanism waned before termination.

Ages. Knowledge of the age of the units of the Keweenawan Supergroup is limited primarily to rocks of the Lake Superior region. Recent U-Pb age dating (e.g., Van Schmus et al., 1982; Davis and Sutcliffe, 1985; Palmer and Davis, 1987; Davis and Paces, 1990; Miller, 1992; Van Schmus, 1992; Paces and Miller, 1993) indicates that volcanism and plutonism occurred between roughly 1,108 and 1,094 Ma, with an intrusion on Michipicoten Island in northeast Lake Superior (Fig. 5) dated at 1,087 Ma. Isotopic age dating of the Midcontinent Rift rocks outside of the Lake Superior region has yielded only a single useful age. Van Schmus et al. (1990) have obtained a U-Pb age of 1,097.5 Ma for a gabbro from the Texaco Poersch #1 drill hole in northeast Kansas. This age lies within the range of published isotopic ages of the Keweenawan volcanic and plutonic rocks of the Lake Superior region, indicating that there is no major age progression along the length of the rift.

The limited accurate isotopic age information for the Keweenawan Supergroup has encouraged the use of paleomagnetic data in stratigraphic correlation (Van Schmus et al., 1982; Green, 1982). The results of extensive paleomagnetic investigations (Halls and Pesonen, 1982) of accurately dated Keweenawan rocks were presented by Palmer and Davis (1987). Their data show rapid changes in the Keweenawan pole positions and a regular progression from the rift's earliest known igneous rocks, the Logan Sills of the Nipigon region north of Lake Superior, to the uppermost sedimentary unit, the Jacobsville Sandstone, lending credence to the use of paleomagnetic parameters for dating the rift rocks. A change in polarity from reversed to normal that has been dated at 1,097 Ma (Davis and Paces, 1990) is widely

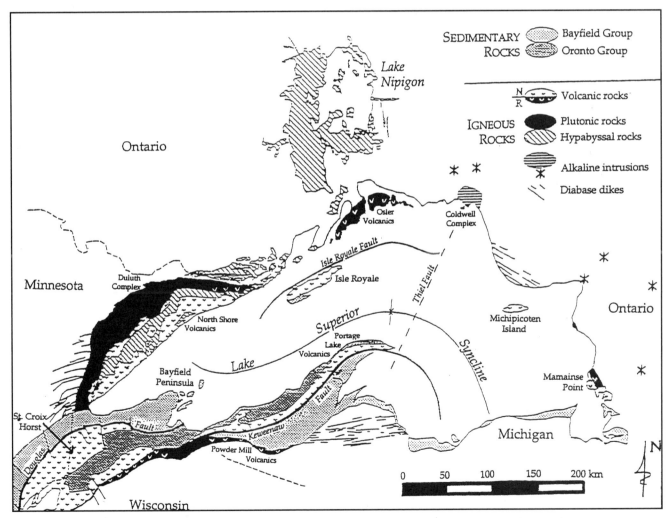

Figure 5. Generalized basement geology of the Lake Superior basin. In legend: N, normal magnetic polarity; R, reversed magnetic polarity. (Modified from Paces and Miller, 1993.)

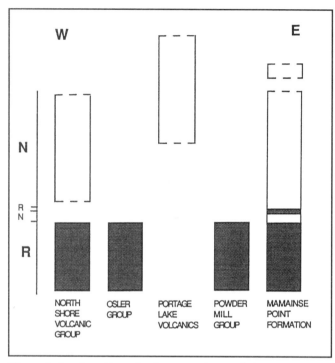

Figure 6. Stratigraphic and paleomagnetic correlation of the major volcanic units around Lake Superior. The locations of these units are shown on Figure 5 (after Klewin and Shirey, 1992). N, normal magnetic polarity; R, reversed magnetic polarity.

observed in the Lake Superior region and, thus, is particularly useful as a stratigraphic correlation tool. However, evidence for another reversal period (Palmer, 1970; Robertson, 1973) of probable short duration in the Mamainse Point sequence (Klewin and Berg, 1991) and in the lower part of the Powder Mill Volcanics (Books, 1972) complicates the use of magnetic polarity as a stratigraphic correlation tool (Fig. 6).

Volumes and rates. Only crude estimates of the volume of volcanic material added to the crust during the Keweenawan rifting can be made with the available information. It is not possible to use the gravity anomaly over the rift to determine the volume of igneous rocks added to the crust because of the destructive interference between the positive anomalies produced by the dense mafic igneous rocks and the negative anomalies derived from both the overlying low-density sedimentary rocks and the thickened crust beneath the rift. Green (1981) estimated a total of 400,000 km³ of both volcanic and intrusive rocks, but subsequent seismic reflection profiling has shown that the volume of volcanic rocks is much greater than previously assumed. Seismic reflection interpretations indicate a mean volume per unit kilometer of strike length of ~500 km³, suggesting that the total volume of volcanics rocks is of the order of 1 million km³. However, this is a minimum estimate because volcanic flows have been eroded from the original rift package. For example, unconformities internal to the volcanic package have been interpreted from seismic reflection profiling (e.g., Chandler et al., 1989; Hinze et al., 1990), which suggest

that volcanic rocks have been removed by erosion. However, more significant erosion of the volcanic rocks is indicated on the margins of the Midcontinent Rift where Keweenawan dikes occur at distances of up to 200 km from the rift axis (Green et al., 1987). These dikes were at one time conduits for volcanic flows and indicate substantial erosion of the volcanic flows on the margins of the rift. As a result, Cannon (1992) estimates that the original volume of volcanic flows was at least 2 million km³. Moreover, this volume does not consider the amount of mantle-derived magma that was trapped in the crust and never extruded. Seismic tomography and seismic and gravity modeling support the existence of mantle derivatives in the crust beneath the rift (e.g., Hinze et al., 1982; Trehu et al., 1991; Hamilton and Mereu, 1993). The amount of mantle-derived material added as intrusions into the crust is unknown, but is less than the volume of the volcanic rocks. In many other rifts, however, the volume of intrusions from the mantle into the lower crust is interpreted to greatly exceed the volume of volcanic rocks (e.g., Mohr, 1983; Ramberg and Morgan, 1984; Neumann et al., 1986). Although the volume of Midcontinent Rift intrusive rocks is poorly constrained by geophysical measurements, the extreme degree of extension may have allowed the majority of mantle derived melts to reach the surface.

Rates of volcanic activity for the Portage Lake Volcanics of the Keweenaw Peninsula of Lake Superior have recently been investigated by Davis and Paces (1990). Their high-precision U-Pb isotopic ages of specific volcanic units, together with measured stratigraphic intervals, indicate an accumulation rate of 1.3 mm/yr over a dated interval of ~2.2 m.y. In addition, they calculate an average interflow repose period of 30 ka and a magma production rate of 0.02 to 0.05 km³/yr. The uncertainty in the production rate reflects the imprecision in estimating the volume of the volcanic rocks. However, as pointed out by Davis and Paces, the estimated range of production rates is of the same order of magnitude as production rates for other rifts and hotspot regimes (e.g., Iceland, Hawaii, and Columbia River Basalts). Davis and Paces (1990) also dated a basalt flow from within the overlying clastic sedimentary rocks (the Lake Shore traps of the Copper Harbor conglomerate) and estimated an accumulation rate of the order of 0.16 mm/yr for the lower-most late-stage sedimentary strata. Subsidence rates are essentially equivalent to accumulation rates based on the reasonable assumption that the surface of the subaerial volcanic flows remained at an approximately constant elevation. These figures, although admittedly poorly constrained, indicate that the rate of subsidence of the rift depression decreased drastically concurrently with the cessation of the major volcanic pulse (Cannon, 1992).

Additional information on magma production rates is provided by Mariano and Hinze (1992, 1994b). Their inversion of magnetic anomaly data of Lake Superior constrained by results obtained from seismic interpretation and gravity modeling suggests that the lower 80% of the volcanic section is reversely magnetized. This indicates that up to 15 km of volcanic flows were extruded during the reversed polarity epoch from 1,108 to

1,097 Ma (Davis and Sutcliffe, 1985; Klewin and Shirey, 1992; Van Schmus, 1992). The resulting accumulation rate, which can be considered a minimum because the time interval for the volcanism is a maximum estimate, is equivalent to the rate obtained by Davis and Paces (1990) for the normally polarized, 1,095-Ma Portage Lake Volcanics.

Paces and Miller (1993) reviewed the isotopic ages of the Keweenawan igneous rocks and found that the ages cluster into rather small age ranges. Although there are alternative interpretations of this clustering, they suggest, on the basis of related stratigraphic, structural, petrologic, and geophysical data, that the clustering indicates magmatism was episodic over the duration of the rifting event. Specifically, their review of the age data suggests major magmatic pulses from 1,109 to 1,106 Ma and from 1,099 to 1,094 Ma with minor substages indicated within the duration of the younger pulse.

Geochemistry. The petrology and geochemistry of the volcanic flows have been reviewed in several recent publications (Green, 1982, 1983, 1987; Van Schmus and Hinze, 1985; Klewin and Shirey, 1992). The flows are bimodal, ranging from picrite to rhyolite, although the rhyolite end of the spectrum is volumetrically much less significant than the rocks of the mafic end of the spectrum. Rhyolites occur primarily as ignimbrites and also as lava domes. The high-alumina olivine tholeiites, which have the most primitive compositions, occur near the tops of the volcanic cycles in the Lake Superior region (Green, 1987; Paces and Bell, 1989; Klewin and Shirey, 1992). Interflow sedimentary units occur within the basaltic sections, becoming more abundant upsection. Occasional thin basalt flows have also been encountered in the overlying sedimentary sections in Michigan (McCallister et al., 1978) and Kansas (Berendsen et al., 1988).

The petrogenesis of the lavas is complex, but recent studies of isotopic systematics of both tholeiitic and rhyolitic volcanic rocks (Brannon, 1984; Dosso, 1984, Klewin, 1989; Paces and Bell, 1989; Nicholson and Shirey, 1990) are beginning to clarify the picture. Paces and Bell (1989) show that in general the amount of crustal contamination in the Portage Lake Volcanics decreases, as indicated by increasing Ni contents, higher in the stratigraphic section (Fig. 7), although there are obvious small-scale cycles in the lava sequence. The primitive tholeiites at the top of sequence have εNd (1095 Ma) that are only slightly depleted (Fig. 8; Paces and Bell, 1989; Nicholson and Shirey, 1990). These near Bulk Earth values indicate that the source of the igneous rocks was enriched with respect to the depleted mantle beneath the Archean crust. Similar results are obtained from samples collected elsewhere along the Midcontinent Rift (Klewin and Shirey, 1992; Van Schmus, 1992). Possible sources of the magmas include subcontinental lithosphere enriched prior to the rifting event, magmas derived from a "fertile" mantle plume, or a mixture of both. The consensus among current workers, based on both geochemical and physical evidence, is that a mantle plume source was involved (e.g., Paces and Bell, 1989; Hutchinson et

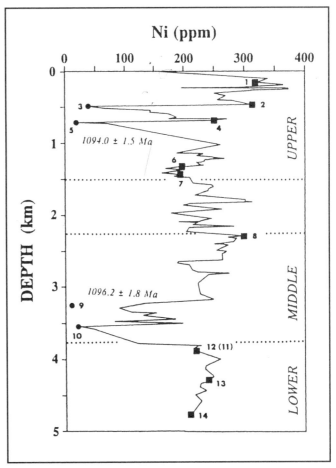

Figure 7. Variation in nickel (Ni) content with stratigraphic depth of the Portage Lake Volcanic section of the Keweenaw Peninsula, Michigan (after Paces and Bell, 1989). The approximate divisions between major cycles are shown as dotted lines. Numbered squares represent olivine tholeiites that plot along major cycle trends, while circles represent more evolved quartz tholeiites to rhyolites that occur at the culmination of minor cycle trends. U-Pb zircon dates are shown near their appropriate stratigraphic positions.

al., 1990; Nicholson and Shirey, 1990; Cannon, 1992; Klewin and Shirey, 1992; Van Schmus, 1992).

Although a mantle plume was the dominant source of the mafic magmas, there were also magmas derived from the upper mantle and the crust. For example, Nicholson and Shirey (1990) identify two groups of Portage Lake Volcanics rhyolites on the basis of Nd and Pb isotopic characteristics. One group is associated with the same source as the basalts, whereas the other is due to partial melting of Archean crust. Berg and Klewin (1988), Klewin and Berg (1991), and Klewin and Shirey (1992) suggest that the early magmas, rich in Mg and Nd, were derived from partial melting of the resident lithosphere at decreasing depths with increasing time, and with fractional crystallization occurring in magma chambers within the crust. Later magmas, typified by the basalts of western Lake Superior and the upper part of the Mamainse Point section in eastern Lake Superior, were derived from partial melting of the rising plume head that had

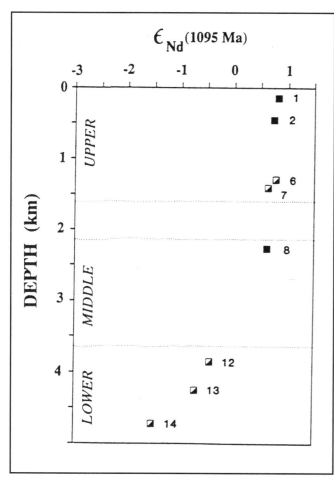

Figure 8. Variation of εNd (1,095 Ma) as a function of stratigraphic depth in the Portage Lake Volcanic section of the Keweenaw Peninsula, Michigan (after Paces and Bell, 1989). The approximate divisions between major cycles are shown as dotted lines. Filled squares have Ni content >300 ppm and half-filled squares have contents between 50 and 250 ppm.

entrained depleted asthenospheric mantle. The combination of sources for the later magmas supports the suggestion of Campbell and Griffiths (1990) and Griffiths and Campbell (1991) regarding the nature of rising mantle plumes.

The volcanic flows have been modified by burial metamorphism (McCallister et al., 1978). In addition, Bornhorst et al. (1988) showed that the permeable flow tops within the Portage Lake Volcanics have experienced prehnite-pumpellyite regional metamorphism due to burial and fluid circulation within the volcanic pile ~35 to 50 m.y. after eruption.

Plutonic rocks

Keweenawan intrusive rocks in the upper crust are volumetrically minor in comparison to the volcanic rocks of the Midcontinent Rift. They range in composition from gabbro to granite and occur as layered complexes, sills, dikes, and isolated alkaline complexes. Weiblen (1982) has reviewed their general characteristics. The layered gabbro complexes, and particularly the Duluth Complex, have been studied extensively. The Duluth Complex consists of two principal phases. The anorthositic phase involved melt and crystal segregation leading to a broad range of rock types, varying from peridotite and anorthosite to felsic lithologies. The troctolitic phase is slightly more primitive than the anorthositic phase (Green, 1987). Recent high-precision U-Pb isotopic dating of the Duluth Complex shows that rocks of the two phases have indistinguishable crystallization ages of roughly 1,099 (+0.4/-0.9) Ma (Miller, 1992). As a result, evolutionary patterns for the melts of the Duluth Complex, based upon the traditional view that the troctolitic phase is younger, are being reevaluated.

Weiblen (1982) has also reviewed the mafic sills and dikes that occur throughout the Lake Superior region, but especially in the Lake Nipigon area north of Lake Superior. The earlier, smaller plutonic bodies have quartz tholeiite compositions and are magnetically reversely polarized, as are the early stage volcanic rocks. Later bodies have olivine tholeiite compositions and are magnetically normally polarized.

Alkaline intrusions are the third major plutonic element of the Keweenawan rocks of the Lake Superior region. These rocks display no intrusive relationship to other Midcontinent Rift rocks. The best known example is the Coldwell alkaline intrusion on the north shore of Lake Superior. It is dated at 1,108 ± 1 Ma (Heaman and Machado, 1992), the same age as the Logan sills (Davis and Sutcliffe, 1985), the earliest of the Midcontinent Rift rocks that are related to tholeiitic magmatism.

There is no record of mantle xenoliths within the volcanic and plutonic rocks of the Midcontinent Rift.

Sedimentary units

Four sequences of sedimentary rocks are associated with the Midcontinent Rift. These units originated before, during, and after the Midcontinent Rift volcanism in the Lake Superior region, and consist mainly of clastic rocks varying from conglomerates to shales. They originated from detritus derived from surrounding pre-Keweenawan rocks and synrift igneous rocks. The oldest of these units lies at the base of the volcanic rocks at widely separated outcrops in the western Lake Superior region, and stratigraphic correlations between outcrops are subject to question. These rocks have been described by Ojakangas and Morey (1982) as relatively thin (~100 m) quartzose sandstones with minor basal conglomerates.

Units within the volcanic sequence make up the second sequence of sedimentary rocks and have been discussed by Merk and Jirsa (1982) and Wallace (1981). These rocks were deposited as red interflow sediments ranging from muds to gravels, laid down in fluvial and lacustrine regimes developed on the volcanic rocks. They generally make up only a small proportion of the volcanic sequence (Wallace, 1981) and are derived primarily from rift volcanics, although exceptions to this are numerous.

As volcanism in the rift waned, sedimentary units continued to be deposited in increasing relative proportion to the volcanic flows in a developing basin over the rift and in flanking basins. The immediate post-volcanic unit is the Oronto Group that includes the Copper Harbor Conglomerate, the Nonesuch Shale, and the Freda Sandstone (Daniels, 1982). The Copper Harbor formation is composed of conglomerates and sandstones made up largely of volcanic detritus. Siltstones and shales are less abundant. All units thicken toward the axis of the rift basin within the lake. Deposition took place on alluvial fans and associated braided plains. Lake deposits occur as evidenced by rare stromatolite horizons and the anoxic lake deposits of the overlying Nonesuch Shale. However, recent studies of the Nonesuch Formation suggest a marine embayment as the depositional environment. Hieshima and Pratt (1991) come to this conclusion on the basis of the correlation of total sulfur and organic carbon, the relatively high total sulfur content, and the absence of sedimentological and geochemical evidence for a saline lacustrine environment. Furthermore, Pratt et al. (1991) find molecular fossils and organic matter in the Nonesuch Formation that are indicative of marine microbial communities. A lower unit of the Nonesuch Shale produces seeps of paraffinic, low-sulfur oil. The upper formation of the Oronto Group is the Freda Sandstone, composed primarily of red-to-buff sandstones that are more mature than the sandstones lower in the group.

The final sedimentary unit related to the Midcontinent Rift is the Bayfield Group (Jacobsville Sandstone), which consists primarily of sandstones that increase upward in textural and mineralogical maturity (Kalliokoski, 1982; Morey and Ojakangas, 1982). The sands were deposited largely in meandering river systems, but also by eolian and possibly lacustrine processes on alluvial plains. The rocks of this group occur over the rift basin and in wedges adjacent to the rift that thin gradually onto the rift shoulders. These sediments are believed to have been deposited in broad sag basins, centered over the rift, that extended well beyond the margins of the rift trough. The contact between this unit and earlier rift-rock units does not crop out, although seismic reflection studies in Lake Superior (Cannon et al., 1989) suggest an angular unconformity, at least locally, between this group and previous units. Tectonic activity during deposition of the Jacobsville Sandstone included not only the concurrent basin development, but also uplift along the Keweenaw reverse fault (Kalliokoski, 1982; 1989). The age of the Jacobsville Formation is poorly constrained stratigraphically, but paleomagnetic measurements indicate an age slightly younger than the underlying Oronto Group (Palmer and Davis, 1987).

Lithostratigraphic equivalents of the Oronto and Bayfield Groups are presumably present along the limbs of the rift outside of the Lake Superior region (e.g., Fowler and Kuenzi, 1978; Dickas, 1986; Berendsen et al., 1988, Witzke, 1990). Stratigraphic correlation has been tenuous, however, due to the lack of time markers in the units and limited paleomagnetic measurements. This is illustrated in Figure 9, which gives the general stratigraphic sequence in the M. G. Eischeid #1 drill hole, Carroll County, Iowa. The lower sequence and upper sequence of this figure are similar to the Oronto and Bayfield Groups, respectively, of the Lake Superior region, although compositional differences exist (Witzke, 1990).

GEOPHYSICAL SURVEYS

Geophysical surveys of the Midcontinent Rift, as is the case with many other paleorifts, have played a major role in defining and determining its nature (Hinze et al., 1992). Outcrops of the rift rocks are limited, and only a few, widely and poorly distributed deep drill holes penetrate into the rift through the overlying sedimentary strata. Even the rock exposures in the Lake Superior region have limitations because they represent the upper or marginal rocks of the rift.

Seismic studies

Seismic studies have had an increasing role in the study of the rift in the Lake Superior region and, more recently, along the southern extensions of the Midcontinent Rift. The earliest of the seismic studies were conducted by United States and Canadian geophysicists within and adjacent to Lake Superior as part of the Upper Mantle Project (Steinhart and Smith, 1966). These studies indicated a complex crust consisting of low-velocity surface sedimentary rocks overlying a thick section of high-velocity mafic volcanic rocks, an interpretation compatible with the Keweenawan geology mapped around the perimeter of Lake Superior. In addition, these studies indicated a high-velocity layer, ranging from 6.63 to 6.90 km/s, beneath the volcanic rocks and a crustal thickness in excess of 50 km. On the basis of these results and related geologic and geophysical studies, Smith et al. (1966) concluded that the Lake Superior geologic basin was a paleorift. Continued crustal seismic studies (Cohen and Meyer, 1966; Ocola and Meyer, 1973; Luetgert and Meyer, 1982; Wold et al., 1982) revealed more information on the regional rift-rock structure within the Lake Superior basin and suggested a similar crustal structure in the western limb of the rift extending from Lake Superior to Kansas. Halls (1982) used crustal seismic time-terms to prepare a crustal thickness map of Lake Superior and environs. His results indicate that the thickened crust extends out of the Lake Superior basin along both the eastern and western limbs of the Midcontinent Rift.

Seismic reflection studies of the rift began in the late 1970s and ushered in a new era of understanding the nature of the Midcontinent Rift. COCORP deep seismic profiling across the rift beneath the Michigan basin and in northeastern Kansas revealed the presence of fault-bounded basins of mafic volcanic rocks leading into thick, upper sedimentary rock units (Brown et al., 1982; Serpa et al., 1984; Zhu and Brown, 1986; Woelk and Hinze, 1991). Increasing interest in the Midcontinent Rift as a potential frontier petroleum province, because of the presence of hydrocarbons in the Oronto Group, led to a number of seismic profiles by geophysical contractors. These profiles, with two-way time (TWT) of 5 s or more, were collected over the western limb and Lake Superior segments of the rift. In general, the quality of these data is excellent; thus, those sections made available to the scientific community have been extremely useful in mapping the finer structure of the Midcontinent Rift not recognizable on seismic refraction profiles. The interpretation of one or more of these profiles has been reported by Dickas (1986), Nyquist and Wang (1988), Chandler et al. (1989), Mudrey et al. (1989), and Hinze et al. (1990). Additional seismic reflection surveys with TWT of 20 s were conducted by the Great Lakes International Multidisciplinary Program on Crustal Evolution (GLIMPCE) in 1986 over the western Great Lakes. These surveys have provided unprecedented images of the rift to sub-Moho depths. Results of these studies in Lake Superior and northern Lake Michigan have confirmed the regional structure interpreted from previous geologic and geophysical studies and have provided critical new details concerning the rift and underlying crust (Behrendt et al., 1988, 1990; Cannon and Green, 1989; Cannon et al., 1989, 1991; Hutchinson et al., 1990; Milkereit et al., 1990; Samson and West, 1992, 1994). Perhaps the most remarkable new information concerning the rift is that in eastern Lake Superior the rift basin reaches a depth of ~30 km with roughly 20 km of mafic volcanic flows overlain by 10 km of Keweenawan sedimentary strata. The pre-rift crust was thinned to less than one-third of its original thickness during the rifting event. However, the present crust is thickened along the rift due to the massive addition of the igneous and sedimentary rocks.

Recent studies of wide-angle reflections and refractions obtained during the GLIMPCE seismic reflection experiments have provided information on the velocities and structure of the crust in central Lake Superior and confirm the current interpretations of the rift basin (Chan and Hajnal, 1989; Jefferson et al., 1989; Lutter et al., 1989; Meyer et al., 1989; Milkereit et al., 1990; Morel-à-l'Huissier and Green, 1989; Shay and Trehu, 1989; Trehu et al., 1991; Hamilton and Mereu, 1993). The subsidence required for the observed thickness of the Oronto Group and volcanic flows has led Hutchinson et al. (1990) to suggest a stretching factor (β) of at least three. This estimate considers both thermal relaxation of the asthenosphere and removal of support from a mantle plume as it moved out of the region or dissipated.

Gravity and magnetic surveys

Gravity and magnetic surveys had an important role in the early definition and study of the Midcontinent Rift (Woollard, 1943; Thiel, 1956; Lyons, 1959, 1970; Craddock et al., 1963; Hinze, 1963; Hinze et al., 1966, 1975, 1982; White, 1966; King and Zietz, 1971; Oray et al., 1973; Halls, 1978; McSwiggen et al.,

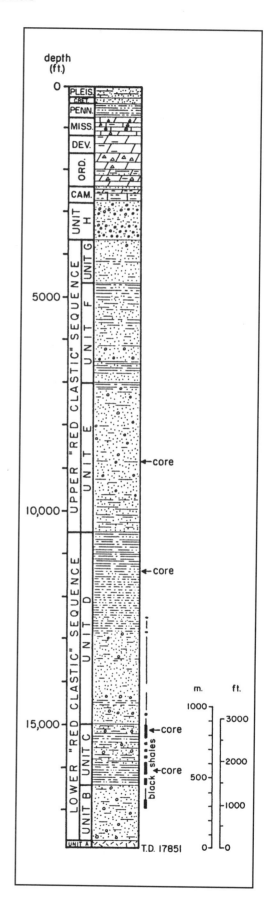

Figure 9. General stratigraphic sequence in the M. G. Eischeid #1 drill hole, Carroll County, Iowa (after Witzke, 1990).

1987) because these methods readily detect the physical property contrasts among the igneous and sedimentary rocks of the rift and the surrounding country rocks. Although many studies involved quantitative modeling, the results of these studies were ambiguous due to the lack of geologic constraints (except for near-surface geology in the Lake Superior region). Nevertheless, the location of the rift as defined by early gravity and magnetic surveys has been corroborated by deep drill holes in Michigan (Sleep and Sloss, 1978), Wisconsin (Dickas, 1992), Minnesota (Morey, 1977; Allen and Chandler, 1993), Iowa (Anderson, 1990a), and Kansas (Berendsen et al., 1988) that have penetrated through the overlying formations into the Midcontinent Rift. Strong density contrasts exist among the country rock (2.7 Mg/m^3), volcanic and plutonic rocks (2.90 to 2.95 Mg/m^3), Oronto Group (2.65 to 2.70 Mg/m^3), and the Bayfield Group (2.35 to 2.45 Mg/m^3). As a result, intense gravity anomalies, both positive and negative, are associated with the rift. In Minnesota the central positive gravity anomaly attains values as much as 100 mGal above background level. In Iowa, marginal, paralleling gravity minima on both sides of the maximum have values more than 80 mGal below the regional level. The resulting intense gravity anomaly of the Midcontinent Rift is the most striking feature of the gravity anomaly map of the central Midcontinent and is the principal method for delineating the extent of the rift (Fig. 1). The related magnetic anomaly of the rift is more complex and less consistent over the interpreted length of the Midcontinent Rift (Fig. 2). Problems in tracing the rift by means of the magnetic anomaly are the result of complicated magnetizations of the volcanic rocks. The remanent magnetization component of the volcanic rocks is of significantly greater magnitude than the induced magnetization component and is either normally or reversely polarized in a different direction than the induced magnetization. Furthermore, the volcanic sequence is deformed, thereby rotating the remanent vector and complicating the magnitude and direction of the total magnetization.

Early modeling (e.g., Thiel, 1956; King and Zietz, 1971) ascribed the complex magnetic anomaly signature and the positive and negative gravity anomalies to upper crustal density and magnetization contrasts related to the rift-rock package. The central positive gravity and magnetic anomalies were interpreted as a horst of mafic volcanic rocks, whereas the flanking anomaly minima were interpreted as originating from low-density, low-magnetization sedimentary strata related to a late-stage sag basin. More recently, models of these anomalies have included mafic intrusions beneath the volcanic rocks and a thickened crust. The mafic intrusions lead to a localized positive gravity effect along the rift axis, adding to the anomaly produced by the volcanic units, whereas the thickened crust produces a broad gravity minimum that complicates the interpretation of the negative anomalies associated with the bordering late-stage sag basins.

Gravity and magnetic anomalies have also been used to postulate extensions of the limbs of the rift beyond the limits in Michigan and Kansas that are shown in Figure 3. Geophysical anomalies in Ohio have been interpreted as a possible extension of the eastern limb of the Midcontinent Rift from Michigan

(e.g., Keller et al., 1982). However, recent COCORP seismic reflection profiling across central Ohio (Pratt et al., 1989) indicates that these anomalies are associated with lower crustal rocks thrust into place along the Grenville Front (Sunwoo, 1989), the western margin of the 1.1 Ga Grenville orogen, rather than by mafic rift-related rocks. The structure of the western limb of the rift has been traced by Yarger (1985) in magnetic data from central Kansas to the Oklahoma border. In a more speculative manner, Nixon (1988) suggests that the rift extends into north-central Oklahoma on the basis of gravity modeling and geologic data. Based on isotopic dating of igneous rocks, surface geology, and geophysical anomalies, Adams and Keller (1994) suggest that the Midcontinent Rift may extend even farther into west Texas and New Mexico.

Gravity and magnetic coverage over the Midcontinent Rift is adequate for regional interpretations. Gridded data sets are publicly available for both gravity anomaly (6-km grid) and total magnetic intensity anomaly (2 km grid) data. Of particular importance are recent high-resolution aeromagnetic surveys of Lake Superior (Teskey et al., 1991) and Minnesota (Chandler, 1991). In Minnesota the high-resolution data permit detailed mapping of lithologic units within the Keweenawan volcanic sequence (Fig. 10) and their structural disturbances (Allen and Chandler, 1992, 1993).

Combined analysis of gravity anomaly and topographic data by Allen et al. (1992) in the Lake Superior region suggests a broad-scale lithospheric feature that may be a remnant of the proposed Keweenaw hotspot (mantle plume; Hutchinson et al., 1990). A broad, low-amplitude topographic high centered on Lake Superior, the so-called Lake Superior Swell (Dutch, 1981), is associated with a regional negative Bouguer gravity anomaly. The elevation/gravity anomaly relationship suggests that the feature is in or approaches isostatic equilibrium. Allen et al. (1992) have shown that the average gravity anomaly increases and the average elevation decreases in a linear manner over a radial distance of 300 to 650 km (Fig. 11) from the approximate center of Lake Superior. Allen et al. (1992) suggest that the correlated gravity and topographic anomalies are vestiges of the magmatic activity associated with the 1.1-Ga rifting event. The mass deficiency may represent a broad underplated region at the base of the crust, a modified upper mantle, or both. Underplating may result from either upper mantle or plume-derived magma that was trapped at the base of the crust. Alternatively, extraction of basaltic melts from either the lithospheric mantle or plume head, which subsequently became attached to the lithosphere, may lead to a less dense upper mantle. The density of the depleted mantle decreases due to an increased MgO/FeO ratio and loss of dense phase garnets (Boyd and McCallister, 1976; Oxburgh and Paramentier, 1977).

Other geophysical studies

In recent years, a few magnetotelluric surveys have been conducted over the Midcontinent Rift (Wunderman, 1986, 1988; Young and Repasky, 1986; Wunderman and Young, 1987a, b;

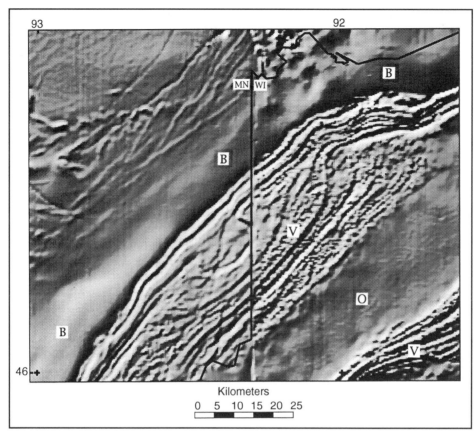

Figure 10. Shaded-relief presentation of high-resolution aeromagnetic data for east-central Minnesota and extreme northwest Wisconsin. The sun illumination is from the north-northwest. Intense northeast-striking magnetic anomalies are associated with Keweenawan volcanic rocks (V), whereas subdued magnetic anomalies are associated with the overlying Oronto Group (O) and the flanking Bayfield Group (B). Linear, northeast-striking anomalies in the northwest corner of the map are associated with Keweenawan dikes cutting Lower Proterozoic basement.

Young et al., 1989). These surveys have been useful for studying the electrically conductive, off-rift sedimentary basins, the gross structure of the rift-rock package, and pre-rift basement lithologic variations. On the basis of different conductivity signatures east and west of the Midcontinent Rift, Wunderman (1988) speculates that the northern segment of the western limb of the rift developed along a 1.8-Ga Penokean suture.

The seismicity of the Midcontinent Rift has been investigated in the Lake Superior region by Frantti (1980); in Minnesota by Mooney (1981), Mooney and Morey (1981), Mooney and Walton (1986), and Chandler and Morey (1989); and in Nebraska and Kansas by Steeples (1980, 1981) and Steeples et al. (1989). In general, seismicity is limited along the Midcontinent Rift with little obvious correlation between known structures and epicenters. A significant exception has been reported by Steeples (1980) who demonstrated that recent microseismicity in Kansas and Nebraska correlates with the Humboldt fault, a fault that parallels the eastern margin of the southern segment of the Midcontinent Rift. In addition, a concentration of microearthquakes occurs

along a related hinge zone fault on the outward margin of the western flanking clastic sedimentary basin. Steeples et al. (1989) also speculate that earthquakes occur near transform faults of the western limb of the rift.

Heat-flow data in the vicinity of the Midcontinent Rift are sparse and poorly distributed (Roy et al., 1972). Available data fall within the range of normal terrestrial heat-flow values. However, we would anticipate a heat-flow minimum over the rift due to the marked attenuation of the granitic crust during the rifting event. Data from Roy et al. (1972) failed to show the expected minimum in a profile over the Midcontinent Rift in Minnesota. Hart and Steinhart (1964) and Steinhart et al., (1967; 1968) published more than 150 heat flow measurements from Lake Superior. These values range from roughly 20 to 60 mW/m^2. They have been interpreted by Drury (1991) and Hart et al. (1994). Of particular interest is a trough of low heat flow values that extends along the northern edge of Lake Superior. This trough roughly correlates with the updip termination of the basaltic flows along the northern edge of the Lake Superior basin.

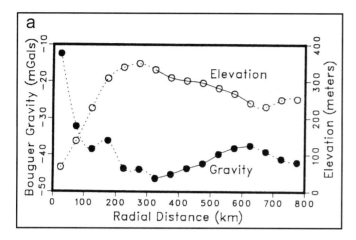

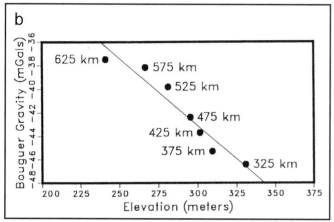

Figure 11. a, Average Bouguer gravity anomaly and average surface elevation versus radial distance from a common point located in central Lake Superior (after Allen et al., 1992). b, Average Bouguer gravity anomaly versus average surface elevation for radial distances between 300 and 650 km from the same common point referenced in a (after Allen et al., 1992).

STRUCTURE

Introduction

The fundamental structure of the Midcontinent Rift in the Lake Superior region, including the asymmetric basin containing a thick section of mafic volcanic rocks overlain by a thick sedimentary sequence, was established a century ago by geologists interested in the copper-bearing Keweenawan rocks (e.g., Irving, 1883). Furthermore, early workers (e.g., Van Hise and Leith, 1911) mapped the significant reverse movement at the base of the Portage Lake Volcanic sequence in northern Michigan. In the following years, steady progress was made in detailed mapping of the surface and near-surface rocks, primarily in the mining districts of Michigan. The Upper Mantle Project in the 1960s led to important results regarding the

structure of both the upper and lower crust of the Lake Superior region (e.g., Steinhart and Smith, 1966). These advances were largely the result of crustal seismic refraction, gravity, and magnetic studies. The interpretations supported inferences from surface geology, but showed that the crustal structure was much more complicated than previously envisioned. A rift origin for the structure was strongly suggested, but caution in acceptance of the rift hypothesis was expressed due to the lack of clear surface evidence for grabens and associated normal faults (White, 1966). Furthermore, it was at this time that the continuity of the structure as a series of segments from Kansas to Lake Superior and across the Michigan basin was established (Hinze et al., 1966, 1975; King and Zietz, 1971; Oray et al., 1973).

A major step forward in the understanding of the structure of the Midcontinent Rift was made in the late 1970s and the 1980s with the acquisition of deeply penetrating seismic reflection data in the Lake Superior region and along both limbs of the rift. These data have been especially useful for constraining the interpretations of ambiguous gravity and magnetic anomaly data. The combined interpretation of seismic reflection, gravity, and magnetic data, constrained by surface geology and the logs of a few deep drill holes, provides the following view of the structure of the Midcontinent Rift.

Near-surface rift structures

Lake Superior basin. In the Lake Superior region, seismic-reflection data, supported by other geophysical and geological data, confirm the existence of a fault-bounded basin filled with a volcanic sequence. The basin is commonly asymmetric with major faulting on only one side, although much of eastern Lake Superior is essentially symmetric (Cannon et al., 1989; Mariano and Hinze, 1992; Samson and West, 1992). Figure 12, a crustal profile transverse to the Midcontinent Rift in western Lake Superior, is a composite of the modified interpretation of GLIMPCE line C (Cannon et al., 1989), an on-shore profile in the Northern Peninsula of Michigan (Hinze et al., 1990), and gravity modeling. The on-shore profile at the southeastern end of the profile is critical because it permits direct correlation between seismic reflections and outcropping geology. Figure 12 shows a thick Keweenawan rock section (~25 km) and a profound thickening of the volcanic rocks suggesting normal growth faults, for example, at the Keweenaw fault (KF). This growth fault is now a high-angle reverse fault due to compressional reactivation during, and perhaps subsequent to, the deposition of the late-stage sedimentary sequence (Bayfield Group/Jacobsville sandstone). Vertical movement on the reverse fault is roughly 5 km on this section, but elsewhere it attains values of up to 7 km. The complementary reverse fault on the northwest side of the rift basin is a growth fault that has experienced only minor late-stage reverse movement. This fault

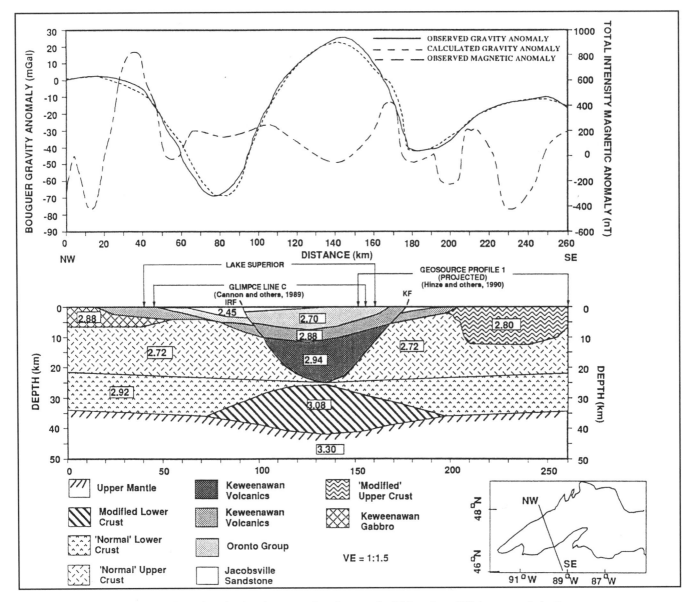

Figure 12. Crustal section (northwest-southeast) across the Midcontinent Rift in western Lake Superior based on seismic reflection profiling and gravity modeling. The seismic reflection interpretation in Lake Superior is based on the interpretations of line C of the GLIMPCE program (Cannon et al., 1989), and the section southeast of the lake across the Keweenaw fault (KF) is derived from seismic reflection and gravity modeling of Hinze et al. (1990; after Hinze et al., 1992). The location of the Isle Royale fault is identified by IRF.

has been interpreted as the Isle Royale fault (IRF), but the extension of the Isle Royale fault from Isle Royale to this profile is equivocal. The structure of the central rift basin within the bounding faults is quite in contrast to the marginal basins, where the basal volcanic rocks and overlying sedimentary strata thin gradually out to the margins. Beneath the rift axis, the pre-rift crust has largely been eliminated during the extensional process or at least heavily intruded by mantle melts. The resulting crust beneath the rift is thickened in contrast to the adjacent Archean crust by intrusions and crustal underplating.

A more symmetric cross section of the Lake Superior rift structure is interpreted along Grant-Norpac, Inc./Argonne line 36 (Fig. 13). This profile extends east-northeast across eastern Lake Superior ~25 km north of the southern shoreline. Figure 14 shows the crustal section derived from combined seismic reflection interpretation and gravity modeling as well as modeling of the magnetic anomaly data (Fig. 15). The volcanic section is roughly 15 km thick with as much as 10 km of overlying Oronto Group and Jacobsville Sandstone. The volcanic and sedimentary units are distinguished on the basis of their reflec-

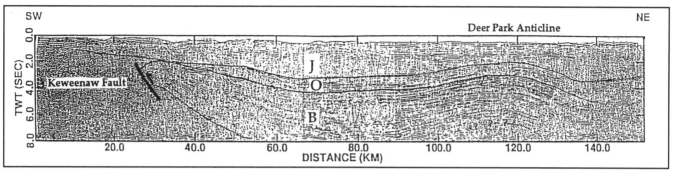

Figure 13. Interpreted migrated record section (southwest-northeast) of Grant-Norpac Inc./Argonne line 36 across the western margin of the Midcontinent Rift in eastern Lake Superior. The Keweenawan rock units are identified as follows: J, Jacobsville sandstone or equivalents; O, Oronto Group; B, Keweenawan basalts (after Mariano and Hinze, 1994a).

tivity (Fig. 13). An anticline, the Deer Park anticline, with a vertical relief of roughly 5 km lies immediately east of the center of the basin. Stratigraphic relations, indicated in the seismic reflection data (Fig. 13), suggest that uplift of the anticline commenced during deposition of the Oronto Group sediments and was terminated before the completion of the Jacobsville Sandstone sedimentation. Presumably, the compression leading to the anticline was contemporaneous with the reverse move-

ment along the southeastern extension of the Keweenaw fault; this fault represents the southwest margin of the volcanic basin in eastern Lake Superior. Reverse movement is also evident in the drag folding of the volcanic units (Fig. 13) at the western margin of the volcanic basin near kilometer 30. Additional structural complexity on this reverse fault has been observed farther northwest toward the Keweenaw Peninsula by Mariano and Hinze (1994a). They interpret the edge of the volcanic

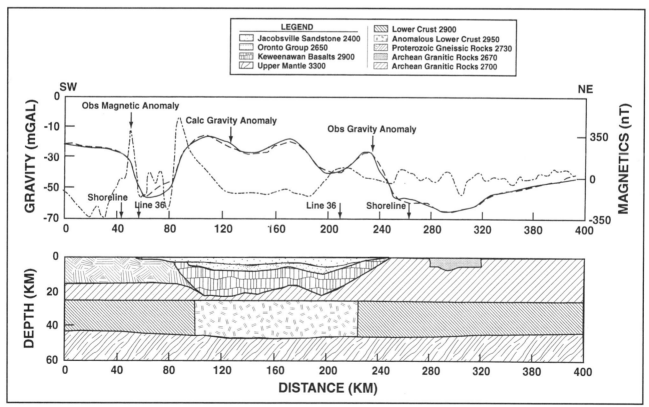

Figure 14. Two and one-half dimensional Bouguer gravity anomaly model of extended seismic reflection line shown in Figure 13 (after Mariano and Hinze, 1994b). OBS is observed anomaly, and CALC is anomaly calculated from illustrated model.

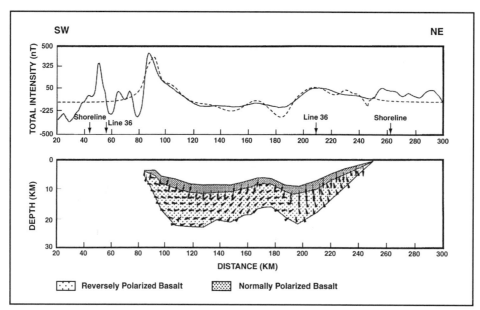

Figure 15. Two-dimensional forward magnetic model of extended seismic reflection line shown in Figures 13 and 14 (after Mariano and Hinze, 1994b). Magnitude of induced magnetization: 2.9 A/m. Magnitude of remanent magnetization: 4.5 A/m. The vectors indicate the direction of total magnetization in the plane of the profile. The essentially nonmagnetic sedimentary strata are not shown.

rocks as a low-angle splay off of the main reverse fault on the basis of the seismic reflection, gravity, and magnetic anomaly data. Previous investigators (Meshref and Hinze, 1970; Behrendt et al., 1988; Cannon et al., 1990) have also observed splays of the main reverse fault that merge with the main fault zone at depth in geophysical/geological models.

A near-vertical fault is indicated at the eastern margin of the Deer Creek anticline, as shown in Figure 13 near kilometer 140, by the disruption of the volcanic reflections. Mariano and Hinze (1994a) have interpreted this as a near-vertical fault, the Crisp Point fault, that developed late in Keweenawan time. Using seismic reflection profiles and gravity and magnetic anomaly data, they have traced the fault northwesterly, parallel to the axis of the anticline to 50 km due south of Michipicoten Island, where the fault turns sharply to the west and continues to the tip of the Keweenaw Peninsula.

As shown in Figure 15, only the upper 20% of the volcanic section is normally magnetically polarized. Similar values are modeled northwest to the Portage Lake Volcanics on the Keweenaw Peninsula. As a result, we infer that the normally polarized volcanic rocks are equivalent to the Portage Lake Volcanics and that the reversely polarized lower basalts are older than the reversal at 1,097 Ma. Inversion of the magnetic anomaly by Mariano and Hinze (1994b) confirms that the magnetization of the basalts extends to depths in excess of 20 km. Thus, the magnetization of the volcanic rocks has not been significantly altered by burial metamorphism or hydrothermal alteration.

South of Michipicoten Island in central eastern Lake Superior in the vicinity of the Crisp Point fault, the gravity and magnetic anomalies are inversely correlated, in contrast to their common direct correlation over much of Lake Superior. The inverse correlation coincides with a low-reflectivity zone on GLIMPCE line F (Fig. 16). This unit is interpreted by Mariano and Hinze (1994b) as a magnetic, low-density felsic volcanic unit that was extruded as the basaltic volcanism terminated.

Western and eastern limbs. The structures of the eastern and western limbs of the Midcontinent Rift are not as well known as the Lake Superior segment due to a cover of Phanerozoic sedimentary strata and fewer deeply penetrating seismic reflection profiles. Over the past decade, however, several new seismic profiles have become available over the limbs of rift in Michigan, Minnesota, Wisconsin, Iowa, and Kansas. These data, together with gravity and magnetic modeling and direct information from an occasional deep drill hole, have shown that the stratigraphic units and structure of the limbs of the Midcontinent Rift are essentially identical to those in the Lake Superior segment (Hinze et al., 1992). The interpretations of Chandler et al. (1989) in Minnesota and Wisconsin and Anderson (1990b) in Iowa have been particularly informative in understanding the nature of the western limb. Farther south along the western limb in Kansas, drilling has shown that Keweenawan igneous rocks (Berendsen et al., 1988; Van Schmus et al., 1990) occur above late-stage clastic sedimentary rocks that formed in sag basins over the rift. This inversion of the normal Keweenawan sequence was shown by Woelk and Hinze (1991), using reprocessed COCORP seismic reflection data and gravity and magnetic modeling, to be a result of high-angle reverse movement, along an originally normal fault, which thrust older igneous rocks over the younger sedimentary rocks (Fig. 17). This interpretation of the Kansas segment of the rift and the iso-

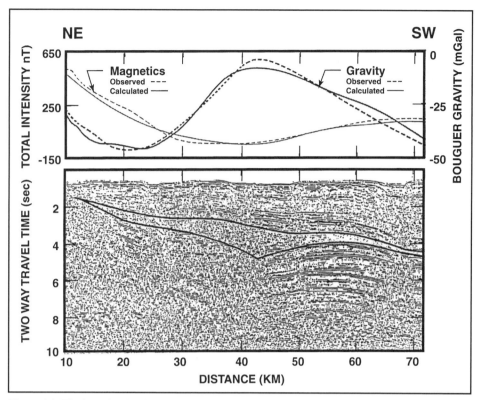

Figure 16. Final stack seismic reflection record section detail of GLIMPCE line F in east-central Lake Superior showing the outline of the interpreted felsic flow used in the forward gravity and magnetic models. The observed and calculated Bouguer gravity and total intensity magnetic anomalies are shown (after Mariano and Hinze, 1994b). The felsic unit has a density of 2,600 kg/m^3, an induced magnetization of 2.9 A/m, and a normally polarized remanent magnetization of 7.0 A/m. The felsic unit lies above basaltic volcanic flows and beneath Oronto Group sedimentary strata.

topic age of 1,097.5 Ma for the igneous rocks, which falls within the narrow age range of Keweenawan igneous rocks in the Lake Superior region, confirms the general consistency in structural style, tectonic evolution, and age along the length of the rift system.

The extension of the Midcontinent Rift from southeastern Lake Superior across northern Michigan (Fox, 1988; Cannon et al., 1991) into southern Michigan is evident in the continuity of the gravity anomaly (Oray et al., 1973; Hinze et al., 1975). However, the available basement drillhole data and the seismic reflection profiling show no evidence of late-stage rift sedimentary rocks along the margins of the rift basin over much of its length in southern Michigan. Furthermore, a significant length of the rift has no correlative magnetic anomaly, suggesting either destructive interference between normally and reversely polarized basalts within the rift basin or a lack of basalts. The latter scenario requires that the positive gravity anomaly is produced solely by deep intrusions that have a minimal magnetic signature. Reverse faulting is equivocal in the Michigan segment of the rift, although seismic reflection profiling (Zhu and Brown, 1986; Fox, 1988) shows that the Keweenawan formations were folded (Fig. 18). The stratigraphic relations between the volcanic flows and the overlying

Keweenawan sedimentary strata are inconclusive regarding the age of the folding because of the strong multiple reflections from the Phanerozoic sedimentary strata of the overlying Michigan basin. These data, however, indicate that the folding was largely completed by the end of the deposition of the late-stage rift sediments. The folding is interpreted to be the result of late-stage compressional stresses, perhaps associated with thrusting within the Grenville orogen. The structural relations between the eastern extremity of the rift in Michigan and the western limit of the coeval (1.1-Ga) Grenville orogeny remains an open question, but the geophysical anomalies suggest that the extreme eastern end of the rift, near the Canadian border, is buried by Grenvillian rocks thrust to the west. Similar thrusting is observed elsewhere along the Grenville Front (Green et al., 1988, 1989; Pratt et al., 1989).

Transverse structures

The asymmetry of the individual rift segments alternates in the Lake Superior portion of the rift (Mudrey and Dickas, 1988; Cannon et al., 1989; Dickas and Mudrey, 1989). Similar patterns of asymmetry have been observed in the western limb of the rift (Chandler et al., 1989) and in the East African Rift System

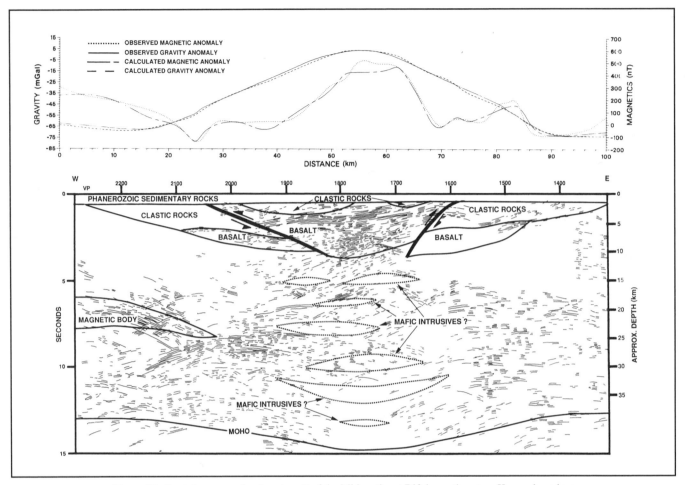

Figure 17. Crustal cross section (west-east) of the Midcontinent Rift in northeastern Kansas based on gravity modeling constrained by magnetic anomaly modeling and seismic reflection profiling (after Woelk and Hinze, 1991).

(Bosworth et al., 1986; Rosendahl, 1987). This asymmetry, however, is not universal in the Midcontinent Rift. Some segments, including eastern Lake Superior and Iowa, are symmetric to only slightly asymmetric. Each segment has its own volcanic and sedimentological depositional patterns. The segmented pattern of en echelon grabens, which developed early in the history of the rift, may have been controlled by old zones of weakness that were reactivated by the rifting process (e.g., Weiblen and Morey, 1980; Mooney and Morey, 1981; Klasner et al., 1982).

One of the most profound cross structures is the Thiel fault that strikes north-northeast to south-southwest across Lake Superior from immediately east of the tip of the Keweenaw Peninsula. The fault occurs along the southeast edge of a ridge that separates two volcanic basins in Lake Superior (Mariano and Hinze, 1994a). Gravity and magnetic modeling of the Thiel fault suggests that little if any thrusting occurred along it during the late-stage compressional event. Numerous other cross faults have been recognized (e.g., Davidson, 1982; Mariano and Hinze, 1994a; Manson and Halls, 1994) that do not transect the entire rift and are probably late-stage features. Perhaps the most

profound of these cross faults are the ancestral faults in the Archean rocks east of Lake Superior that were reactivated to produce the anomalous structure associated with the basalt exposures on Mamainse Point. Many faults along the entire length of the rift have, at least locally, been reactivated by post-Keweenawan stresses, as evidenced by related structures and stratigraphic variations within overlying Phanerozoic sedimentary strata (e.g., Middleton, 1991; Carlson, 1992).

Deep-rift structures

The nature of the crust beneath the rift basins has been a major concern in Midcontinent Rift studies. Seismic refraction profiling and analysis of gravity and magnetic data provided only limited information concerning the deep-rift structures. However, these data did indicate a thickening of the crust to 55 km and a possible densification of the lower crust (e.g., Hinze et al., 1982). The recent availability of wide-angle seismic reflection data along GLIMPCE line A, a roughly north-south profile in central Lake Superior, has resulted in a major

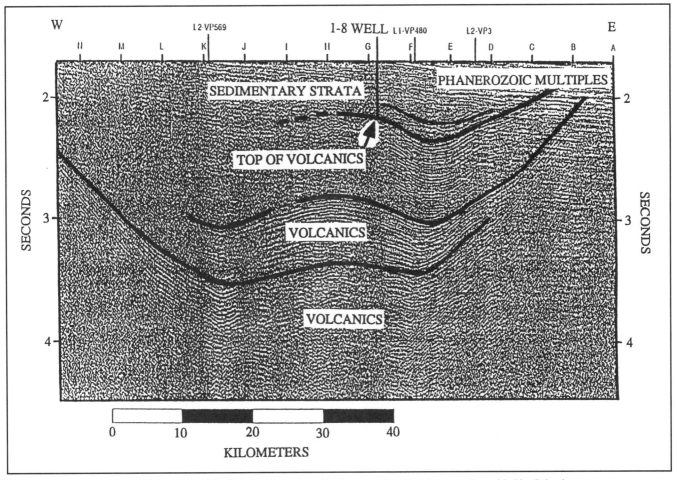

W   L2-VP569   1-8 WELL   L1-VP480   L2-VP3   E

H   M   L   K   J   I   H   G   F   E   D   C   B   A

SEDIMENTARY STRATA   PHANEROZOIC MULTIPLES

TOP OF VOLCANICS

VOLCANICS

VOLCANICS

SECONDS

0   10   20   30   40

KILOMETERS

Figure 18. Interpreted final stack of seismic reflection record section (west-east) provided by Seismic Reflections, Inc., that parallels the Midcontinent Rift in central Southern Peninsula of Michigan (after Fox, 1988). The drill hole shown is the McClure-Sparks #1-8, which encountered Keweenawan basalts at ~2.2 s (~5.3 km) below late-stage Keweenawan clastic sedimentary strata.

breakthrough in the study of deep-rift structures. Trehu et al. (1991) interpreted the seismic data using an iterative, forward modeling approach together with gravity modeling. Their interpretion shows a thickened crust and a lower crust of higher density and higher velocity beneath the rift, similar to the models portrayed in Figures 12 and 14. Hamilton and Mereu (1993) performed two-dimensional travel-time tomographic inversion on data also from GLIMPCE line A. Their results confirm the presence of a thickened crust with a broad zone of high-velocity (7.0 to 7.2 km/s) lower crust beneath the axis of the rift. This roughly 50-km-wide lower crustal zone correlates with a higher density lower crust beneath the rift in eastern Lake Superior interpreted by Mariano and Hinze (1994b) using gravity modeling constrained by seismic data. The high-density, high-seismic velocity lower crust is believed to be a result of a combination of intrusion and underplating.

Samson and West (1992) reprocessed the seismic reflection data of GLIMPCE line F in eastern Lake Superior and found a lack of reflectors beneath the central graben. They sug-

gested that this transparent zone shows that large-scale crustal remnants or mantle intrusions are absent beneath the graben; if they were present, reflections would be observed in the profiling. They also found a complex crust-mantle transition zone in the reflection data that they interpreted as being consistent with underplating of the crust during the rifting process. Green (1991) has performed teleseismic tomographic inversion on the western limb of the rift in Wisconsin and Minnesota. He found that the lower crust in the range from 28 to 46 km depth shows higher velocities than in the adjacent crust. He also found a slightly lower upper mantle velocity beneath the rift. Thus, the overall result from a variety of studies is that the crust is thickened beneath the rift and that the seismic velocity and density of the lower crust are increased probably due to mantle-derived intrusions and crustal underplating.

Faults penetrating the entire crust have not been observed in the seismic data; rather, deep reflections are parallel to the boundaries of the lower crust. These reflections may originate from ductile shear zones involved in the thinning of the crust during the extension process or from mantle intrusions. There is no evidence of a laterally displaced anomalous lower crust beneath the Midcontinent Rift as would be anticipated in rifts that involved simple shear (Wernicke, 1985).

Vertical movements

Vertical movements are a significant component of the continental rifting process. Evidence indicates that the Midcontinent Rift commenced with sag basin development on a positive topographic surface. Soon thereafter, normal faults developed along one or both of the margins of the basins. This subsidence was due to a combination of extension and mantle plume dynamics as the head of the plume rose to and spread out beneath the base of the lithosphere (e.g., Campbell and Griffiths, 1990; Griffiths and Campbell, 1991). Extension produced faulting in the brittle, upper crust that, in some segments of the Midcontinent Rift, was completely removed. Concurrently, the lower crust was thinned by ductile shear. The presence of major unconformities within the volcanic rock package and variations in the thickness of the rift rocks indicate that subsidence was neither continuous nor uniform. Subsidence generally kept pace with volcanism, with only the oldest of the flows being deposited subaqueously.

Following the volcanism and the primarily extensional phase of the rifting, graben development ceased and broad, sag basins developed along the axis of the rift. The subsidence was caused by a combination of thermal collapse and loading by the mass of the volcanic flows. Sedimentation kept pace with subsidence, and sediments increased in maturity during the history of the basins. The presence of marine affinities in the Oronto Group (Hieshima and Pratt, 1991; Pratt et al., 1991) indicates that the water levels in the basins were at sea level.

Uplifts are a significant element of modern rifts (Ramberg and Morgan, 1984) and, by analogy, probably existed along the Midcontinent Rift. Uplifts in modern rifts take two forms: broad, domed areas and flank uplifts. Broad domal uplifts occur at an early stage in the rifting process, when active rifting predominates as an ascending mantle plume spreads out at the base of the lithosphere. This pattern of uplift is consistent with the model studies of plumes by Griffiths and Campbell (1991). Strong evidence for regional upwarp, such as observed with the East African Rift, is subtle in the Midcontinent Rift; however, Keweenawan rocks were deposited on a continental terrane, indicating that the region was a positive topographic feature. Flank uplifts along rifts are produced by buoyancy forces associated with lithospheric thinning. Characteristically, they are larger in amplitude but less extensive than domal uplifts, and occur after the rifting is initiated. Evidence for this form of uplift in the Midcontinent Rift is suggested by the sedimentary rocks and their pattern of deposition in the subsiding basins. Merk and Jirsa (1982) discuss the relative amount of uplift around the Lake Superior basin in Keweenawan time based of their analysis of interflow sedimentary rocks.

Evidence of an isolated Keweenawan uplift in northern Wisconsin and Michigan has been presented by Peterman and Sims (1988). They mapped a region of anomalously low (ca. 1.1 Ga) Rb-Sr ages in the older rocks of the Wisconsin arch. They interpret these ages as representing closure of the Rb-Sr isotopic system as the region was arched upward, rapidly eroded, and cooled beneath the blocking temperature. They suggest that the uplift was a forebulge caused by loading of the crust by the dense igneous rocks of the Midcontinent Rift. However, flexural modeling of the rift (Peterman and Sims, 1988; Nyquist and Wang, 1989) does not satisfy the amount of uplift required by the model of Peterman and Sims.

A final stage of vertical uplift is evident along the length of the Midcontinent Rift from Kansas (Woelk and Hinze, 1991) to Lake Superior to northern Michigan (Fox, 1988). Vertical movements, evidenced by the high-angle reverse faults interpreted as reactivated normal faults, have 5 km or more of vertical movement (Chandler et al., 1989; Cannon et al., 1989; Hinze et al., 1990). Other indications of vertical movement include anticlines within the rift-rock package (e.g., Fox, 1988; Mariano and Hinze, 1994a) and drag folding along reverse faults (Mariano and Hinze, 1994a). These vertical movements were concurrent (Kalliokoski, 1982) with the late-stage sedimentary basin subsidence and continued after the subsidence was terminated. Reverse movement along the Keweenaw fault of northern Michigan has been dated at ca. 1,060 Ma (Cannon et al., 1991).

Uplifts appear anomalous in a rift regime, although they are observed in continental rifts that have undergone regional compression (Milanovsky, 1981). The uplifts along the Midcontinent Rift are unusual in that they occur throughout the rift, regardless of rift orientation. We suggest that this origin is related to the concurrent regional maximum horizontal compressional stresses associated with the east-to-west thrusting within the Grenville orogen immediately to the east of the Midcontinent Rift (Van Schmus and Hinze, 1985). Van Schmus (1992) indicates that the peak of the orogeny was from 970 to 1,030 Ma (Schärer and Gower, 1988). These regional stresses, acting upon the thermally weakened crust of the rift, could have reactivated the normal faults as high-angle reverse faults and also produced arching. Regional compression may also explain the sudden cessation of extension and the fact that the rift did not proceed to split the proto–North American continent, despite the nearly complete rupture of the crust.

Currently, the region centered around Lake Superior is a topographic high in isostatic response to the underplated crust or depleted upper mantle (Allen et al., 1992) associated with the rifting event. This topographic anomaly is believed to be a vestige of a broad domal uplift that was initiated at the end of the rifting process.

EVOLUTION OF THE MIDCONTINENT RIFT

During the past decade, the acceptance of a rift origin for the Lake Superior basin and the Midcontinent geophysical anomaly has been a critical step forward in understanding the tectonic evolution of the Midcontinent Rift. However, this recognition fails to clarify the mechanics of the causative process and the ultimate origin of the rift. Numerous processes have been suggested as the cause of continental rifts. In general, these processes fall into one of two categories. There are active mechanisms in which a thermal perturbation of the mantle, a plume, causes a disturbance within the overlying lithosphere and passive mechanisms, where plate interactions cause regional horizontal extension in the lithosphere (Baker and Morgan, 1981). Proposed origins of the Midcontinent Rift have invoked both mechanisms.

The overlap in time between the nearby Grenville orogeny and the 1,100-Ma Midcontinent Rift has suggested a cause and effect relation related to a passive mechanism (e.g., Hinze and Wold, 1982; Van Schmus and Hinze, 1985; Gordon and Hempton, 1986). Cambray (1988) has suggested a passive origin wherein the opening and closing vectors of the Midcontinent Rift are restricted to an approximately north-south direction. However, the consistency of structures along the length of the rift and the lack of strike-slip motion on the limbs of the rift do not support the largely transtensional stresses on the limbs of the rift predicted by this model.

Numerous authors have suggested an active mechanism involving an upper mantle perturbation associated with an asthenospheric mantle plume or hotspot (e.g., Hinze et al., 1972; White, 1972; Green, 1983; Hutchinson et al., 1990; Cannon and Hinze, 1992; Klewin and Shirey, 1992). The early papers on this subject were largely speculative, without strong, quantitative support. This situation has changed, however, with the availability of improved deep seismic reflection data and intensive isotopic and geochemical analyses of the rift's volcanic rocks. These investigations suggest that the Midcontinent Rift is analogous to rifts associated with passive continental margins and mantle plumes (hotspots). The observed volume (Behrendt et al., 1988; Cannon et al., 1989; Chandler et al., 1989; Cannon, 1992) and rate (Davis and Paces, 1990); Cannon, 1992) of volcanism, the lack of an associated pervasive dike swarm with a consistent strike (Cannon and Hinze, 1992), the inferred extension (Chandler, 1983; Hutchinson et al., 1990), and the isotopic character of the Keweenawan volcanic rocks (Paces and Bell, 1989; Hutchinson et al., 1990; Nicholson and Shirey, 1990; Klewin and Shirey, 1992) support a mantle plume origin for the Midcontinent Rift. Hutchinson et al., (1990) discuss the geochemical and geophysical evidence for the origin of the Midcontinent Rift with a hot asthenospheric plume, which they term the "Keweenaw hotspot," and the geologic implications of this interpretation.

Klewin and Shirey (1992), however, suggest that the lithospheric origin for the early basalts of the Midcontinent Rift, based on their isotopic and trace element geochemistry, indicate that the plume was not responsible for the rifting event. Furthermore, the spatial and temporal proximity of the Midcontinent Rift and the Grenville orogeny indicates a possible connection and a passive mechanism perhaps associated with some still

undefined plate boundary forces. The significance of the relation between the Midcontinent Rift and the Grenville orogeny remains elusive, as does the source of the deviatoric tension that caused the Midcontinent Rift. None of the proposed mechanisms for tensional stresses, such as bending or membrane stresses or drag on the base of the lithosphere, provides a satisfactory explanation for the available evidence. Perhaps instabilities in the head of the asthenospheric mantle plume cause complex convection patterns leading to local tension that is capable of producing the observed geometry of the Midcontinent Rift. Resolution of this problem awaits further studies.

When discussing the evolution of the Midcontinent Rift, it is useful to portray the development in a series of cross sections of the lithosphere through time. We present our current ideas of the evolution of the rift in Figure 19. These cross sections are simplified, and no attempt is made to illustrate the plan view of the rift. The diagrams generally represent the culmination of a particular episode in the development of the rift, and absolute ages are applied to the diagram to the extent possible. A brief description of these episodes follows. As shown in Figure 19a, the rift is initiated with a broad, domal isostatic uplift over an ascending asthenospheric plume at ca. 1,110 Ma. Convectional movements within the head of the mantle plume may cause extension or they may have an external origin. As illustrated in Figure 19b, at ca. 1,108 Ma, extension reached a magnitude that, together with thermal weakening of the lithosphere, initiated a thinning or necking of the crust and upper mantle. As the head of the mantle plume approached the base of the lithosphere, it spread laterally, and the crest of the uplift subsided while the distal portions continued to rise. Clastic sediments were deposited in this basin at ca. 1,108 Ma. Shortly thereafter, portions of the upper mantle were melted by the thermal pulse of the rising plume. The resulting tholeiitic magmas rose through developing extensional cracks in the lithosphere. Evidence suggests that this early stage volcanism and the associated feeder dikes were much more widespread than the present lateral extent of the rift.

The head of the plume reached the base of the lithosphere (Fig. 19c) and spread out laterally (ca. 1,100 Ma). Continued extension led to additional thinning of the lithosphere, both crust and mantle, and focusing of the graben development along the present axis of the rift. Concurrent with the graben development, enormous volumes of basalts were extruded into the depression. The magmas were derived from lithospheric melting and an increasing amount of decompression melting within the plume (White and McKenzie, 1989). This was the period of major extension of the brittle crust with graben development and volcanism that kept pace with subsidence. During the following ~6 m.y. (Fig. 19d), the plume continued to spread laterally, and melts from the plume became the predominant source of volcanic flows and intrusions. The melts intruded the crust and were extruded regionally over the rift, but primarily into the grabens and half grabens that continued to develop in the colder, brittle crust over the continually thinning litho-

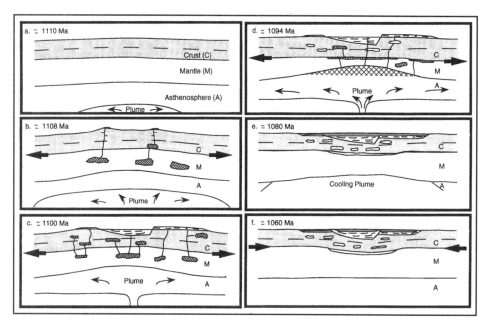

Figure 19. Diagrammatic sketches of lithospheric cross sections illustrating the magmatic/tectonic evolution of the Midcontinent Rift System from inception (a) to termination (f). Sketches are not to scale and mark different time intervals. Arrows indicate orientation of principal horizontal stress. Crosshatched areas are regions where magmas originate within the lithosphere and mantle plume. Lined areas in and adjacent to the rift basin are volcanic rocks and overlying dotted patterns are late-stage sedimentary rocks.

sphere. Regional crustal underplating occurred with crystallization of melts ponded at the base of the crust and within the lower crust. With this last surge of magmatic/tectonic activity, the mantle plume became inactive, extension terminated, and igneous activity sharply declined. However, volcanic activity did continue in a minor way to at least 1,087 Ma.

As the head of the plume cooled (Fig. 19e), its upper portions were attached to the thinned lithosphere and the lithosphere sagged under the load of the massive extrusions and intrusions along the axis of the grabens. The margins of the rift gave rise to locally derived sediments, but the margins were soon covered by the sediments of the Oronto Group and then by the more mature sediments of the Bayfield Group/Jacobsville Sandstone in the expanding axial basins. The extension was replaced by regional compression, probably from the Grenville orogen (Fig. 19f), that reactivated zones of weakness within the thermally softened lithosphere of the rift zone. Most noticeably, the graben-normal faults were reactivated as high-angle reverse faults that were initiated during and following deposition of the Bayfield Group/Jacobsville Sandstone. In addition, regional folding of the rift rocks occurred at this time. The lithospheric structures portrayed in Figure 19f are much as they are today in many segments of the Midcontinent Rift. Subsequent uplift, however, has led to partial, or in some regions complete, erosion of both the volcanic and clastic sedimentary rocks adjacent to the rift and the upper rift-rocks along the axis of the Midcontinent Rift. The region remains a topographic high due to regional underplating of the crust or modification of the upper mantle. The portrayed series of

"snapshots" of the lithosphere of the Midcontinent Rift region at about 1.1 Ga is speculative and highly generalized; modifications to the sections are caused by local igneous events, deflections of regional stresses, and variations in the strength and other physical properties of the lithosphere.

CONCLUSIONS

At 1,100 Ma, a major disruption of the proto–North American continent led to a nearly complete rupture of the crust along the arcuate, ~2,000-km-length of the Midcontinent Rift System. The majority of the rift's igneous rocks were derived from a melting mantle plume and the overlying lithosphere over a short span of less than 20 m.y. The relation of the plume to the extension resulting the rift is presently unclear. The Midcontinent Rift is a profound disruption of the crust, as indicated by its anomalous geophysical signatures. The geophysical data, together with limited outcrops of the rift rocks and a few deep drill holes, suggest that the pre-rift crust was thinned to less than one-third of its original thickness. The crust beneath the rift has been thickened, however, due to the addition of as much as 20 km of mantle-derived volcanic rocks and as much as 10 km of post-volcanic clastic sedimentary strata. The original rift grabens and their volcanic and sedimentary fill have been modified by regional compression, perhaps associated with the roughly contemporaneous Grenville orogenic activity. The most profound of these modifications are the reactivation of the normal, graben faults as high-angle reverse faults with more than 5 km of vertical displacement.

During the past decade, geophysical studies, particularly deeply penetrating seismic reflection profiling, and isotopic analysis of the rift's volcanic rocks have provided much new insight into the Midcontinent Rift and its structural and magmatic evolution. Many problems, however, remain to be solved. For example, of special concern is the lack of information concerning the degree and form of modification of the lower crust beneath the rift. In addition, little is known about the manifestations of the rift in the upper mantle. Other significant questions concern the nature and source of the tensional stresses. Were they generated entirely by the mantle plume head, or were other mechanisms involved? Was there a significant uplift produced by the head of the mantle plume? Also, is there evidence of a plume trail migrating to or from Lake Superior? Furthermore, what is the relation of the origin, development, and termination of the Midcontinent Rift to the evolution of the Grenville orogen?

Future research of the Midcontinent Rift should be directed at questions such as these. In addition, continued detailed investigations are needed to gain a more comprehensive knowledge of the geochemistry, magnetostratigraphy, and absolute ages of the rift rocks. Geochemical and isotopic studies should provide a clearer picture of the magmatic evolution of the rift and plume through both time and space. Additional isotopic dating and paleomagnetic measurements will hopefully result in a better understanding of the relative ages of the rift units along the entire length of the Midcontinent Rift. Also, efforts should be directed toward mapping the finer structures within the rift and expanding and refining the gross structural interpretations. Of particular interest are the transitional, transverse zones within the Midcontinent Rift, along which the rift structure changes abruptly and dramatically. Are these transition zones related to ancestral structures? How did the rift evolve along them?

During the next decade, detailed investigations of these topics, at specific sites along the rift, will hopefully provide a stronger framework for developing an integrated model of the entire rift. In the future, a larger number of investigators will need to be concerned with integrating information from the individual locations along the rift. Integrated models should consider the relative ages of the individual rift segments and their structural relations to one another and lead to a better understanding of the spatial and temporal evolution of the Midcontinent Rift.

ACKNOWLEDGMENTS

We acknowledge the useful interaction with our colleagues and the constructive criticism of numerous specialists working on the Midcontinent Rift. Financial support for this overview was provided in part by National Science Foundation Grant Number EAR-8617315.

REFERENCES CITED

Adams, D. C., and Keller, G. R., 1994, Possible extension of the Midcontinent Rift in west Texas and New Mexico: Canadian Journal of Earth Sciences, v. 31, p. 709–720.

Allen, D. J., and Chandler, V. W., 1992, The utility of high-resolution aeromagnetic data for investigating the Midcontinent Rift in east-central Minnesota [abs.]: Institute on Lake Superior Geology, 38th, Proceedings and Abstracts, Hurley, Wisconsin, p. 3–5.

Allen, D. J., and Chandler, V. W., 1993, Magnetic properties of the Midcontinent Rift volcanics at Osseo, Minnesota: Clues to the origin of aeromagnetic anomalies [abs.]: Institute on Lake Superior Geology, 39th, Proceedings and Abstracts, Eveleth, Minnesota, p. 2–3.

Allen, D. J., Hinze, W. J., and Cannon, W. F., 1992, Drainage, topographic, and gravity anomalies in the Lake Superior region: Evidence for a 1100 Ma mantle plume: Geophysical Research Letters, v. 19, p. 2119–2122.

Anderson, R. R., editor, 1990a, The Amoco M. G. Eischeid #1 Deep Petroleum Test Carroll County, Iowa: Iowa Department of Natural Resources Special Report Series Number 2, 185 p.

Anderson, R. R., 1990b, Interpretation of geophysical data over the Midcontinent Rift System in the area of the M. G. Eischeid #1 Petroleum Test, Carroll County, Iowa, in Anderson, R. R., ed., The Amoco M. G. Eischeid #1 deep petroleum test Carroll County, Iowa: Iowa Department of Natural Resources Special Report Series No. 2, p. 27–38.

Baker, B. N., and Morgan, P., 1981, Continental rifting: Progress and outlook: Eos (Transactions, Americal Geophysical Union), v. 62, p. 585–586.

Behrendt, J. C., Green, A. G., Cannon, W. F., Hutchinson, D. R., Lee, M., Milkereit, B., Agena, W. F., and Spencer, C., 1988, Crustal structure of the Midcontinent Rift System: Results from GLIMPCE deep seismic reflection profiles: Geology, v. 16, p. 81–85.

Behrendt, J. C., Hutchinson, D. R., Lee, M. W., Agena, W. F., Trehu, A., Thornber, C. R., Cannon, W. F., and Green, A., 1990, Seismic reflection (GLIMPCE) evidence of deep crustal and upper mantle intrusions and magmatic underplating associated with the Midcontinent Rift System of North America: Tectonophysics, v. 173, p. 617–626.

Berendsen, P., Borcherding, R. M., Doveton, J., Gerhard, L., Newell, K. D., Steeples, D., and Watney, W. L., 1988, Texaco Poersch #1, Washington County, Kansas—Preliminary geologic report of the pre-Phanerozoic rocks: Kansas Geological Survey Open-file Report 88-22, 116 p.

Berg, J. H., and Klewin, K. W., 1988, High-MgO lavas from the Keweenawan Midcontinent Rift near Mamainse Point, Ontario: Geology, v. 16, p. 1003–1006.

Books, K. G., 1972, Paleomagnetism of some Lake Superior rocks: U.S. Geological Survey Professional Paper 760, 42 p.

Bornhorst, T. J., Paces, J. B., Grant, N., Obradavich, J. D., and Huber, N. K., 1988, Age of native copper mineralization, Keweenaw Peninsula, Michigan: Economic Geology, v. 83, p. 619–625.

Bosworth, W., Lambiase, J., and Keisler, R., 1986, A new look at Gregory's Rift: The structural style of continental rifting: Eos (Transactions, American Geophysical Union), v. 67, p. 577–583.

Boyd, F. R., and McCallister, R. H., 1976, Densities of fertile and sterile garnet peridotites: Geophysical Research Letters, v. 15, p. 673–675.

Brannon, J., 1984, Geochemistry of successive lava flows of the Keweenawan North Shore Volcanic Group [Ph.D. thesis]: St. Louis, Washington University, 312 p.

Brown, L., Jensen, L., Oliver, J., Kaufman, S., and Steiner, D., 1982, Rift structure beneath the Michigan basin from COCORP profiling: Geology, v. 10, p. 645–649.

Campbell, I. H., and Griffiths, R. W., 1990, Implications of mantle plume structure for the evolution of flood basalts: Earth and Planetary Science Letters, v. 99, p. 79–93.

Cambray, F. W., 1988, A tectonic model for the Mid-Continent Rift System [abs.]: Institute on Lake Superior Geology, 34th, Proceedings and Abstracts, Marquette, Michigan, p. 17.

Cannon, W. F., 1992, The Midcontinent Rift in the Lake Superior region with emphasis on its geodynamic evolution: Tectonophysics, v. 213, p. 41–48.

Cannon, W. F., and Green, A. G., 1989, A model for development of the Midcontinent Rift in the Lake Superior region [abs.]: Eos (Transactions, American Geophysical Union), v. 70, p. 272.

Cannon, W. F., and Hinze, W. J., 1992, Speculations on the origin of the North American Midcontinent Rift: Tectonophysics, v. 213, p. 49–55.

Cannon, W. F., and 11 others, 1989, The Midcontinent Rift beneath Lake Superior from GLIMPCE seismic reflection profiling: Tectonics, v. 8, p. 305–332.

Cannon, W. F., Peterman, Z. E., and Sims, P. K., 1990, Structural and isotopic evidence for Middle Proterozoic thrust faulting of Archean and Early Proterozoic rocks near the Gogebic Range, Michigan and Wisconsin [abs.]: Institute on Lake Superior Geology, 36th, Proceedings and Abstracts, Thunder Bay, Ontario, p. 11–12.

Cannon, W. F., Lee, M. W., Hinze, W. J., Schulz, K. J., and Green, A. G., 1991, Deep crustal structure of Precambrian crust beneath northern Lake Michigan, Midcontinent North America: Geology, v. 19, p. 207–210.

Carlson, M. P., 1992, Tectonic implications and influence of the Midcontinent Rift System in Nebraska and adjoining areas [abs.]: International Basement Tectonics Conference, 10th, Duluth, Minnesota, p. 143–145.

Chan, W. K., and Hajnal, Z., 1989, Some features of the Midcontinent Rift from high density refraction signals [abs.]: Eos (Transactions, American Geophysical Union), v. 70, p. 274.

Chandler, V. W., 1983, Correlation of magnetic anomalies in east-central Minnesota and northwestern Wisconsin: Constraints on magnitude and direction of Keweenawan rifting: Geology, v. 11, p. 174–176.

Chandler, V. W., 1991, Aeromagnetic anomaly map of Minnesota: Minnesota Geological Survey State Map. Series 5-17, scale 1:500 000.

Chandler, V. W., and Morey, G. B., 1989, Seismic history of Minnesota and its geologic significance: An update: Seismological Research Letters, v. 60, p. 79–86.

Chandler, V. W., McSwiggen, P. L., Morey, G. B., Hinze, W. J., and Anderson, R. L., 1989, Interpretation of seismic reflection, gravity and magnetic data across the Middle Proterozoic Midcontinent Rift System in western Wisconsin, eastern Minnesota, and central Iowa: American Association of Petroleum Geologists Bulletin, v. 73, p. 261–275.

Cohen, T. J., and Meyer, R. P., 1966, The Midcontinent gravity high: Gross crustal structure, in Steinhart, J. S., Smith, T. J., eds., The Earth Beneath the continents: American Geophysical Union Monograph 10, p. 141–165.

Craddock, C., Thiel, E. C., and Gross, B., 1963, A gravity investigation of the Precambrian of southeastern Minnesota and western Michigan: Journal of Geophysical Research, v. 68, p. 6015–6032.

Daniels, P. A., Jr., 1982, Upper Precambrian sedimentary rocks: Oronto Group, in Wold, R. J., and Hinze, W. J., eds., Geology and tectonics of the Lake Superior basin: Geological Society of America Memoir 156, p. 107–133.

Davidson, D. M., Jr., 1982, Geological evidence relating to interpretations of the Lake Superior basin structure, in Wold, R. J., and Hinze, W. J., eds., Geology and tectonics of the Lake Superior basin: Geological Society of America Memoir 156, p. 5–14.

Davis, D. W., and Sutcliffe, R. H., 1985, U-Pb ages from the Nipigon plate and northern Lake Superior: Geological Society of America Bulletin, v. 96, p. 1572–1579.

Davis, D. W., and Paces, J. B., 1990, Time resolution of geologic events on the Keweenaw Peninsula and implications for development of the Midcontinent Rift System: Earth and Planetary Science Letters, v. 97, p. 54–64.

Dickas, A. B., 1986, Comparative Precambrian stratigraphy and structure along the Mid-Continent Rift: American Association of Petroleum Geologists Bulletin, v. 70, p. 225–238.

Dickas, A. B., 1992, Preliminary report, geology and geophysics of the Terra/Patrick #7-22 Wildcat hydrocarbon test, Midcontinent Rift, Keystone Township, Bayfield County, Wisconsin [abs.]: Institute on Lake Superior Geology, 38th, Proceedings and Abstracts, Hurley, Wisconsin, p. 25–26.

Dickas, A. B., and Mudrey, M. G., Jr., 1989, Structural differentiation of the Midcontinent Rift System [abs.]: Eos (Transactions, American Geophysical Union), v. 70, p. 275.

Dosso, L., 1984, The nature of the Precambrian subcontinental mantle: Isotopic study (Sr, Nd, Pb) of Keweenawan volcanism of the north shore of Lake Superior [Ph.D. thesis]: Minneapolis, University of Minnesota, 221 p.

Drury, M. J., 1991, Heat flow in the Canadian Shield and its relation to other geophysical parameters, in Cermak, V., and Rybach, L., eds., Terrestrial heat flow and the lithosphere structure: Berlin, Springer-Verlag, p. 317–337.

Dutch, S. I., 1981, Isostasy, epeirogeny, and the highland rim of Lake Superior: Zeitschrift für Geomorphologie. Neue Folge, Suppl., v. 40, p. 27–41.

Fowler, J. H., and Kuenzi, W. D., 1978, Keweenawan turbidites in Michigan (deep borehole red beds): A founded basin sequence developed during evolution of a protoceanic rift systems: Journal of Geophysical Research, v. 83, p. 5833–5843.

Fox, A. J., 1988, An integrated geophysical study of the southeastern extension of the Midcontinent Rift System [M.S. Thesis]: West Lafayette, Purdue University, 112 p.

Frantti, G. G., 1980, Seismicity and crustal studies in the Lake Superior Precambrian province [abs.]: Eos (Transactions, American Geophysical Union), v. 61, p. 96.

Gordon, M. B., and Hempton, M. R., 1986, Collision-induced rifting: The Grenville orogeny and the Keweenawan Rift of North America: Tectonophysics, v. 127, p. 1–25.

Green, A. G., Milkereit, B., Davidson, A., Spencer, C., Hutchinson, D. R., Cannon, W. F., Lee, M. W., Agena, W. F., Behrendt, J. C., and Hinze, W. J., 1988, Crustal structure of the Grenville front and adjacent terranes: Geology, v. 16, p. 788–792.

Green, A. G., and 9 others, 1989, A GLIMPCE of the deep crust beneath the Great Lakes: American Geophysical Union Monograph 51, p. 65–80.

Green, J. C., 1981, Pre-Tertiary plateau basalts, in Basaltic Volcanism Study Project, Basaltic volcanism on the terrestrial planets: New York, Pergamon Press, p. 30–77.

Green, J. C., 1982, Geology of Keweenawan extrusive rocks, in Wold, R. J., and Hinze, W. J. eds., Geology and tectonics of the Lake Superior basin: Geological Society of America Memoir 156, p. 47–55.

Green, J. C., 1983, Geologic and geochemical evidence for the nature and development of the Middle Proterozoic (Keweenawan) Midcontinent Rift of North America: Tectonophysics, v. 94, p. 413–437.

Green, J. C., 1987, Plateau basalts of the Keweenawan North Shore Volcanic Group, in Biggs, D. L., ed., Centennial Field Guide Voulme 3: North-Central Section of the Geological Society of America: Geological Society of America, p. 59–62.

Green, J. C., Bornhorst, T. J., Chandler, V. W., Mudrey, M. G., Jr., Myers, P. R., Pesonen, L. V., and Wilband, J. T., 1987, Keweenawan dykes of the Lake Superior region: Evidence for the evolution of the Middle Proterozoic Midcontinent Rift of North America, in Halls, H. C., and Fahrig, W. F., eds., Mafic dyke swarms: Geological Association of Canada Special Paper 34, p. 289–302.

Green, W. V., 1991, Lithospheric seismic structure of the Cenozoic Kenya Rift and the Precambrian Midcontinent Rift from teleseismic tomography [Ph.D. thesis]: Madison, Wisconsin, University of Wisconsin-Madison, 179 p.

Griffiths, R. W., and Campbell, I. H., 1991, Interaction of mantle plume heads with the Earth's surface and onset of small-scale convection: Journal of Geophysical Research, v. 96, p. 18295–18310.

Halls, H. C., 1969, Compressional wave velocities of Keweenawan rock specimens from the Lake Superior region: Canadian Journal of Earth Science, v. 6, p. 555–568.

Halls, H. C., 1978, The late Precambrian central North America rift system—A survey of recent geological and geophysical investigations, in Neumann, E. R., and Ramberg, I. B., eds., Tectonics and geophysics of continental rifts: NATO Advanced Study Institute, Series C: Boston: Reidel, v. 37, p. 111–123.

Halls, H. C., 1982, Crustal thickness in the Lake Superior region, in Wold, R. J., and Hinze, W. J., eds., Geology and tectonics of the Lake Superior basin: Geological Society of America Memoir 156, p. 239–244.

Halls, H. C., and Pesonen, L. J., 1982, Paleomagnetism of Keweenawan rocks, in

Wold, R. J., and Hinze, W. J., eds., Geology and tectonics of the Lake Superior basin: Geological Society of America Memoir 156, p. 173–202.

Hamilton, D. A., and Mereu, R. F., 1993, 2-D tomographic imaging across the North American Midcontinent Rift System: Geophysiccal Journal International, v. 112, p. 344–358.

Hart, S. R., and Steinhart, J. S., 1964, The heat flow and thermal properties of Lake Superior: Carnegie Institution of Washington Year Book 63, p. 326–328.

Hart, S. R., Steinhart, J. S., and Smith, T. J., 1994, Terrestrial heat flow in Lake Superior: Canadian Journal of Earth Science, v. 31, p. 698–708.

Heaman, L. M., and Machado, N., 1992, Timing and origin of the Midcontinent Rift alkaline magmatism, North America: Evidence from the Coldwell Complex: Contributions to Mineralogy and Petrology, v. 110, p. 289–303.

Hieshima, G. B., and Pratt, L. M., 1991, Sulfur/carbon ratios and extractable organic matter of the Middle Proterozoic Nonesuch Formation, North American Midcontinent Rift: Precambrian Research, v. 54, p. 65–79.

Hinze, W. J., 1963, Regional gravity and magnetic anomaly maps of the Southern Peninsula of Michigan: Michigan Geological Survey Report of Investigation 1, 26 p.

Hinze, W. J., and Kelly, W. C., 1988, Scientific drilling into the Midcontinent Rift System: Eos (Transactions, American Geophysical Union), v. 69, p. 1649, 1656–1657.

Hinze, W. J., and Wold, R. J., 1982, Lake Superior geology and tectonics—overview and major unsolved problems, in Wold, R. J, and Hinze, W. J, eds., Geology and tectonics of the Lake Superior basin: Geological Society of America Memoir 156, p. 273–280.

Hinze, W. J., O'Hara, N. W., Trow, J. W., and Secor, G. B., 1966, Aeromagnetic studies of eastern Lake Superior, in Steinhart, J. S., and Smith, T. J., eds., The Earth beneath the continents: American Geophysical Union Monograph 10, p. 95–110.

Hinze, W. J., Roy, R. F., and Davidson, D. M., Jr., 1972, The origin of late Precambrian rifts [abs.]: Geological Society of America Abstracts with Program, v. 4, p. 723.

Hinze, W. J., Kellogg, R. L., and O'Hara, N. W., 1975, Geophysical studies of basement geology of Southern Peninsula of Michigan: American Association of Petroleum Geologists Bulletin, v. 59, p. 1562–1584.

Hinze, W. J., Wold, R. J., and O'Hara, N. W., 1982, Gravity and magnetic anomaly studies of Lake Superior, in Wold, R. J., and Hinze, W. J, eds., Geology and tectonics of the Lake Superior basin: Geological Society of America Memoir 156, p. 203–222.

Hinze, W. J., Braile, L. W., and Chandler, V. W., 1990, A geophysical profile of the southern margin of the Midcontinent Rift System in western Lake Superior: Tectonics, v. 9, p. 303–310.

Hinze, W. J., Allen, D. J., Fox, A. J., Sunwoo, D., Woelk, T., and Green, A. G., 1992, Geophysical investigations and crustal structure of the North American Midcontinent Rift System: Tectonophysics, v. 213, p. 17–32.

Hutchinson, D. R., White, R. S., Cannon, W. F., and Schulz, K. J., 1990, Keweenawan hot spot: Geophysical evidence for a 1.1 Ga mantle plume beneath the Midcontinent Rift System: Journal of Geophysical Research, v. 95, p. 10869–10884.

Irving, R. D., 1883, The copper-bearing rocks of Lake Superior: U.S. Geological Survey Monograph 5, 464 p.

Jefferson, T. J., Karl, J. H., and Meyer, P. J., 1989, A model of the velocity distribution across the Midcontinent Rift: GLIMPCE Line A [abs.]: Eos (Transactions, American Geophysical Union), v. 70, p. 274.

Kalliokoski, J., 1982, Jacobsville Sandstone, in Wold, R. J., and Hinze, W. J., eds., Geology and tectonics of the Lake Superior basin: Geological Society of America Memoir 156, p. 147–155.

Kalliokoski, J., 1989, Jacobsville Sandstone and tectonic activity [abs.]: Institute on Lake Superior Geology, 35th, Proceedings and Abstracts, Duluth, Minnesota, p. 39–41.

Keller, G. R., Bland, A. E., and Greenberg, J. K., 1982, Evidence for a major late Precambrian tectonic event (rifting) in the eastern midcontinent region, United States: Tectonics, v. 1, p. 213–222.

King, E. R., and Zietz, I., 1971, Aeromagnetic study of the Midcontinent gravity high of central United States: Geological Society of America Bulletin, v. 82, p. 2187–2208.

Klasner, J. S., Cannon, W. F., and Van Schmus, W. R., 1982, The pre-Keweenawan tectonic history of southern Canadian Shield and its influence on formation of the Midcontinent Rift, in Wolds, R. J., and Hinze, W. J., eds., Geology and tectonics of the Lake Superior basin: Geological Society of America Memoir 156, p. 27–46.

Klewin, K. W., 1989, Polybaric fractionation in an evolving continental rift: Evidence from the Keweenawan Mid-continental Rift: Journal of Geology, v. 97, p. 65–76.

Klewin, K. W., and Berg, J. H., 1991, Petrology of the Keweenawan Mamainse Point lavas, Ontario: Petrogenesis and continental rift evolution: Journal of Geophysical Research, v. 96, p. 457–474.

Klewin, K. W., and Shirey, S. B., 1992, The igneous petrology and magmatic evolution of the Midcontinent Rift System: Tectonophysics, v. 213, p. 33–40.

Lippus, C. S., 1988, The seismic properties of mafic volcanic rocks of the Keweenawan Supergroup and their implications [M.S. thesis]: West Lafayette, Purdue University, 160 p.

Luetgert, J. H., and Meyer, R. P., 1982, Structure of the western basin of Lake Superior from cross structure refraction profiles, in Wold, R. J., and Hinze, W. J., eds., Geology and tectonics of the Lake Superior Basin: Geological Society of America Memoir 156, p. 245–255.

Lutter, W. J., Trehu, A. M., Nowack, R. L., and Shay, J. T., 1989, Inversion for crustal structure using first arrival refraction data from the 1986 Lake Superior GLIMPCE experiment [abs.]: Eos (Transactions, American Geophysical Union), v. 70, p. 274.

Lyons, P. L., 1950, A gravity map of the United States: Tulsa Geological Society Digest, v. 18, p. 33–43.

Lyons, P. L., 1959, The Greenleaf anomaly, a significant gravity feature, in Hambleton, W. M. ed., Symposium on geophysics in Kansas: Kansas Geological Survey Bulletin 137, p. 105–120.

Lyons, P. L., 1970, Continental and oceanic geophysics, in Johnson, H., and Smith, B. L., eds., The megatectonics of continents and oceans: New Brunswick, New Jersey, Rutgers University Press, p. 147–166.

Manson, M. L., and Halls, H. C., 1994, Post-Keweenawan compressional faults in the eastern Lake Superior region and their tectonic significance: Canadian Journal of Earth Science, v. 31, 640–651.

Mariano, J., and Hinze, W. J., 1992, Magnetic and seismic reflection interpretation of the Midcontinent Rift in eastern Lake Superior [abs.]: Eos (Transactions, American Geophysical Union), v. 73, p. 194.

Mariano, J., and Hinze, W. J., 1994a, Structural interpretation of the Midcontinent Rift in eastern Lake Superior from seismic reflection and potential-field studies: Canadian Journal of Earth Science, v. 31, p. 619–628.

Mariano, J., and Hinze, W. J., 1994b, Gravity and magnetic models of the Midcontinent Rift in eastern Lake Superior: Canadian Journal of Earth Science, v. 31, p. 661–674.

McCallister, R. H., Boctor, N. Z., and Hinze, W. J., 1978, Petrology of the spilitic rocks from the Michigan Basin drill hole, 1978: Journal of Geophysical Research, v. 83, p. 5825–5831.

McSwiggen, P. L., Morey, G. B., and Chandler, V. W., 1987, New model of Midcontinent Rift in eastern Minnesota and western Wisconsin: Tectonics, v. 6, p. 677–685.

Merk, G. P., and Jirsa, M. A., 1982, Provenance and tectonic significance of the Keweenawan interflow sedimentary rocks, in Wold, R. J, and Hinze, W. J., eds., Geology and tectonics of the Lake Superior basin: Geological Society of America Memoir 156, p. 97–106.

Meshref, W. M., and Hinze, W. J., 1970, Geologic interpretation of aeromagnetic data in western Upper Peninsula of Michigan: Michigan Department of Natural Resources, Geological Survey Report of Investigation 12, 25 p.

Meyer, P. J., Karl, J. H., and Jefferson, T. J., 1989, Beamsteering applied to the GLIMPCE wide-angle data: Line A [abs.]: Eos (Transactions, American Geophysical Union), v. 70, p. 276.

Middleton, M. D., 1991, A preliminary study of rejuvenated movement along a Precambrian fault, St. Croix County, Wisconsin [abs.]: Institute on Lake Superior Geology, 37th, Proceedings and Abstracts, Eau Claire, Wisconsin, p. 70–71.

Milanovsky, E. E., 1981, Aulacogens of ancient platforms: problems of their origin and tectonic development: Tectonophysics, v. 73, p. 213–248.

Milkereit, B., Green, A. G., Lee, M. W., Agena, W. F., and Spencer, D., 1990, Pre- and post-stack migration of GLIMPCE reflection data: Tectonophysics, v. 173, p. 1–13.

Miller, J. D., Jr., 1992, The need for a new paradigm regarding the petrogenesis of the Duluth Complex [abs.]: Institute on Lake Superior Geology, 38th, Proceedings and Abstracts, Hurley, Wisconsin, p. 65–67.

Mohr, P., 1983, The Morton-Black hypothesis for the thinning of continental crust—revisited in western Afar: Tectonophysics, v. 94, p. 509–528.

Mooney, H. M., 1981, Minnesota earthquakes—1979: Earthquake Notes, v. 52, p. 57–58.

Mooney, H. M., and Morey, G. B., 1981, Seismic history of Minnesota and its geological significance: Bulletin of the Seismological Society of America, v. 71, p. 199–210.

Mooney, H. M., and Walton, M., 1986, Seismicity and tectonic relationships for Upper Great Lakes Precambrian Shield Province: U.S. Nuclear Regulatory Commission Final Report, July 1981—December 1982, NUREG/CR-3150, 47 p.

Morel-à-l'Huissier, P., and Green, A. G., 1989, Modeling the GLIMPCE 1986 seismic refraction data recorded along seismic reflection Line A [abs.]: Eos (Transactions, American Geophysical Union), v. 70, p. 277.

Morey, G. B., 1977, Revised Keweenawan subsurface stratigraphy, southwestern, Minnesota: Minnesota Geological Survey Report of Investigation 16, 67 p.

Morey, G. B., and Ojakangas, R. W., 1982, Keweenawan sedimentary rocks of eastern Minnesota and northwestern Wisconsin, *in* Wold, R. J., and Hinze, W. J., eds., Geology and tectonics of the Lake Superior basin: Geological Society of America Memoir 156, p. 135–146.

Morey, G. B., and Van Schmus, W. R., 1988, Correlation of Precambrian rocks of the Lake Superior region, United States: U.S. Geological Survey Professional Paper 1241-F, 31 p.

Mudrey, M. G., Jr., and Dickas, A. B., 1988, Midcontinent Rift model based upon Gregory Rift tectonic and sedimentation geometries [abs.]: Institute on Lake Superior Geology, 34th, Proceedings and Abstracts, Marquette, Michigan, p. 73–75.

Mudrey, M. G., Jr., McGinnis, L. D., Ervin, C. P., Nyquist, J., Dickas, A. B., Morey, G. B., Green, A. G., and Sexton, J. L., 1989, Structure of the Midcontinent Rift System in western Lake Superior: Results from 8-sec reflection seismic data and gravity and magnetic anomalies [abs.]: Eos (Transactions, American Geophysical Union), v. 70, p. 278.

Neumann, E.-R., Pallesen, S., and Andersen, P., 1986, Mass estimates of cumulates and residues after anatexis in the Oslo graben: Journal of Geophysical Research, v. 91, p. 11629–11640.

Nicholson, S. W., and Shirey, S. B., 1990, Evidence for a Precambrian mantle plume: A Sr, Nd, and Pb isotopic study of the Midcontinent Rift System in the Lake Superior region: Journal of Geophysical Research, v. 95, p. 10851–10868.

Nicholson, S. W., Cannon, W. F., and Schulz, K. J., 1992, Metallogeny of the Midcontinent Rift System of North America: Precambrian Research, v. 58, p. 355–386.

Nixon, G. A., 1988, An analysis of geophysical anomalies in north-central Oklahoma and their relationship to the Midcontinent geophysical anomaly [M.S. thesis]: Norman, University of Oklahoma, 118 p.

Nyquist, J. E., and Wang, H. F., 1988, Flexural modeling of the Midcontinent Rift: Journal of Geophysical Research, v. 93, p. 8852–8868.

Nyquist, J. E., and Wang, H. F., 1989, Flexural modeling of the Midcontinent Rift and the Goodman Swell [abs.]: Eos (Transactions, American Geophysical Union), v. 70, p. 275.

Ocola, L. C., and Meyer, R. P., 1973, Central North American rift system, 1. Structure of the axial zone from seismic and gravimetric data: Journal of Geophysical Research, v. 78, p. 5173–5194.

Ojakangas, R. W., and Morey, G. B., 1982, Keweenawan pre-volcanic quartz sandstones and related rocks of the Lake Superior region, in Wold, R. J., and Hinze, W. J., eds., Geology and tectonics of the Lake Superior

basin: Geological Society of America Memoir 156, p. 85–96.

Oray, E., Hinze, W. J., and O'Hara, N. W., 1973, Gravity and magnetic evidence for the eastern termination of the Lake Superior syncline: Geological Society of America Bulletin, v. 84, p. 2763–2780.

Oxburgh, E. R., and Parmentier, E. M., 1977, Compositional and density stratification in oceanic lithosphere—causes and consequences: Journal of the Geological Society of London, v. 133, p. 343–355.

Paces, J. B., and Bell, K., 1989, Non-depleted sub-continental mantle beneath the Superior Province of the Canadian Shield: Nd-Sr isotopic and trace element evidence from Midcontinent Rift basalts: Geochimica et Cosmochimia Acta, v. 53, p. 2023–2035.

Paces, J. B., and Miller, J. D., Jr., 1993, Precise U-Pb ages of Duluth Complex and related mafic intrusions, northeastern Minnesota, U.S.A.: Geochronological insights to physical, petrogenetic, paleomagnetic and tectono-magmatic processes associated with the 1.1 Ga Midcontinent Rift System: Journal of Geophysical Research, v. 98, 13997–14013.

Palmer, H. C., 1970, Paleomagnetism and correlation of some middle Keweenawan rocks, Lake Superior: Canadian Journal of Earth Science, v. 7, p. 1410–1436.

Palmer, H. C., and Davis, D. W., 1987, Paleomagnetism and U-Pb geochronology of volcanic rocks from Michipicoten Island, Lake Superior, Canada: Precise calibration of the Keweenawan polar wander track: Precambrian Research, v. 37, p. 157–171.

Peterman, Z. E., and Sims, P. K., 1988, The Goodman Swell: A lithospheric flexure caused by crustal loading along the Midcontinent Rift System: Tectonics, v. 7, p. 1077–1090.

Pratt, L. M., Summons, R. E., and Hieshima, G. B., 1991, Sterane and Triterpane biomarkers in the Precambrian Nonesuch Formation, North American Midcontinent Rift: Geochimica et Cosmochimica Acta, v. 55, p. 911–916.

Pratt, T., Culotta, R., Hauser, E., Nelson, D., Brown, L., Kaufman, S., Oliver, J., and Hinze, W., 1989, Major Proterozoic basement features of the eastern Midcontinent of North America revealed by recent COCORP profiling: Geology, v. 17, p. 505–509.

Ramberg, I. B., and Morgan, P., 1984, Physical characteristics and evolutionary trends of continental rifts: Proceedings, International Geological Congress, 27th, Moscow: Utrecht, Netherlands, VNU Science Press, v. 7, p. 165–216.

Robertson, W. A., 1973, Pole positions from the Mamainse Point lavas and their bearing on a Keweenawan pole path and polarity sequence: Canadian Journal of Earth Science, v. 10, p. 1541–1555.

Rosendahl, B. R., 1987, Architecture of continental rifts with special reference to East Africa: Annual Reviews of Earth and Planetary Science, v. 15, p. 445–503.

Roy, R. F., Blackwell, D. D., and Decker, E. R., 1972, Continental heat flow, in Robertson, E. C., ed., The nature of the solid Earth: New York, New York, McGraw-Hill, p. 506–543.

Samson, C., and West, G. F., 1992, Crustal structure of the Midcontinent Rift System in eastern Lake Superior from controlled-amplitude analysis of GLIMPCE deep reflection seismic data: Canadian Journal of Earth Science, v. 29, p. 636–649.

Samson, C., and West, G. F., 1994, Detailed basin structure and tectonic evolution of the Midcontinent Rift System in eastern Lake Superior from reprocessing of GLIMPCE deep reflection seismic data: Canadian Journal of Earth Science, v. 31, p. 629-639.

Schärer, U., and Gower, C. F., 1988, Crustal evolution in eastern Labrador: Constraints from precise U-Pb ages: Precambrian Research, v. 38, p. 405–421.

Serpa, L., Setzer, T., Farmer, H., Brown, L., Oliver, J., et al., 1984, Structure of the southern Keweenawan Rift from COCORP surveys across the Midcontinent geophysical anomaly in northeastern Kansas: Tectonics, v. 3, p. 367–384.

Shay, J. T., and Trehu, A. M., 1989, Structure of the Midcontinent Rift beneath Lake Superior [abs.]: Eos (Transactions, American Geophysical Union), v. 70, p. 274.

Sleep, N. H., and Sloss, L. L., 1978, A deep borehole in the Michigan basin: Journal of Geophysical Research, v. 83, p. 5815–5824.

Smith, T. J., Steinhart, J. S., and Aldrich, L. T., 1966, Crustal structure under Lake Superior, *in* Steinhart, J. S., and Smith, T. J., eds., The Earth beneath the continents: American Geophysical Union Monograph 10, p. 181–197.

Steeples, D. W., 1980, Microearthquakes recorded by the Kansas Geological Survey: Kansas Geological Survey Report Investigation 37, 25 p.

Steeples, D. W., 1981, Kansas earthquakes—1979: Earthquake Notes, v. 52, p. 42–48.

Steeples, D. W., Hildebrand, G. M., Bennett, B. C., Miller, R. O., Chung, Y., and Knapp, R. W., 1989, Kansas-Nebraska seismicity studies using the Kansas-Nebraska microearthquake network: U.S. Nuclear Regulatory Commissions Final Report, NUREG/CR-5045, 71 p.

Steinhart, J. S., and Smith, T. J., eds., 1966, The Earth beneath the continents: American Geophysical Union Monograph 10, 663 p.

Steinhart, J. S., Hart, S. R., and Smith, T. J., 1967, Heat flow: Carnegie Institution of Washington Yearbook 66, p. 52–57.

Steinhart, J. S., Hart, S. R., and Smith, T. J., 1968, Heat flow: Carnegie Institution of Washington Yearbook 67, p. 130–140.

Sunwoo, D., 1989, An integrated geophysical study of the Grenville Front in Ohio [M.S. thesis]: West Lafayette, Purdue University, 94 p.

Teskey, D. J., Thomas, M. D., Gibb, R. A., Dods, S. D., Kucks, R. P., Chandler, V. W., Fadaie, K., and Phillips, J. O., 1991, High resolution aeromagnetic survey of Lake Superior: Eos (Transactions, American Geophysical Union), v. 72, p. 81, 85–86.

Thiel, E. C., 1956, Correlation of gravity anomalies with the Keweenawan geology of Wisconsin and Minnesota: Geological Society of America Bulletin, v. 67, p. 1079–1100.

Trehu, A., and 15 others, 1991, Imaging the Midcontinent Rift beneath Lake Superior using large aperture seismic data: Geophysical Research Letters, v. 18, p. 625–628.

Van Hise, C. R., and Leith, C. K., 1911, The geology of the Lake Superior region: U.S. Geological Survey Monograph 52, 641 p.

Van Schmus, W. R., 1992, Tectonic setting of the Midcontinent Rift System: Tectonophysics, v. 213, p. 1–15.

Van Schmus, W. R., and Hinze, W. J., 1985, The Midcontinent Rift System: Annual Reviews of Earth and Planetary Science, v. 13, p. 345–383.

Van Schmus, W. R., Green, J. C., and Halls, H. C., 1982, Geochronology of Keweenawan rocks of the Lake Superior region: A summary, *in* Wold, R. J., and Hinze, W. J., eds., Geology and tectonics of the Lake Superior basin: Geological Society of America Memoir 156, p. 165–171.

Van Schmus, W. R., Bickford, M. E., and Zietz, I., 1987, Early and Middle Proterozoic provinces in the central United States, *in* Kroner, A., ed., Proterozoic lithospheric evolution: Geodynamic Series, v. 17, p. 43–68.

Van Schmus, W. R., Martin, M. W., and Berendsen, P., 1990, Age and isotopic composition of Nd and Pb for subsurface samples of the 1100 Ma old Midcontinent Rift, central United States: Geological Society of America, Abstracts with Program, v. 21, p. A174.

Wallace, H., 1981, Keweenawan geology of the Lake Superior basin, *in* Campbell, F. H. A., ed., Proterozoic basins of Canada: Geological Survey of Canada, Paper 81-10, p. 399–417.

Weiblen, P. W., 1982, Keweenawan intrusive igneous rocks, *in* Wold, R. J., and Hinze, W. J., eds., Geology and tectonics of the Lake Superior Basin: Geological Society of America Memoir 156, p. 57–82.

Weiblen, P. W., and Morey, G. B., 1980, A summary of the stratigraphy, petrology, and structure of the Duluth Complex: American Journal of Science, v. 280A, p. 88–133.

Wernicke, B., 1985, Uniform-sense normal simple shear of the continental lithosphere: Canadian Journal of Earth Science, v. 22, p. 108–125.

White, W. S., 1966, Tectonics of the Keweenawan basin, western Lake Superior region: U.S. Geological Survey Professional Paper 524-E, p. E1–E23.

White, W. S., 1972, Keweenawan flood basalts and continental rifting [abs.]: Geological Society of America Abstracts with Program, v. 4, p. 532–534.

White, R., and McKenzie, D., 1989, Magmatism at rift zones: The generation of volcanic continental margins and flood basalts: Journal of Geophysical Research, v. 94, p. 7685–7729.

Witzke, B. J., 1990, General stratigraphy of the Phanerozoic and Keweenawan sequence, M. G. Eischeid #1 drillhole, Carroll County, Iowa, *in* Anderson, R. R., ed., The Amoco M. G. Eischeid #1 Deep Petroleum Test, Carroll County, Iowa: Iowa Geological Survey Special Report Series No. 2, p. 39–58.

Woelk, T. S., and Hinze, W. J., 1991, Model of the Midcontinent Rift System in northeastern Kansas: Geology, v. 19, p. 277–280.

Wold, R. J., and Hinze, W. J., 1982, Introduction, *in* Wold, R. J., and Hinze, W. J., eds., Geology and tectonics of the Lake Superior basin: Geological Society of America Memoir 156, p. 1–4.

Wold, R. J., Hutchinson, D. R., and Johnson, T. C., 1982, Topography and surficial structure of Lake Superior bedrock as based on seismic reflection profiles, in Wold, R. J., and Hinze, W. J., eds., Geology and tectonics of the Lake Superior basin: Geological Society of America Memoir 156, p. 239–244.

Woollard, G. P., 1943, Transcontinental gravitational and magnetic profile of North America and its relation to geologic structure: Geological Society of America Bulletin, v. 54, p. 747–790.

Woollard, G. P., and Joesting, H. R., 1964, Bouguer gravity anomaly map of the United States: U.S. Geological Survey, scale 1:25 000 000.

Wunderman, R. L., 1986, Structure of the Keweenaw Rift and adjacent crust from a magnetotelluric survey in central Minnesota-Wisconsin [abs.]: Institute on Lake Superior Geology, 32nd, Proceedings and Abstracts, Wisconsin Rapids, Wisconsin, p. 91–92.

Wunderman, R. L., 1988, Crustal structure across the exposed axis of the Midcontinent Rift and adjacent flanks, based on magnetotelluric data, central Minnesota–Wisconsin: A case for crustal inhomogeneity and possible reactivation tectonics [Ph.D. thesis]: Houghton, Michigan Technological University, 209 p.

Wunderman, R. L., and Young, C. T., 1987a, Evidence for widespread basement décollement and related crustal asymmetry associated with the western limb of the Midcontinent Rift [abs.]: Institute on Lake Superior Geology, 33rd, Proceedings and Abstracts, Wawa, Ontario, p. 85–86.

Wunderman, R. L., and Young, C. T., 1987b, Is the path of the Keweenawan Midcontinent Rift controlled by an older continental suture? [abs.]: Geological Society of America Abstracts with Program, v. 19, p. 900.

Yarger, H. L., 1985, Kansas basement study using spectrally filtered aeromagnetic data, *in* Hinze, W. J., ed., The utility of regional gravity and magnetic anomaly maps: Tulsa, Oklahoma, Society of Exploration Geophysicists, p. 213–232.

Young, C. T., and Repasky, T. R., 1986, A magnetotelluric transect of the Jacobsville Sandstone in northern Michigan: Geological Society of America Bulletin, v. 97, p. 711–716.

Young, C. T., Adams, D., Loukili, A., Wunderman, R., 1989, Electrical conductivity models of the Lake Superior region (Midcontinent USA) [abs.]: Eos (Transactions, American Geophysical Union), v. 70, p. 274.

Zhu, T., and Brown, L., 1986, Consortium for Continental Reflection Profiling Michigan surveys: reprocessing and results: Journal of Geophysical Research, v. 91, p. 11477–11495.

MANUSCRIPT ACCEPTED BY THE SOCIETY JANUARY 16, 1996

Geological Society of America
Special Paper 312
1997

Segmented structure of the Middle Proterozoic Midcontinent Rift System, North America

A. B. Dickas
Department of Geosciences, University of Wisconsin, Superior, Wisconsin 54880
M. G. Mudrey, Jr.
Wisconsin Geological and Natural History Survey, University of Wisconsin (Extension), 3817 Mineral Point Road, Madison, Wisconsin 53705

ABSTRACT

We interpret recently acquired on-shore and off-shore reflection seismic data and deep mineral-exploration drill-holes to show that the Middle Proterozoic Midcontinent Rift System is similar to the East African Rift, where a series of opposed half grabens, principally infilled with rift-derived igneous and sedimentary rock stratigraphic packages, are separated by accommodation structures. Employing the East African Rift as a structural model, we suggest that the Midcontinent Rift System (MRS) is composed of both first-order and second-order rift segments. Five first-order segments, termed zones, constitute the entire MRS extending from Kansas to lower Michigan by way of the Lake Superior Basin. These zones are identified on the basis of major breaks in gravity and magnetic patterns, seismic geometry, lithologic packages, and terrane composition. First-order segments are themselves divided, by the degree of local structural maturity, into second-order rift segments. Within the Superior Zone, geologically and geophysically the most thoroughly documented zone of the MRS, four such second-order segments are identified and described.

This interpretation suggests that stratigraphic and structural correlation within the MRS may not be valid within the framework of a regional symmetric model. Instead, correlation must consider isolated igneous and sedimentary packages created unequally and in direct response to local rift developments.

INTRODUCTION AND HISTORIC BACKGROUND

The 1.1-Ga Middle Proterozoic Midcontinent Rift System (MRS) is a major intracontinental, thermo-tectonic structure that has been traced by regional gravity (Lyons and O'Hara, 1982) and magnetic data (King and Zietz, 1971), subsurface drilling, and outcrop control over a length of 2,000 km within the central United States (Fig. 1). Outcrops are known only in the Lake Superior region and define a structural basin that has been filled by plateau lava and sedimentary rock comprising the Keweenawan Supergroup (Fig. 2). Basalt, underlain by thin quartz sandstone and interbedded with polymictic red-bed units

(Merk and Jirsa, 1982), was extruded along the rift during the short geologic time interval from ca. 1,109 Ma to ca. 1,094 Ma (Paces and Davis, 1988). Beginning during the late stages of volcanic activity, a suite of maturing upward, clastic sedimentary rock was deposited.

In 1943 George Woollard reported a major positive gravity anomaly near Clay Center, Kansas, and subsequently (Woollard, 1943) reported the existence of a long narrow belt of positive gravity anomalies stretching at least 1,200 km from Lake Superior southwestward into Kansas (Fig. 1). Thiel (1956, p. 1088) called this feature the "midcontinent gravity high" and noted the strong spatial relationship of this gravity anomaly to

Dickas, A. B., and Mudrey, M. G., Jr., 1997, Segmented structure of the Middle Proterozoic Midcontinent Rift System, North America, *in* Ojakangas, R. W., Dickas, A. B., and Green, J. C., eds., Middle Proterozoic to Cambrian Rifting, Central North America: Boulder, Colorado, Geological Society of America Special Paper 312.

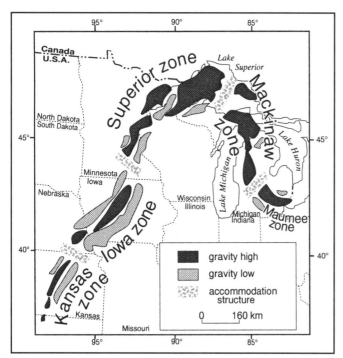

Figure 1. Simplified map of the Midcontinent Gravity High and the Mid-Michigan Gravity High. The Midcontinent Gravity High extends from Kansas northeast to the central section of the Lake Superior Basin. The Mid-Michigan Gravity High extends from the eastern section of the Lake Superior Basin southeast to the Michigan-Ohio border. The rock sequences causing these geophysical trends constitute the Midcontinent Rift System (MRS). Also shown, as hachured bars cutting across the axis of the MRS, is the location of the four accommodation features of first-order structural significance, as proposed in this study. These accommodation structures subdivide the MRS into five zones, named on the basis of principle geographic location. Gravity trend modified from Lyons and O'Hara (1982).

Keweenawan rock in the Lake Superior area. His geophysical modeling was limited but led him to state "the high anomaly gradients bounding the lava mass suggest faults between high-density lava and low-density sandstone and shale" (Thiel 1956, p. 1089).

By the 1960s, a broad consensus emerged that the MRS was an aborted, structurally symmetric, intracontinental rift of major proportions, extending from the Oklahoma–Kansas state line northeast to Lake Superior, and then southeast under the Michigan Basin, terminating in the vicinity of Detroit, Michigan. Within the southwest Lake Superior Basin, field geology identified the center of the rift as the St. Croix Horst, flanked by half-graben basins (Fig. 2). Summaries of variations of this tectonic model may be found in Steinhart and Smith (1966), Wold and Hinze (1982), and Dickas (1986a).

A major revision in the tectonic interpretation of the MRS emerged in the 1980s when Serpa et al. (1984) evaluated a deep reflection seismic profile extending across the southern part of this rift in Kansas. Their model, which was compatible with other well-studied modern rifts, incorporated asymmetric basins, fault block rotations, and fanning of sedimentary lay-

ers. Dickas (1986b) suggested that these structural model differences were a reflection of differing stages of tectonic maturity. In the mid-1980s, some data from reflection seismic programs conducted for petroleum exploration along the MRS became available. Interpretation of these data suggest that an asymmetric, segmented structure may well be the correct model for the MRS, particularly for the Lake Superior portion of this trend.

FUNDAMENTAL RIFT UNIT

Seismic analysis of the East African Rift System (Rosendahl and Livingstone, 1983) led to new ideas regarding phased development of intracontinental rifts. These concepts resulted from observations that Lakes Nyasa (Malawi) and Tanganyika (Tanzania) were "divided into natural structural units with a certain dimensional repeatability" (Lorber, 1984, p. 15). This geometric duplication was identified as the "fundamental rift unit." A review of the literature permits us to formulate a composite picture of a typical "fundamental rift unit."

Geometry

In plan view the fundamental unit is approximately a parallelogram with length of 50 to 190 km, and widths from 20 to 60 km. Length to width ratios vary from 2 to 4. A worldwide study of fundamental rift units (Reynolds, 1984) indicates an average length of 119 km and an average width of 40 km, resulting in a typical length to width ratio of 3.

In cross-section view the fundamental unit is commonly triangular, defining a half-graben. Reynolds (1984) reports this half-graben structure in the Suez, Rio Grande, and Viking Rifts; Logatchev et al. (1983) in the Baikal Rift; Schuepbach and Vail (1980) in the Greenland margin; and Bosworth et al. (1986) in the North Sea Graben, Connecticut River Rift, and the Rhine-graben. Although displaying geometries of first-order rift segments, rift units should also possess unique second-order geometries, expressed in their structural, morphological, and depositional histories.

Longitudinal structures

When the half-graben structure is initiated, it is bordered on its deep side by a major curvilinear, normal, listric fault system, and on the shallow side by a monocline, or series of normal, small displacement step faults. These structural elements are consistent with those denoted in the "taphrogeosyncline" of Kay (1951).

Terminal structures

Fundamental units are bounded by complex structures striking obliquely to the rift axis. These structures, named "accommodation zones" by Bosworth et al. (1986), are vari-

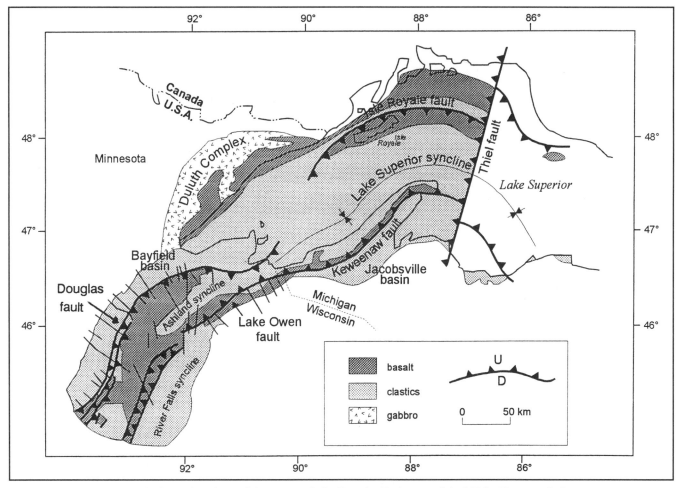

Figure 2. General geology of the western Lake Superior Basin part of the MRS. The central, St. Croix Horst, originally a graben, is flanked by half-graben basins and bordered by regional fault systems. The strikes of these faults parallel the rift axis, but fault dip directions are inward toward the rift axis. After Dickas (1986a).

ously reported in the literature as including: interbasinal highs, ridges, deeply buried "humps," basement sills, "antiform" highs, transverse structures, transform faults, and regions of uplifted basement. In active rifts terminal structures are often coincident with changes in boundary fault orientation, hot springs, and high heat flow (Reynolds, 1984), the preferential occurrence of intrusive and extrusive igneous activity (McConnell, 1980), and geomorphic discontinuities (Rosendahl and Livingstone, 1983).

Relation to rift system

Rift systems often are composed of amalgamated fundamental units consisting of half grabens (Scott and Rosendahl, 1989). Along the rift trend, juxtaposed half grabens typically display alternating structural and depositional polarities (orientations; Reynolds, 1984). Separate half-graben basins may thus contain differing spatial, lithic, and temporal stratigraphic packages.

DATA BASE IN LAKE SUPERIOR REGION

Encouraged by discoveries in Precambrian rocks in Siberia, Australia, and China, the petroleum industry began conducting extensive reflection seismic programs in the Lake Superior region in late 1983 (Dickas, 1997). In 1985 the Great Lakes International Multidisciplinary Program on Crustal Evolution (GLIMPCE), a consortium formed to improve the information data base on crustal structure in the Great Lakes region, recorded 630 km of seismic data in Lake Superior to a time-depth of 20 seconds (Behrendt et al., 1988; Cannon et al., 1989). Portions of GLIMPCE profiles A and C (Fig. 3) have been of particular value to this study and are briefly described here.

GLIMPCE seismic profile A extends from the northern apex of Lake Superior southward to the south shore in the vicinity of Marquette, Michigan (Fig. 3). Within the central third of this section (between shot points 1000 and 2800), Cannon et al. (1989) interpreted a thick sequence of seismic reflectors as Middle Proterozoic volcanic (pre–Portage Lake and

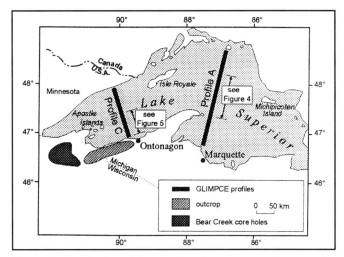

Figure 3. Distribution of data used in this report to reinterpret the structural geology of the MRS.

Portage Lake) and interbedded sedimentary rock lying above Archean basement and directly overlain by the conformable Oronto Group, consisting of conglomerate, sandstone, and shale (Fig. 4). A prominent southward increase in thickness of the pre–Portage Lake and Portage Lake section during the Middle Proterozoic is displayed in this interpretation. The southern border of this stratigraphic wedge is the Keweenaw fault, which crops out along the Keweenaw Peninsula to the west of the profile. The northern border is marked by the Isle Royale fault. The southerly increase in section thickness suggests that this sector of the Keweenaw fault was tectonically active, while the corresponding Isle Royale fault was passive, during the Middle Proterozoic, allowing ponding of basaltic infill against the topographic ramp caused by the Keweenaw fault scarp.

GLIMPCE seismic profile C was recorded from the Minnesota shoreline of Lake Superior southeast to offshore Ontonagon, Michigan (Fig. 3). Cannon et al. (1989) again document a stratigraphic wedge created by geometric fanning within the Portage Lake and pre–Portage Lake sequences (between shot points 1000 and 1800; Fig. 5). However, the direction of thickening along profile C is reversed from that observed in profile A, showing an increase to the north. Thus during the Middle Proterozoic, the infill history of this particular rift sector was directly influenced by displacement along the Isle Royale, and not the Keweenaw, fault.

Additional data supplement information obtained from GLIMPCE profiles. A late 1950s Bear Creek Mining Company exploration program included the drilling of 48 shallow core holes (maximum depth of 1,005 m) in northwestern Wisconsin (Fig. 3). This program unsuccessfully sought cupriferous mineralization within the Nonesuch Formation, the central unit of the Middle Protozeroic Oronto Group. As Nonesuch deposition did not extend throughout the geographical extent of this drilling program, only 15 of 48 cores intersected a Nonesuch section (Table 1). Isopach mapping of these 15 Nonesuch Formation sections discloses increasing thickness to the south (Fig. 6). Because the Oronto Group is regionally conformable with the underlying Portage Lake Volcanics, as evidenced by outcrop study, this southerly increase in Nonesuch Formation thickness suggests the older volcanic sequence also thickens to the south. This sedimentary and volcanic rock wedge was formed by Middle Proterozoic activity along the Lake Owen fault, located to the immediate south (Fig. 2). In contrast the Douglas fault, the northern boundary of this half-graben shaped stratigraphic package, was apparently passive during this time.

The Nonesuch Formation is exposed exclusively along the

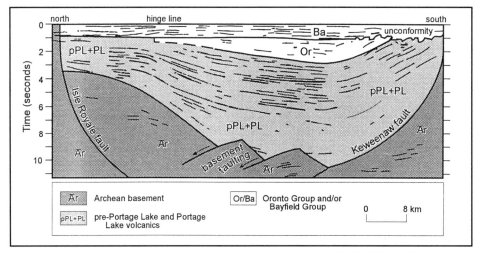

Figure 4. Line diagram interpretation of seismic reflection data along GLIMPCE profile A, located in the eastern section of Lake Superior. The profile displays that portion of the central MRS structure bounded by the Isle Royale fault (SP 2825) and the Keweenaw fault (SP 1030) (see Fig. 3 for location). Vertical scale is in seconds of two-way travel time. Note southward increase in thickness of pre–Portage Lake and Portage Lake volcanic rocks and conformable Oronto Group sedimentary strata. The angular unconformity shown is local in extent. Modified from Cannon et al., 1989.

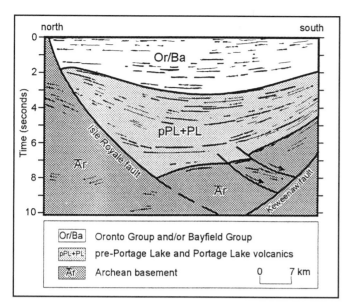

Figure 5. Line diagram interpretation of seismic reflection data along GLIMPCE profile C, located in western Lake Superior. The south half of profile C is shown from the Isle Royale fault (SP 1000) to the south end of profile C (SP 1850; see Fig. 3 for location). Note that the northward direction of increase in rock thickness is opposite to that displayed in GLIMPCE profile A (Fig. 4). Vertical scale is in seconds of two-way travel time. Modified from Cannon et al., 1989.

southwest shoreline of Lake Superior in the vicinity of the Wisconsin-Michigan border (Fig. 3). Thirteen measured sections of this formation have been collected, corrected for structural dip, and mapped (Table 1). The resulting isopach pattern displays a distinct and uniform increase in thickness to the north, under Lake Superior (Fig. 6). This thickness orientation agrees with that interpreted from the GLIMCPE seismic profile C, but is the opposite of that interpreted from the GLIMCPE profile A and the isopach data from Bear Creek cores. This reversal in direction of increasing isopach thickness, as seen on GLIMCPE profiles A and C, suggests the development in the western Lake Superior Basin of axially juxtaposed fault blocks of similar strike but opposing structural dip orientations.

STRUCTURAL DIFFERENTIATION OF THE MIDCONTINENT RIFT SYSTEM

The structural and tectonic history of the Midcontinent Rift System (MRS) presented here is based on a hierarchical classification of rift elements proposed by Rosendahl (1987), whereby individual rift units join to form rift zones, which in turn form rift branches that finally, in sum, create a rift system. Rosendahl (1987) demonstrated this descriptive scheme by adopting it to the morphology and structure of the East African Rift System, with "system" representing the largest rift scale (Table 2).

The entire 2,000-km trend of the Midcontinent Rift, extending from Kansas to the Lower Peninsula of Michigan via

the Lake Superior Basin (Fig. 1), constitutes the Midcontinent Rift System. In scale, this system is comparable to the East African Rift System (Table 2). The Midcontinent Rift System is subdivided into two branches, which join within the central Lake Superior Basin (Klasner et al., 1982). These branches were originally identified by their geophysical characteristic as the Midcontinent Gravity High (Thiel, 1956), which trends northeast-southwest and displays a strong potential field signature, and the Mid-Michigan Gravity High (Hinze et al., 1975), which trends southeast-northwest and is identified by a subdued gravity anomaly caused by thick overlying Phanerozoic strata.

We subdivide the MRS into five first-order zones, separated by terminal (accommodation) structures and differentiated on the basis of rift-geometry and tectonics. Clockwise, from the southwest termination of the system in Kansas, we name the zones Kansas, Iowa, Superior, Mackinaw, and Maumee (Fig. 1). The geologic characteristics of these segments (Table 3) suggest that structural maturity, as measured by amount of rift extension, type and depth of faulting, and degree of structural asymmetry, decreases both southeast and southwest of the Superior Zone (Dickas, 1986b; McSwiggen et al., 1987).

The least mature appear to be the Maumee and Mackinaw Zones, which jointly compose the eastern arm of the Midcontinent Rift System. The Maumee Zone is a symmetrical, unfaulted extensional structure (Brown et al., 1982) associated with a low-amplitude gravity signature and Precambrian "redbed" stratigraphy (Sleep and Sloss, 1978). The Mackinaw Zone is also similar in geometry, but differs from the Maumee Zone because it is associated with listric normal faults (Behrendt et al., 1988), suggestive of a higher stage of structural development than the unfaulted Maumee Zone.

The Kansas Zone, which forms the most southwesterly segment of the western branch of the MRS, is isolated from its counterpart in Iowa by a granitic terrain in southeast Nebraska (Muehlberger et al., 1967). This zone narrows to the south to a region of deep crust feeder dikes (Anderson, 1992). The Kansas Zone was originally identified, by interpretation of a COCORP seismic survey, as an extensional asymmetric basin plunging to the west and bounded by easterly dipping normal faults (Serpa et al., 1984). A deep test of the hydrocarbon potential of this segment of the MRS, however, drilled a Precambrian igneous and underlying sedimentary clastic rock sequence that is stratigraphically transposed to lithologic sequences present to the north in Iowa (Berendsen et al., 1988) and Wisconsin (Dickas, 1986a). Woelk and Hinze (1991) have reinterpreted the Kansas COCORP line as displaying a westerly plunging structure bounded by reverse faults, traceable in depth to ~10 km.

The Iowa Zone is bounded by accommodation structures defined by an interruption of the central gravity maxima displayed by the Midcontinent Gravity High in southeastern Nebraska (at its contact with the Kansas Zone), and a left-lateral shift in the same gravity anomaly in south-central Minnesota (at its contact with the Superior Zone). Further evidence for these terminal (accommodation) structures include the pres-

TABLE 1. MEASURED DRILL CORE AND OUTCROP SECTIONS OF THE NONESUCH FORM-TION, NORTHWESTERN WISCONSIN, AND ADJACENT UPPER PENINSULA OF MICHIGAN*

Location	Thickness (m)	Source of Data[†]
19-47N-10W DO-5, Wisconsin	72	Bear Creek
15-47N-10W DO-6, Wisconsin	67	Bear Creek
03-45N-10W DO-8, Wisconsin	128	Bear Creek
34-47N-11W DO-11, Wisconsin	42+	Bear Creek
08-46N-10W DO-13, Wisconsin	88	Bear Creek
11-46N-10W DO-14, Wisconsin	96	Bear Creek
06-45N-06W WC-2, Wisconsin	109	Bear Creek
15-45N-06W WC-3, Wisconsin	117	Bear Creek
36-47N-08W WC-9, Wisconsin	147	Bear Creek
36-46N-09W WC-13, Wisconsin	120	Bear Creek
32-48N-09W WC-14, Wisconsin	63	Bear Creek
08-45N-06W WC-16, Wisconsin	77+	Bear Creek
03-45N-07W WC-17, Wisconsin	99	Bear Creek
06-45N-09W WC-18, Wisconsin	137	Bear Creek
03-45N-08W WC-22, Wisconsin	133	Bear Creek
17-45N-02W Brownstone Falls, Wisconsin	113	Elmore, 1981
18-46N-01W Potato River Falls, Wisconsin	92	Elmore, 1981
30-47N-01E Parker Creek, Wisconsin	114	Elmore, 1981
20-47N-01E Montreal River, Wisconsin	99	Elmore, 1981
10-49N-46W Black River, Michigan	20+	Elmore, 1981
30-50N-45W Presque Isle River, Michigan	183	Elmore, 1981
03-50N-42W Porcupine Mountains, Michigan	175	Daniels, 1982
13-51N-42W Big Iron River, Michigan	229	Elmore, 1981
13-51N-42W Big Iron River, Michigan	179	White and Wright, 1954
01-50N-43W Little Iron River, Michigan	57+	Elmore, 1981
01-50N-43W White Pine Area, Michigan	185	Daniels, 1982
23-55N-34W Swedetown, Michigan	138+	Elmore, 1981
10-56N-34W Calumet Area, Michigan	215	Daniels, 1982

*All data have been corrected for structural dip.
[†]All Bear Creek localities refer to drill core; remainder are outcrop.

ence of plutonic intrusions in southeast Minnesota and northeastern Iowa (Chandler et al., 1989), and the documentation of the Belle Plaine fault system of southern Minnesota (Sloan and Danes, 1962). Analyses of exploration company seismic profiles from central Iowa (Anderson, 1992) identify a central horst associated with deep structures and bounded by multiple systems of reactivated reverse faults, one of which can be traced to a depth of 20 km (Chandler et al., 1989). A combination of seismic and gravity modeling suggests these deep middle and lower crust structures are dominated by a solid, mafic intrusive core averaging 35 km in width and extending to a depth of at least 40 km (Chandler et al., 1989). This well-developed rift core is in contrast to the less-developed rift core in the Kansas Zone, modeled by Woelk and Hinze (1991) as composed of isolated mafic intrusive pillows.

The Superior Zone appears to be the most mature of all proposed MRS tectonic segments. It is bounded by terminal (accommodation) structures that are associated with large-scale, regional faulting, and it includes the area of maximum

MRS width development of 150 km (Klasner et al., 1982). The Superior Zone is characterized by a central horst bounded by reverse fault systems that can be studied in outcrop, the Duluth Gabbro Complex (Miller, 1989), and a mafic intrusive rift core that in model analysis averages 50 km in width and extends to a depth of approximately 50 km (Van Schmus and Hinze, 1985). The Belle Plaine fault system, a major northwest-trending structural zone located in southern Minnesota, forms an accommodation structure juxtaposing the Superior and the Iowa Zones (Chandler et al., 1989). The accommodation structure bounding the northeastern edge of the Superior Zone, identified as the Thiel fault (Hinze et al., 1975), geographically coincides with a region of crustal thickening known as the Trans-Superior tectonic structure (Klasner et al., 1982).

The fundamental rift unit concept (Rosendahl and Livingstone, 1983) can be applied to the Lake Superior Basin, the only sector of the MRS with outcrops of rift-related igneous and sedimentary rocks. Here we identify four such units, named Manitou, Ontonagon, Brule, and Chisago (Fig. 7), averaging

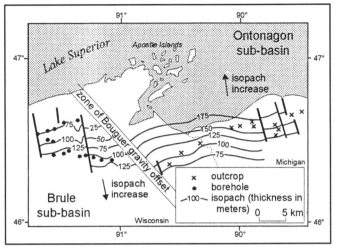

Figure 6. Isopach map of the Nonesuch Formation of the upper Keweenawan Oronto Group, southwestern shoreline of Lake Superior. The large dots are exploratory drill sites in northwestern Wisconsin; for details see Table 1. The x patterns represent thicknesses of Nonesuch Formation outcrops from the Wisconsin-Michigan border area; see Table 1 for details. Note the different directions of thickening shown by black arrows, suggesting these two sets of isopach data are associated with differing rift units, separated by an accommodation structure. The location of such an accommodation structure is indicated by a zone of Bouguer gravity offset, as identified from mapping by Craddock et al. (1969) and Klasner et al. (1979). The Ontonagon and Brule subbasins shown here are two of the four rift units composing the Superior Zone of the MRS.

TABLE 2. A COMPARISON OF EAST AFRICA RIFT SYSTEM AND MIDCONTINENT RIFT SYSTEM REGIONAL STRUCTURAL ELEMENTS*

Element	African Model	U.S. Model
System	East African Rift	Midcontinent Rift
Branch	Eastern	Mid-Michigan Gravity High[†]
	Western	Midcontinent Gravity High[†]
Zone	Gregory (Kenya)	Kansas[§]
	Stefanie (Chow Bahir)	Iowa[§]
500-700 km	Tanganyika	Superior[§]
in length	Usno-Omo-Kibish	Mackinaw[§]
	Malawi	Maumee[§]
Unit**	Turkana	Manitou[§]
	Baringo-Bogoria[‡]	Ontonagon[§]
80-160 km	Nakura-Naivasha[‡]	Brule[§]
in length	Magadi-Natron[‡]	Chisago[§]
l/w ratio 2-4/1		

*After the rift scale classification by Rosendahl, 1987. The architectural elements listed here should be considered typical, rather than correlative, examples.
[†]Known principally as a geophysical entity.
[§]Terminology as proposed in this chapter.
**Fundamental rift building structural block.
[‡]After Bosworth et al., 1986.

150 km in length by 60 km in width and having a length/width ratio of 2.5. In other geometric and geologic aspects these units are consistent with the description of the typical fundamental unit presented earlier in this chapter. Cross-sectional geometry and stratigraphic polarity characterizing these units has been clearly identified (Figs. 4 and 5) and verified by seismic analysis by McSwiggen et al. (1987), Nyquist and Wang (1988), Chandler et al. (1989), Cannon et al. (1989), and McGinnis et al. (1989), and by surface and subsurface isopach mapping (Fig. 6). Igneous and sedimentation wedges of alternating isopach patterns distinguish juxtaposed rift units.

Three (accommodation) structures of second-order tectonic significance are identified within the Lake Superior Zone (Table 3). The structure separating the Chisago from the Brule Unit (Fig. 7) is located on the basis of a constriction in the regional gravity pattern in Anoka County, Minnesota (Craddock et al., 1969), the recent identification of the Bloomer fault of Wisconsin by Mudrey et al. (1987), and the inclusion of a "very deep—large excess of mass" into a potential field model used by Chandler et al. (1989, p. 271) to interpret a seismic profile across this section of the Midcontinent Rift System. The accommodation structure located to the northeast, separating the Brule from the Ontonagon Unit and isolating the opposing Nonesuch Formation core and outcrop dips mapped in northwest Wisconsin (Mudrey and Dickas, 1988), has been recognized as "lineament D" by Klasner et al. (1982), and as the Mineral Lake fault by Mudrey and Brown (1988). Additional evidence for this structure includes a major bifurcation in the regional gravity field (Klasner et al., 1979), and vertical Oronto Group outcrops in the on-land sector of this feature. The third accommodation structure, separating the Ontonagon from the Manitou Unit, is located along the trend of a principle left lateral shift of the Lake Superior Bouguer gravity field (Klasner et al., 1979), and separates the middle and upper Keweenawan sedimentary and igneous rock packages that are apposed to each other both by structural dip and thickening patterns, as seen on GLIMPCE seismic profiles A and C (Figs. 4 and 5).

The Superior Zone rift units (Fig. 7) have a distinctive stratigraphic thickening polarity, as identified by seismic, subsurface, and outcrop studies; southerly in the Brule and Manitou units and northerly in the Ontonagon and Chisago units. These rift units are further distinguished by limited spatial distribution of the Duluth Complex intrusion (Miller, 1989), and differing Oronto Group organic petrologies (Elmore et al., 1989). This lack of continuity of rift trend patterns is an argument for the presence of the half-graben basins and their opposing structural and stratigraphic polarity.

The increased burial depth of rift rocks beneath Phanerozoic strata in both directions away from the Lake Superior area and the absence of published geophysical data along the rift prevent such detailed differentiation outside the Superior Zone. The possibility, however, of such differentiation must be con-

**TABLE 3. PRESENTATION OF GEOLOGICAL, GEOPHYSICAL, AND GEOCHEMICAL
INFORMATION TO DIVIDE THE MIDCONTINENT RIFT SYSTEM INTO ZONES
SEPARATED BY ACCOMMODATION STRUCTURES
OF FIRST-ORDER TECTONIC SIGNIFICANCE***

Rift Division	Supporting Evidence
Maumee Zone	Sag basin on COCORP seismic lines. McClure #1-8 Sparks well stratigraphy.
First-order accommodation structure	Gravity "dogleg," Montcalm Co., Michigan. First Seismic Corp. seismic interpretation.
Mackinaw Zone	GLIMPCE profile F extension faulting. McClure #1 Beaver Island test. Coalescing of Midcontinent Rift structure.
First-order accommodation structure	Thiel fault previously identified. Fault break on GLIMPCE line G. Major change in MRS strike. Interrupted gravity and magnetic patterns.
Superior Zone	Structural Detail[†] ———→
First-order accommodation structure	Interrupted gravity and magnetic patterns. Granitic intrusions, southeast Minnesota. Belle Plaine (transform?) fault system. Borehole and gravity control, southern Minnesota
Iowa Zone	Identified Iowa central horst structure. Presence of extension and contraction faulting. Published regional well control.
First-order accommodation structure	Broken gravity and magnetic trends. Granite terrane, southeastern Nebraska. Local well control. South-central magnetic lineament.
Kansas Zone	Extension faulting on COCORP seismic lines. Asymmmetric central graben. Texaco #1 Noel Poersch test and stratigraphy.

Manitou Unit	S isopach. GLIMPCE A.
Second-order accommodation structure	Altered gravity trend.
Ontonagon Unit	N isopach GLIMPCE C. Duluth Complex intrusion. Remanent magnetic reversals. N outcrop thickening.
Second-order accommodation structure	Terminated gravity trend. Transverse basement ridge. Archean fault trend. Vertical outcrop dips.
Brule Unit	Different organic petrologies. S isopach Bear Creek Cores. S isopach Petty-Ray Seismic.
Second-order accommodation structure	Bloomer fault (field evidence). Constricted gravity trend.
Chisago Unit	W isopach Petty-Ray Seismic.

*Left side of table. Because of the greater wealth of such information available within the Superior
Zone, this rift segment is further divided into four rift units, separated by accommodation structures of
second-order tectonic significance (right side of table).
†N, S, and W refer to compass directions of isopach and outcrop thickening.

sidered in evaluating the structural and stratigraphic histories of other zones and units of the Midcontinent Rift System. Such consideration could explain the contrasting estimates for Precambrian hydrocarbon reported in recent studies. Hatch and Morey (1985) reported early thermal maturity and pre-Phanerozoic loss of hydrocarbon from Middle Proterozoic strata on the central horst within the Chisago Unit of the Supe-rior Zone. Kelly and Nishioka (1985), in contrast, document the presence of indigenous, live crude oil dating to at least 1,047 Ma associated with similar age strata of the Ontonagon Unit within the same MRS zone. Considering the differing geologic development of these fundamental rift units, the economic prospects for the intervening Brule Unit could mimic either adjacent unit.

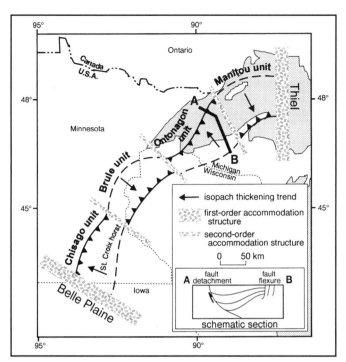

Figure 7. Structural interpretation of the Superior Zone of the Midcontinent Rift System. The Superior Zone is bounded by the Belle Plaine (southern Minnesota) and the Thiel (eastern Lake Superior Basin) faults and is subdivided into four rift units by three accommodation structures of second-order tectonic significance. Note each of the four identified rift units possesses a sediment thickening direction opposite to that of an adjacent rift unit. The barbed border of each rift unit represents the location of upper Keweenawan faults bounding the thick wedges of volcanic and associated sedimentary rocks. The dashed line identifies the position of a hinge line that became active during the later contractional phase of the rift system, creating the present-day horst structure.

CONCLUSIONS

The adaptability of the fundamental rift unit concept to the MRS, especially in the region of the Lake Superior Basin, is demonstrated by the analysis of newly acquired geologic and geophysical data. This analysis shows that the MRS, through its history of evolution, followed tectonic and sedimentation processes similar to those operating in the present-day East African Rift System. We suggest that an East Africa Rift type of structural development controlled differential infilling of individual MRS units, particularly in the Lake Superior Basin. Our analysis includes only that time from the inception of rifting through the early development of the MRS during pre-Oronto and Oronto time. We suggest that tectonic adjustments continued into post-Oronto time, however these later adjustments were not necessarily in response to the same stresses that led to initial development of the MRS.

The recognition of the principle of individual tectonic development of fundamental rift units, jointly composing larger

rift zones and branches, requires that any determined geologic history of one rift unit does not necessarily offer tectonic insight into the development of adjacent rift units. Thus the exploration of the MRS, whether for mineralization or hydrocarbons, should be based on structural, sedimentation, and maturation studies on a basin-to-basin basis.

ACKNOWLEDGMENTS

The authors would like to thank Raymond R. Anderson, Iowa Department of Natural Resources Geological Survey Bureau, and Pieter Berendsen, Kansas Geological Survey, for offering insightful suggestions. We are also indebted to two anonymous reviewers for their helpful comments. Finally, we would like to acknowledge financial support for this research provided by the Cooperative Extension division of the University of Wisconsin-Extension.

REFERENCES CITED

Anderson, R. R., 1992, The Midcontinent Rift of Iowa [Ph.D. thesis]: Iowa City, The University of Iowa, 323 p.

Behrendt, J. C., Green, A. G., Cannon, W. F., Hutchinson, D. R., Lee, M. W., Milkereit, B., Agena, W. F., and Spencer, C., 1988, Crustal structure of the Midcontinent Rift System: Results from GLIMPCE deep seismic reflection profiles: Geology, v. 16, p. 81–85.

Berendsen, P., Borcherding, R. M., Doveton, J., Gerhard, L., Newell, K. D., Steeples, D., and Watney, W. L., 1988, Texaco Poersch #1, Washington County, Kansas—preliminary geologic report of the pre-Phanerozoic rocks: Kansas Geological Survey Open-File Report 88-22, 116 p.

Bosworth, W., Lambiase, J., and Keisler, R., 1986, A new look at Gregory's Rift: The structural style of continental rifting: Eos (Transactions, American Geophysical Union), v. 67, no. 29, p. 577, 582, 583.

Brown, L., Jensen, L., Oliver, J., Kaufman, S., and Steiner, D., 1982, Rift structures beneath the Michigan Basin from COCORP profiling: Geology, v. 10, p. 645–649.

Cannon, W. F., and 11 others, 1989, The North American Midcontinent Rift beneath Lake Superior from GLIMPCE seismic reflection profiling: Tectonics, v. 8, p. 305–332.

Chandler, V. W., McSwiggen, P. L., Morey, G. B., Hinze, W. J., and Anderson, R. R., 1989, Interpretation of seismic reflection, gravity, and magnetic data across Middle Proterozoic Mid-continent Rift System in northwestern Wisconsin, eastern Minnesota, and central Iowa: American Association of Petroleum Geologists Bulletin, v. 73, p. 261–275.

Craddock, C., Mooney, H. M., and Kolehmainen, V., 1969, Simple Bouguer gravity map of Minnesota and northwestern Wisconsin: Minnesota Geological Survey Miscellaneous Map Series Map M-10, scale 1:1,000,000.

Daniels, P. A., 1982, Upper Precambrian sedimentary rocks, Oronto Group, Michigan–Wisconsin, in Wold, R. J., and Hinze, W. J., eds., Geology and tectonics of the Lake Superior Basin: Geological Society of America Memoir 156, p. 107–133.

Dickas, A. B., 1986a, Comparative Precambrian stratigraphy and structure along the Mid-Continent Rift: American Association of Petroleum Geologists Bulletin, v. 70, p. 225–238.

Dickas, A. B., 1986b, Seismologic analysis of arrested stage development of the Midcontinent Rift, in Mudrey, M. G., Jr., ed., Precambrian petroleum potential, Wisconsin and Michigan: Geoscience Wisconsin, Vol. 11, p. 45–52.

Dickas, A. B., 1997, Exploration for hydrocarbon along the Midcontinent Rift

System trend of Wisconsin and the Lake Superior Basin: 1983 through 1991: Wisconsin Geological and Natural History Survey (in press).

Elmore, R. D., 1981, The Copper Harbor Conglomerate and Nonesuch Shale: Sedimentation in a Precambrian intracontinental rift, Upper Michigan: [Ph.D. thesis]: University of Michigan, 192 p.

Elmore, R. D., Milavec, G. J., Imbus, S. W., and Engel, M., 1989, The Precambrian Nonesuch Formation of the North American Mid-Continent rift, sedimentology and organic geochemical aspects of lacustrine deposition: Precambrian Research, p. 1–23.

Hatch, J. R., and Morey, G. B., 1985, Hydrocarbon source rock evaluation of Middle Proterozoic Solor Church Formation, North American Mid-Continent Rift System, Rice County, Minnesota: American Association of Petroleum Geologists Bulletin, v. 69, p. 1208–1216.

Hinze, W. J., Kellogg, R. L., and O'Hara, N. W., 1975, Geophysical studies of basement geology of Southern Peninsula of Michigan: American Association of Petroleum Geologists Bulletin, v. 59, p. 1562–1584.

Kay, M., 1951, North American geosynclines: Geological Society of America Memoir 48, 143 p.

Kelly, W. C., and Nishioka, G. K., 1985, Precambrian oil inclusions in late veins and the role of hydrocarbons in copper mineralization at White Pine, Michigan: Geology, v. 13, p. 334–337.

King, E. R., and Zietz, I., 1971, Aeromagnetic study of the Midcontinent Gravity High of central United States: Geological Society of America Bulletin, v. 82, p. 2187–2207.

Klasner, J. S., Wold, R. J., Hinze, W. J., Bacon, L. O., O'Hara, N. W., and Berkson, J. M., 1979, Bouguer gravity anomaly map of the northern Michigan–Lake Superior region: U.S. Geological Survey Geophysical Investigations Map GP-930, scale 1:1,000,000

Klasner, J. S., Cannon, W. F., and Van Schmus, W. R., 1982, The pre-Keweenawan tectonic history of southern Canadian Shield and its influence on formation of the Midcontinent Rift, *in* Wold, R. J., and Hinze, W. J., eds, Geology and tectonics of the Lake Superior Basin: Geological Society of America Memoir 156, p. 27–46.

Logatchev, N. A., Zorin, Y. A., and Rogozhina, V. A., 1983, Cenozoic continental rifting and geologic formations as illustrated by the Kenya and Baikal rift zones: Geotectonics, v. 17, p. 83–92.

Lorber, P. M., 1984, The Kigoma Basin of Lake Tanganyika: Acoustic stratigraphy and structure of an active continental rift [M.S. thesis]: Duke University, 73 p.

Lyons, P. L., and O'Hara, N. W., chairpersons, 1982, Gravity anomaly map of the United States, exclusive of Alaska and Hawaii: Society of Exploration Geophysicists, 2 maps, each 1:2,500,000.

McConnell, R. B., 1980, A Precambrian origin for the proto-rift dislocation belt of eastern Africa, *in* Geodynamic evolution of the Afro-Arabian Rift System: Rome, Italy, Accademia Nazionale dei Lincei, 705 p.

McGinnis, L. D., and 10 others, 1989, Geophysical and tectonic study of sedimentary basins and the upper crust beneath Lake Superior [abs.]: American Geophysical Union Annual Meeting, p. 272.

McSwiggen, P. L., Morey, G. B., and Chandler, V. W., 1987, New model of the Midcontinent Rift in eastern Minnesota and western Wisconsin: Tectonics, v. 6, p. 677–685.

Merk, G. P., and Jirsa, M. A., 1982, Provenance and tectonic significance of the Keweenawan interflow sedimentary rocks, *in* Wold, R. J., and Hinze, W. J., eds., Geology and tectonics of the Lake Superior Basin: Geological Society of America Memoir 156, p. 97–105

Miller, J. D., 1989, Middle Keweenawan intrusive igneous history, *in* Dickas, A. B., ed., Lake Superior Basin segment of the Midcontinent Rift System: International Geological Congress, 28th, Field Trip Guidebook T344, p. 13–15.

Mudrey, M. G., Jr., and Brown, B. A., 1988, Summary of field mapping, Superior sheet: Wisconsin Geological and Natural History Survey Open-File Reports 88-7, 39 p.

Mudrey, M. G., Jr., and Dickas, A. B., 1988, Midcontinent Rift model based upon Gregory Rift tectonic and sedimentation geometries [abs.]: Institute on Lake Superior Geology, 34th, Proceedings and Abstracts, v. 34, p. 73–75.

Mudrey, M. G., Jr., LaBerge, G. L., Myers, P. E., and Kordua, W. S., 1987, Bedrock geology of Wisconsin; northwest sheet: Wisconsin Geological and Natural History Survey, Map 87-11, scale 1:250,000, 2 sheets.

Muehlberger, W. R., Denison, R. E., and Lidiak, E. G., 1967, Basement rocks in continental interior of United States: American Association of Petroleum Geologists Bulletin, v. 51, p. 2351–2380.

Nyquist, J. E., and Wang, H. F., 1988, Flexural modeling of the Midcontinent Rift: Journal of Geophysical Research, v. 93, p. 8852–8868.

Paces, J. B., and Davis, D. W., 1988, Implications of high precision U-Pb age dates on zircons from Portage Lake volcanic basalts on Midcontinent Rift subsidence rates, lava flow response periods and magma production rates: Institute on Lake Superior Geology, 34th, Abstracts, p. 85–86.

Reynolds, D. J., 1984, Structural and dimensional repetition in continental rifts [M.S. thesis]: Durham, North Carolina, Duke University, 175 p.

Rosendahl, B. R., 1987, Architecture of continental rifts with special reference to East Africa: Annual Review Earth and Planetary Science, v. 15, p. 445–503.

Rosendahl, B. R., and Livingstone, D. A., 1983, Rift lakes of East Africa—New seismic data and implications for future research: Episodes, v. 1983, p. 14–19.

Schuepbach, M. A., and Vail, P. R., 1980, Evolution of outer highs on divergent continental margins, *in* Continental tectonics: National Academy of Sciences, 197 p.

Scott, D. L., and Rosendahl, B. R., 1989, North Viking Graben: An East African perspective: American Association of Petroleum Geologists Bulletin, v. 73, p. 155–165.

Serpa, L., Setzer, R., Farmer, H., Brown, L., Oliver, J., Kaufman, S., and Sharp, J., 1984, Structure of the southern Keweenawan Rift from COCORP surveys across the Midcontinent geophysical anomaly in northeastern Kansas: Tectonics, v. 3, p. 367–384.

Sleep, N. J., and Sloss, L. L., 1978, A deep borehole in the Michigan Basin: Journal of Geophysical Research, v. 83, p. 5815–5819.

Sloan, R. E., and Danes, Z. F., 1962, A geologic and gravity survey of the Belle Plaine area, Minnesota: Proceedings, Minnesota Academy of Science, v. 30, p. 49–52.

Steinhart, J. S., and Smith, T. J., eds., 1966, The earth beneath the continents: American Geophysical Union Geophysical Monograph 10, 663 p.

Thiel, E., 1956, Correlation of gravity anomalies with the Keweenawan geology of Wisconsin and Minnesota: Geological Society of America Bulletin, v. 67, p. 1079–1100.

Van Schmus, W. R., and Hinze, W. J., 1985, The Midcontinent Rift System: Annual Review of Earth and Planetary Sciences, v. 13, p. 345–383.

White, W. S., and Wright, J. C., 1954, The White Pine copper deposit, Ontonagon County, Michigan: Economic Geology, v. 49, p. 675–716.

Woelk, T. S., and Hinze, W. J., 1991, Model of the Midcontinent Rift System in northeastern Kansas: Geology, v. 19, p. 277–280.

Wold, R. J., and Hinze, W. J., eds., 1982, Geology and tectonics of the Lake Superior Basin: Geological Society of America Memoir 156, 280 p.

Woollard, G. P., 1943, Transcontinental gravitational and magnetic profile of North America and its relation to geologic structure: Geological Society of America Bulletin 54, p. 747–790.

Manuscript Accepetd by the Society January 16, 1996

Geological Society of America
Special Paper 312
1997

Integrated geophysical modeling of the
North American Midcontinent Rift System:
New interpretations for western Lake Superior,
northwestern Wisconsin, and eastern Minnesota

David J. Allen* and William J. Hinze
Department of Earth and Atmospheric Sciences, Purdue University, West Lafayette, Indiana 47907
Albert B. Dickas
Department of Geosciences, University of Wisconsin, Superior, Wisconsin 54880
M. G. Mudrey, Jr.
Wisconsin Geological and Natural History Survey, University of Wisconsin (Extension), 3817 Mineral Point Road, Madison,
Wisconsin 53705

ABSTRACT

Integrated geophysical investigations of the North American Midcontinent Rift System have resulted in a new understanding of the structure, stratigraphy, and evolution of this 1,100-Ma aborted continental rift. Interpretation of seismic reflection, gravity and magnetic anomaly, seismic refraction, rock physical property, and geologic data has identified a great degree of structural heterogeneity of the rift system in eastern Minnesota, northwestern Wisconsin, and western Lake Superior. In the western Lake Superior region, two ridges of pre-rift basement rocks are identified by pinch out of the rift's volcanic strata and lower portion of the overlying sedimentary sequence. In addition, regional rift reverse faults terminate above these ridges. Three-dimensional gravity modeling, constrained by seismic reflection profiles, suggests that both ridges are underlain by and composed of a belt of granitic rocks within the buried Archean greenstone-granite province beneath the rift basin, suggesting a significant influence of ancestral structures throughout the evolution of the rift system. Magnetic modeling indicates that magmatism did not occur uniformly along the length of the rift. In Minnesota and Wisconsin, the great majority of the rift's volcanic rocks are normally polarized and probably younger than the recorded ca. 1,098-Ma magnetic reversal, in contrast to western Lake Superior, where the lower half of the volcanic sequence is reversely polarized and was probably erupted before ca. 1,098-Ma. Gravity modeling suggests that the mass deficiency associated with crustal thickening along the rift is probably compensated by the positive effect of dense rift intrusions in the lower crust. The volume of magma trapped in the lower crust may be similar to that erupted into the rift basin.

*Present address: Exxon Exploration Company, 222 Benmar, Houston, Texas 77060.

Allen, D. J., Hinze, W. J., Dickas, A. B., and Mudrey, M. G., Jr., 1997, Integrated geophysical modeling of the North American Midcontinent Rift System: New interpretations for western Lake Superior, northwestern Wisconsin, and eastern Minnesota, *in* Ojakangas, R. W., Dickas, A. B., and Green, J. C., eds., Middle Proterozoic to Cambrian Rifting, Central North America: Boulder, Colorado, Geological Society of America Special Paper 312.

The spatial distribution of pre-Keweenawan ridges, termination of regional fault systems, and other structural detials more completely define the relation between regional rift tectonics and localized structural and stratigraphic asymmetry. The nature of accommodation structures dividing the Lake Superior basin into individual unit, previously identified by gravity, magnetic, and isopach mapping, is further clarified by seismic reflection analysis and potential field modeling.

INTRODUCTION

The most striking anorogenic geologic feature of the basement rocks of the North American craton is the Middle Proterozoic (Keweenawan) Midcontinent Rift System (MRS) . The 1,100-Ma MRS is a major Precambrian magmatic and sedimentary province and tectonic disruption of the lithosphere that extends ~2,000 km from Kansas through Lake Superior and into southern Michigan (Fig. 1; Thiel, 1956; Hinze et al., 1966; King and Zietz, 1971; Hinze et al., 1975; Dickas, 1986a). The MRS is one of the world' s great continental rifts and is unique among paleorifts in its extensive length and enormous volume of igneous and sedimentary rocks.

Rocks of the Midcontinent Rift System are exposed only in the Lake Superior region; elsewhere, they are buried beneath Phanerozoic platform sedimentary strata. Even in the Lake Superior region, the rift rocks are poorly exposed due to a thick mantle of Pleistocene glacial deposits and the waters of Lake Superior. Consequently, geophysical surveys play a critical role in investigating the structure and stratigraphy of the MRS (Hinze et al., 1992).

This study focuses on the application of geophysical techniques to investigate critical problems concerning the structure, stratigraphy, and evolution of the Midcontinent Rift System in western Lake Superior, northwestern Wisconsin, and eastern Minnesota. Seismic reflection and refraction, gravity and magnetic anomaly, rock physical property, and geologic data are interpreted in an integrated manner to develop two- and three-dimensional quantitative geophysical models of the rift. These models lead to a clearer picture of the rift's structural and stratigraphic relations, resulting in an improved understanding of the temporal and spatial evolution of the MRS.

GEOLOGIC ELEMENTS

The rocks of the Midcontinent Rift System in the Lake Superior region constitute the 1.1-Ga Keweenawan Supergroup (Morey and Van Schmus, 1988) The supergroup consists of two major suites: a lower, predominantly volcanic sequence, and an overlying clastic sedimentary package. Seismic reflection studies indicate total rift fill thicknesses of ~30 km in Lake Superior (e.g., Behrendt et al., 1988; Cannon et al., 1989).

The igneous sequence consists mainly of basaltic volcanic flows with lesser proportions of predominantly mafic intrusions and interflow clastic sedimentary strata. In western Lake Superior, seismic reflection data image a maximum of 16 km of volcanic flows beneath as much as 9 km of post-volcanic sedimentary strata (Allen, 1994). The volcanic and plutonic rocks of the Lake Superior region have U-Pb ages that range between 1,109 and 1,087 Ma (Davis and Sutcliffe, 1985; Palmer and Davis, 1987; Davis and Paces, 1990; Heaman and Machado, 1992; Miller, 1992; Van Schmus, 1992). Moreover, Paces and Miller (1993) suggest that most of the magmatism in the Lake Superior region occurred during two major pulses, from 1,109 to 1,106 Ma and from 1,099 to 1,094 Ma. The consensus among current investigators, based on both geochemical and physical evidence, is that magmas were derived from both lithospheric and mantle plume sources (e.g., Berg and Klewin, 1988; Paces and Bell, 1989; Hutchinson et al., 1990; Nicholson and Shirey, 1990; Klewin and Berg, 1991; Cannon, 1992; Klewin and Shirey, 1992; Van Schmus, 1992; Miller and Chandler, this volume). Early magmas may have been derived from the lithosphere, by partial melting due to heat provided by an ascending mantle plume, and also by assimilation of lithospheric material by plume-derived magmas as surface conduits developed. Later magmas are geochemically primitive and were most probably derived from partial melting of the plume head, with minimal lithospheric contamination as the magmas rose to the surface.

As volcanism and plutonism waned, sedimentary strata were deposited in the subsiding volcanic basins and in flanking sedimentary basins floored principally by pre-Keweenawan basement rocks. The textural and compositional maturity of the Keweenawan sedimentary sequence increases upsection, from the older, relatively immature Oronto Group (Daniels, 1982) to the younger, relatively mature Bayfield Group (Kalliokoski, 1982; Morey and Ojakangas, 1982). The Oronto Group strata contain a substantial proportion of rock fragments, derived primarily from Keweenawan igneous rocks along the margins of the volcanic basins. The Bayfield Group, on the other hand, is composed almost entirely of quartz and feldspar, suggesting that volcanic source rocks were either removed or buried, and that these sediments were derived from erosion of older Archean granitic rocks (Ojakangas and Morey, 1982) and, possibly, reworking of Oronto Group clastic sedimentary rocks. Progressing upsection, the Oronto Group consists of the Copper Harbor Conglomerate, Nonesuch Formation, and Freda Sandstone, which have maximum exposed thicknesses of 1,800 m, 200 m, and 3,600 m (Daniels, 1982). The thickness of the overlying Bayfield Group is poorly constrained because the Oronto-Bayfield contact is neither well-understood nor well-exposed. A deep drill hole in northern Michigan, however, penetrated more than 1,000 m of Bayfield Group sandstones (Bacon, 1966), and

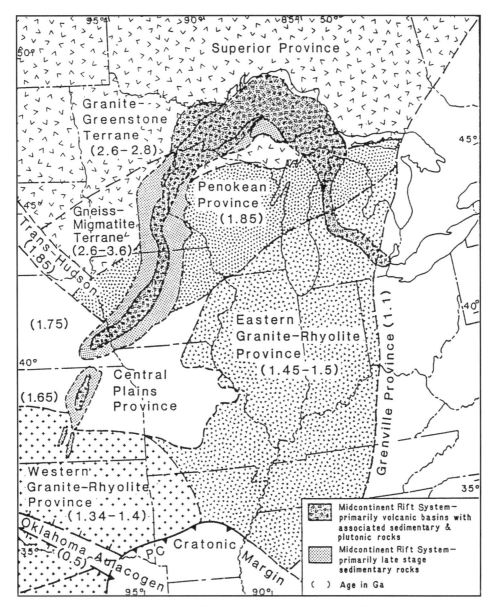

Figure 1. Basement provinces of the Midcontinent of North America (modified from Van Schmus et al., 1982), showing the position of the Midcontinent Rift System (after Hinze and Kelly, 1988).

recent geophysical interpretations (Allen, 1994) suggest even greater thicknesses in the flanking sedimentary basins along the margins of the MRS.

The present distribution of Keweenawan igneous and sedimentary rocks (Fig. 2) is largely the result of late-stage compressional stresses that resulted in structural inversion of the MRS. These stresses may have been related to convergence between the Grenville Province allochthon (Fig. 1) and the proto–North American continent within which the MRS developed (e.g., Cannon, 1994), although the relation between these provinces remains elusive. Reverse faults (Fig. 2) thrust older volcanic rocks and Oronto Group strata over younger Bayfield Group rocks and delineate the margins of uplifted volcanic-

rooted horsts (Thiel, 1956; White, 1966b; Dickas, 1986b), such as the St. Croix Horst of Minnesota and Wisconsin (SCH, Fig. 2). Thrusting was in part concurrent with deposition of the Bayfield Group, but continued after Keweenawan sedimentation ended (Kalliokoski, 1969). Based on Rb-Sr closure ages Cannon et al. (1990) suggest that thrusting occurred at ca. 1060 ± 20 Ma.

Although the pre-rift crust was attenuated during the rifting process, the present crust is relatively thick, as much as 55 km (Halls, 1982; Hamilton and Mereu, 1993), due to the great thickness of Keweenawan volcanic and sedimentary strata. In addition, the lower crust beneath the rift basin has been intruded and underplated (Behrendt et al., 1990; Green, 1991;

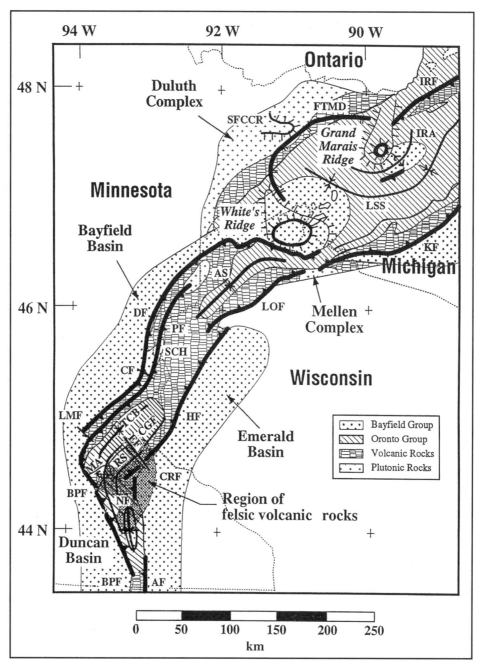

Figure 2. Geologic map of the Midcontinent Rift System, western Lake Superior, northwestern Wisconsin and eastern Minnesota, showing lithologies and major structures. Bold hachured lines within White's ridge and Grand Marais ridge encompass regions where Keweenawan volcanic rocks are absent. Reverse faults: KF, Keweenaw fault; LOF, Lake Owen fault; HF, Hastings fault; CRF, Castle Rock fault; NF, Northfield fault; AF, Austin fault; IRF, Isle Royale fault; DF, Douglas fault, PF, Pine fault; CGF, Cottage Grove fault. Cross-faults: BPF, Belle Plaine fault; LMF, Lake Minnetonka fault; CF, Chisago fault. Folds: IRA, Isle Royale anticline; LSS, Lake Superior syncline; AS, Ashland syncline; MA, Montgomery anticline; RS, Rice syncline; TCB, Twin Cities Basin. Other features: SCH, St. Croix Horst; SFCCR, Schroeder-Forest Center crustal ridge; FTMD, Finland Tectono-Magnetic Discontinuity normal fault.

Trehu et al., 1991; Hamilton and Mereu, 1993), and the upper mantle may have also been altered during the rifting process (Allen et al., 1992).

GEOPHYSICAL ELEMENTS AND METHODOLOGY

Seismic reflection data are of enormous value to investigations of the MRS because of the layered nature of the rift's volcanic and sedimentary strata. Figure 3 shows the locations of the seismic profiles relevant to this study, superimposed on a Bouguer gravity anomaly map. In western Lake Superior, the data set consists of eleven eight-second profiles (Fig. 3) obtained from Grant-Norpac, Inc., and released by Argonne National Laboratory (McGinnis and Mudrey, 1991) and one twenty-second profile obtained as part of the Great Lakes Multidisciplinary Program on Crustal Evolution (GLIMPCE; Behrendt et al., 1988). Twenty-one five- or six-second proprietary profiles in Wisconsin and Minnesota (Fig. 3) provide information concerning the southwestern extension of the MRS from western Lake Superior. As shown in Figure 3, only three of the onshore profiles in Wisconsin and Minnesota have been previously interpreted. In addition, just one prior investigation (Sexton and Henson, 1994) has focused on the application of the Grant-Norpac/Argonne profiles to the investigation of the rift's structure in western Lake Superior.

Seismic refraction data provide important information concerning the gross structure of the lithosphere along the MRS, and are especially valuable for investigating lower crustal structure, as only one of the seismic reflection profiles of this study (GLIMPCE line C, Fig. 3) images the Moho. Refraction data indicate that the crust is anomalously thick and dense along the rift axis (Halls, 1982; Trehu et al., 1991; Hamilton and Mereu, 1993), probably the result of deep mafic intrusions beneath the rift basin.

Gravity data provide important information concerning the MRS because of the intense density contrasts among the rift's igneous and sedimentary rocks and pre-Keweenawan basement rocks. Positive gravity anomalies, such as over the St. Croix Horst in Minnesota and Wisconsin, over the Duluth Complex in northeastern Minnesota, and in portions of western Lake Superior (Fig. 3; compare with Fig. 2), are associated with thick sections of mafic igneous rocks (density ~2,950 kg/m^3, Allen, 1994). The negative anomalies that flank the St. Croix Horst in Minnesota and Wisconsin (Fig. 3) are associated with marginal basins that contain mainly low-density (~2,400 kg/m^3; Halls, 1969) Bayfield Group sandstones overlying thin sequences of older Oronto Group strata and volcanic rocks. The gravity anomaly signature of the rift in western Lake Superior (Fig. 3) is remarkably different and more complicated than the signature in Wisconsin and Minnesota. Of particular interest are two negative, roughly circular anomalies located in Lake Superior and on the Bayfield Peninsula of Wisconsin (Fig. 3). These intense anomalies are not well understood, although Hinze et al. (1982) and Dickas (1986c) suggest that they might be produced by relatively

thick sections of low-density Bayfield Group sedimentary rocks, thinning of the rift's dense volcanic sequence, or low-density sources beneath the rift basin. White (1966b) suggested that the southern (Bayfield Peninsula) negative anomaly is produced by a ridge of pre-Keweenawan basement rocks above which the rift's volcanic rocks are relatively thin or absent. Until this study, however, no quantitative models have been proposed.

Magnetic data are useful for investigating both the structure and stratigraphy of the rift's igneous rocks. The Keweenawan volcanic strata are associated with intense, short-wavelength, linear magnetic anomalies that strike parallel to the bedding of the volcanic flows (Fig. 4). The correlation between the volcanic strata and the magnetic anomalies is the result of vertical (stratigraphic) variations in the magnetization of the dipping Keweenawan volcanic section (Allen and Chandler, 1993) and represents a powerful tool for mapping the structure of the MRS (Allen and Chandler, 1992; Allen, 1994). Paleomagnetic data are useful for estimating the relative ages of the rift's rocks (Green, 1982; Halls and Pesonen, 1982; Van Schmus et al., 1982; Palmer and Halls, 1986; Palmer and Davis, 1987). A change in polarity from reversed to normal, dated at ca. 1,098 Ma (Davis and Paces, 1990; Paces and Miller, 1993), is widely observed in the Lake Superior region and is therefore useful as a stratigraphic correlation tool. However, evidence for another reversal period (Palmer, 1970; Robertson, 1973) of probable short duration in the Mamainse Point volcanic sequence in the eastern Lake Superior region (Klewin and Berg, 1991) and possibly in the lower portion of the Powder Mill Volcanic Group of northern Michigan (Books, 1972) complicates the use of magnetic polarity as a stratigraphic correlation tool.

By themselves, the seismic, gravity, and magnetic data are important tools for investigating the structure and stratigraphy of the MRS. When interpreted together, however, they reveal additional information that is not contained in any of the individual data sets. For example, the magnetic data can be used to precisely locate faults and contacts where seismic and geologic data are inconclusive or unavailable. The seismic reflection data then become very useful for placing constraints on configurations of igneous and sedimentary bodies of quantitative gravity models. Moreover, the gravity data can also help to resolve ambiguities in interpretation of the seismic reflection data. The gravity models are further constrained by seismic refraction data, which provide evidence concerning the configuration of the lower crust and upper mantle, information that typically is not provided by seismic reflection profiles. Finally, using the body configurations that evolve from the combined seismic interpretation and gravity modeling, quantitative magnetic models are constructed to evaluate the magnetic properties of the rift's volcanic and plutonic rocks. By determining the approximate proportions of normally and reversely polarized rocks in the Keweenawan igneous section, it is possible to estimate the amounts of magma that may have cooled before and after the magnetic reversal at ca. 1,098 Ma, and therefore determine the relative ages of different segments of the MRS.

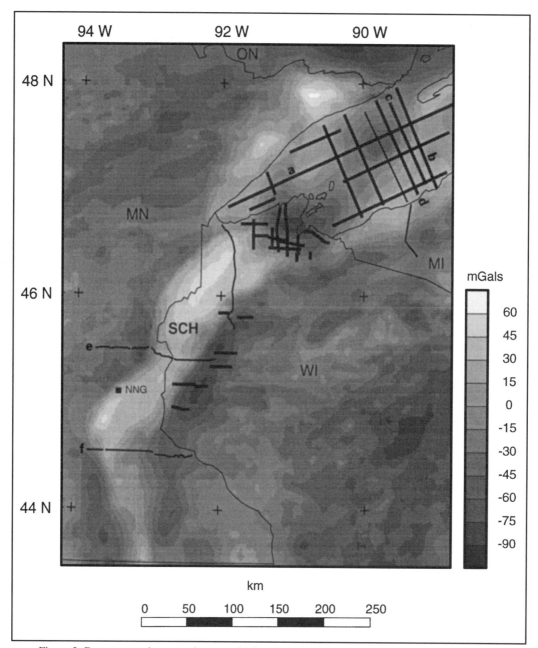

Figure 3. Bouguer gravity anomaly map with locations of seismic profiles used in this study. Thin lines indicate profiles that have been discussed in detail in journal articles. Thin line in Lake Superior is GLIMPCE profile C. Thick lines show recent profiles that have not previously been interpreted in detail. Lines identified as a, b, c, d, e, and f represent locations of Figures 7, 8, 9, 11, 16, and 17, respectively. SCH, St. Croix Horst; NNG, Northern Natural Gas deep drill hole (not discussed in this chapter).

GEOLOGIC INTERPRETATION

In the early 1980s, the identification of the MRS as a frontier hydrocarbon province marked the beginning of a modern era of interpretation of the structure and evolution of the MRS. This era of interpretations, based on seismic reflection profiles, gravity and magnetic data, and isotopic studies of the Keweenawan volcanic rocks of the rift, contrasts with earlier interpretations that relied principally on geologic field information, regional seismic refraction profiles, and limited gravity and magnetic data. Over the past decade, studies of the MRS have become more definitive with the increasing availability of seismic reflection, gravity, and magnetic data, both public and proprietary (Dickas, 1986a; Behrendt, et al., 1988; Berendsen et al., 1988;

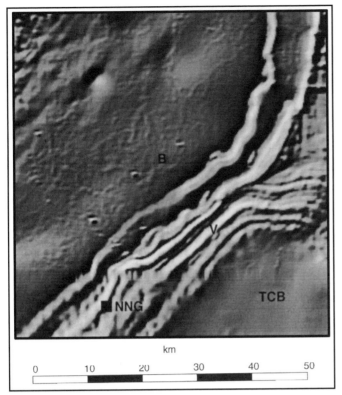

km

0   10   20   30   40   50

Figure 4. Shaded relief aeromagnetic map of a portion of the Midcontinent Rift System in east-central Minnesota. Northwestern illumination. TCB, Twin Cities Basin; V, Keweenawan volcanic strata; B, Bayfield Basin; NNG, Northern Natural Gas deep drill hole (not discussed in this chapter).

Cannon et al., 1989; Anderson, 1990; Hinze et al., 1992). Within the western Lake Superior region of the MRS, additional interpretational details are given by Allen (1994).

Stratigraphic relations, western Lake superior region

Interpretations of seismic reflection profiles in and near western Lake Superior are summarized in Figures 5 and 6. These maps indicate respectively the thicknesses of the Portage Lake and Chengwatana Volcanic Groups (equivalent to the total thickness of the Keweenawan igneous sequence, except in northeastern Minnesota and within ~20 km of the Minnesota shoreline) and the Keweenawan sedimentary sequence (Oronto and Bayfield Groups combined). Interpretations of offshore profiles are linked to exposed Keweenawan geology in Michigan by an onshore profile (Hinze et al., 1990) that terminates just 4 km southeast of GLIMPCE profile C (Fig. 3). The seismic interpretations in western Lake Superior are enhanced because all horizons must be consistent at points of profile intersections, greatly reducing the ambiguity associated with interpreting a single seismic profile.

Figure 5 indicates the Keweenawan volcanic sequence is thickest in Lake Superior along the Lake Superior syncline (5 s

two-way travel time ~16 km) and in Wisconsin along the St. Croix Horst (4 s two-way travel time ~13 km), but is very thin or absent above two ridges of pre-Keweenawan (Archean) basement rocks. Both ridges coincide with intense negative gravity anomalies (compare Figs. 3 and 5), confirming previous interpretations (Weber and Greenacre, 1966; White, 1996a; Wold and Ostenso, 1966; Dickas, 1986c) that the negative anomalies are produced, at least in part, by thinning of the dense Keweenawan igneous rocks. Figure 7 is a long profile extending northeast from the southwestern corner of Lake Superior (see Fig. 3 for location), traversing White's ridge (to the southwest) and the Grand Marais ridge (to the northeast). This and other profiles indicate the rift's volcanic strata thin gradually to a feather edge onto both ridges (Fig. 7), with no evidence for substantial normal or reverse faulting. Hence, these ridges, composed of Archean rock, remained positive Keweenawan topographic features as the adjacent volcanic basins gradually subsided.

As shown in Figure 6, the thickest section of Keweenawan sedimentary strata in western Lake Superior occurs along the Lake Superior syncline (3.5 s two-way travel time ~9 km). The synclinal axis bends west and then northwest toward Minnesota, and is not continuous with the Ashland syncline in Wisconsin (Fig. 6). Rather, the two synclines are separated by White's ridge, which remained a positive topographic feature after volcanism ceased. In fact, the Copper Harbor Conglomerate and Nonesuch Formation actually pinch out, and the Freda Sandstone lies directly above the pre-Keweenawan basement along the crests of both White's ridge and the Grand Marais ridge (Fig. 7). (The dashed line in Figure 7 represents a prominent reflection within the lower Freda Sandstone, correlated offshore by means of the onshore profile in Michigan [Fig. 3] that traverses the Oronto Group.)

Seismic data also indicate the great majority of Keweenawan sedimentary strata in western Lake Superior belong to the Oronto Group. For example, Figure 8 shows nearly horizontal reflections just northwest of exposed, gently-dipping (5° NW) Freda Sandstone (Oronto Group) along the Michigan shoreline, indicating that the overlying Bayfield Group, if present, must be very thin along this profile. Similar seismic-geologic relations along other profiles reveal that the Bayfield Group is relatively thin or absent throughout most of western Lake Superior.

As suggested by Figures 7 and 8, Oronto Group sedimentary strata typically lie conformably above the volcanic flows. In several locations in western Lake Superior, however, unconformable relations are observed (Fig. 9), indicating local uplift and erosion of the Keweenawan volcanic flows prior to their burial beneath the Oronto Group. These unconformities occur along the margins of White's ridge and the Grand Marais ridge (Fig. 10), and coincide approximately with regions where the Copper Harbor Conglomerate and Nonesuch Formation are absent (Fig. 10). The ridges, therefore, were uplifted and eroded after volcanism waned, and

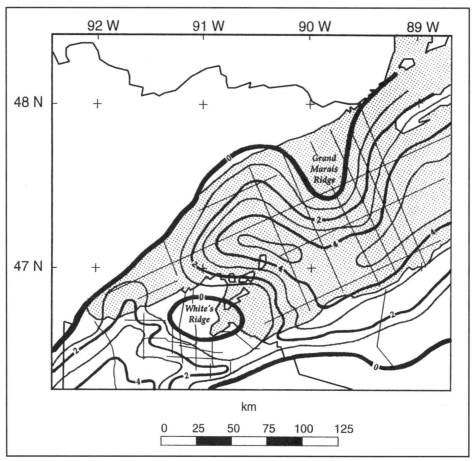

Figure 5. Contour map, thickness of the Portage Lake and Chengwatana Volcanic Groups. Thin lines indicate seismic reflection profiles. Contour interval is 1 s two-way travel time (1 s ~ 3.3 km). Dotted pattern identifies Lake Superior.

remained positive features until the Freda Sandstone was deposited on top of them.

Structural relations, western Lake Superior region

The structure of the MRS in western Lake Superior is different from that of the St. Croix Horst to the southwest and of central Lake Superior to the east in that major reverse faulting is absent. As shown in Figure 6, the Isle Royale reverse fault terminates ~40 km southwest of the Isle. Displacement across the fault decreases, from ~3 km along the seismic profile nearest the Isle (Fig. 8) to <1 km along the adjacent profile to the southwest, and the fault terminates before crossing the next seismic profile (Fig. 11; see Fig. 3 for profile location). Along this profile, however, the Keweenawan rocks are deformed (Fig. 11), indicating that late-stage contraction was accommodated by folding instead of thrusting. The Isle Royale anticline (near 65 km along Fig. 11), extends parallel to the Lake Superior syncline (Fig. 6) and is faulted along the southern margin of the Grand Marais ridge (Fig. 6). This small fault is imaged only along GLIMPCE profile C and is therefore a local feature,

in contrast to the interpretation by Sexton and Henson (1994) who name this feature the Ojibwa fault.

In Wisconsin, proprietary seismic reflection data indicate that displacement across the Douglas fault decreases to the east (Figs. 6 and 12). East of 91°W, the fault flattens to a décollement within the Oronto Group, and the Douglas fault then diminishes to a monocline east of Ashland, Wisconsin (Fig. 12). This interpretation is confirmed by the dissipation of the gravity anomaly (second vertical derivative) signature of the Douglas fault in northwestern Wisconsin (Fig. 13). Moreover, the seismic data in Lake Superior show no evidence for an eastern extension of the Douglas fault, and are incompatible with interpretations (e.g., Sexton and Henson, 1994) that suggest the fault extends into the lake.

Nature of the Pre-Keweenawan basement ridges

White's Ridge represents a significant discordant structure within the MRS. This ridge separates the deep volcanic basin of western Lake Superior from the deep basin along the St. Croix Horst (Fig. 5), and delineates the accommodation structure

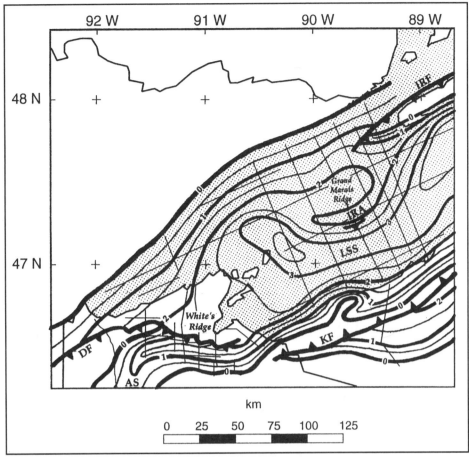

Figure 6. Contour map, base of Oronto Group. Thin lines indicate seismic reflection profiles. Contour interval is 0.5 s two-way travel time (1 s ~ 2.5 km). DF, Douglas Fault; KF, Keweenaw fault; IRF, Isle Royale fault; AS, Ashland syncline; LSS, Lake Superior syncline; IRA, Isle Royale anticline. Dotted pattern identifies Lake Superior.

between the Brule and Ontonagon rift units as identified by Dickas and Mudrey (1989, this volume) and Cannon et al. (1989). The rift widens abruptly as it crosses White's Ridge (Figs. 1 and 2), and the rift basin is much deeper in western Lake Superior than along the western limb (Figs. 5 and 6). Southwest of White's Ridge, the rift structure consists of a central basaltic horst bound by reverse faults and flanked by clastic sedimentary basins. Northeast of the ridge, however, major reverse faulting is absent, and flanking sedimentary basins are less common (Fig. 2).

The Grand Marais ridge, a counterpart to White's ridge (Fig. 7), is located immediately southwest of Isle Royale (Fig. 2). This structure delineates the accommodation feature separating the Ontonagon and Manitou rift units (Dickas and Mudrey, this volume).

Gravity modeling provides critical insight into the nature of White's ridge and the coeval Grand Marais ridge. Figure 14 is a residual gravity anomaly map, prepared by subtracting the (three-dimensionally) calculated effect of the Keweenawan volcanic and sedimentary strata from the observed gravity anomaly (Fig. 3). The prominent northeast-striking negative residual

anomaly (Fig. 14) is robust and not greatly altered as the model densities of the volcanic, sedimentary, and pre-Keweenawan basement rocks are varied within reasonable limits (Allen, 1994). The negative anomaly encompasses both pre-Keweenawan ridges (compare Figs. 5 and 14), and must be produced by a mass deficiency within the middle or upper crust, because the moderate to steep gravity gradients (Fig. 14) preclude sources at greater depths. In addition, the anomaly cannot be produced by variation in the density of the Keweenawan volcanic rocks, because the anomaly extends across the basement ridges where the volcanic sequence is absent. Moreover, the anomaly is probably not the result of low-density (Bayfield Group) sedimentary rocks because more than 2.5 km of such strata are necessary to properly model the anomaly, much greater than the maximum thickness of these rocks (~1.0 km) allowed by seismic-geologic relations (e.g., Fig. 8).

Therefore, the negative residual anomaly is likely produced by a mass deficiency within the pre-Keweenawan basement beneath the rift basin. Beyond the margins of the MRS, negative anomalies of comparable amplitude, width, and orientation (G, Fig. 14) are associated with granitic belts in the Archean

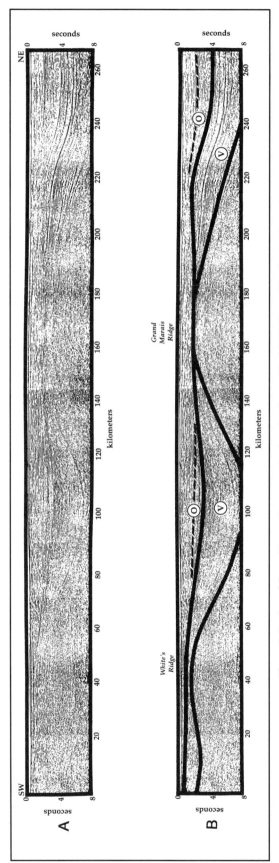

Figure 7. Portion of Grant-Norpac/Argonne seismic reflection profile LS-08 (line a, Fig. 3). A, migrated seismic section; B, interpretation. V, volcanic flows; O, Oronto Group sedimentary strata. Dashed line indicates a prominent reflection within the lower Freda Sandstone.

greenstone-granite province (Fig. 1). A similar feature may be present within the buried Archean basement beneath the rift basin, an interpretation that is supported by quantitative gravity modeling by Allen (1994). A relatively thick section (~10 to 15 km) of buried pre-Keweenawan crustal rocks is required to model the observed residual anomaly (Allen, 1994). Consequently, previous models involving a nearly complete separation of the continental crust along the MRS (e.g., Cannon, 1992) may require modification in western Lake Superior. Perhaps the large volume of Archean rocks preserved beneath the Keweenawan sequence indicates that the rift basin in western Lake Superior formed (in part) by subsidence of the crust, and that extension may have been accommodated by the injection of dikes or by ductile stretching of the crust rather than by extensive normal faulting.

The proposed Archean granitic belt constitutes the pre-Keweenawan basement along most of White's ridge and the Grand Marais ridge (compare Figs. 5 and 14), and apparently played an influential role throughout the evolution of the MRS. First, the trend of the MRS (southwesterly in easternmost western Lake Superior and along the St. Croix Horst) is disrupted (northwesterly) near the interpreted granitic belt (compare Figs. 6 and 14). Hence, this pre-Keweenawan feature may have influenced the development of the MRS by acting as a buttress. A minimally fractured and nonfoliated mass of granite should have relatively great strength compared with the surrounding crustal rocks and would probably resist deformation (Klasner et al., 1982). Perhaps tensional stresses were more effective and the rift propagated more easily across the buttress (northwesterly) than along its axis (southwesterly), resulting in the dramatic change in strike of the MRS and the local absence of volcanic strata along the crests of the ridges. Second, post-volcanic uplift and erosion of the ridges (Figs. 9 and 10) may have been an isostatic response, as the granitic rocks are considerably less dense (~2,650 kg/m^3) than the surrounding basaltic volcanic flows (density ~2,950 kg/m^3). This uplift may also reflect an early pulse of contraction associated with the initial stages of the Grenvillian collision east of the MRS (Fig. 1). In fact, such contraction might have also been responsible for the termination of volcanism. Finally, the Douglas and Isle Royale faults terminate as they approach White's ridge and the Grand Marais ridge, respectively, and the Isle Royale anticline follows the southern margin of the Grand Marais ridge (Fig. 6). Apparently, the pre-Keweenawan basement ridges continued to play active roles during late-stage contraction along the MRS, probably by acting as buttresses that resisted contractional deformation, resulting in the absence of reverse faulting in most of western Lake Superior.

Geologic and geophysical data support the existence of a third pre-Keweenawan basement ridge in the western Lake Superior region. The Schroeder–Forest Center crustal ridge (Miller and Chandler, this volume) extends west-northwest–east-southeast beneath the Duluth Complex and North Shore Volcanic Group of northeastern Minnesota (Fig. 2) and may actually be a

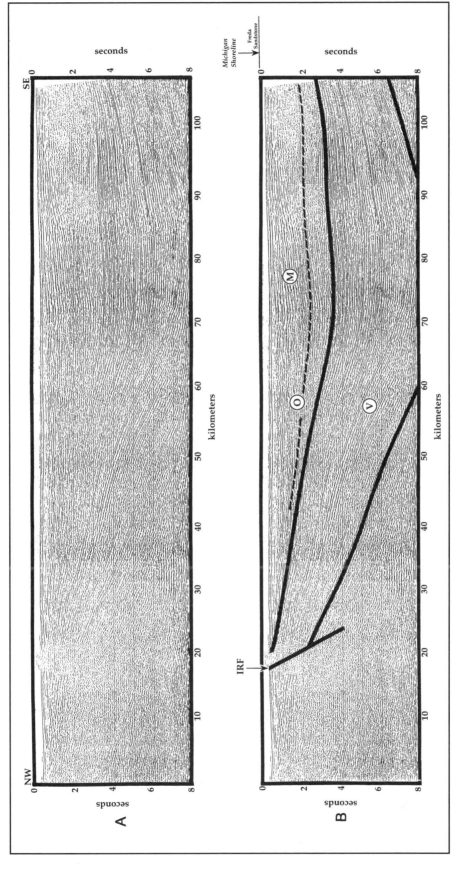

Figure 8. Grant-Norpac/Argonne seismic reflection profile LS-43 (line b, Fig. 3). A, migrated seismic section; B, interpretation. V, volcanic flows; O, Oronto Group sedimentary strata; M, lake-bottom multiple reflections; IRF, Isle Royale fault. Dashed line indicates a prominent reflection within the lower Freda Sandstone.

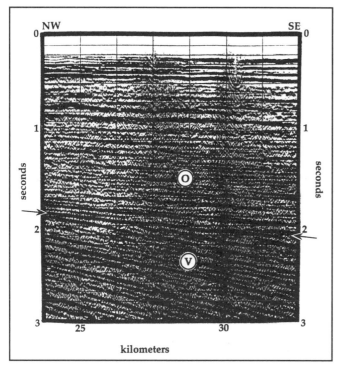

Figure 9. Portion of Grant-Norpac/Argonne migrated seismic reflection profile LS-45 (line c, Fig. 3). Arrows delineate unconformable contact between volcanic flows (V) and Oronto Group strata (O).

northwestern extension of White's ridge (Allen, 1994). The crustal ridge is inferred by an abundance of Archean xenoliths within the overlying Keweenawan intrusive rocks (Boerboom, 1994), a pronounced thinning of the Keweenawan Beaver Bay Complex (Miller and Chandler, this volume), a paleotopographic high during the deposition of the North Shore Volcanic Group (Jirsa, 1984), discontinuities within the North Shore Volcanic Group (Grout et al., 1959; Green, 1972, 1983; Miller and Chandler, this volume), and also a thinning of the Early Proterozoic Biwabik and Gunflint iron formations (Morey, 1972a, b). Moreover, the local negative gravity anomaly (saddle) associated with this feature (compare Figs. 2 and 3) also suggests that the Keweenawan mafic igneous rocks are relatively thin in this area, consistent with interpretations by White (1966b), Ferderer (1982), and Chandler (1990).

Stratigraphic relations, western limb of the Midcontinent Ridge System

The stratigraphy of the marginal basins that flank the St. Croix Horst in Wisconsin and Minnesota (Fig. 2) must be inferred geophysically due to poor exposure of Keweenawan rocks and a lack of deep drill holes. The dense coverage of seismic reflection profiles (Fig. 3) along the eastern flanking basin and associated negative gravity anomaly (which attains its maximum negative values near Emerald, Wisconsin) permits a detailed investigation of the Keweenawan stratigraphy in this area. Historically, this basin has been referred to as the River Falls syncline because

overlying Paleozoic rocks are deformed in such a manner. The seismic data, however, indicate that the Keweenawan structure is not synclinal, but is instead a wedge-shaped, westward-deepening basin (Fig. 15), more appropriately named the Emerald Basin (Allen, 1994). To reproduce the intense negative gravity anomaly associated with this basin (Fig. 3), the sedimentary strata must belong predominantly to the low-density Bayfield Group (density ~2,400 kg/m^3) or the lithostratigraphic-equivalent Fond du Lac Formation and Hinckley Sandstone. Older, intermediate-density (~2,650 kg/m^3) Oronto Group strata, if present, compose a very thin basal layer (e.g., Fig. 16). Similar combined analysis of seismic reflection and gravity data in Minnesota indicates that the Bayfield Basin, the counterpart to the Emerald Basin situated west and northwest of the St. Croix Horst (Dickas, 1986a), also contains mainly Bayfield-equivalent strata (Fig. 16). However, Oronto Group strata are relatively thick in the southern Emerald Basin along the eastern margin of the Midcontinent Rift System (MRS) in southeastern Minnesota (Fig. 17). This is the only location where these strata are regionally prevalent in a Keweenawan flanking sedimentary basin; elsewhere, the distribution of the Oronto Group roughly coincides with that of the underlying volcanic strata. Perhaps the southern Emerald Basin subsided preferentially due to the combined loads of the volcanic basins to both the northwest and southwest (e.g., Figs. 2 and 18).

Gravity modeling, constrained by three long seismic reflection profiles across the St. Croix Horst in Wisconsin and Minnesota (Fig. 3), indicates that the Keweenawan volcanic rocks typically have an average density (~2,950 kg/m^3) consistent with basaltic volcanic flows (Fig. 16). In southeastern Minnesota, however, the relatively low average density of the upper volcanic sequence along the southernmost seismic profile (Fig. 17; see Fig. 3 for profile location) indicates a large proportion of low-density felsic volcanic flows or the prevalence of interflow clastic sedimentary strata. In fact, felsic rocks are encountered in deep drill holes north of this profile, just west of the Minnesota-Wisconsin border (FV, Fig. 18). In addition, combined gravity and magnetic modeling by Allen (1994) supports the interpretation that felsic rocks are abundant in the Keweenawan igneous sequence over a large area in southeastern Minnesota (Fig. 2). The prevalence of felsic rocks implies that the rift magmas in this area may have been derived (to a large degree) from melting of the crust or assimilation of crustal material by plume-derived magmas. Notably, the felsic rocks occur where the MRS bends to the southeast (Fig. 2). Tensional stress (which was probably oriented northwest-southeast) may have been less effective in opening the north-northwest-striking segment of the MRS in southeastern Minnesota, and plume-derived magmas may have encountered greater resistance while rising through the lithosphere. Consequently, plume magmas may have incorporated a large amount of crustal material; additional magmas may have been derived from crustal melting due to the heat provided by trapped plume magmas.

As shown in Figure 18, the Keweenawan volcanic sequence is thickest north of 46°N and south of 45°N. These areas correspond to the Ashland syncline and Twin Cities

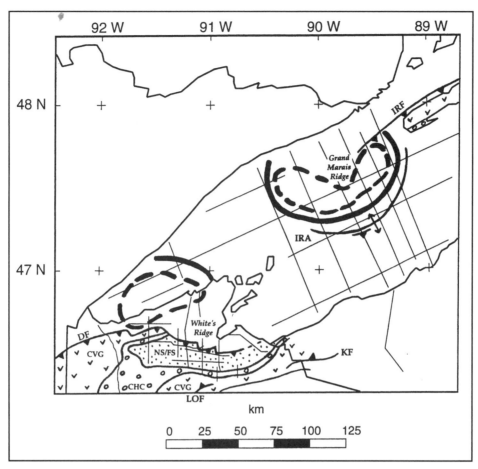

Figure 10. Map of the MRS indicating significant stratigraphic relations of the Oronto Group in northwestern Wisconsin and western Lake Superior. Thin lines indicate seismic reflection profiles. Thick dashed lines enclose areas where Oronto Group strata lie unconformably above volcanic flows. Thick solid lines delineate pinch-out of the Copper Harbor Conglomerate (CHC) and Nonesuch Shale (NS) onto pre-Keweenawan basement ridges. FS, Freda Sandstone; CVG, Chengwatana Volcanic Group; DF, Douglas fault; LOF, Lake Owen fault; KF, Keweenaw fault; IRF, Isle Royale fault; IRA, Isle Royale anticline.

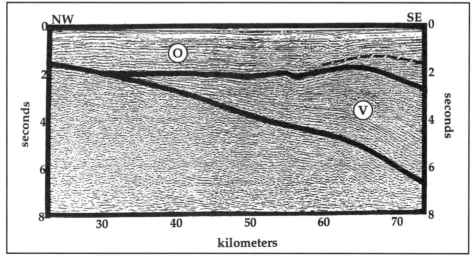

Figure 11. Portion of Grant-Norpac/Argonne migrated seismic reflection profile LS-47 (line d, Fig. 3). V, volcanic flows; O, Oronto Group sedimentary strata. Dashed line indicates a prominent reflection within the lower Freda Sandstone.

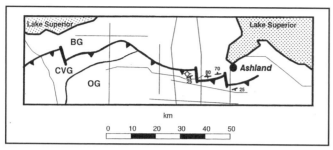

Figure 12. Map of the Douglas fault in northwestern Wisconsin. Thin lines indicate seismic reflection profiles. CVG, Chengwatana Volcanic Group; OG, Oronto Group; BG, Bayfield Group. Strike-dip symbols indicate attitude of the Freda Sandstone, the youngest formation of the Oronto Group. Well location marks the site of the deep Midcontinent Rift Terra-Patrick #7–22 hydrocarbon test, drilled to a total depth of 1,514 m in 1992.

Basin, respectively, and are associated with the thickest sections of overlying Oronto Group strata. The intervening segment of the St. Croix Horst (near 45°30′N), however, is characterized by a relatively thin volcanic sequence (Fig. 18) without an overlying section of Keweenawan sedimentary strata (Fig. 2). This distribution of rift igneous and sedimentary rocks may reflect varying degrees of magmatism and sedimentation along the MRS. Alternatively, variation in the thickness

of the Keweenawan section may be the result of nonuniform uplift and erosion of the St. Croix Horst (e.g., Fig. 22).

Structural relations, western limb of the MRS

Shaded relief magnetic anomaly maps (Figs. 19 and 20) are very useful for investigating the structure of the western limb of the MRS due to the intense magnetization contrasts between the rift's volcanic and sedimentary rocks and also within the volcanic sequence (Allen and Chandler, 1993). Magnetic anomaly data are often the primary means by which rift structures are identified due to a lack of exposure of the Keweenawan sequence and relatively few long seismic reflection profiles (Fig. 3). Magnetic data, in conjunction with gravity, seismic, and geologic data, reveal several structural complexities pertaining to the reverse faults along the rift's western limb in Wisconsin and Minnesota.

Reverse faulting. The southeastern margin of the St. Croix Horst (Fig. 2) has traditionally been interpreted as a single reverse fault, the Lake Owen fault (White, 1966b). Geologic evidence concerning this fault is limited, and the fault is delineated primarily on the basis of gravity and magnetic data. The Lake Owen fault is associated with a pronounced magnetic anomaly signature (Fig. 19) and an intense, paired positive-negative second vertical derivative gravity anomaly (Fig. 13). As indicated by

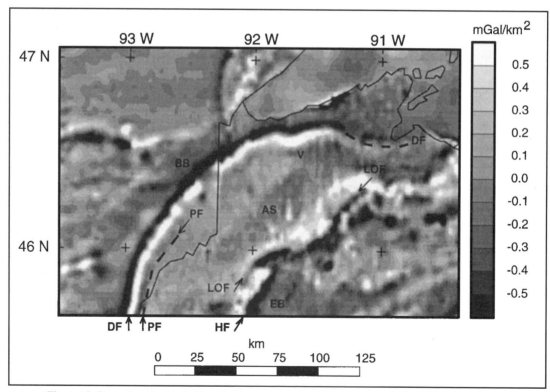

Figure 13. Second vertical derivative Bouguer gravity anomaly map, northern St. Croix Horst region. DF, Douglas fault (dashed line is extension of the fault based on gravity and seismic reflection data); LOF, Lake Owen fault; HF, Hastings fault; PF, Pine Fault; AS, Ashland syncline; V, abrupt thinning of the outcrop belt of volcanic rocks; BB, Bayfield Basin; EB, Emerald Basin.

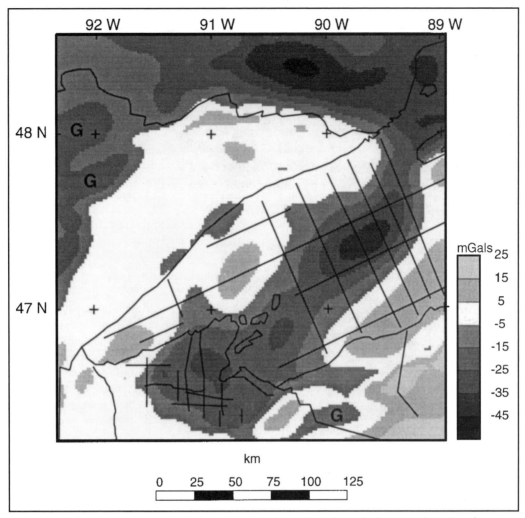

Figure 14. Residual gravity anomaly map prepared by subtracting calculated gravity values from observed gravity values. Values between -5 and +5 mGal are indicated by white. G, negative anomalies associated with Archean granitic rocks. Thin lines indicate seismic reflection profiles. Calculation based on densities of 2,400, 2,650, 2,950, and 2,750 kg/m³ for the Bayfield Group, Oronto Group, volcanic strata, and average pre-Keweenawan basement, respectively, and a regional base level of -45 mGal (e.g., Allen et al., 1992).

Figures 13 and 19, the Lake Owen fault terminates just south of 46°N, approximately 15 km west of the northeastern termination of the Hastings fault, a reverse fault extending southwest into Minnesota. The relation between the Hastings and Lake Owen faults is unclear. These faults might represent fault splays that merge at depth. Alternatively, the two faults may have been a single reverse fault that was displaced ~15 km by a left-lateral strike-slip fault, although there is no evidence that such a fault disrupts the interior of the St. Croix Horst or the pre-Keweenawan rocks to the southeast. Moreover, a left-lateral strike-slip fault would be inconsistent with the predominance of right-lateral Keweenawan strike-slip faulting in northwestern Wisconsin (e.g., Fig. 12; Dickas and Mudrey, 1991).

The discontinuity between the Lake Owen and Hastings faults coincides with the northern margin of the Emerald Basin

(Fig. 15). The northern margin of the basin is clearly delineated by exposure of the pre-Keweenawan Barron Quartzite (Fig. 15; Mudrey et al., 1987) and further identified by termination of the negative gravity anomaly (Fig. 3) associated with the basin. This correlation suggests that the Emerald Basin may be genetically related to the Hastings fault. Perhaps, the Emerald Basin represents a foreland basin that subsided under the weight of the adjacent St. Croix Horst. The textural and compositional maturity of the Bayfield Group (i.e., the predominant strata in this basin), however, suggests a pre-Keweenawan provenance for the basin, combined with possible erosion of Oronto Group strata, and minimal contribution from the uplifted volcanic rocks. The absence of a flanking sedimentary basin southeast of the Lake Owen Fault (Fig. 2) suggests that this fault may have experienced only minimal reverse motion.

The Lake Owen–Hastings fault discontinuity occurs near the Great Lakes Tectonic Zone (GLTZ), a suture that extends northeast across Minnesota (Fig. 21) and into northern Wisconsin (Sims et al., 1980). The GLTZ divides the northern, ~2.7-Ga greenstone-granite terrane from the southern, ~3.5-Ga gneiss-migmatite terrane (Fig. 21; Sims et al., 1980). The Lake Owen-Hastings fault discontinuity also occurs near the Niagara fault, the southern boundary of the Archean Superior Province with the Early Proterozoic Penokean magmatic terrane of central Wisconsin (Fig. 1; Sims et al., 1989). Perhaps differences in crustal strength, thickness, and rheological properties between adjacent pre-Keweenawan terranes caused the reverse faults to become segmented along the southeastern margin of the St. Croix Horst.

To the southwest in Minnesota, the Hastings fault terminates at the Empire fault, a northwest-southeast cross structure within the MRS (Fig. 20). Reverse faulting southwest of the Empire fault occurs along the Castle Rock fault (Fig. 20), a structure imaged by seismic reflection data (Fig. 17), and along the Northfield fault (Fig. 22), which is inferred by the presence of volcanic rocks at shallow depth (Allen, 1994). South of the Northfield fault, reverse faulting is equivocal (Fig. 22),

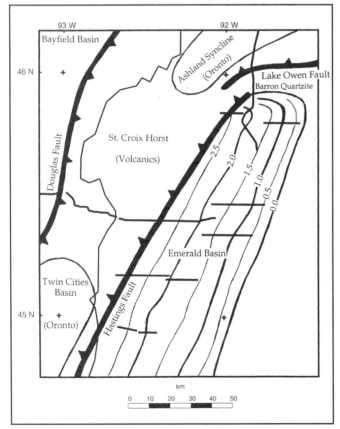

Figure 15. Contour map, base of the Keweenawan sedimentary strata within the Emerald basin. Thick lines indicate seismic reflection profiles. Contour interval is 0.5 s two-way travel time (1 s ~ 2.0 km).

although gravity and magnetic modeling by Allen (1994) indicates that displacement would be relatively minor (<1 km).

The western margin of the St. Croix Horst is delineated by two parallel reverse faults, the Douglas fault and the Pine fault (Sims and Zietz, 1967; Allen and Chandler, 1992). The Pine fault is identified by an abrupt discontinuity in the strike of the short-wavelength linear magnetic anomalies (Fig. 19) associated with the Keweenawan volcanic flows. The Douglas and Pine faults are disrupted by the Chisago fault, a minor cross fault north of the Twin Cities Basin (Fig. 20), and either terminate at or are substantially disrupted by the Lake Minnetonka fault, another cross fault along the western margin of the St. Croix Horst (Fig. 20). Southwest of the Lake Minnetonka fault, magnetic anomaly data indicate reverse faulting, although it is unclear whether this is a southwestern extension of the Pine fault or a northern extension of the Belle Plaine fault (PF?, Fig. 20).

Nature of the Belle Plaine fault. The northwest-striking Belle Plaine fault is a major cross structure that extends from the southwestern termination of the St. Croix Horst to the northern termination of the Iowa Horst (Figs. 2 and 20; Gibbs et al., 1984; Chandler et al., 1989). The Belle Plaine fault delineates the southern margin (accommodation structure) of the Superior zone (Dickas and Mudrey, this volume). Knowledge of the fault's structure is based primarily on gravity and magnetic data due to a relatively thick (~500 m) section of lower Paleozoic sedimentary strata. As shown in Figure 20, the Belle Plaine fault is not a single fault, but rather a series of en echelon fault segments.

Deformation of Paleozoic strata along the trace of the Belle Plaine fault near the city of Belle Plaine (BP, Fig. 20; Sloan and Danes, 1962) and in Waseca County (WC, Fig. 20; Bloomgren, 1993) indicates post-Ordovician reactivation of the fault. Stratigraphic relations within Paleozoic rocks suggest that the northeast side of the fault was uplifted, at most ~300 m, during Phanerozoic time. The amplitudes of the associated gravity and magnetic anomalies, however, indicate a much greater magnitude of displacement across the fault during Keweenawan time (Allen, 1994). The abrupt increase in the thickness of the volcanic sequence across the Belle Plaine fault (Fig. 17 and 18) suggests this fault may have originally developed as a normal growth fault.

In addition, the Belle Plaine fault may have been reactivated as a reverse fault during late-stage contraction along the MRS. For example, the magnetic data indicate a juxtaposition of Keweenawan volcanic and sedimentary strata just west of the Montgomery anticline (Fig. 20), suggesting that volcanic rocks were thrust west over younger sedimentary strata. The axis of the Montgomery Anticline is nearly parallel to the Belle Plaine fault (Fig. 20), suggesting that this fold may be a rollover structure produced by dragging within the hanging wall of the Belle Plaine fault, an interpretation also suggested by seismic reflection data (Fig. 17).

The MRS bends abruptly to the southeast along the Belle Plaine fault. Sloan and Danes (1962) suggested that the fault

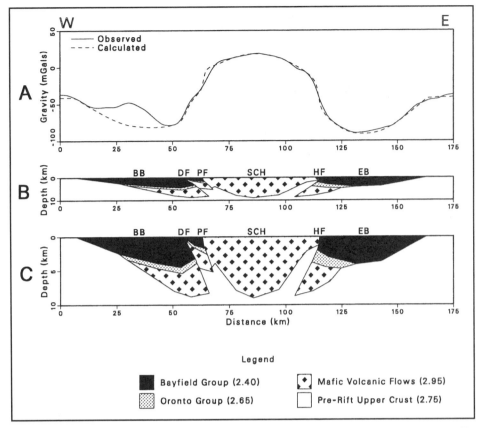

Figure 16. Two-dimensional gravity model of the Midcontinent Rift System along the seismic profile across the central St. Croix Horst in Minnesota and Wisconsin (line, Fig. 3). A, observed and calculated profiles; B, model, vertical exaggeration 1:1; C, model, vertical exaggeration 3:1. Densities are expressed in g/cm^3 (1 g/ cm^3 = 1,000 kg/m^3). SCH, St. Croix Horst; DF, Douglas fault; PF, Pine fault; HF, Hastings fault; BB, Bayfield Basin; EB, Emerald Basin. The positive gravity anomaly near 30 km is a local, north-northwest-striking feature (see Fig. 3) that is probably produced by a positive density contrast in the pre-Keweenawan basement beneath the Bayfield Basin.

experienced ~150 km of left-lateral strike-slip motion, and that the St. Croix and Iowa Horsts were originally a single, continuous feature. Beyond the margins of the MRS, however, geophysical and geologic data do not suggest the presence of major strike-slip faulting disrupting older pre-Keweenawan terranes. Moreover, it is not possible to obtain a continuous gravity anomaly signature by restoring the proposed strike-slip motion along the fault. In fact, the continuity of the rift's central positive gravity anomaly in southeastern Minnesota (Fig. 3) suggests that the MRS merely bends to the southeast.

It is unclear why the MRS is deflected in this area. However, older structures within the pre-Keweenawan terranes are parallel to and along strike with the Belle Plaine fault. In central Minnesota, for example, the western terminations of both the Animikie Basin outliers and the Penokean fold and thrust belt occur ~150 km northwest of the MRS, and on-line with the strike of the Belle Plaine fault (Fig. 21). In addition, V. W. Chandler (personal communication, 1993) speculates that the Belle Plaine fault may occur along the eastern margin of the Archean gneiss-migmatite terrane of southwestern Minnesota

(Fig. 1). On the basis of an intense magnetic anomaly just west of the Belle Plaine fault, McSwiggen et al. (1987) propose that the MRS was deflected east of a strong, pre-Keweenawan granitic intrusion. In any event, it appears likely that the development of the rift in southeastern Minnesota was influenced by ancestral zones of weakness in the older Proterozoic and Archean basement.

Magnitude of uplift. Quantitative gravity and magnetic modeling (e.g., Figs. 16 and 17) is useful for investigating the structure of the MRS at depth. Figure 22 summarizes the results of combined gravity and magnetic modeling along eight profiles across the MRS in Minnesota and Wisconsin (Allen, 1994). Reverse faulting is not uniform along the rift system. Across the Belle Plaine fault, vertical displacement (Fig. 22) is typically ~1 km or less, and never greater than ~3 km. Uplift of the central volcanic basin increases along the St. Croix Horst in a northeast direction, from ~1 km across the Northfield fault, to ~3 km across the Castle Rock fault, and to ~8 km across the Hastings fault (Fig. 22). Reverse displacement across the Lake Owen fault cannot be estimated (or even proven) because the

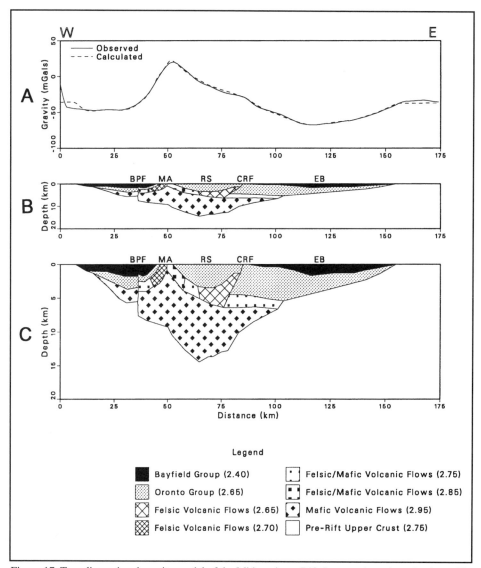

Figure 17. Two-dimensional gravity model of the Midcontinent Rift System along the seismic profile across the southern St. Croix Horst in Minnesota (line f, Fig. 3). A, observed and calculated profiles; B, model, vertical exaggeration 1:1; C, model, vertical exaggeration 3:1. Densities are expressed in g/cm^3 = 1,000 kg/m^3). BPF, Belle Plaine fault; CRF, Castle Rock fault; MA, Montgomery anticline; RS, Rice syncline; EB, Emerald Basin.

fault apparently thrusts Keweenawan volcanic strata over older Keweenawan and pre-Keweenawan rocks.

Along the western margin of the St. Croix Horst, displacement across the Douglas-Pine fault system increases to the northeast in Minnesota (Fig. 22), but then decreases to the east-northeast in Wisconsin, as required by the termination of the Douglas fault (Fig. 12 and 13). Perhaps, cross structures, such as the Empire fault, Lake Minnetonka fault, Chisago fault, and transition zone between the Hastings and Lake Owen faults (Figs. 19 and 20), developed to accommodate differential uplift between adjacent segments of the St. Croix Horst. Maximum uplift of the horst apparently occurred within its central portion between the

Ashland syncline and Twin Cities Basin (Fig. 2 and 22), suggesting that Oronto Group strata may have been deposited and later eroded from this portion of the St. Croix Horst.

The remarkable difference in the magnitude of uplift between the southeastern Minnesota segment of the MRS and the St. Croix Horst (Fig. 22) may be related to the orientation of the rift segments with respect to late-stage compressional stresses. For example, Cannon (1994) interprets that these stresses were oriented southeast-northwest, perpendicular to the St. Croix Horst but nearly parallel to the MRS in southeastern Minnesota, offering a possible explanation for the relatively minor reverse displacement across the Belle Plaine fault.

Figure 22 also indicates interpreted normal faults along the western limb of the MRS. Normal faulting is suggested by an abrupt increase in the thickness of the Keweenawan volcanic sequence, such as in Figures 16 (across the Pine fault) and 17 (across the Belle Plaine fault). The interpreted normal faults often coincide with known reverse faults, suggesting the reverse faults (in many instances) are reactivated normal growth faults.

Lower-crustal structure

Seismic refraction data (Halls, 1982; Hamilton and Mereu, 1993) indicate that the crust is as thick as ~55 km along the axis of the MRS in the Lake Superior region, in contrast to the surrounding craton where it is ~40 km thick. This amount of crustal thickening would be expected to produce a ~50-mGal,

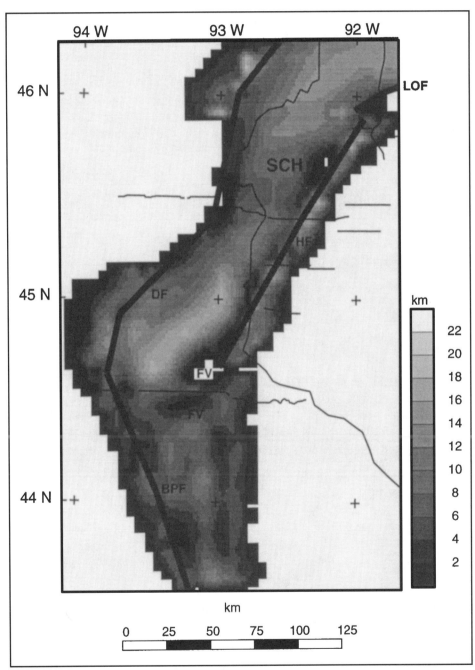

Figure 18. Thickness of Chengwatana Volcanic Group based on three-dimensional gravity modeling. Thin lines indicate seismic reflection profiles. BPF, Belle Plaine fault; DF, Douglas fault; HF, Hastings fault; LOF, Lake Owen fault; SCH, St. Croix Horst; FV, felsic volcanic rocks.

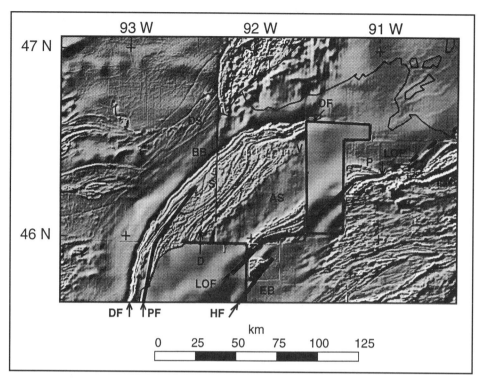

Figure 19. Shaded relief aeromagnetic map, northern St. Croix Horst region. Northwestern illumination. DF, Douglas fault; LOF, Lake Owen fault (dashed line is possible extension of the fault; Cannon et al., 1993); HF, Hastings fault; PF, Pine fault; AS, Ashland syncline; V, abrupt thinning of the outcrop belt of volcanic rocks; BB, Bayfield Basin; EB, Emerald Basin; DS, Keweenawan dike swarm; D, interpreted dike; S, interpreted sill-like intrusion (also shown in Fig. 2); P, pinch-out of volcanic strata. Heavy rectangular lines denote aeromagnetic survey boundaries.

long-wavelength negative gravity anomaly (Allen, 1994). The residual gravity anomaly map (Fig. 14), however, contains no such broad negative anomaly. Moreover, no permutation of (reasonable) model densities for the various Keweenawan units results in a residual regional negative anomaly (Allen, 1994). Consequently, the negative gravitational effect of the thickened crust in western Lake Superior is likely compensated by a positive mass imbalance, probably due to dense lower-crustal intrusions or crustal underplating. This interpretation is consistent with tomographic inversion of seismic refraction data in central Lake Superior (Hamilton and Mereu, 1993), which image a zone of high velocity (and presumably high density) in the lower crust along the rift axis.

To properly compensate the regional negative effect of the thickened crust in western Lake Superior, Allen (1994) suggests (based on gravity modeling) that between 30 to 50% of the Keweenawan rift magma was trapped in the lower crust, while the remaining 50 to 70% was erupted. Estimates for other rifts, however, suggest that extrusive rocks typically compose just 5 to 35% of the mantle-derived igneous rocks (Ramberg, 1976; Mohr, 1983; Crisp, 1984; Ramberg and Morgan, 1984; Shaw, 1985; Neumann et al., 1986; Riciputi and Johnson, 1990). Perhaps the higher proportion of Keweenawan magma

that reached the surface is indicative of a more rapid rate of extension of the MRS.

Information concerning the lower-crustal structure of the MRS in Wisconsin and Minnesota is relatively limited. Refraction data, however, indicate ~5 km of crustal thickening along the axis of the St. Croix Horst (Ocola and Meyer, 1973), an interpretation that is consistent with the results of teleseismic tomographic inversion by Green (1991). In addition, Green (1991) detected a zone of high velocity between 28 and 46 km, suggesting that the lower crust may have been intruded by Keweenawan magmas. Hence, the lower-crustal structure of the MRS along its western limb may be similar to that in Lake Superior.

Relative ages along the Midcontinent Rift System

As indicated by magnetic modeling along eight profiles in Minnesota and Wisconsin (Allen, 1994), the Keweenawan volcanic sequence of the St,. Croix Horst and in southeastern Minnesota is predominantly normally polarized. Reversely polarized rocks are either absent or form relatively thin layers along the base of the volcanic basin (e.g., Fig. 23). Consequently, the majority of the Keweenawan igneous sequence along the western limb of the Midcontinent Rift System (MRS) is probably younger than the ca. 1,098 Ma magnetic reversal documented in the Lake Superior region (Paces and Miller, 1993). This interpretation is consistent with the magnetic modeling results of Chandler et al. (1989) in Wisconsin and Minnesota, and Woelk (1989) in Kansas who also interpreted large proportions of normally polarized volcanic rocks.

Northeast of White's ridge, however, magnetic modeling (Allen 1994) indicates that roughly half of the volcanic sequence in western Lake Superior is reversely polarized and probably older than ca. 1,098 Ma. In addition, Mariano and Hinze (1994) discovered that ~80% of the volcanic sequence in eastern Lake Superior is reversely polarized. Consequently, major magmatic pulses did not occur at the same time along the length of the MRS. Instead, major magmatic activity began first in eastern Lake Superior, then in western Lake Superior, and finally along the rift's western limb.

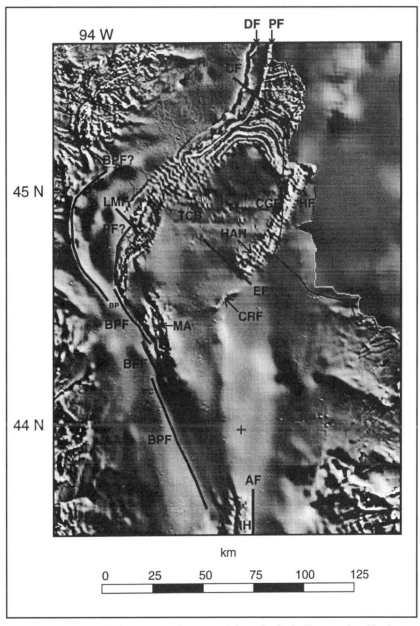

Figure 20. Shaded relief aeromagnetic map, southern St. Croix Horst region. Northeastern illumination. Reverse faults: DF, Douglas fault; PF, Pine fault; HF, Hastings fault; CGF, Cottage Grove fault; CRF, Castle Rock fault; AF, Austin fault. Cross-faults: BPF, Belle Plaine fault; CF, Chisago Fault; LMF, Lake Minnetonka fault; EF, Empire fault. Other structural features: TCB, Twin Cities Basin; MA, Montgomery anticline; IH, Iowa Horst; HAH, Hudson-Afton Horst; R, reversely polarized intrusion. BP, city of Belle Plaine; WC, Waseca County.

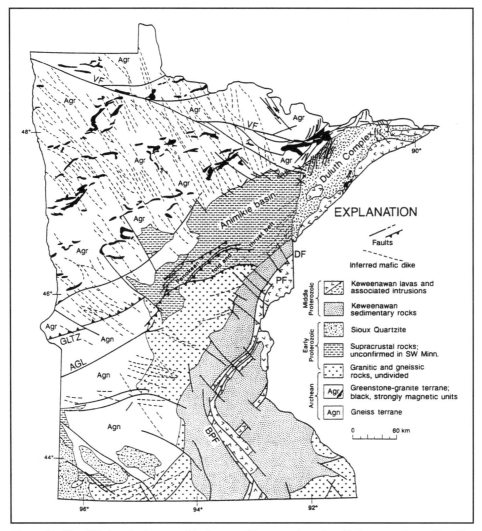

Figure 21. Simplified geologic map of Archean and Proterozoic rocks of Minnesota (from Chandler and Schaap, 1991). BPF, Belle Plaine fault; DF, Douglas fault; PF, Pine fault; VF, Vermillion fault; GLTZ, Great Lakes Tectonic Zone; AGL, Appleton Geophysical Lineament.

The presence of reversely polarized Keweenawan igneous rocks in Kansas (Van Schmus et al., 1990), in Nebraska (Marshall, 1994), and in the lower portions of some magnetic models across the St. Croix Horst (e.g., Fig. 23), however, indicates that at least some magmatic activity occurred in these areas before ca. 1,098 Ma. Hence, magmatism may have been contemporaneous along the length of the MRS, but at any one time occurred at different rates along the various rift segments.

CONCLUSIONS

Integrated interpretation of geologic and geophysical data, collected mainly for hydrocarbon exploration along the Midcontinent Rift System during the 1980s, has produced a new understanding of the structure, stratigraphy, and evolution of the rift in western Lake Superior, northwestern Wisconsin, and eastern Minnesota. Ancestral features, contained within pre-Keweenawan basement terranes, played influential roles throughout the history of the rift system. Such features are responsible for the remarkable structural and stratigraphic heterogeneity of the MRS in the Superior zone (the greater western Lake Superior region).

The most significant structure identified by this study is an Archean granitic belt extending between and including White's ridge and the Grand Marais ridge. These ridges, acting as local accommodation features, greatly influenced the distribution of rift volcanic and post-volcanic sedimentary rocks in western Lake Superior.

In west-central Wisconsin, seismic reflection and gravity data delineate the geographic extent, depth, and stratigraphy of the asymmetric Emerald Basin, formerly suspected only as a downward continuation of a poorly understood synclinal structure of Paleozoic age.

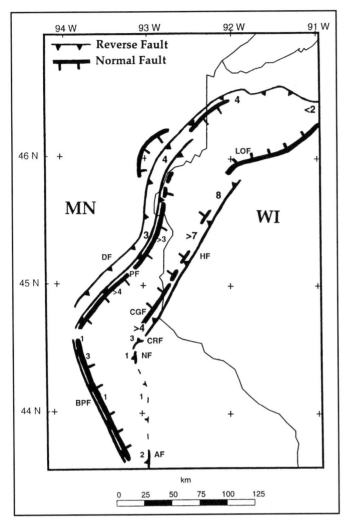

Figure 22. Map of the MRS in eastern Minnesota and northwestern Wisconsin indicating reverse faults (thin barbed lines) and interpreted normal faults (thick hachured lines). BPF, Belle Plaine fault; DF, Douglas fault; PF, Pine fault; AF, Austin fault; NF, Northfield fault; CRF, Castle Rock fault; CGF, Cottage Grove fault; HF, Hastings fault; LOF, Lake Owen fault. Estimates of vertical displacement (km) across the reverse faults are indicated in their appropriate locations on the hanging wall. Dashed line represents an equivocal reverse fault between the AF and NF.

ACKNOWLEDGMENTS

Proprietary seismic reflection data were generous provided the authors by Texaco Exploration and Production, Inc., Amoco Production Company, and Petty-Ray Geophysical Division of Geosource, Inc. The authors thank the reviewers, Val Chandler of the Minnesota Geological Survey and James Robertson of the Wisconsin Geological and Natural History Survey, for their useful suggestions.

The senior author wishes to thank V. W. Chandler of the Minnesota Geological Survey and L. W. Braile, S. D. King, D. W. Levandowski, and Y.-N. Shieh of Purdue University for their helpful guidance in this project. Financial assistance was provided by Amoco Production Company, ARCO Oil and Gas Company, Chevron Oil Company, the Minnesota Geological Survey, Purdue University, University of Wisconsin-Superior (Extension), and the Wisconsin Geological and Natural History Survey.

REFERENCES CITED

Allen, D. J., 1994, An integrated geophysical investigation of the Midcontinent Rift System: western Lake Superior, Minnesota and Wisconsin [Ph.D. thesis]: Purdue University, 267 p.

Allen, D. J., and Chandler, V. W., 1992, The utility of high-resolution aeromagnetic data. for investigating the Midcontinent Rift in east-central Minnesota [abs.]: Proceedings and Abstracts, Institute on Lake Superior Geology, Hurley, Wisconsin, v. 38, part 1, p. 3–5.

Allen, D. J., and Chandler, V. W., 1993, Magnetic properties of the Midcontinent Rift volcanics at Osseo, Minnesota: Clues to the origin of aeromagnetic anomalies [abs]: Proceedings and Abstracts, Institute on Lake Superior Geology, Eveleth, Minnesota, v. 39, part 1, p. 2-3.

Allen, D. J., Hinze, W. J., and Cannon, W. F., 1992, Drainage, topographic, and gravity anomalies in the Lake Superior region: evidence for a 1100 Ma mantle plume: Geophysical Research Letters, v. 19, p. 2119–2122.

Anderson, R. R., editor, 1990, The Amoco M. G. Eischeid #1 Deep Petroleum Test Carroll County, Iowa: Iowa Department of Natural Resources Special Report Series No. 2, 185 p.

Bacon, L. O., 1966, Geologic structure east and south of the Keweenaw fault on the basis of geophysical evidence, in Steinhart, J. S., and Smith, T. J., eds., The Earth beneath the continents: American Geophysical Union Monograph 10, p. 42–55.

Behrendt, J. C., Green, A. G., Cannon, W. F., Hutchinson, D. R., Lee, M., Milkereit, B., Agena, W. F., and Spencer, C., 1988, Crustal structure of the Midcontinent Rift System: Results from the GLIMPCE deep seismic reflection profiles: Geology, v. 16, p. 81–85.

Behrendt, J. C., Hutchinson, D. R., Lee, M. W., Agena, W. F., Trehu, A., Thornber, C. R., Cannon, W. F., and Green, A., 1990, Seismic reflection (GLIMPCE) evidence of deep crustal and upper mantle intrusions and magmatic underplating associated with the Midcontineht Rift System of North America: Tectonophysics, v. 173, p. 617–626.

Berendsen, P., Borcherding, R. M., Doveton, J., Gerhard, L., Newall, K. D., Steeples, D., and Watney, W. L., 1988, Texaco Poersch #1, Washington County, Kansas—preliminary geologic report of the pre-Phanerozoic rocks: Kansas Geological Survey Open-File Report 82–22, 116 p.

Berg, J. H., and Klewin, K. W., 1988, High-MgO lavas from the Keweenawan Midcontinent near Mamainse Point, Ontario: Geology, v. 16, p. 1003–1006.

Bloomgren, B. A., 1993, Bedrock geology of Waseca County, Minnesota: Minnesota Geological Survey Miscellaneous Map Series M-73, scale 1:62 500 and smaller.

Boerboom, T. J., 1994, Archean crustal xenoliths in a Keweenawan hypabyssal sill, northeastern Minnesota. White was right! [abs.]: Proceedings and Abstracts, Institute on Lake Superior Geology, Houghton, Michigan, v. 40, p. 5–6.

Books, K. G.,1972, Paleomagnetism of some Lake Superior Keweenawan rocks: U.S. Geological Survey Professional Paper 760, 42 p.

Cannon, W. F., 1992, Process rates during Midcontinent rifting: clues to the origin of the Midcontinent Rift System [abs.]: Proceedings and Abstracts, Institute on Lake Superior Geology, Hurley, Wisconsin, v. 38, part 1, p. 17–19.

Cannon, W. F., 1994, Closing of the Midcontinent Rift—a far-field effect of Grenvillian compression: Geology, v. 22, p. 155–158.

Cannon, W. F. , 1989, The Midcontinent Rift beneath Lake Superior from GLIMPCE seismic reflection profiling: Tectonics, v. 8, p. 305–332.

Cannon, W. F., Peterman, Z. E., and Sims, P. K., 1990, Structural and isotopic evidence for Middle Proterozoic thrust faulting of Archean and Early Proterozoic rocks near the Gogebic Range, Michigan and Wisconsin [abs.]:

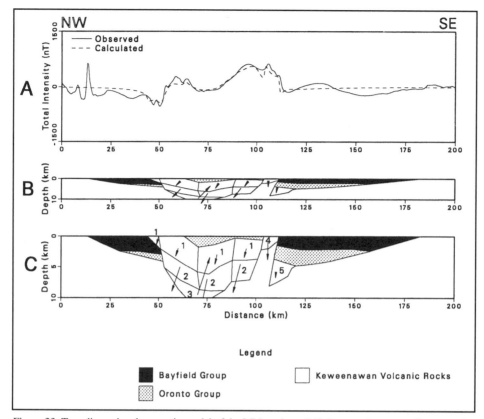

Figure 23. Two-dimensional magnetic model of the Midcontinent Rift System along a profile across the Twin Cities Basin in Minnesota. A, observed and calculated profiles; B, model, vertical exaggeration 1:1; C, model, vertical exaggeration 3:1. Arrows denote direction of total magnetization in the plane of the profile; lengths of arrows are proportional to the magnetization. Magnetic properties [susceptibility (SI)/remanent magnetization (A/m) (direction)/Koenigsberger ratio (Q)]: 1, 0.0075/1.20 (N) /3.4; 2, 0.0239/3.80(N)/3.4; 3, 0.0.591/9.40(R)/3.4; 4, 0.0754/1.50(R)/ 0.4; 5, 0.0126/0.00/0.0. N, normal polarization; R, reversed polarization.

Proceedings and Abstracts, Institute on Lake Superior Geology, Thunder Bay, Ontario, v. 36, part l, p. 11–12.

Cannon, W. F., Nicholson, S. W., Zartman, R. E., Peterman, Z. E., and Davis, D. W., 1993, The Kallander Creek volcanoes—A remnant of a Keweenawan central volcano centered near Mellen, Wisconsin [abs.]: Proceedings and Abstracts, Institute on Lake Superior Geology, Eveleth, Minnesota, v. 39, part 1, p. 20–21.

Chandler, V. W., 1990, Geologic interpretation of gravity and magnetic data over the central part of the Duluth Complex, northeastern Minnesota: Economic Geology, v. 85, p. 816–829.

Chandler, V. W., and Schaap, B. D., 1991, Bouger gravity anomaly map of Minnesota: Minnesota Geological Survey State Map Series S-16, scale 1:500,000.

Chandler, V. W., McSwiggen, P. L., Morey, G. B., Hinze, W. J., and Anderson, R. L., 1989, Interpretation of seismic reflection, gravity and magnetic data across the Middle Proterozoic Midcontinent Rift System in western Wisconsin, eastern Minnesota, and central Iowa: American Association of Petroleum Geologists Bulletin, v. 73, p. 26l–275.

Crisp, J. A., 1984, Rates of magma emplacement and volcanic output: Journal of Volcanology and Geothermal Research, v. 20, p. 177–211.

Daniels, P. A., Jr., 1982, Upper Precambrian sedimentary rocks: Oronto Group, *in* Wold, R. J., and Hinze, W. J., eds., Geology and tectonics of the Lake Superior basin: Geological Society of America Memoir 156, p. 107–133.

Davis, D. W., and Sutcliffe, R. H., 1985, U-Pb ages from the Nipigon plate and northern Lake Superior: Geological Society of America Bulletin, v. 96,

p. 1572–1579.

Davis, D. W., and Paces, J. B., 1990, Time resolution of geologic events on the Keweenaw Peninsula and implications for development of the Midcontinent Rift System: Earth and Planetary Science Letters, v. 97, p. 54–64.

Dickas, A. B., 1986a, Comparative Precambrian stratigraphy and structure along the Midcontinent Rift: American Association of Petroleum Geologists Bulletin, v. 70, p. 225–238.

Dickas, A. B., 1986b, Nature of the northern boundary of the St. Croix Horst, *in* Mudrey, M. G., ed., Precambrian petroleum potential, Wisconsin and Michigan: Wisconsin Gelogic and Natural History Survey, Geoscience Wisconsin, v. 11, p. 39–44.

Dickas, A. B., 1986c, Interpretation of western Lake Superior gravity low, *in* Mudrey, M. G., ed., Precambrian petroleum potential, Wisconsin and Michigan: Wisconsin Gelogic and Natural History Survey, Geoscience Wisconsin, v. 11, p. 53–58.

Dickas, A. B., and Mudrey, M. B. , Jr., 1989, Structural differentiation of the Midcontinent Rift System [abs.]: Eos, (Transactions, American Geophysical Union) v. 70, p. 275.

Dickas, A. B., and Mudrey, M. B. , Jr., 1991, Geology and petrography of the Amnicon pluton, Douglas County, Wisconsin: Proceedings, Institute on Lake Superior Geology, 37th, Eau Claire, Wisconsin, p. 41–43.

Ferderer, R. J., 1982, Gravity and magnetic modelling of the southern half of the Duluth Complex, northeastern Minnesota [M.A. thesis]: Bloomington, Indiana, Indiana University, 99 p.

Gibbs, A. K., Payne, B., Setzer, T., Brown, L. D., Oliver, J. E., and Kaufman, S.,

1984, Seismic-reflection study of the Precambrian crust of central Minnesota: Geological Society of America Bulletin, v. 95, p. 280–294.

Green, J. C., 1972, North Shore Volcanic Group, in Sims, P. K., and Morey, G. B., eds., Geology of Minnesota: A centennial volume: Minnesota Geological Survey, p. 294–332.

Green, J. C., 1982, Geology of Keweenawan extrusive rocks, in Wold, R. J., and Hinze, W. J., eds., Geology and tectonics of the Lake Superior Basin: Geological Society of America Memoir 156, p. 47–55.

Green, J. C., 1983, Geologic and geochemical evidence for the nature and development of the Middle Proterozoic (Keweenawan) Midcontinent Rift of North America: Tectonophysics, v. 94, p. 413–437.

Green, W. V., 1991, Lithospheric seismic structure of the Cenozoic Kenya rift and the Precambrian Midcontinent rift from teleseismic tomography [Ph.D. thesis]: Madison, Wisconsin, University of Wisconsin-Madison, 179 p.

Grout, F. F., Sharp, R. P., and Schwartz, G. M., 1959, The geology of Cook County, Minnesota: Minnesota Geological Survey Bulletin, v. 39, 163 p.

Halls, H. C., 1969, Compressional wave velocities of Keweenawan rock specimens from the Lake Superior region: Canadian Journal of Earth Science, v. 39, 163 p.

Halls, H. C., 1982, Crustal thickness in the Lake Superior region, in Wold, R. J., and Hinze, W. J., eds., Geology and tectonics of the Lake Superior basin: Geological Society of America Memoir 156, p. 239–244.

Halls, H. C., and Pesonen, L. J., 1982, Paleomagnetism of Keweenawan rocks, in Wold, R. J., and Hinze, W. J., eds., Geology and tectonics of the Lake Superior basin: Geological Society of America Memoir 156, p. 173–302.

Hamilton, D. A., and Mereu, R. F., 1993, 2-D tomographic imaging across the North American Midcontinent Rift System: Geophysical Journal International, v. 112, p. 344–358.

Heaman, L. M., and Machado, N., 1992, Timing and origin of the Midcontinent Rift alkaline magmatism, North America: Evidence from the Coldwell Complex: Contributions to Mineralogy and Petrology, v. 110, p. 289–303.

Hinze, W. J., O'Hara, N. W., Trow, J. W., and Secor, G. B., 1966, Aeromagnetic studies of eastern Lake Superior, in Steinhart, J. S., and Smith, T. J., eds., The Earth beneath the continents: American Geophysical Union Monograph 10, p. 95–110.

Hinze, W. J., Kellogg, R. L., and O'Hara, N. W., 1975, Geophysical studies of basement geology of southern peninsula of Michigan: American Association of Petroleum Geologists Bulletin, v. 59, . p. 1562–1584.

Hinze, W. J., Wold, R. J., O'Hara, N. W., 1982, Gravity and magnetic anomaly studies of Lake Superior, in Wold, R. J., and Hinze, W. J., eds., Geology and tectonics of the Lake Superior basin: Geological Society of America Memoir 156, p. 203–222.

Hinze, W. J., and Kelly, W. C., 1988, Scientific drilling into the Midcontinent Rift System: Eos, (Transactions, American Geophysical Union), v. 69, p. 1649, 1656–1657.

Hinze, W. J., Braile, L. W., and Chandler, V. W., 1990, A geophysical profile of the southern margin of the Midcontinent Rift System in western Lake Superior: Tectonics, v. 9, p. 303–310.

Hinze, W. J., Allen, D. J., Fox, A. J., Sunwood, D., Woelk, T., and Green, A. G., 1992, Geophysical investigations and crustal structure of the North American Midcontinent Rift System: Tectonophysics, v. 213, p. 17–32.

Hutchinson, D. R., White, R. S., Cannon, W. F., and Schulz, K. J., 1990, Keweenaw hot spot: Geophysical evidence for a 1.1 Ga mantle plume beneath the Midcontinent Rift System: Journal of Geophysical Research, v. 95, p. 10869–10884.

Jirsa, M. A., 1984, Interflow sedimentary rocks in the Keweenawan North Shore Volcanic Group, northeastern Minnesota: Minnesota Geological Survey Report of Investigations 30, 20 p.

Kalliokoski, J., 1982, Jacobsville sandstone, in Wold, R. J., and Hinze, W. J., eds., Geology and tectonics of the Lake Superior basin: Geological Society of America Memoir 156, p. 147–155.

Kalliokoski, J., 1989, Jacobsville Sandstone and tectonic activity [abs.]: Proceedings and Abstracts, Institute on Lake Superior Geology, Duluth, Minnesota, v. 35, part 1, p. 39–41.

King, E. R., and Zietz, I., 1971, Aeromagnetic study of the Midcontinent gravity high of central United States: Geological Society of America Bulletin, v. 82, p. 218–208.

Klasner, J. S., Cannon, W. F., and Van Schmus, W. R., 1982, The pre-Keweenawan tectonic history of southern Canadian Shield and its influence on formation of the Midcontinent Rift, in Wold, R. J., and Hinze, W. J., eds., Geology and tectonics of the Lake Superior basin: Geological Society of America Memoirs 156, p. 27–46

Klewin, K. W., and Berg, J. H., 1991, Petrology of the Keweenawan Mamainse Point lavas, Ontario: Petrogenesis and continental rift evolution: Journal of Geophysical Research, v. 96, p. 457–474.

Klewin, K. W., and Shirey, S. B., 1992, The igneous petrology and magmatic evolution of the Midcontinent Rift System: Tectonophysics, v. 213, p. 33–40.

Mariano, J., and Hinze, W. J., 1994, Gravity and magnetic models of the Midcontinent Rift in eastern Lake Superior: Canadian Journal of Earth Science, v. 31, p. 661–674.

Marshall, L. P., 1994, Geochemistry of the Midcontinent Rift System basalt in the subsurface of Nebraska [M.S. thesis]: Pittsburgh, Pennsylvania, University of Pittsburgh, 165 p.

McGinnis, L. D., and Mudrey, M. G., Jr., 1991, Seismic reflection profiling and tectonic evolution of the Midcontinent Rift in Lake Superior: Wisconsin Geological and Natural History Survey Miscellaneous Paper 91–2, 11 p.

McSwiggen, P. L., Morey, G. B., and Chandler, V. W., 1987, New model of Midcontinent Rift in eastern Minnesota and western Wisconsin: Tectonics, v. 6, p. 677–685.

Miller, J. D., Jr., 1992, The need for a new paradigm regarding the petrogenesis of the Duluth Complex [abs.]: Proceedings and Abstracts, Institute on Lake Superior Geology, Hurley, Wisconsin, v. 38, part 1, p. 65–67.

Mohr, P., 1983, The Morton-Black hypothesis for the thinning of the continental crust; revisited in western Afar: Tectonophysics, v. 94, p. 509–528.

Morey, G. B., 1972a, Mesabi range, in Sims, P. K, and Morey, G. B., eds., Geology of Minnesota: A centennial volume: Minnesota Geological Survey, p. 204–217.

Morey, G. B., 1972b, Gunflint range, in Sims, P. K., and Morey, G. B., eds., Geology of Minnesota: A centennial volume: Minnesota Geological Survey, p. 218–225.

Morey, G. B., and Ojakangas, R. W., 1982, Keweenawan sedimentary rocks of eastern Minnesota and northwestern Wisconsin, in Wold, R. J., and Hinze, W. J., eds., Geology and tectonics of the Lake Superior basin: Geological Society of America Memoir 156, p. 135–146.

Morey, G. B., and Van Schmus, W. R., 1988, Correlation of Precambrian rocks of the Lake Superior region, United States: U.S. Geological Survey Professional Paper 1241-F, 31 p.

Mudrey, M. G., Jr., LaBerge, G. L., Myers, P. E., and Cordua, W. S., 1987, Bedrock geology of Wisconsin, northwest sheet: Wisconsin Geological and Natural History Survey Map 87-1la, scale 1:250,000.

Neumann, E.-R., Pallesen, S., and Andersen, P., 1986, Mass estimates of cumulates and residues after anatexis in the Oslo graben: Journal of Geophysical Research, v. 91, p. 11629–11640.

Nicholson, S. W., and Shirey, S. B., 1990, Evidence for a Precambrian mantle plume: a Sr, Nd, Pb isotopic study of the Midcontinent Rift System in the Lake Superior region: Journal of Geophysical Research, v. 95, p. 10851–10868.

Ocola, L. C., and Meyer, R. P., 1973, Central North American rift system, 1. structure of the axial zone from seismic and gravimetric data: Journal of Geophysical Research, v. 78, p. 5173–5194.

Ojakangas, R. W., and Morey, G. B., 1982, Keweenawan pre-volcanic quartz sandstones and related rocks of the Lake Superior region, in Wold, R. J., Hinze, W. J., eds., Geology and tectonics of the Lake Superior basin: Geological Society of America Memoir 156, p. 85–96.

Paces, J. B., and Bell, K., 1989, Non-depleted sub-continental mantle beneath the Superior Province of the Canadian Shield: Nd-Sr isotopic and trace element evidence from the Midcontinent Rift basalts: Geochimica et Cosmochimica Acta, v. 53, p. 2023–2035.

Paces, J. B., and Miller, J. D., 1993, Precise U-Pb ages of Duluth Complex and related mafic intrusions, northeastern Minnesota, U.S.A.: geochronological insights to physical, petrogenetic, paleomagnetic and tectono-magmatic processes associated with the 1.1 Ga Midcontinent Rift System: Journal of Geophysical Research, v. 98, p. 13997–14013.

Palmer, H. C., 1970, Paleomagnetism and correlation of some Middle Keweenawan rocks, Lake Superior: Canadian Journal of Earth Science, v. 7, p. 1410–1436.

Palmer, H. .C., and Halls, H. C., 1986, Paleomagnetism of the Powder Mill group, Michigan and Wisconsin: a reassessment of the Logan Loop: Journal of Geophysical Research, v. 91, p. 11571–11580.

Palmer, H. C., and Davis, D. W., 1987, Paleomagnetism and U-Pb geochronology of volcanic rocks from Michipicoten Island, Lake Superior, Canada: precise calibration of the Keweenawan polar wander track: Precambrian Research, v. 37, p. 157–171.

Ramberg, I. B., 1976, Gravimetry interpretation of the Oslo graben and associated igneous rocks: Norges Geologiske Undersökelse, v. 325, p. 1–194.

Ramberg, I. B., and Morgan, P., 1984, Physical characteristics and evolutionary trends of continental rifts, *in* Proceedings, 27th International Geological Congress, Volume 7: Amsterdam, Verenigde Nederlandse Uitgeversbedrijver Science Press, p. 165–216.

Riciputi, L. R., and Johnson, C. M., 1990, Nd and Pb isotope variations in the multicyclic central caldera cluster of the San Juan volcanic field, Colorado, and implications for crustal hybridization: Geology, v. 18, p. 975–978.

Robertson, W. A., 1973, Pole positions from the Mamainse Point lavas and their bearing on a Keweenawan pole path and polarity sequence: Canadian Journal of Earth Science, v. 10, p. 1541–1555.

Sexton, J. L., and Henson, H., Jr., 1994, Interpretation of seismic reflection and gravity profile data in western Lake Superior: Canadian Journal of Earth Science, v. 31, p. 652–660.

Shaw, H. R., 1985, Links between magma-tectonic rate balances, plutonism, and volcanism: Journal of Geophysical Research, v. 90, p. 11275–11288.

Sims, P. K., and Zietz, I., 1967, Aeromagnetic and inferred Precambrian paleogeologic map of east-central Minnesota and part of Wisconsin: U.S. Geological Survey Miscellaneous Investigations Series Map I-2185, scale 1:500 000.

Sims, P. K., Card, K. D., Morey, G. B., and Peterman, Z. E., 1980, The Great Lakes tectonic zone—A major crustal structure in central North America: Geological Society of America Bulletin, v. 91, p. 690–698.

Sims, P. K., Van Schmus, W. R., Schulz, K. J., and Peterman Z. E., 1989, Tectono-stratigraphic evolution of the Early Proterozoic Wisconsin magmatic terranes of the Penokean orogen: Canadian Journal of Earth Science, v. 26, p. 2145–2158.

Sloan, R. E., and Danes, Z. F., 1962, A geologic and gravity survey of the Belle Plaine area, Minnesota: Minnesota Academy of Science Proceedings, v. 30, p. 49–52.

Thiel, E. C., 1956, Correlation of gravity anomalies with the Keweenawan geology of Wisconsin and Minnesota: Geological Society of America Bulletin, v. 67, p. 1079–1100.

Trehu, A., and 15 others, 1991, Imaging the Midcontinent Rift beneath Lake Superior using large aperture seismic data: Geophysical Research Letters, v. 18, p. 625–628.

Van Schmus, W. R., 1992, Tectonic setting of the Midcontinent Rift System: Tectonophysics, v. 213, p. 1–15.

Van Schmus, W. R., Green, J. C., and Halls, H. C., 1982, Geochronology of Keweenawan rocks of the Lake Superior region: a summary, *in* Wold, R. J., and Hinze, W. J., eds., Geology and tectonics of the Lake Superior basin: Geological Society of America Memoir 156, p. 165–171.

Van Schmus, W. R., Martin, M. W., Sprowl, D. R., Geissman, J., and Berendsen, P., 1990, Age, Nd and Pb isotopic composition, and magnetic polarity for subsurface samples of the 1100 Ma Midcontinent Rift [abs.]: Geological Society of America Abstracts with Program, v. 21, p. A174.

Weber, J. R., and Goodacre, A. K., 1966, A reconnaissance underwater gravity survey of Lake Superior, *in* Steinhart, J. S., and Smith, T. J., eds., The Earth beneath the continents: American Geophysical Union Geophysical Monograph 10, p. 56–65.

White, W. S., 1966a, Geologic evidence for crustal structure in the western Lake Superior basin, *in* Steinhart, J. S., and Smith, T. J., eds., The Earth beneath the continents: American Geophysical Union Monograph 10, p. 28–41.

White, W. S., 1966b, Tectonics of the Keweenawan basin, western Lake Superior region: U.S. Geological Survey Professional Paper 524-E, p. El–E23.

Woelk, T. S., 1989, An integrated geophysical study of the Midcontinent Rift System in northeastern Kansas [M.S. thesis]: West Lafayette, Indiana, Purdue University, 75 p.

Wold, R. J., and Ostenso, N. A., 1966, Aeromagnetic, gravity, and sub-bottom profiling studies in western Lake Superior, *in* Steinhart, J. S., and Smith, T. J., eds., The Earth beneath the continents: American Geophysical Union Geophysical Monograph 10, p. 66–94.

MANUSCRIPT ACCEPTED BY THE SOCIETY JANUARY 16, 1996

Geological Society of America
Special Paper 312
1997

Geology, petrology, and tectonic significance
of the Beaver Bay Complex, northeastern Minnesota

James D. Miller, Jr., and Val W. Chandler
Minnesota Geological Survey, 2642 University Avenue, St. Paul, Minnesota 55114

ABSTRACT

The Beaver Bay Complex (BBC) is a hypabyssal, multiple-intrusive igneous complex exposed over a 600 km² area in northeastern Minnesota. It was emplaced into comagmatic volcanic rocks at 1.1 Ga during the development of the Midcontinent Rift System. Thirteen intrusive units have been identified that represent a minimum of six major intrusive events. With the exception of a body of granophyre, most intrusions were formed from gabbroic to dioritic parental magmas with successive intrusions generally involving less evolved compositions. Compositional variations within intrusive units developed as a result of in situ magmatic differentiation, assimilation of footwall rocks, and composite intrusions of evolved magma from deeper staging chambers. Over the exposed extent of the BBC, intrusion shapes appear to have been controlled by a shallow crustal ridge that trends northwesterly across the BBC. The focus of emplacement appears to have migrated toward the rift axis and toward higher stratigraphic levels with time perhaps reflecting plate drift and thickening of the volcanic pile. In the larger context of the Midcontinent Rift System, geologic, geochronologic, geophysical, and geochemical data consistently indicate that the BBC, particularly the youngest Beaver River diabase dike and sheet network, acted as a magma conduit and structural boundary to the development and infilling of the western end of the Portage Lake Volcanic basin during the main stage of rift volcanism and graben formation.

INTRODUCTION

The intrusive mafic igneous rocks that form pronounced highlands in the area around the North Shore community of Beaver Bay were recognized by the earliest geologic studies of northeastern Minnesota, being termed the Beaver Bay Group by Irving (1883) and the Beaver Bay Diabase by Winchell (1899). The unique occurrence of large blocks of nearly pure anorthosite as xenoliths in some of the diabase intrusions has generated a continuing interest in these rocks (Lawson, 1893; Grout and Schwartz, 1939; Morrison et al., 1983), not all of it scientific in nature. About 1906, a fledgling abrasives company developed a short-lived quarry, crushing, and transport facility in the area to extract the anorthosite, mistaking it for corundum. Despite this

setback, the company, Minnesota Mining and Manufacturing (3M), survived.

The term "Beaver Bay Complex" was used by Grout and Schwartz (1939) to include intrusive rocks extending some 20 km inland from shoreline exposures between Split Rock Point and Little Marais (Fig. 1A and Fig. 3, in a following section). Although various "facies" of intermediate to mafic intrusive rock were recognized by Grout and Schwartz, all were classified as diabase. Only anorthosite inclusions and granophyre bodies (red rock) were distinguished as distinctive units of the Beaver Bay Complex (BBC). The boundary between the BBC "diabase" and deeper seated Duluth Complex "gabbros", though obscured by thick glacial debris, was thought to be transitional. The multiple-intrusive nature of the BBC was first doc-

Miller, J. D., Jr., and Chandler, V. W., 1997, Geology, petrology, and tectonic significance of the Beaver Bay Complex, northeastern Minnesota, *in* Ojakangas, R. W., Dickas, A. B., and Green, J. C., eds., Middle Proterozoic to Cambrian Rifting, Central North America: Boulder, Colorado, Geological Society of America Special Paper 312.

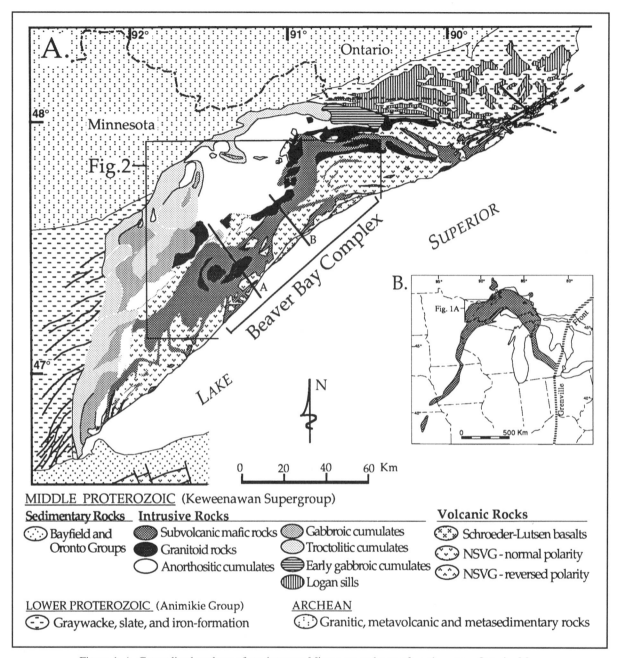

Figure 1. A. Generalized geology of northeastern Minnesota and part of northwestern Ontario. Map units lacking outline represent geology inferred from aeromagnetic data and limited exposure. Shoreline extent of the Beaver Bay Complex and the area portrayed in Figure 2 are noted. Lines A and B correspond to the interpretive cross sections shown in Figure 8. B. Regional extent of igneous rocks of the Midcontinent Rift System based largely on gravity data.

umented by Gehman (1957) with his detailed mapping and petrologic study of the area between Split Rock Point and Beaver Bay, although more recent studies (Miller, 1988; Shank, 1989) have revised his interpretations (see following discussion). Another detailed mapping and petrologic study of a 25 km² area north of Finland by Stevenson (1974) delineated several more intrusive units of the BBC, including a well-differentiated mafic layered intrusion. A detailed mapping study of a

small part of the BBC grew out of site preparation for the Mile-post 7 tailings basin, which was constructed for the taconite milling operation near Silver Bay (Green, 1982b). With the exception of these few areas, however, the detailed geology and petrogenesis of the BBC was, until recently, poorly understood.

In 1993, the Minnesota Geological Survey completed an eight-year program to map the bedrock geology of the Beaver Bay Complex at a 1:24,000 scale and to characterize its transition into the Duluth Complex. The project was funded in part through the U.S. Geological Survey's COGEOMAP program. All or parts of eleven 7.5 minute quadrangles, covering over 900 square kilometers, were mapped and some 2,100 hand samples collected. Studies of these handsamples have included petrographic observations of over 1,500 thin sections, electron microprobe analysis of about 250 samples producing more than 3,500 individual analyses, more than 230 whole rock analyses, and about 250 specific gravity and magnetic susceptibility measurements. Five geologic maps have been produced from the project (Miller, 1988; Miller et al., 1989, 1993, 1994; Boerboom and Miller, 1994). This chapter summarizes the general geology and petrology of the BBC based on the results of this mapping and related studies. We will show the BBC to be composed of at least 13 major intrusive units that demonstrate various styles of emplacement, a broad range of parental magma compositions, and unique crystallization histories. In the larger tectonic context of the Midcontinent Rift System, we conclude that the creation of the BBC was closely associated with graben formation and active volcanism characteristic of the main phase of rifting.

REGIONAL SETTING

The intrusive and volcanic geology of northeastern Minnesota represents the northwestern flank of the Middle Proterozoic Midcontinent Rift System (MRS, Fig. 1). The MRS is exposed only in the Lake Superior region but can be traced by its gravity and magnetic anomaly pattern along an arcuate path from Kansas to the Lower Peninsula of Michigan (Fig. 1B). In general, the MRS is composed of a lower sequence of predominantly flood basalts and related intrusions and an upper red bed sequence of largely fluvial sediment, which together have been broadly folded into asymmetric basins due to syndepositional downwarping and late contractional deformation. Recent seismic profiles across Lake Superior reveal basalt flow thicknesses as great as 20 km in some axial portions of the rift (Cannon et al., 1989). The large volume of lava calculated to have infilled the rift has led to the conclusion that the source of the magmatism, and perhaps the tectonism, was a mantle plume (Hutchinson et al., 1990; Cannon and Hinze, 1992). The effect of contemporaneous Grenvillian orogenesis on the tectonic evolution of the MRS is debatable, though most would agree that Grenvillian deformation was the likely cause of late reversed motion evident on many originally graben-bounding faults (Fig. 1B). Over most of the MRS, this late tectonism resulted in the creation of a central horst of volcanic rock flanked by deep sedimentary basins.

The Middle Proterozoic geology of northeastern Minnesota is unique relative to other parts of the MRS because of the predominance of intrusive rocks over volcanic rocks. Except for the Lake Superior shoreline, which is composed of 6–8 km of largely basalt flows (Green, 1972), most of northeastern Minnesota (~60%) is apparently underlain by intrusive igneous rocks (Fig. 1A). The bulk of these are troctolitic, gabbroic, anorthositic, and granitic rocks comprising the multiple intrusions of the crescent-shaped Duluth Complex (Weiblen

and Morey, 1980; Miller and Weiblen, 1990). Most Duluth Complex intrusions were emplaced in the vicinity of the unconformity between the pre-rift basement of Early Proterozoic and Archean rocks and the early rift basalt flows. A significant volume of magma also was emplaced at medial levels of the volcanic edifice with the greatest concentration of these intrusions forming the Beaver Bay Complex. Gravity, aeromagnetic, and seismic data over northeastern Minnesota indicate that the mafic intrusive rocks evident at the surface are deeply rooted to depths of at least 10 km (Ferderer, 1982; Chandler, 1990).

Recent high precision U-Pb dating of zircons from intrusive rocks of northeastern Minnesota (Paces and Miller, 1993) indicate prolonged and episodic intrusive activity between 1,107 and 1,096 Ma, roughly half the period of overall magmatic activity of the MRS (1,109–1,086 Ma). The earliest intrusive activity is indicated by an 1,106.9 ± 0.6 Ma age for reversely polarized, layered gabbros along the northern tier of the Duluth Complex (early gabbroic cumulates, Fig. 1A). However, four ages between 1,099.3 ± 0.3 and 1,098.6 ± 0.5 of samples representing the main troctolitic and anorthositic series of the Duluth Complex indicate that the bulk of the complex was emplaced after an 8-m.y. hiatus. Two samples from the BBC give indistinguishable ages of 1,096.1 ± 0.8 Ma and 1,095.8 ± 1.2 Ma. The slightly older age is from the Sonju Lake intrusion and the younger age is from a Silver Bay intrusion (see following discussion). The later age represents the youngest recognized intrusive component of the BBC. However, the onset of BBC magmatism and its coevality relative to the Duluth Complex is unknown because attempts to date the oldest intrusive phases of the BBC have been unsuccessful (Paces and Miller, 1993).

GEOLOGY OF THE BEAVER BAY COMPLEX

Recent geological mapping reveals that the historical definition of the Beaver Bay Complex (Grout and Schwartz, 1939) must be expanded to at least include all intrusive igneous rocks exposed in the arcuate area extending from the lakeshore between Split Rock Point and Grand Marais and inland as much as 20 km (Fig. 1A). The northwest boundary of the BBC is taken to generally correspond to the northwestern margin of Houghtaling Creek troctolite with its associated flanking intrusions (Wilson Lake ferrogabbro, Dam Five Lake gabbronorite; see Fig. 4). The troctolite is a 2- to 3-km-wide, northeast-trending, keel-shaped intrusion that is easily recognizable in outcrop, has a large exposure area, and has a distinctive, easily traceable aeromagnetic signature (Fig. 2). In contrast to the dikes and thin sheeted bodies of typically non-cumulate gabbroic rocks that characterize much of the BBC, the predominant rock types to the northwest of the Houghtaling Creek troctolite are structurally complex, coarse-grained gabbroic anorthosite cumulates, homogeneous granophyric granite, and volcanic rocks that are strongly hornfelsed and brecciated by variably contaminated mafic intrusions. This assemblage is typical of the roof zone of the Duluth Complex elsewhere (Weiblen and Morey, 1980).

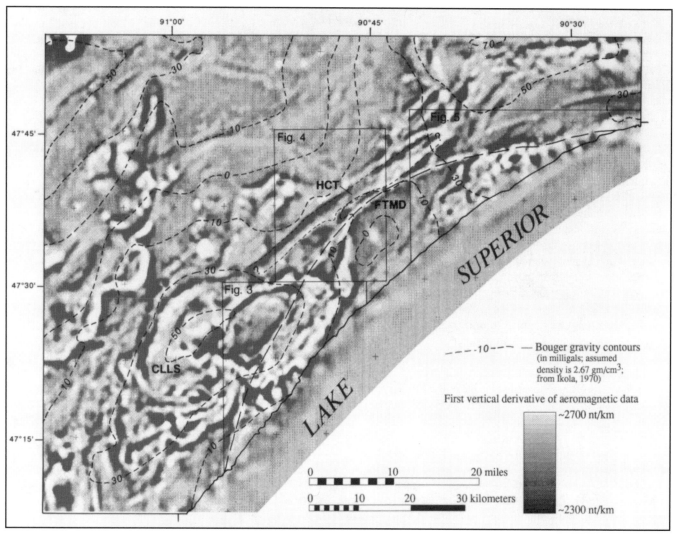

Figure 2. Gray-scale image of the first vertical derivative (reduced to pole for Keweenawan rem-
anance) of aeromagnetic data (Chandler, 1983) and contoured Bouger gravity data (Ikola, 1970) over
the Beaver Bay Complex. Dashed lines are Bouger gravity contours from Ikola (1970). Coverage of
geologic maps (Figs. 3 to 5) are outlined and some geologic features referenced in the text are
denoted: HCT, Houghtaling Creek troctolite; FTMD, Finland tectono-magmatic discontinuity;
CLLS, Cloquet Lake Layered Series.

Aeromagnetic data and reconnaissance mapping indicate
an extensive belt of subvolcanic intrusions extends to the northeast and southwest of what is herein defined as the BBC
(Fig. 1A). Beyond the western limit of outcrop, which runs
approximately north-northeast from Split Rock Point (Figs. 3,
4), interpretations of aeromagnetic anomalies (Fig. 2) and limited drill core suggest the presence of a large, nested, multiple-
intrusive layered complex, informally termed the Cloquet Lake
layered series (Chandler, 1990; Meints et al., 1993). The aeromagnetic anomaly pattern of the Houghtaling Creek troctolite
seems to merge into the lower part of the layered series (Fig. 2)
suggesting a cogenetic relationship. Although this layered complex can be considered a southwestern extension of the BBC,
for all practical purposes, the western boundary of the BBC is

the limit of exposure. Beyond its easternmost extent of outcrop,
the aeromagnetic anomaly associated with the Houghtaling
Creek troctolite suggests that the dike pinches out and inter-
sects intrusions trending in a more northerly direction (Fig. 2).
These latter intrusions appear to emanate from a more sharply
curved mass of hypabyssal sheets and thicker layered bodies
that are well exposed in the vicinity of Sawbill and Brule Lakes

Figure 3. Geology of the southern Beaver Bay Complex. Generalized
and slightly modified from 1:24,000 scale geologic maps of the Silver
Bay and Split Rock Point NE 7.5 minute quadrangles (Miller, 1988),
the Illgen City quadrangle (Miller et al., 1989), the Finland and Doyle
Lake quadrangles (Miller et al., 1993), and unpublished data from the
Little Marais quadrangle.

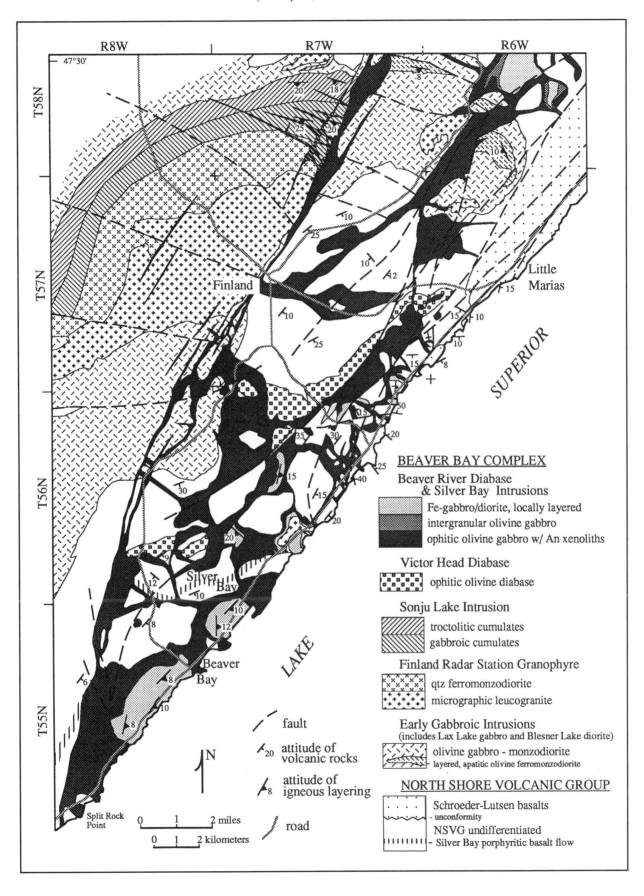

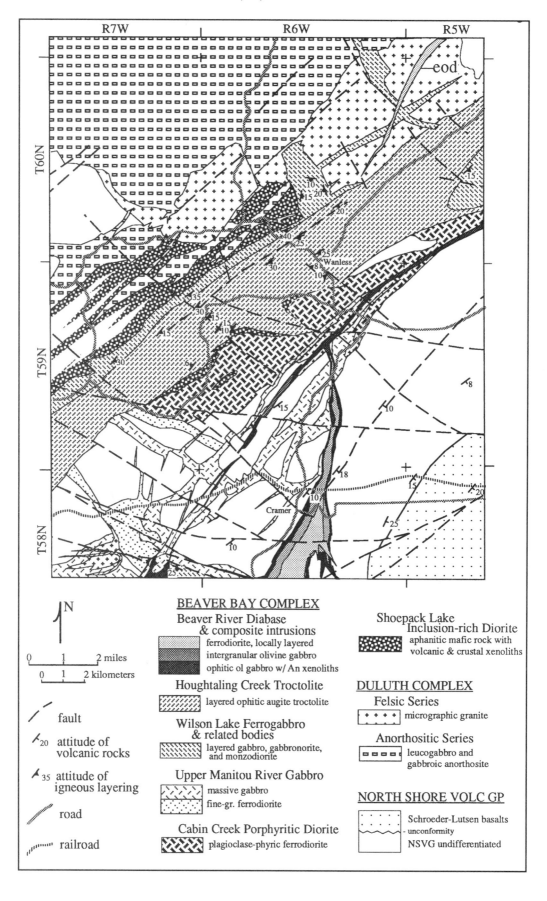

Figure 4. Geology of the northern Beaver Bay Complex. Generalized from 1:24,000 scale geologic maps of the Cabin Lake and Cramer 7.5 minute quadrangles (Miller et al., 1994) and the Silver Island Lake, Wilson Lake, and Toohey Lake 7.5 minute quadrangles (Boerboom and Miller, 1994).

———————————◄———————————

(Green, 1982a; Burnell, 1976). Although more study of this northern complex is necessary to compare its composition, intrusive setting, and age of emplacement to the BBC, its geometry and location are sufficiently different to warrant its distinction as a separate intrusive suite.

Within the exposed area of the BBC as defined above, thirteen major intrusive map units have been distinguished which represent a minimum of six distinct intrusive events. This numerical disparity reflects the composite nature of most intrusive units and the uncertain correlation of units over the extensive area of BBC exposure. Because of this uncertainty, it is useful to discuss the exposed geology of the BBC in terms of three areas—southern, northern, and eastern. The southern BBC is contained within the triangular area enclosed by the western extent of exposure, the shoreline extending northeast

from Split Rock Point, and a northern limit at about 47°30' latitude (Fig. 3). This area includes the Split Rock Point NE, Silver Bay, Illgen City, Doyle Lake, Finland, and Little Marias 7.5-minute quadrangles. The northern BBC extends north of 47°30' to the Houghtaling troctolite and is bounded to the east and west by the limit of exposure (Fig. 4). This area is covered by the Cabin Lake, Cramer, Silver Island Lake, Wilson Lake, and Toohey Lake 7.5-minute quadrangles. The eastern BBC corresponds to nearshore exposures between Schroeder and Grand Marais and extends inland about 10 km (Fig. 5). Before giving a description of BBC intrusive units in these three areas, we will first describe the geology of the volcanic rocks into which they were emplaced.

North Shore Volcanic Group

Basalt, minor felsic flows, and thin interflow sedimentary rocks are exposed along nearly the entire length of Minnesota's north shore of Lake Superior to a cumulative thickness of 7–10 km (Green, 1972) and are collectively termed the North Shore Volcanic Group (NSVG). In detail, the NSVG can be broken down into formational groupings of distinctive compo-

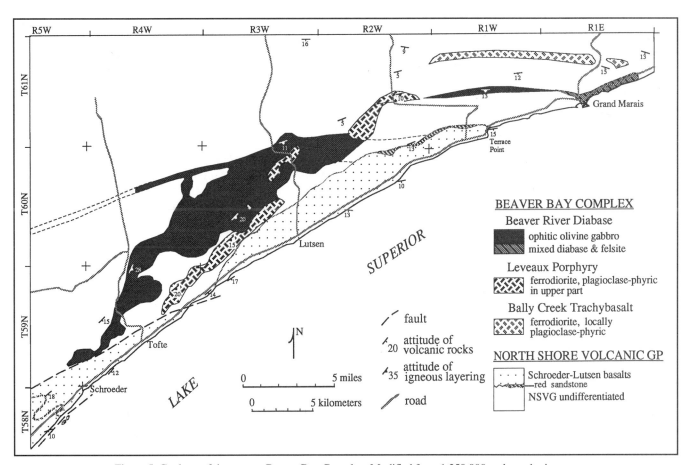

Figure 5. Geology of the eastern Beaver Bay Complex. Modified from 1:250,000 scale geologic map of the Two Harbors 1° × 2° quadrangle (Green, 1982a). Dashed lines indicate the projection of aeromagnetic anomalies, apparently associated with the Beaver River diabase, in areas of no exposure.

sitions and physical attributes. The lava sequence exposed between Split Rock Point and Grand Marais is the statigraphically highest part of the NSVG and measures about 3–3.5 km thick. In increasing stratigraphic order, the following informal volcanic sequences have been recognized by Green (1972, 1992) progressing northeast from Split Rock Point to the top of the sequence at Tofte.

- *Gooseberry River basalts.* Mix of ophitic olivine tholeiitic basalts and transitional basalts; includes a rhyolite flow and a strongly plagioclase porphyritic basalt flow.
- *Baptism River lavas.* Mixed basaltic lavas bounded below by an aphyric rhyolite (Silver Beaver felsite) and above by a thick porphyritic rhyolite (Palisade Head rhyolite; Green and Fitz, 1993).
- *Bell Harbor lavas.* Mostly intermediate basalts (quartz tholeiites, basaltic andesites, and transitional basalts), some olivine tholeiite flows, one thin felsic flow, and several interflow siltstone, sandstone, and polymict conglomerate units.
- *Schroeder-Lutsen basalts.* Except for a thick transitional basalt flow (Manitou basalt) and a volcanic breccia unit (Pork Bay breccia), this approximately 1.2-km-thick unit is composed entirely of olivine tholeiitic basalt flows. The base of this unit is locally in angular unconformity with older flows and is commonly marked by a polymict conglomerate.

The lavas underlying the Schroeder-Lutsen basalts to the northeast of Tofte toward Grand Marais (Fig. 2) are andesitic and felsic (Good Harbor Bay andesite and Grand Marias rhyolite) and do not obviously correlate with the Bell Harbor lavas. The northeastern exposure of the base of the Schroeder-Lutsen basalts (at Terrace Point) is conformable above a thick red sandstone unit, which in turn conformably overlies the andesite flow (Fig. 5).

The regional structure of the NSVG flows defines a broad, half-saucer-shaped basin dipping gently (<20°) lakeward. From east-southeasterly dips along the southern stretch of the lakeshore, flow contacts gradually shift to become south dipping along the northern limb of the basin (Fig. 1A). In the vicinity of the BBC, however, this regional structure is significantly disrupted (Figs. 3–5). The cause for most of this disruption is attributable to deformation (block faulting and rotation) that accommodated the emplacement of the various BBC intrusions, particularly the dikes and sills of the Beaver River diabase. With the notable exception of the Schroeder-Lutsen basalts, all volcanic sequences noted above are intruded and deformed by the BBC. Because of this deformation, the correlation of volcanic sequences across this area is commonly difficult. Another source of deformation is from shore-parallel faults that cut volcanic and intrusive rocks alike. Local offset evidence and slickenside measurements indicate both normal and reversed oblique-slip displacement along these faults (Witthuhn, this volume; Green, 1992). Some of this motion could be related to the late-stage contractional deformation evident throughout the rift.

Southern Beaver Bay Complex

Although being the focus of most earlier studies of the BBC, the total number and sequence of intrusions comprising the southern BBC was incompletely known prior to recent mapping (Miller, 1988; Miller et al., 1989, 1993). From oldest to youngest, five major intrusive units have been identified in the southern BBC—the Lax Lake gabbro/Blesner Lake diorite, the Finland granophyre, the Sonju Lake intrusion, the Beaver River diabase, and the Silver Bay intrusions (Fig. 3)—as well as several smaller bodies.

A complex suite of hydrothermally altered, gabbroic to monzodioritic rocks, which represent the earliest intrusions in the southern BBC, are found in two areas—one southwest of Finland (T56N, R8W), termed the *Lax Lake gabbro,* and another, the *Blesner Lake diorite,* to the northeast (T58N, R7–6W, Fig. 3). The nearly identical lithologies and contact relationships within the Lax Lake and Blesner Lake units (Miller et al., 1993) suggest that they originally formed a continuous intrusive mass that was separated by younger intrusions. Because of a paucity of internal structure, it is difficult to determine the shape of this early gabbroic mass, although a shallow south-dipping lensoid sheet seems likely from its outcrop pattern and displaced volcanic rocks. The variation in rock types define a crude concentric zonation with the least evolved rock type, ophitic olivine diabase, at the margins and passing inward to intergranular gabbro, apatitic olivine ferrodiorite, ferromonzodiorite, and eventually to quartz ferromonzodiorite. In the Blesner Lake diorite, the apatitic ferrodiorite locally becomes well laminated (Fig. 3) and somewhat resembles the upper differentiates of the Sonju Lake intrusion. Locally abrupt changes in rock type and the lack of cumulate textures (save for the laminated ferrodiorite) imply that these varied lithologies resulted more from multiple injections of progressively more evolved magma than from post-emplacement differentiation. Moreover, moderate to strong hydrothermal alteration and an irregular distribution of interstitial granophyre ubiquitous in all rock types suggest that these early gabbroic intrusions were strongly contaminated by hydrothermal fluids and felsic melts generated from the water-bearing volcanic rocks into which the intrusions were emplaced.

Exposed over a 40 km^2 area in the northwestern part of the southern BBC is a distinctive, oval-shaped mass of leucogranite and quartz ferromonzodiorite, termed the Finland Radar Station granophyre (Miller et al., 1993), or simply the *Finland granophyre.* The main phase of the granophyre is a homogeneous, salmon-colored, micrographic leucogranite with abundant miarolitic cavities. To the north, this granitic phase, which contains less than 5% Fe-silicates and oxides, abruptly grades to a quartz ferromonzodiorite characterized by 5–20% prismatic, iron-rich pyroxene, amphibole, and locally olivine and 20–40% micrographic felsic matrix. Although the intrusive relationships between the Finland granophyre and the older Lax Lake gabbro along its southern margin and the younger Beaver River diabase at its eastern extent are well established by several exposures of sharp contacts, the genetic relationship of the granophyre to the Sonju Lake layered intrusion along its gradational northern boundary is more problematic. Stevenson (1974) considered the monzodioritic and granitic phases of the granophyre to be intrusive into the Sonju Lake intrusion although he cites no evidence for that interpretation. Given the parallel zonation of the two phases of the granophyre with the strike of cumulate units of the Sonju Lake intrusion (Fig. 3; also see Miller et al., 1993) and generally smooth compositional variations across these units (Fig. 6B), a comagmatic relationship resulting from crystallization differentiation could be envisioned. Such an interpretation is inconsistent, however, with the large amount of granophyre relative to the layered mafic rocks apparent from the geologic map (Fig. 3). Modeling of gravity and aeromagnetic data across these units (Miller et al., 1990) confirm that the volume of granophyre is at least as great as the volume of underlying mafic rocks and thus is too great to be a differentiate of the layered intrusion. Also, ongoing isotopic studies indicate that the granophyre has an isotopic composition distinct from the mafic rocks and therefore could not have evolved from them (J. Vervoort, personal communication, 1995). A more plausible interpretation is that the mafic magma that produced the Sonju Lake intrusion

underplated the lower density Finland granophyre and perhaps caused it to partially melt. Petrologic and isotopic studies are underway to determine the extent of assimilation across the mafic-felsic contact.

The *Sonju Lake intrusion* is the most completely differentiated mafic layered intrusion recognized in the Keweenawan magmatic system (Stevenson, 1974; Weiblen, 1982). The intrusion is a shallow (15–30°) south- to southeast-dipping, 1,200-m-thick sheet, which has an apparent strike length of over 15 km (based on its aeromagnetic signature, Fig. 2) extending west from its abrupt truncation by the Beaver River diabase (Fig. 3). Starting with a picritic diabase basal contact zone against older gabbroic rock, an upwardly differentiated sequence of cumulate rocks indicates a cumulus mineral paragenesis of $Ol \rightarrow Pl + Ol \rightarrow Pl + Cpx + Ol \rightarrow Pl + Cpx + Ox(-Ol) \rightarrow Pl + Cpx + Ox + Ap + Ol$. This cumulate stratigraphy displays smooth, logarithmic cryptic variations of upwardly decreasing Mg values of mafic silicates and An contents of plagioclase. These characteristics are consistent with its formation by closed-system fractional crystallization of a moderately evolved tholeiitic basaltic magma (Table 1). An orthogonal set of mafic-felsic hybrid dikes intruding the upper part and eastern margin of the Sonju Lake intrusion (Fig. 3) appear to have resulted from mixing of residual felsic magma in

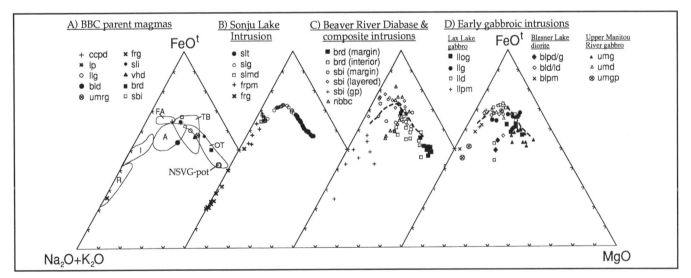

Figure 6. AFM diagrams (Na$_2$O+K$_2$O-FeO-MgO, wt.%) of Beaver Bay Complex (BBC) compositions. (A) Comparison of BBC parental compositions (Table 1) to major North Shore Volcanic Group (NSVG) lava compositions, OT, olivine tholeiite; TB, transitional basalt; A, andesite; FA, ferroandesite; I, icelandite; R, rhyolite, (after Green, 1983). Beaver Bay Complex intrusions are shown as symbols that are listed on Table 1. NSVG-pot is the primitive "NSVG olivine tholeiite" composition listed in Table 1. (B) Calculated liquid line of descent of the Sonju Lake intrusion through troctolitic (slt), gabbroic (slg), and monzodiorite (slmd) intervals of the layered sequence. Also plotted are whole rock compositions of Finland granite (frg) and quartz ferromonzodiorite (frpm). (C) Whole rock composition plots of Beaver River diabase (brd, ophitic margins and coarse subophitic interiors distinguished), Silver Bay intrusions (sbi, coarse marginal facies, layered ferrogabbroic cumulate interiors, and granophyric compositions distinguished), and composite intrusions from the northern BBC (nbbc). Sonju Lake intrusion differentiation trend (dashed line) is also shown. (D) Whole rock composition plots of gabbroic (llog, llg, blpd/g, umg), dioritic (lld, bld/ld, umd), and ferromonzodioritic (llpm, blpm, umgp) compositions from the Lax Lake gabbro, Blesner Lake diorite, and Upper Manitou River gabbro are plotted. Abbreviations correspond to map units described in Miller (1988) and Miller et al. (1993, 1994). Sonju Lake intrusion differentiation trend (dashed line) is also shown.

TABLE 1. PARENTAL COMPOSITIONS OF BEAVER BAY COMPLEX INTRUSIONS

Unit*	Cabin Creek Porph. Diorite	Leveaux Porphyry	Lax Lake Gabbro	Blesner Lake Diorite	Upper Manitou River Gabbro	Finland Radar St. Granophyre	Sonju Lake Intrusion	Victor Head Diabase	Beaver River Diabase	Silver Bay Intrusions	North Shore Volcanic Group
Phase	Aphyric Facies	Aphyric Facies	Average Gabbro	Average Gabbro	Average Gabbro	Average Granite	Bulk Comp.§	Average Diabase	Marginal Diabase	Marginal Gabbro	Olivine Tholeiite**
Analyses	3	2†	5	5	12	9	84	3	12	12	14
wt. %											
SiO_2	53.5	53.5	47.7	49.5	48.3	73.9	48.4	49.7	48.0	49.5	49.5
TiO_2	2.30	2.45	2.27	1.70	1.64	0.30	2.22	1.71	1.31	3.25	0.80
Al_2O_3	12.5	13.5	16.2	18.6	16.3	12.5	13.9	15.9	16.7	12.4	18.2
FeO^t	14.7	13.4	12.5	9.4	11.6	3.0	14.6	11.2	10.5	15.4	8.4
MnO	0.20	0.16	0.19	0.14	0.18	0.05	0.21	0.15	0.17	0.23	0.13
MgO	3.2	3.4	4.8	3.63	5.7	0.1	7.9	5.3	7.8	3.6	8.8
CaO	6.8	7.1	9.7	10.21	10.2	0.6	9.1	9.5	10.4	7.8	11.4
Na_2O	2.81	2.83	2.74	3.80	2.58	3.60	2.54	2.57	2.30	2.85	2.38
K_2O	2.02	1.21	0.78	0.70	0.56	4.90	0.68	0.68	0.35	1.17	0.19
P_2O_5	0.41	0.42	0.24	0.14	0.14	0.07	0.30	0.25	0.14	0.47	0.06
LOI	1.83	1.70	1.17	2.85	0.99	1.00	0.18	1.39	1.71	1.42	
Total	99.2	99.7	99.6	100.73	98.2	100.0	99.9	98.3	99.4	98.1	99.8
Mg no.	29.5	31.3	40.5	40.8	46.7	5.6	49.1	45.7	56.9	29.3	65.3
An	29.3	38.8	50.4	45.6	54.2	6.2	50.1	52.5	60.1	35.4	63.1
ppm											
Sc	31	29		33	28	1.4	35	31	26	38.9	26.9
Cr	43	35	144	124	111	1.8	194	107	156	29.1	361
Co	40	39	63	41	49	0.7	76	60	57	60.5	48.8
Ni	30	38	77	37	77	9.3	185	85	185	34	277
Rb	55	27.5	28	28	22	157	21	23	13	40	4
Sr	231	296	208	368	242	66	235	257	245	209	207
Ba	475	551	268	179	151	1,029	173	333	137	370	58
Y	37		30	22	21	114	20	27	17	47.7	13
Zr	244	155	152	138	100	736	115	160	95	324	60
La	31.7	40		17.6	13.7	87.8	14.9	28.8	10.3	16.9	4.16
Ce	69.5	87		38.6	32	187	33.8	58	23.2	38	10.3
Sm	7.9	9.1		5.1	4.1	16.7	4.2	6.0	3.3	5.0	1.91
Eu	2.4	2.9		1.8	1.4	2.8	1.6	1.9	1.27	1.8	0.83
Tb	1.3	1.5		0.9	0.7	2.6	0.8	0.63	0.8	0.41	
Yb	3.5	4.0		2.8	2.2	9.8	2.1	3.1	2.03	3.7	1.38
Lu	0.51	0.58		0.42	0.31	1.41	0.32	0.53	0.28	0.57	0.21

*Rock types arranged in approximate order of decreasing age.
†One of the two analyses averaged is of the lower nonporphyritic unit reported by Green, 1972.
§Composition is a weighted average of 82 whole rock analyses through the Sonju Lake intrusion.
**Average composition of 14 unaltered, primitive olivine thoeliite basalts containing less than 3 percent glass (Brannon, 1984, Table 5.5).

the upper differentiates of the layered intrusion with mafic magma forming the adjacent Beaver River diabase.

The *Beaver River diabase* is the most areally extensive intrusive phase of the entire BBC and is found in contact with all other BBC units, except the Houghtaling Creek troctolite. In the southern BBC, it occurs as a series of dikes, sills, and sheets of ophitic olivine gabbro that grades into coarser and more subophitic to intergranular gabbros in the medial portions of thicker sheets. The western and northern extent of the Beaver River diabase is marked by a continuous dike (or dike set) traceable in outcrop and by its aeromagnetic signature along a 110-km-long arcuate path (Figs. 2–5). Within the concavity of this bounding

dike are several other large, often bifurcating dikes and several large sheets dipping gently southeast. The complex of sheets holding up table-top highlands in the southern BBC (Fig. 3) may be part of an originally continuous, nearly horizontal sheet that has locally been eroded through to expose volcanic rocks forming the footwall. An aphanitic mafic rock with abundant inclusions of anorthosite, granite, basalt, and felsite locally occurs at the margins of some diabase intrusions and probably represents brecciation, engulfment, and incipient assimilation attending the initial intrusion of the diabase.

The most intriguing feature of the Beaver River diabase is the presence of large (as much as several hundred meters in diameter), rounded to angular inclusions of nearly pure *anorthosite*. These bodies, which locally display brecciation and recrystallization textures (Morrison et al., 1983), are particularly common in the upper and lower margins of the larger diabase sheets. With the exception of Lawson (1893), who thought the anorthosite to be the tops of deeply rooted Archean mountains around which Keweenawan rocks had flowed, most workers recognized their xenolithic nature, although their source has been in doubt. The lack of any discernable chill of the diabase against the anorthosite suggests that the inclusions were derived from a hot (deep?) source. Grout and Schwartz (1939) believed that the anorthositic gabbros of the Duluth complex supplied the inclusions, though they noted several inconsistencies such as the coarser nature and lighter color of the inclusions. A petrographic and geochemical analysis of the inclusions by Morrison et al. (1983) suggested from rather ambiguous isotopic data that the inclusions were probably derived from an older (≤1.9 Ma) deep crustal source. X-ray diffraction analyses (Miller, 1980, unpublished data) indicates that plagioclase from the inclusions has a significantly higher state of structural disorder compared to plagioclase from the Duluth Complex anorthositic rocks, further confirming a deep crustal source to the inclusions. However, if a plagioclase crystal mush origin for Duluth Complex anorthositic rocks is correct (Miller and Weiblen, 1990), a corollary of such a model is that significant amounts of Keweenawan anorthosite, generated by plagioclase flotation under high pressure, should have formed in the deep crust prior to BBC magmatism at 1,096 Ma (the two dated anorthositic rocks of the Duluth Complex have 1,099 Ma ages; Paces and Miller, 1993). Under deep crustal conditions, such plagioclase cumulates would probably be distinctive in texture and composition from their shallow crustal counterparts. Moreover, the ambiguous isotopic compositions of the inclusions may indicate that anorthosite-forming Keweenawan magmas were contaminated by older crust rather than older anorthosite being contaminated by interaction with Keweewawan magmas, as concluded by Morrison et al. (1983).

The youngest intrusions recognized in the southern BBC are a suite of small, isolated bodies of varied shape and size, collectively termed the *Silver Bay intrusions*. The bodies are intermediate in composition and form composite intrusions in Beaver River diabase dikes and sheets. The largest of these, located southwest of Beaver Bay (Fig. 3), is a gently southeast-dipping, concentrically zoned, lensoid body emplaced into the upper part of a thick diabase sheet. The two major zones of the body—an outer ring of coarse, vari-textured ferromonzodiorite and an inner core of strongly laminated and locally layered olivine ferrogabbro, gabbronorite, and ferromonzodiorite—were interpreted by Gehman (1957) to represent two distinct intrusions. However, recent mapping (Miller, 1988) and petrologic studies (Shank,1989; Chalokwu and Miller, 1992) concluded that the outer unit is a marginal facies to the inner sequence of iron-rich cumulates. Petrologic studies of this intrusion and a smaller body near Silver Bay reveal an in situ

differentiation trend toward extreme iron enrichment. In the southern BBC, six such zoned bodies and several smaller massive bodies (Fig. 3) span a considerable range of intermediate compositions suggesting that these bodies were derived from a fractionating magma chamber at various stages of differentiation. Geochemical modeling demonstrates that fractional crystallization of the Beaver River diabase could have generated the magma compositions that gave rise to the various Silver Bay intrusions, although this differentiation apparently took place at a deeper crustal level.

Several minor intrusions also comprise the southern BBC, although their temporal and comagmatic relationships to the major intrusive units are unclear. The most extensive of these is a 50- to 65-m-thick sill of fine- to medium-grained, ophitic olivine diabase that can be traced over a strike-length of at least 17 km (Fig. 3) and is informally referred to as the *Victor Head diabase*. Although clearly older than the Beaver River diabase that intrudes it, the Victor Head diabase is not in contact with any older units. Another small, unnamed intrusion of uncertain affinity is a fine-grained intergranular ferrodiorite that cuts the Gooseberry River basalts and the southern Lax Lake gabbro (Miller,1988; Green, 1982b). Also, several isolated masses of micrographic and locally spherulitic leucogranite are intrusive into the Blesner Lake diorite and a Silver Bay intrusion (Fig. 3). The former mass may represent satellite intrusions that emanate from the Finland granophyre. The Silver Bay body may have formed by localized melting of felsic volcanic rocks that locally form the footwall to the Beaver River diabase and Silver Bay intrusions.

Northern Beaver Bay Complex

Prior to recent mapping, very little was known about this part of the BBC (Green, 1982a). Our mapping now shows that northern BBC is characterized by the convergence of intrusive units into a narrow band of northeast trending dikes (Fig. 4). Intrusive relationships indicate five major emplacement events forming (1) Shoepack Lake inclusion-rich diorite, (2) the Cabin Creek porphyritic diorite, (3) the Upper Manitou River gabbro, (4) the Houghtaling Creek troctolite and earlier marginal intrusions, and (5) the Beaver River diabase and its composite intrusions. The intrusive activity represented by these units apparently spanned a greater time period than recognized in either the southern or eastern BBC. Except for the Beaver River diabase, correlation of intrusive units between the northern and southern BBC is not definitive. Based on relative timing and compositional similarities, it would seem that the Upper Manitou River gabbro is comagmatic with the early gabbroic intrusions to the south and that the Houghtaling Creek troctolite is perhaps correlative with the Sonju Lake intrusion. However, as discussed in the style of emplacement section below, possible interpretations of the aeromagnetic data suggest that the Houghtaling Creek may be older than the Upper Manitou River gabbro. The earlier Shoepack Lake inclusion-rich diabase and the Cabin Creek porphyritic diorite have no apparent equivalents in the southern BBC.

The earliest intrusive phase of the northern BBC is a strongly contaminated, inclusion-laden mafic rock termed the *Shoepack Lake inclusion-rich diorite.* Although contact relationships with other BBC rocks are obscure, this rock type is intruded by both granophyre and gabbroic anorthositic rocks of the Duluth Complex (Fig. 4). The rock is characterized by angular inclusions of varied size and irregular shape in a black, dense aphanitic matrix. The aphanitic matrix is commonly amygdaloidal, contains various amounts of feldspar and quartz xenocrysts, and locally shows evidence of incomplete mixing with a felsic melt. Felsic volcanic rocks are the most common inclusion type, although mafic volcanics and gabbroic rocks are locally present. More significant is the occurrence of granitic gneiss, biotite schist, and amphibolite inclusions, presumably of Archean age (Boerboom, 1994). As interpreted from similar inclusion-rich mafic rocks observed in the margins of some Beaver River diabase dikes in the southern BBC, this rock probably represents brecciation and assimilation attending the initial stages of emplacement of a mafic magma into a newly developed intrusion conduit. That this rock type is found on both sides of the Houghtaling Creek troctolite indicates that the troctolite utilized this same conduit at a later stage of BBC magmatism.

The next intrusive phase is represented by a plagioclase-phyric ferrodiorite termed the *Cabin Creek porphyritic diorite.* It is intruded by all younger BBC intrusive phases (Fig. 4). Subequant to tabular, labradoritic plagioclase phenocrysts as large as 4 cm across and rare cognate anorthosite inclusions comprise about 30% of the rock on average. However, some phases of this unit are aphyric and others contain as much as 50% phenocrysts. A consistently fine-grained matrix is composed of various mixtures of andesine, Fe-augite, Ti-magnetite, and a granophyric mesostasis yielding bulk compositions ranging from ferrodiorite to quartz ferromonzonite. Because of a lack of alignment of the phenocrysts or any gradations in texture, it is difficult to ascertain the exact shape of the Cabin Creek body. Its general map distribution suggests that, at least along its eastern extent, it is a thick sheet somewhat conformable with volcanic rocks. The composition and texture of plagioclase phenocrysts (An_{60} average) and ferrodioritic matrix comprising the Cabin Creek porphyritic diorite are very similar to anorthositic rocks associated with the Duluth Complex on the northwestern side of the Houghtaling Creek troctolite (Fig. 4; Scott Creek leucogabbro, Katydid Lake gabbroic anorthosite, and Outlaw Lake gabbroic anorthosite of Boerboom and Miller, 1994). Like the plagioclase crystal mushes that Miller and Weiblen (1990) postulated to be parental to the anorthositic rocks of the Duluth Complex, dioritic magmas charged with variable concentrations of intratelluric plagioclase were probably parental to the Cabin Creek porphyritic diorite. If it can be shown that the porphyritic diorite of the BBC and the gabbroic anorthosites of the Duluth Complex are indeed cosanguine, this may help explain a paradox of the anorthositic rocks. Miller and Weiblen (1990) noted that the calculated abundances of trapped liquid component comprising the anorthositic rocks (5-30 wt.%) are much less

than what could have transported the plagioclase to its site of emplacement and therefore speculated that some of this transporting magma was expelled during or after emplacement. Perhaps the porphyritic diorite with its many aphyric facies represents such an expelled liquid.

The *Upper Manitou River gabbro* is a complex suite of mostly gabbroic rocks that forms a crudely orthogonal pattern of dikes and sills throughout much of the northern BBC (Fig. 4). Except for a narrow northeast-trending dike cutting the Cabin Creek porphyritic diorite, most of the unit is intrusive into volcanic rocks dominated by rhyolitic to icelanditic lava flows. The major rock type of the Upper Manitou River gabbro is a medium- to coarse-grained, massive, ophitic to intergranular, mildly granophyric gabbro. The more ophitic phases of the gabbro are typically olivine-bearing and, in the vicinity of the Sonju Lake intrusion, the rock is actually an augite troctolite. Near contacts, the gabbro may contain as much as 30% granophyric mesostasis and be hydrothermally altered, probably resulting from assimilation of volatile-rich, felsic wall rock material. In two separate areas of the Upper Manitou River gabbro, the dominant rock type is a fine-grained ferromonzodiorite (Fig. 4). This rock locally resembles an intermediate volcanic rock, however, its predominantly massive character, lack of granoblastic texture, and local coarsening suggest it is intrusive. Contact relationships with more gabbroic phases are unclear, but the strong differences in texture and composition between the gabbroic and monzodioritic phases suggest that they represent composite intrusions of distinct magmas. The range of rock types, lack of internal structure, and relative age of emplacement of the Upper Manitou River gabbro are characteristics consistent with its being comagmatic with the Lax Lake gabbro and Blesner Lake diorite of the southern BBC. However, in contrast to its southern counterparts, the Upper Manitou River gabbro is, on average, more gabbroic than ferromonzodioritic (Table 1, Fig. 6A, D).

Prior to emplacement of the Houghtaling Creek troctolite, an earlier group of layered intrusions were emplaced into the same conduit first created by the Shoepack Lake inclusion-rich diorite. Although grouped together in Figure 4, Boerboom and Miller (1994) distinguished three intrusive units: (1) the *Wilson Lake ferrogabbro,* a generally massive olivine gabbro in two orthogonal dikes and a small stock (within the stock the gabbro develops a cumulate texture and grades upward into granophyric ferromonzodiorite); (2) the *Dam Five Lake gabbronorite,* a well-layered, northeast-trending sequence of gabbronoritic to olivine gabbroic cumulates that forms most of the northwestern footwall of the Houghtaling Creek troctolite; and (3) the *Four-mile Lake ferrogabbro,* a strongly differentiated layered sequence of gabbro, gabbronorite, ferromonzodiorite, quartz monzodiorite, and melanogranophyre that occurs within the Houghtaling Creek troctolite. Although adjacent to one another, the different structural grain and cumulate compositions of the Wilson Lake and Dam Five Lake units suggest that they are separate intrusions. And although nearly conformable with the Houghtaling Creek troctolite, the composition and texture of the Dam Five Lake gabbronorite does not fit with the cumulate stratigraphy and cryptic variation defined by the troctolite. Likewise, the Four-mile Lake ferrogabbro, which is situated in the axial part of the Houghtaling Creek troctolite, is too evolved to be an in situ differentiate of the troctolite. Rather the Four-mile Lake ferrogabbro may be a composite intrusion or, more likely, a detached part of either the Dam Five Lake gabbronorite or the Wilson Lake ferrogabbro.

Except for a narrow contact zone of decussate, coarse-grained olivine gabbro, the *Houghtaling Creek troctolite* is composed mostly of well-laminated olivine-plagioclase cumulates. A keel-shaped body is implied by igneous lamination and rare modal layering, which indicate a gently dipping, asymmetric trough near the surface (Fig. 4), and by its aeromagnetic anomaly (Fig. 2), which indicates a deeply rooted intrusion. Limited cryptic and modal variations occur over 300–500 m of stratigraphic section, which can be inferred from the attitude of lamination and layering across the body. Over this thickness, cumulus olivine ranges from Fo_{68} near the northern margin to Fo_{57} at the highest stratigraphic level and interstitial pyroxene and oxide increase in modal abundance. This is approximately the same compositional and modal variation observed in the troctolitic part of the Sonju Lake intrusion. If these two bodies evolved in a similar fashion, this would suggest that an upper gabbroic part of the Houghtaling Creek troctolite has been eroded away. Again, the narrow thickness over which the Four-mile Lake ferrogabbro is strongly differentiated is not consistent with its being the upper differentiate of the Houghtaling Creek troctolite with its broad cryptic variation. Taking the similarity of the two cumulate systems a step farther, it seems reasonable to speculate that the Houghtaling Creek troctolite represents a feeder dike to the southeast-dipping sheet of the Sonju Lake intrusion.

The youngest intrusive phase of the northern BBC is the *Beaver River diabase* and its composite intrusions. This unit projects north from the southern BBC as two prominent dikes that converge into one in the vicinity of Sec. 11, T59N, R6W (Fig. 4). Although characteristically composed of ophitic olivine gabbro as observed to the south, the Beaver River diabase in the northern BBC is distinctive in two significant ways. First, it only rarely contains anorthosite inclusions. This may be due to the fact that the northern Beaver River bodies are steep-walled dikes, instead of sheets where inclusions could have settled or floated to the margins. A second noticeable difference is the multiple composite intrusions in the axial portion of the dikes, particularly the eastern branch. Composite intrusions, such as the Silver Bay intrusions, are present in the southern BBC, but these tend to be single-pulse, widely dispersed composite bodies in the upper parts of thick diabase sheets. Within the axial part of the eastern Beaver River diabase dike, however, composite intrusions extend over 15 km from the northeastern part of the southern BBC (Fig. 3) to the convergence point of the two branches and contain rocks formed by at least three major

intrusive pulses. The first pulse into both branches of the dike formed the main ophitic olivine gabbro that gradually coarsens inward from a chilled margin with volcanic rocks and earlier intrusions. A second pulse created a medium-grained, intergranular olivine gabbro in all of the eastern and part of the western arms (Fig. 4). In larger diabase sheets to the south, this intergranular phase was also commonly observed, although at this earlier stage of mapping, it was not recognized as a distinct intrusive phase of the diabase. The composite nature of the intergranular gabbro is clearly evident in the northern BBC, however, by the presence of abrupt, transgressive contacts between medium-grained intergranular olivine gabbro and medium- to fine-grained, ophitic olivine gabbro/diabase. Mineral composition changes across these contacts are as much as 25% Fo in olivine and 10% En in augite, both decreasing in the intergranular gabbro. A third pulse created a tear-drop-shaped body of massive ferromonzodiorite grading inward to a well-laminated olivine ferrodiorite in the thickened southern end of the eastern dike (Figs. 3, 4). This phase resembles the large zoned bodies of the Silver Bay intrusions to the south (Fig. 3). In one location (center of T58N, R6W), the ferrodiorite transgresses both the intergranular gabbro and the early ophitic gabbro. Other minor lithologies present in the eastern dike include a medium-grained ophitic gabbro with poikilitic olivine that commonly occurs between the ophitic and intergranular gabbro phases and a leucocratic olivine gabbro that occurs between the intergranular gabbro and the diorite body. These rock types may represent smaller discrete pulses of distinct composition or may be locally differentiated subfacies of the three main phases. In any case, the multiple composite nature of these intrusions is quite evident in this part of the BBC.

Eastern Beaver Bay Complex

The nearshore exposures of hypabyssal intrusive rocks extending from Tofte to Grand Marais and inland some 5–9 km have been investigated by Green (1972, 1982a, 1992), but have not been mapped in the same detail as the southern and northern BBC. And although not previously grouped with the other intrusions of the BBC, aeromagnetic data (Fig. 2) demonstrate the continuity of intrusive units from the northern BBC to this area. Moreover, the rock types present in this eastern area are identical to those found in other areas of the BBC. The eastern BBC is composed of two major intrusive units—the Leveaux porphyry/Bally Creek trachybasalt and the Beaver River diabase with its trademark anorthosite inclusions (Fig. 5). These two units were emplaced adjacent and subparallel to one another, however, chilled intrusive contacts and inclusion relationships indicate an older age for the Leveaux porphyry.

The *Leveaux porphyry* (Green, 1972, 1992) occurs as a thick (50–100 m), southeast-dipping (10–20°) sill of plagioclase porphyritic ferrodiorite forming prominent cuesta-like hills above the shoreline between Tofte and Lutsen (Fig. 5). Also, small outliers of the Leveaux porphyry form Bear and Gull

Islands 3 km southwest of Schroeder. The sill is composed of an upper porphyritic and a lower aphyric facies, which grade into one another over a 1- to 3-m-thick interval (Green, 1972). The upper porphyritic half contains as much as 40% coarse (1–2 cm) labradoritic plagioclase phenocrysts in a brown, fine- to medium-grained matrix composed of felty plagioclase, subprismatic augite, coarse Fe-Ti oxide, acicular apatite, and a 5–15% granophyre mesostasis. Locally, olivine or pigeonite is also present. The blocky to rectangular labradorite phenocrysts are typically unaltered and show subtle compositional zoning (An$_{63-55}$). The lower aphyric portion is identical in mineralogy and texture to the matrix of the porphyritic unit. Green (1972) suggested that the two facies may have formed by plagioclase phenocryst flotation subsequent to emplacement. The *Bally Creek trachybasalt* is a locally plagioclase porphyritic, intergranular, granophyric pigeonite ferrodiorite that occurs in a 30-m-thick sill north of Grand Marias (Green, 1972). It is compositionally similar to the Leveaux porphyry (Fig. 5) but cannot be structurally connected to it. The nearly identical mineralogy, texture, and composition of the Leveaux porphyry, Bally Creek trachybasalt, and the Cabin Creek porphyritic diorite and their intrusive-stratigraphic relationship to other BBC units strongly suggest a comagmatic relationship.

In the eastern BBC, the *Beaver River diabase* generally has the same lithologic and structural characteristics found throughout this most extensive of BBC intrusions, though mapping is insufficient to delineate possible composite intrusions evident elsewhere. The aeromagnetic anomaly pattern, the map distribution, and the shallow attitude of rare olivine layering indicate that the diabase intrusion is a subhorizontal to slightly south-dipping sheet fed by a feeder dike along its northern margin. Abundant anorthosite inclusions are found at the southwestern end of the sheet near Tofte. The most notable of these inclusions is the 300-m-wide block that holds up Carlton Peak. Projection of the diabase east of the R3W–R2W township line (Fig. 5) is uncertain. The aeromagnetic anomaly associated with the diabase strikes due east (dashed line) to where it blends in with the positive anomaly associated with the Schroeder-Lutsen basalts. This may indicate that the diabase sheet projects beneath these basalts or that the diabase and basalts are juxtaposed by faulting. Unfortunately, exposure is lacking in this area. A thin, conformable sill of ophitic diabase that is similar to the Beaver River diabase but lacks anorthosite inclusions, forms Murphy Mountain, north of Terrace Point (Fig. 5). At Grand Marais and on trend with the Murphy Mountain sill, ophitic diabase is found in complex contact with felsic lavas (Green, 1972, 1982a).

PETROLOGY OF BEAVER BAY COMPLEX MAGMAS

The texture and mineralogy of most rocks comprising BBC intrusions imply that their whole rock compositions approximate the magma compositions from which they crystallized, a characteristic consistent with their hypabyssal setting. Obvious exceptions to this are the cumulates of Sonju Lake intrusion, the Houghtaling Creek troctolite, and some Silver Bay intrusions; the plagioclase porphyritic phases of the Cabin Creek and Leveaux porphyritic diorites; and the olivine enriched portions of the Beaver River diabase. Nevertheless, the parent magma compositions of most BBC intrusions can be straightforwardly estimated from average whole rock compositions of the more homogeneous intrusions (e.g., Finland granophyre, Victor Head diabase, aphyric phase of the Leveaux porphyry, and Cabin Creek porphyritic diorite) or from the average compositions of noncumulate marginal phases of the various composite or differentiated intrusions (e.g., Lax Lake gabbro, Blesner Lake diorite, Upper Manitou River gabbro, Beaver River diabase, Silver Bay intrusions). These estimates are summarized in Table 1. Because the picritic basal margin of the strongly differentiated Sonju Lake intrusion does not represent a liquid composition, its parent magma composition has been calculated from the weighted averages of 82 whole rock analyses profiling its 1,200-m thickness. Although the Houghtaling Creek troctolite has a marginal phase of coarse-grained, decussate olivine gabbro, its bulk composition appears to represent a mixture of magma and cumulus minerals; for example, TiO$_2$ < 0.5%, and Mg# > 67.

Excluding the uniquely felsic Finland granophyre, the range of BBC parent magma compositions (Table 1) is similar to the olivine tholeiite and transitional basalt compositions that dominate the NSVG (Fig. 6A). The tightly clustered trend of BBC parent magma compositions evident on an AFM diagram (Alk-Fe-Mg; Fig. 6A) and the systematic variation of other elemental abundances suggest that all mafic BBC magmas evolved from a common olivine tholeiitic primary magma type. (The obvious deviation of the marginal gabbro phase of the Blesner Lake diorite, bld, Fig. 6A, from the trend defined by other BBC compositions reflects the strongly altered and alkali metasomatized character of this unit.) Such a primary composition, which is approximated by the most primitive, high-Al olivine tholeiites of the NSVG (Table 1), is thought to have given rise to most MRS magmas, especially in later stages of magmatism (Green, 1983; Miller and Weiblen, 1990; Klewin and Shirey, 1992). That even the most primitive of the BBC intrusions, the Beaver River diabase, is significantly evolved from a primitive olivine tholeiite composition (Table 1) indicates that all BBC parent magmas were generated in turn by magmatic differentiation of such a primary composition in deeper staging chambers. Petrologic modeling of some BBC intrusions and other hypabyssal bodies that intruded the NSVG (Jerde, 1991) suggests that most magmas experienced multistage, polybaric fractionation between their extraction from the mantle and their subvolcanic emplacement. Although the presently available radiometric ages do not indicate an overlap of magmatic activity between the BBC and the Duluth Complex (Paces and Miller, 1993), additional dates and more detailed petrologic studies may show that some Duluth Complex intrusions acted as the final levels of staging and differentiation of some BBC-bound magmas. Such consanguinity seems most apparent between the anorthositic series of the Duluth Complex and the porphyritic diorites of the BBC.

The intrusive-stratigraphic variations of BBC mafic rocks implies a general progression toward less evolved parental magma compositions with successively younger intrusions (Fig. 6A; Table 1), a trend also recognized in the lithostratigraphic variations of NSVG compositions (Green, 1983). Excluding granitic rocks and composite intrusions, the early intrusions of porphyritic diorite are the most evolved parental compositions (Mg# 24–32) in the BBC. Although the next cycle of mafic intrusions produced a compositionally diverse suite of gabbroic to monzodioritic rocks (Lax Lake gabbro, Blesner Lake diorite, and Upper Manitou River gabbro), initial intrusions were of less evolved gabbroic composition (Mg# 40–46). The next mafic intrusion cycle is characterized by the moderately evolved parent magma of the Sonju Lake intrusion (Mg# 49), the Victor Head diabase (Mg# 46), and perhaps the Houghtaling Creek troctolite, presuming its parental composition is similar to the Sonju Lake intrusion. The last intrusive cycle began with the intrusion of the least evolved Beaver River diabase (Mg# ~57). The significance of this compositionally inverted sequence of intrusion is probably the same as for comparable variations in the NSVG, namely, that it reflects progressively shorter crustal residence times of magmas in their passage from the mantle to the upper crust.

The variation of rock types within individual mafic intrusive units is the result of combinations of three processes—internal differentiation, assimilation, and composite intrusion. All intrusions experienced some degree of internal magmatic differentiation during crystallization, although for most, this was accomplished by bouyant migration of residual magma out of marginal solidification zones in conductively cooled bodies. This type of differentiation resulted in gradual changes in mineralogy (e.g., increased modes of Fe-Ti oxide and granophyre at the expense of mafic silicates), texture (e.g., transition from ophitic to intergranular), and composition (e.g., increases in Si, Fe, Ti, Na, K, and incompatible trace elements). These gradational changes are evident in the early gabbroic intrusions (Lax Lake gabbro, Blesner Lake diorite, and Upper Manitou River gabbro), the dikes and sills of the Beaver River diabase, and more subtly, in the early porphyritic ferrodiorite bodies and the Victor Head diabase. More extreme differentiation driven by crystal fractionation is evident in the intrusions hosting cumulate rocks—the Houghtaling Creek troctolite, the layered ferrogabbro bodies of the Silver Bay intrusions (Shank, 1989), and, most notably, the Sonju Lake intrusion. Stratigraphic variations in cumulus mineralogy and composition through the 1-km-thick Sonju Lake intrusion indicate that strong magmatic differentiation was driven by closed-system, fractional crystallization and accumulation of cumulus minerals at the floor of the sheetlike body. Calculations of the liquid line of descent of the Sonju Lake intrusion indicate a Fenner trend of differentiation characterized by strong iron enrichment (Fig. 6B), typical of shallow,

tholeiitic mafic intrusions (Wager and Brown, 1968). As discussed above, although the quartz ferromonzodiorite and granite of the Finland granophyre appear to have compositions consistent with their being upper differentiates of the Sonju Lake intrusion (Fig. 6B), isotopic data and volume estimates indicate that such comagmatism is improbable.

Another source of compositional diversity within several BBC intrusive units is composite intrusions of progressively more differentiated magma compositions. Such a multiple intrusive process is implied by abrupt lithologic variations among the early gabbroic intrusions, but this phenomenon is most definitely evident with the emplacement of Silver Bay intrusions into medial portions of Beaver River diabase dikes and sheets. Although the range of rock compositions comprising the Beaver River and Silver Bay intrusive suites defines a continuum suggestive of progressive internal differentiation (Fig. 6C), such a continuum is rarely evident when composite intrusions are viewed individually. Most contacts between ophitic olivine diabase and monzodioritic rocks of the Silver Bay intrusions are sharp and characterized by a dramatic change in texture and mineralogy and a marked compositional hiatus. For example, three stages of intrusion evident in one small outcrop area (<15 m^2, T58N, R6W, Sec. 15) display jumps in olivine and pyroxene composition from Fo_{54} / En_{69} in ophitic diabase, to Fo_{29} / En_{57} in intergranular gabbro, to Fo_{18} / En_{40} in ferrodiorite. The range of compositions defined by noncumulate rocks of the various Silver Bay intrusions (SBI margin, Fig. 6C; average composition, Table 1) implies that composite intrusions were emplaced at many different times over an extended period. Petrologic modeling (Shank, 1989; Chalokwu and Miller, 1992) demonstrates that most of the compositional continuum defined by the Beaver River–Silver Bay composite system can be accounted for by fractional crystallization (in a deeper staging chamber) of a single parent magma corresponding to the most primitive Beaver River diabase compositions (Table 1).

A third source of compositional diversity within many BBC intrusions is assimilation of felsic melts. Most such melts appear to have been generated by anatexis of somewhat hydrated volcanic footwall rocks, although in at least one instance where Beaver River diabase intrudes the Sonju Lake intrusion (Fig. 3), felsic melts were also derived from older, incompletely crystallized intrusions. Because of strong viscosity contrasts, assimilation of felsic melt by mafic magmas was commonly incomplete as now manifest by irregular segregations of granophyre in hydrothermally altered gabbroic to dioritic rocks. Nearly all BBC intrusions show some evidence of felsic contamination, with the most extreme example being the heterogeneous Shoepack Lake inclusion-rich diabase. Where felsic melts have been more thoroughly mixed into the mafic intrusions, it is difficult to discriminate such contributions from those generated by in situ magmatic differentiation or by composite intrusions of highly evolved magmas generated in deeper systems. For example, it is impossible to estimate from field relations and geochemistry how much of the granophyre forming the roof zone of the Sonju Lake intrusion was contributed by melting of the Finland granophyre versus that generated by fractional crystallization of the layered intrusion (Fig. 6B). A different problem is posed by the strongly altered quartz ferromonzonitic rocks that grade inward from ferrodioritic to gabbroic rocks in the Lax Lake gabbro and Blesner Lake diorite (Miller, 1988; Miller et al., 1993). Because magmatic differentiation alone cannot account for the exceptional scatter of compositions characterizing these bodies (Fig. 6D), it is likely that much of the quartz ferromonzonite formed by infiltration of hydrous felsic melts from the volcanic footwall into the core of the body. Overall, the varied lithologies of the early gabbroic intrusions seems to reflect, most completely, the effects of all three processes that have given rise to the compositional diversity of BBC magmas.

STRUCTURAL STYLE OF EMPLACEMENT

The shapes of BBC intrusions differ noticeably over the arcuate extent of the BBC. Whereas dikelike bodies are most common in the northern BBC, a mix of dikes and sheetlike intrusions, the latter especially common among the earlier intrusions, are present in the southern and eastern areas. One explanation for this difference in intrusion shape is that the northern BBC represents a deeper level of exposure such that only the feeders to higher level sheetlike bodies are exposed. However, another possible reason for these differences may be related to the control of deeper crustal structure on the location and style of emplacement. Such a possibility is implied by geologic and geophysical evidence which indicate that a large mass of pre-Keweenawan crust lies close to the surface below the northern BBC.

In the pioneering geophysical study of western Lake Superior, White (1966) concluded that the saddle in the Bouger gravity over northeastern Minnesota (Fig. 2) represents an extension of a basement ridge that projects northward from the Bayfield Peninsula and acted as a divide to two thick basins of lava accumulation. Recently, Allen's (Allen et al., 1994) interpretation of integrated gravity, aeromagnetic, and seismic data over western Lake Superior confirmed White's conclusion that a buried crustal ridge of granitic composition lies beneath the Bayfield Peninsula, which they termed White's Ridge, and also identified a previously unrecognized block to the southwest of Isle Royale, which they termed the Grand Marais Ridge (Fig. 7). Their study did not closely evaluate the geophysical data over northeastern Minnesota, however. On that score, Davidson (1972), in attempting to explain the abundance of granophyre bodies in the roof zone of the Duluth Complex, speculated that the gravity saddle represented the buried projection of the granitic rocks of the Archean Giants Range batholith (Fig. 7). Modeling gravity and magnetic data along an east-west profile transecting the southern BBC, Ferderer (1982) interpreted a shallow, low-density crustal block just offshore of

Little Marais. Most recently, Chandler (1990) concluded that the saddle represents a structural thinning of Keweenawan intrusions over granitic crust, a preponderance of low-density anorthositic rock, or both. More direct geologic evidence that older crust exists at shallow to moderate depths beneath the northern BBC has been the recent discovery of migmatite, granite gneiss, amphibolite, and biotite schist inclusions in Shoepack Lake inclusion-rich diorite occurring in the Wanless area (Fig. 4; Boerboom, 1994; Boerboom and Miller, 1994). We conclude, therefore, that not only is a basement ridge, most likely the southeastern extension of the Giants Range batholith, the probable cause of the northwest-trending gravity minimum, but also suggest that this feature exerted significant structural control on the focus of both volcanic and intrusive activity in northeastern Minnesota. We term this crustal structure the Schroeder-Forest Center crustal ridge (S-FC, Fig. 7). It is interesting to note that this structural feature may have also affected Early Proterozoic geology as evidenced by the thinning of the Biwabik and Gunflint iron-formations toward their truncation by the Duluth Complex (Fig. 7; Morey, 1972a, b).

Several lines of evidence confirm White's (1966) conclusion that a basement high separated the accumulation of the many NSVG lava flows into two basins or at least into an asymmetric one. Although the gentle, arcing dips of lava flows exposed along the North Shore seem to define a single, saucer-shaped basinal structure, the volcanic sequences forming the northeastern and southwestern limbs of the NSVG basin (Fig. 1A) do not obviously correlate, especially where the sequence crosses the BBC (Green, 1972, 1983). Our detailed mapping confirms that the volcanic stratigraphy cannot be directly traced across the BBC. Grout et al. (1959) noted that the number of flows in the southwestern limb is more than three times the number in the northeastern limb. Moreover, cumulative thickness estimates of NSVG flows lying between two major time boundaries—(1) the major normal-reverse magnetic polarity reversal below, and (2) the unconformity with the Schroeder-Lutsen basalts above (Fig. 1A)—indicates that the southwestern segment is at least twice the thickness of the northeastern segment (Green, 1972). Jirsa (1984) documented an increased occurrence of interflow conglomerate and sandstone units between Lutsen and Silver Bay, which he attributed to moderate to locally significant relief during volcanism. Felsic volcanic rocks, perhaps indicative of crustal melting or diminished magmatic activity, are notably abundant in the area overlying the proposed crustal ridge in the northern BBC (Miller et al., 1994; Boerboom and Miller, 1994).

Geologic and geophysical evidence also suggest that the Schroeder–Forest Center crustal ridge had a strong effect on the form and distribution of BBC intrusions, mainly by deflecting magmatism into two intrusive "basins." The narrow focusing of northeast-trending dikes in the northern BBC suggests that repeated magma intrusions followed the same narrow conduit through the lower crust. This conduit may mark the abrupt southeastern margin of the proposed crustal ridge or perhaps may follow an early formed and often-used fault through the ridge (Figs.

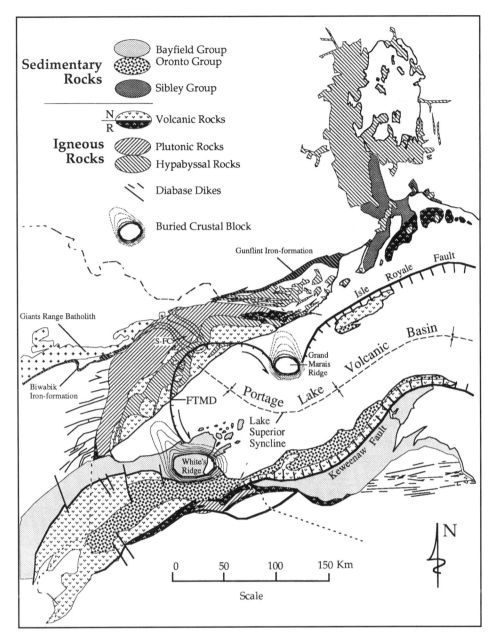

Figure 7. Keweenawan geology and tectonic interpretation the western Lake Superior region showing the extent of the Portage Lake volcanic basin (modified after Allen et al., 1994). Heavy hatched lines denote growth fault boundaries of the Portage Lake volcanic basin with length of hatchures indicating relative amounts of displacement. Note that the throw on the Finland tectono-magmatic discontinuity (FTMD) is considerably less than on the Keweenawan and Isle Royale structures and that a component of strike-slip is probable along the southeast-trending extensions of the discontinuity. The positioning of White's ridge and Grand Marais crustal blocks are based on modelling of aeromagnetic, gravity, and seismic data by Allen et al. (1994). The heavy upper contours of these crustal blocks represent areas of no volcanic rock cover. The position and form of the Schroeder– Forest Center crustal ridge (S-FC) is schematically inferred from gravity and geologic data presented in this report. The significance of the positions of the Archean Giants Range batholith and the Lower Proterozoic Biwabik and Gunflint iron-formations are discussed in the text.

7, 8B). The aeromagnetic anomaly map of the area (Fig. 2) shows that the narrow anomalies associated with the northern BBC dikes merge into two broad areas of cusped anomaly patterns to the northeast and southwest. These magnetic anomalies correspond to two Bouger gravity highs of greater than 50 milli-gals (Fig. 2). The northern anomaly area is known to contain thick mafic intrusions and granophyre (Davidson, 1972; Green, 1982a), but the details of the geology and the relationship of these intrusive bodies to the BBC and Duluth Complex are poorly understood.

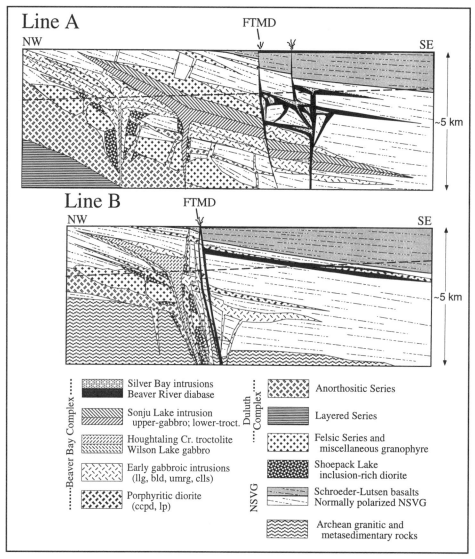

Figure 8. Schematic interpretive cross sections across the southern (Line A) and northern (Line B) Beaver Bay Complex showing the relative structural and intrusive relationships of the major volcanic and intrusive units. Cross section traces roughly correspond to Lines A and B in Figure 1A. The tops of the cross sections portray the paleosurfaces at the time of Schroeder-Lutsen basalt/Beaver River diabase magmatism (~1,096 Ma). The dashed lines indicate the approximate present erosion levels, but do not take into account the effects of post-magmatic faulting.

Gravity and magnetic modeling over the southern anomaly area, of which the southern BBC forms the eastern flank, indicate that mafic intrusive rocks extend to depths of between 10 km (White, 1966) and 20 km (Ferderer, 1982). Strategic shallow bedrock drilling in the unexposed area of the southern anomalies (Meints et al., 1993) indicates that, at the surface, the area is composed of a nested layered complex of sheetlike mafic intrusions and lensoidal granophyre bodies, termed the Cloquet Lake layered series (Figs. 2, 8A). Within this layered series, the anomaly patterns indicate that successively younger intrusions were emplaced above and east of previous intrusions (Fig. 8). The nested mode of emplacement implied by the Cloquet Lake layered series is also evident in southern BBC intru-

sions, which occupy the eastern flank of the southern anomaly. The Sonju Lake intrusion appears to have separated the Lax Lake gabbro and Finland granophyre from similar rocks now forming its northern footwall and from unexposed rocks to the west (Figs. 3, 8A). Subsequently and farther east, the Beaver River diabase and its composite Silver Bay intrusions truncated the eastern extension of the Sonju Lake intrusion and displaced the Blesner Lake diorite from the Lax Lake gabbro.

With the possible exception of the Houghtaling Creek troctolite, an eastward migration of intrusion is also evident in the dikes of the northern BBC though over a more narrow width (Figs. 4, 8B). The aeromagnetic anomaly associated with the Houghtaling Creek troctolite appears to either merge into the lower section of the Cloquet Lake layered series or be cut by it (Fig. 2). If either case were true, this would argue for the Houghtaling Creek being older than the early gabbroic intrusions. On the other hand, this appearance of vergence may instead reflect continuity between Cloquet Lake layered series and the older gabbroic intrusions (Dam Five Lake gabbronorite, Wilson Lake ferrogabbro), which the Houghtaling Creek troctolite intrudes. Moreover, the anomaly of the troctolite does not seem to be affected by the northwest-trending intrusions of the Upper Manitou River gabbro (Figs. 2, 4), which might be expected were the gabbro younger than the troctolite. And as mentioned previously, the composition and internal structure of the Houghtaling Creek troctolite is more consistent with its being correlative with the Sonju Lake intrusion. Age dating may resolve this problem.

The eastward and up-section migration of successive BBC intrusions (Fig. 8) has several important implications for the tectonic evolution of the BBC and the MRS. The shift in the focus of intrusion toward the rift axis may indicate a shift in the location of the mantle source or deep crustal staging chambers feeding the BBC or, more likely, it is a consequence of plate separation over a fixed source of magma. Similarly, the emplacement of sheetlike BBC intrusions into progressively higher levels of the volcanic pile can be explained in a couple of ways. Assuming that depth of emplacement is controlled by the level at which hydrostatic equilibrium is attained (Bradley, 1965), migration-up stratigraphy could be attributable to the lower density of successively more primitive magmas or may represent magma ponding at a constant depth in a continuously thickening volcanic pile. A complicating factor to evaluating the controlling process on the depth of emplacement is the inhomogeneity of the volcanic country rock. Given that many sheet intrusions were emplaced beneath felsic lava flows or, in the case of the Sonju Lake intrusion, a granophyre body, the volcanic stratigraphy appears to exert significant local control within a more general depth range of approximate hydrostatic equilibrium.

TECTONIC SIGNIFICANCE OF THE BEAVER BAY COMPLEX

Within the last decade, acquisition of deep seismic reflection data across Lake Superior, radiogenic isotope data, and high-resolution U-Pb ages of Keweenawan volcanic and intrusive rocks have greatly improved our understanding of the tectono-magmatic evolution of the MRS. The general geodynamic model emerging from this new database (Cannon, 1992) concludes that the MRS is a plume-generated intracontinental rift that evolved in several stages. Early stages of volcanic activity between 1,109 and about 1,100 Ma were characterized by slow to moderate rates of flood basalt volcanism into broad, gently subsiding basins. Isotopic data suggest that basaltic magmas were formed from mixtures of plume and lithospheric mantle melts (Klewin and Shirey, 1992; Nicholson and Shirey, 1990). Beginning with the emplacement of the main phase of the Duluth Complex at 1,099 Ma, the rates of intrusion/erup-

tion and rift subsidence dramatically increased and, throughout most of the MRS, volcanism became focused into central asymmetric grabens. Magmas generated during this main stage of igneous activity were derived from high degrees of melting from an undepleted mantle plume (Paces and Bell, 1989; Nicholson and Shirey, 1990). In the western Lake Superior region, this phase of volcanic activity is represented by the Portage Lake Volcanics (Davis and Paces, 1990). Around 1,094 Ma, volcanic activity began to wane and eventually ceased by 1,086 Ma. Continued subsidence with the thermal collapse of the mantle plume resulted in the rift becoming infilled by fluvial and lacustrine sediment that locally reached thicknesses of 8 km. The final phase of MRS tectonism involved the creation of a central horst throughout most of the rift by reverse motion along originally normal, graben-bounding faults, with as much as 10 km of vertical displacement. The cause of this horizontal maximum principal compressive stress has been attributed to the increased influence of Grenvillian tectonics with the waning of dilational stresses imparted by the mantle plume (Cannon and Hinze, 1992).

Within this geodynamic model, the 1,096-Ma ages recently determined for both the Sonju Lake intrusion (1,096.1 ± 0.8) and a Silver Bay intrusion (1,095.8 ± 1.2, Paces and Miller, 1993) indicate that most BBC intrusive activity was contemporaneous with the period of prolific volcanism and graben formation in the western Lake Superior region of the rift. Davis and Paces (1990) acquired a similar age (1,096.2 ± 1.8 Ma) from a basaltic flow near the base of the nearly 5-km-thick section of Portage Lake Volcanics exposed on the Keweenaw Peninsula of Michigan (Fig. 7). Our interpretations of seismic reflection profiles across western Lake Superior (GLIMPCE line C, Cannon et al., 1989; Grant-Norpac lines 57, 53, 47, 45, and 43, unpublished data) agree with others (Thompson et al., 1990; Allen et al., 1994) that the base of the Portage Lake Volcanics correlates with the top of the NSVG in Minnesota, specifically the Schroeder-Lusten basalts. Such a correlation is consistent with the predominance of olivine tholeiite basalts in both sequences. Moreover, the local development of an angular unconformity at the base of the Schroeder-Lutsen basalts lends further support to the hypothesis that the Portage Lake Volcanics represents a major change in the geodynamic evolution of the MRS by the onset of rapid graben formation (Cannon, 1992). The existence of an unconformity at the base of the Portage Lake Volcanics has not been recognized elsewhere because the sequence is truncated against the Keweenaw and Isle Royale faults (Fig. 7). This stratigraphic correlation of volcanic units implies that the emplacement of the BBC was contemporaneous with the formation of the Schroeder-Lutsten basalts. Indeed, the composition, composite nature, and volume of the Beaver River diabase provide consistent evidence that at least this extensive intrusive component of the BBC contributed to the rapid volcanic infilling and associated structural subsidence of the Portage Lake volcanic basin during the main stage of rifting.

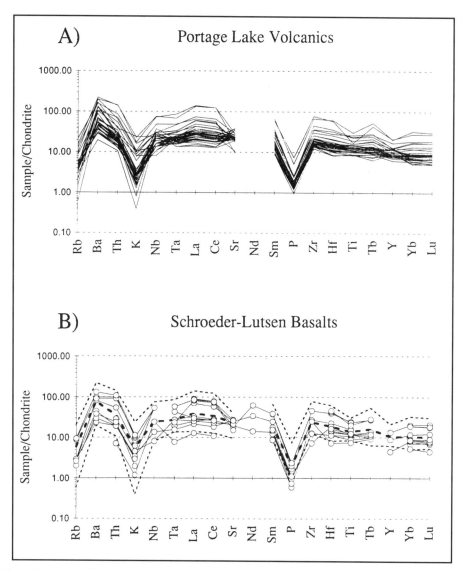

Figure 9. Incompatible trace element spidergrams comparing the chondrite-normalized compositions of (A) Portage Lake Volcanics (Paces, 1988); (B) Schroeder-Lutsen basalts (Green, 1986); (C) ophitic olivine diabase forming the margins of Beaver River diabase intrusions, and (D) various rock types from differentiated interiors of Beaver River diabase intrusions and from composite intrusions. The maximum, minimum, and average abundances of the Portage Lake Volcanic compositions are noted by dashed lines in plots B–D. All analyses normalized to C1 chondrite compositions (Sun and McDonough, 1989). Many data sets are incomplete and were derived from various laboratories by various methods.

Figure 9 shows that the geochemical variability of the Portage Lake Volcanics (Paces, 1988) is very similar to that defined by the few analyses of Schroeder-Lutsen basalts (Green, 1986). Moreover, the span of compositions defined by the Beaver River diabase and its composite intrusions fall within the same range and with generally the same trace element pattern. Because similar geochemical characteristics are shared by most Keweenawan tholeiitic mafic rocks, however, the compositional similarities of the Portage Lake Volcanics, the Schroeder-Lutsen basalts, and the Beaver River diabase allow, but do not prove, a comagmatic relationship.

Another aspect of the Beaver River diabase consistent with its acting as a conduit to surface volcanics is its multiple intrusive history. The three composite intrusions identified in the eastern dike of the Beaver River diabase of the northern BBC (Fig. 4) are distinguished by obvious breaks in rock type. However, many more subtle changes in texture, mode, and mineralogy are evident within the body that probably indicate numerous magma intrusions with only minor compositional differences between successive pulses. Petrologic modeling indicates that the parental magmas of the Silver Bay composite intrusions in the southern BBC (Figs. 4) could have been derived from further differentia-

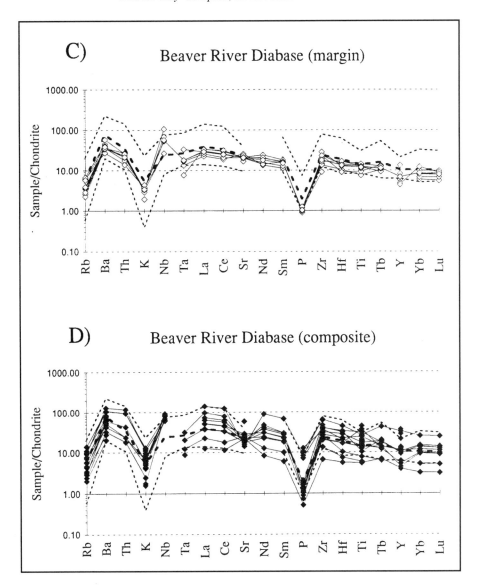

tion of primitive Beaver River diabase magma in deeper staging chambers (Shank, 1989; Chalokwu and Miller, 1992). The broad compositional range defined by these many composite bodies (Fig. 6C) suggests that their emplacement occurred in multiple events over a protracted period of time.

Finally, the large volume of magma that created the extensive dikes and sill system of the Beaver River diabase, which is more than 100 km in length, is consistent with the increased volume of volcanism represented by the Portage Lake Volcanics (Davis and Paces, 1990; Cannon, 1992). Not knowing the amount of diabase at depth nor that eroded away, the total volume of rock forming the Beaver River diabase cannot be reasonably estimated. However, given the presence of many large anorthosite xenoliths of deep crustal origin, feeder dikes as much as several hundred meters wide must have existed to mid- or lower-crustal depths at some point in their intrusive history. If the dikes indeed vented to the surface, any such volume

calculation would grossly underestimate the actual volume of magma that passed through the Beaver River diabase dikes since mafic dikes tend to collapse as magma pressure decreases (Shaw, 1980; Gudmundsson, 1990). Although many well-delineated dikes are considerably less than a few hundred meters across, incomplete exposure and uncertainties about the attitude and width of some diabase intrusions preclude a determination of whether all dikes are presently too narrow to have accommodated the larger anorthosite xenoliths. Nevertheless, it seems likely that very large volumes of magma were involved in the creation of the Beaver River diabase system.

Certain aspects and explanations of the structural style of magma emplacement are also congruous with the BBC being related to the development and infilling of a rapidly subsiding, axial rift basin. The apparent eastward migration in the focus of intrusion (Fig. 8) is in line with migration of volcanic activity into a more centrally located basin during the main phase of rift volcanism. Also, the interpretation that the upward migration of successive sheet intrusions (Fig. 8) represents emplacement at near constant depths in an ever-thickening volcanic pile is consistent with high rates of eruption and rapid subsidence that characterized the main stage of rift volcanism (Cannon, 1992). Finally, the fact that the arcuate pattern of the BBC and especially the Beaver River diabase (Fig. 1A) completely encircles the exposure area of the Schroeder-Lusten basalts is consistent with the BBC defining the structural edge of the basin into which the volcanics accumulated (Fig. 8).

The most convincing evidence that the BBC, particularly, the Beaver River diabase, acted as a basin-bounding, lava-supplying system is the rift-parallel normal faulting that attended the emplacement of the diabase dikes and sheets. The orientations of volcanic sequences intruded by the Beaver River diabase indicate that dike and sheet emplacement was accommodated by complex block faulting and rotation of the volcanic country rock on a local and regional scale. The most pronounced and obvious fault displacement is associated with the arcuate diabase dike (locally dike set) that bounds the western and northern extent of the Beaver River diabase and runs the entire length of the BBC (Figs. 1–5). This structure, termed the Finland tectono-magmatic discontinuity or FTMD (Miller et al., 1993), is easily recognizable on aeromagnetic images of the BBC because it truncates the anomaly patterns of many earlier intrusions (Fig. 2).

The sense of displacement across the FTMD is clearly downdrop of the east (rift) side, but the amount of throw seems to vary along its length. Near the southern end of the FTMD, the offset of a distinctive plagioclase-porphyritic basalt flow (Fig. 3) implies a vertical displacement of 1.5 ± 0.3 km. Farther north, the presumed correlation of the Lax Lake gabbro and Blesner Lake diorite implies vertical displacement in the range of 2.5–6.5 km based on 15–18 km of apparent horizontal offset across the FTMD and assuming the bodies have sheetlike forms that conform with the shallow (10–20°) southern dips of their bounding volcanic rocks (Fig. 3). This correlation and sense of displacement predicts, however, that the Finland granophyre and Sonju Lake intrusion should occur downsection (north) of the Blesner Lake diorite on the southeast side of the FTMD. The consistently southern dip of all structural features in the Blesner Lake diorite and the volcanics north and south of it, preclude the possibility that the lack of the granophyre and layered intrusion exposure is due to a flattening out or reversal of the regional structure. It is possible, however, that the Finland granophyre pinches out dramatically east of the FTMD or that it remains largely buried beneath the Blesner Lake diorite. A small mass of granophyre intruding the Blesner Lake diorite 3–4 km east of the FTMD (Fig. 3) may represent an apophysis of such a buried body. However, the persistent thickness and east-west strike of well-layered cumulate rocks adjacent to the FTMD indicates that a rapid thinning of the Sonju Lake intrusion is unlikely. The only hint of Sonju Lake cumulates north of the Blesner Lake diorite is a south-dipping sequence of well-laminated apatitic olivine ferrodiorite cumulates that occur

along the north margin of the Blesner Lake diorite (Figs. 3, 4). Assuming these units are correlative implies a vertical displacement on the FTMD of about 1.5–2 km and requires not only that Finland granophyre abruptly pinches out, but also that a major northwest-oriented fault zone cuts out the lower sequence of the layered intrusion (Fig. 4). The rare occurrence of a west-northwest-trending Beaver River diabase dike where such an unexposed fault zone should exist suggests that the structure was an accommodation zone to the emplacement of the two main north-northeast-trending diabase dikes (Figs. 3, 4). In the northern BBC, a normal sense of displacement across the FTMD is implied by the translation of the Cabin Creek porphyritic diorite and the Upper Manitou River gabbro, though the amount of throw is difficult to estimate because of the uncertain shape of these bodies. Volcanic units do not obviously correlative across either arm of the Beaver River diabase in northern BBC suggesting a significant but unknown amount of displacement.

We conclude that the BBC played a significant, and heretofore unrecognized, role in the tectono-magmatic evolution of the MRS in the western Lake Superior region. The eastward progression of successive BBC intrusions culminating with the emplacement of the extensive Beaver River diabase system is consistent with current tectonic models suggesting that the main stage of rift volcanism became focussed into central basins bounded by growth faults. Evidence that the Isle Royale and Keweenaw faults were the main growth faults of the Portage Lake volcanic basin is indicated in seismic profiles by a dramatic thinning of the Portage Lake Volcanics sequences outboard of these faults (Cannon et al, 1989; Allen et al., 1994). However, it must now be recognized that the Beaver River diabase system and its FTMD structure played a similar, albeit less prominent role as a growth fault at the western end of the Portage Lake volcanic basin (Fig. 7). Unlike the Isle Royale and Keweenaw structures, however, the FTMD did not experience later reverse motion, thereby preserving evidence of its role as a magma conduit to surface lavas.

Recognizing the FTMD as the western structural boundary of the Portage Lake volcanic basin fits well with the Allen et al. (1994) model of the MRS in western Lake Superior area. Based on interpretations of gravity, aeromagnetic, and seismic data, Allen and coworkers identified two large, isolated blocks of pre-Keweenawan crust, one situated beneath the Bayfield Peninsula (White's ridge) and the other southwest of Isle Royale (Grand Marais ridge), that strongly control the shape of the Portage Lake volcanic basin (Fig. 7). Their model shows that the main rift axis centered between the Keweenaw Peninsula and Isle Royale, which they termed the Lake Superior syncline, curves around the Grand Marais ridge to the northwest where it projects into the cusp of the FTMD. At its deepest, they estimate that the graben structure is filled with 18 km of volcanics, both Portage Lake Volcanics and older flows, and 9 km of sedimentary rocks (mostly Oronto Group). The volcanics pinch out against the two crustal ridges and are absent over some areas of both. White's ridge divides the Portage Lake volcanic basin from another deep trough to the southwest (the Ashland syncline), which hosts the Chengwatana Volcanic Group. The position of the Schroeder–Forest Center ridge, which Allen et al. (1994) did not recognize, apparently controlled the northwestern extent of the basin. If principal tensional stresses were oriented northwest-southeast, displacement along the FTMD where it scissors into the White's and Grand Marais ridges would be expected to be oblique, if not strike-slip. This configuration greatly expands earlier estimates of the size of the Portage Lake volcanic basin and the volume of magma that filled it (e.g., Davis and Paces, 1990).

The buried crustal blocks identified by Allen et al. (1994) may also explain the FTMD's escape from the reversed motion, which reactivated the Keweenaw and Isle Royale growth faults and created a central horst along most of the MRS. Allen et al. (1994) concluded from investigations of gravity data and seismic profiles that the main reverse faults bounding the northwest side of the central horst, the Isle Royale and Douglas (Fig. 7), are deflected and terminated along the southeast sides of these crustal blocks. Although this contradicts the Cannon et al. (1989) interpretation that the Douglas fault continues as far northeast as the implied position of the Grand Marais ridge (Fig. 7), Allen et al. (1994) had the benefit of a more complete network of seismic profiles covering the western Lake Superior area. As shown in Figures 3, 4, and 5, many northeast-trending faults cutting all units occur along the North Shore, however, these are as commonly normal as reverse (Witthuhn, this volume). One reverse fault of potentially significant displacement is implied by the occurrence of Leveaux porphyry on two islands just offshore of Schroeder (Fig. 5). Although Green (1992) suggests that the Leveaux cuts at a low angle through the Schroeder-Lutsen basalts, a 15–20° southeast dip of plagioclase lamination (Green, 1982a) implies that the sill must project beneath the basalt and therefore must be upthrust against them at Bear and Gull Islands. In any case, no reverse faults are recognized in northeastern Minnesota more than a few kilometers from the shoreline.

ACKNOWLEDGMENTS

Partial support for field mapping, petrographic services, and geochemical analyses was provided between 1985 and 1993 by the U.S. Geological Survey's COGEOMAP program (USDI-14-08-0001-A0391, A0436, A0531, A0649, A0796, A0846). Additional support was provided by State Special Appropriations from the Minnesota State Legislature to the Minnesota Geological Survey. Processing of the aeromagnetic data was accomplished by software provided by the U.S. Geological Survey. The authors wish to acknowledge the vital contributions of the many students who assisted in field mapping over the eight years of the COGEOMAP project: Susan Brink, Jim Dunlap, Dusty Early, Eric Jerde, Dan Holm, Robbie Lamons, Nancy Nelson, Rich Patelke, Dean Peterson, Colin Reichhoff, Jayne Reichhoff, Bernie Saini-Eidukat, Steve Shank, Mike Spicuzza, and Ed Venzke. Special thanks to Terry Boerboom for his field acumen and his assistance in managing much of the project, and to John Green for his guidance, assistance, and encouragement over these past eight years. Technical reviews by John Green and Bill Cannon were helpful and greatly appreciated.

REFERENCES CITED

Allen, D. J., Hinze, W. J., Dickas, A. B., and Mudrey, M. G., Jr., 1994, Geophysical investigations of the Midcontinent Rift System: A new model for western Lake Superior and northern Wisconsin: Proceedings, Annual Institute on Lake Superior Geology, 40th Part 1, Program and Abstracts, p. 1–2.

Boerboom, T. B., 1994, Archean crustal xenoliths in a Keweenawan hypabyssal sill, northeastern Minnesota. White was right!: Proceedings, Annual Institute on Lake Superior Geology, 40th, Part 1, Program and Abstracts, p. 5–6.

Boerboom, T. B., and Miller, J. D., Jr., 1994, Bedrock geologic map of the Silver Island Lake, Wilson Lake, and western Toohey Lake quadrangles, Lake and Cook Counties, Minnesota: Minnesota Geological Survey Miscellaneous Map Series, M-81, scale 1:24,000.

Bradley, J., 1965, Intrusion of major dolerite sills: Royal Society of New Zealand Transactions, Geology, v. 3, p. 27–55.

Brannon, J. C., 1984, Geochemistry of sucessive lava flows of the Keweenawan North Shore Volcanic Group [Ph. D. thesis]: St. Louis, Missouri, Washington University, 312 p.

Burnell, J. R., 1976, Petrology and structural relations of the Brule Lake intrusions, Cook County, Minnesota [M.S. thesis]: Duluth, Minnesota, University of Minnesota, 105 p.

Cannon, W. F., 1992, The Midcontinent Rift in the Lake Superior region with emphasis on its geodynamic evolution: Tectonophysics, v. 213, p. 41–48.

Cannon, W. F., and Hinze, W. J., 1992, Speculations on the origin of the North American Midcontinent Rift: Tectonophysics, v. 213, p. 49–55.

Cannon, W. F., and 11 others, 1989, The North American Midcontinent Rift beneath Lake Superior from GLIMPCE seismic reflection profiling: Tectonics, v. 8, p. 305–332.

Chandler, V. W., 1983, Aeromagnetic map of Minnesota, Cook and Lake Counties: Minnesota Geological Survey Aeromagnetic Map Series, A-1, scale 1:250,000.

Chandler, V. W., 1990, Geologic interpretation of gravity and magnetic data over the central part of the Duluth Complex, northeastern Minnesota: Economic Geology, v. 85, p. 816–829.

Chalokwu, C. I., and Miller, J. D., Jr., 1992, Differentiation of Keweenawan Silver Bay Intrusions, Beaver Bay Complex, northeastern Minnesota: Geological Society of America Abstracts with Programs, v. 24, no. 7, p. A86.

Davidson, D. M., 1972, Eastern part of the Duluth Complex, *in* Sims, P. K., and Morey, G. B., eds., Geology of Minnesota—A centennial volume: Minnesota Geological Survey, p. 354–360.

Davis, D. W., and Paces, J. B., 1990, Time resolution of geologic events on the Keweenaw Peninsula and implications for development of the Midcontinent Rift System: Earth and Planetary Science Letters, v. 97, p. 54–64.

Ferderer, R. J., 1982, Gravity and magnetic modelling of the southern half of the Duluth Complex, northeastern Minnesota [M.A. thesis]: Bloomington, Indiana, Indiana University, 99 p.

Gehman, H. M, 1957, The petrology of the Beaver Bay Complex, Lake County, Minnesota [Ph.D. thesis]: Minneapolis, Minnesota, University of Minnesota, 300 p.

Green, J. C., 1972, North Shore Volcanic Group, *in* Sims, P. K., and Morey, G. B., eds., Geology of Minnesota—A centennial volume: Minnesota Geological Survey, p. 294–332.

Green, J. C., 1982a, Geologic map atlas of Minnesota, Two Harbors sheet: Minnesota Geological Survey, scale 1:250,000.

Green, J. C., 1982b, Geology of the Milepost 7 area, Lake County, Minnesota: Minnesota Geological Survey Report of Investigations 26, 12 p.

Green, J. C., 1983, Geologic and geochemical evidence for the nature and development of the Middle Proterozoic (Keweenawan) Midcontinent Rift of North America: Tectonophysics, v. 94, p. 413–437.

Green, J. C., 1986, Lithogeochemistry of Keweenawan igneous rocks: St. Paul, Minnesota Department of Natural Resources, Division of Minerals, Project 241-4, 94 p.

Green, J. C., 1992, Geologic map of the north shore of Lake Superior, Lake and Cook Counties, Minnesota: Part 1. Little Marias to Tofte: Minnesota Geological Survey Miscellaneous Map Series, M-71, scale 1:24,000.

Green, J. C., and Fitz, T. J., III, 1993, Extensive felsic lavas and rheoignimbrites in the Keweewawan Midcontinent Rift plateau volcanics, Minnesota: petrographic and field recognition: Journal of Volcanology and Geothermal Research, v. 54, p. 177–196.

Grout, F. F., and Schwartz, G. M, 1939, The geology of anorthosites of the Minnesota Coast of Lake Superior: Minnesota Geological Survey Bulletin, v. 28, 119 p.

Grout, F. F., Sharp, R. P., and Schwartz, G. M, 1959, The geology of Cook County Minnesota: Minnesota Geological Survey Bulletin, v. 39, 163 p.

Gudmundsson, A., 1990, Dyke emplacement at divergent plate boundaries, *in* Parker, A. J., Rickwood, P. C., and Tucker, D. H., eds., Mafic dykes and emplacement mechanisms: Rotterdam, Balkema, p. 47–62.

Hutchinson, D. R., White, R. S., Cannon, W. F., and Shultz, K. F., 1990, Keweenawan hot spot: Geophysical evidence for a 1.1 Ga mantle plume beneath the Midcontinent Rift System: Journal of Geophysical Research, v. 95, p. 10869–10844.

Ikola, R. J., 1970, Simple Bouger gravity map of Minnesota, Two Harbors sheet: Minnesota Geological Survey Miscellaneous Map Series, M-9, scale 1:250,000.

Irving, R. D., 1883, The copper-bearing rocks of Lake Superior: U.S. Geological Survey, Monograph 5, 464 p.

Jerde, E. A., 1991, Geochemistry and petrology of hypabyssal rocks associated with the Midcontinent Rift, northeastern Minnesota [Ph.D. thesis]: Los Angeles, California, University of California, 305 p.

Jirsa, M. A., 1984, Interflow sedimentary rocks in the Keweenawan North Shore Volcanic Group, northeastern Minnesota: Minnesota Geological Survey Report of Investigations 30, 20 p.

Klewin, K. W., and Shirey, S. B., 1992, The igneous petrology and magmatic evolution of the Midcontinent Rift System: Tectonophysics, v. 213, p. 33–40.

Lawson, A. C., 1893, The anorthosytes of the Minnesota Coast of Lake Superior: Geological and Natural History Survey of Minnesota, Bulletin 8, Part 1, p. 1–23.

Meints, J. P., Jirsa, M. A., Chandler, V. W., and Miller, J. D., Jr., 1993, Scientific core drilling in parts of Itasca, St. Louis, and Lake Counties, northeastern Minnesota, 1989–1991: Minnesota Geological Survey Information Circular 37, 159 p.

Miller, J. D., Jr., 1988, Geologic map of the Split Rock Point NE and Silver Bay quadrangles, Lake County, Minnesota: Minnesota Geological Survey, Miscellaneous Map Series, M-68, scale 1:24,000.

Miller, J. D., Jr., and Weiblen, P. W., 1990, Anorthositic rocks of the Duluth Complex: Examples of rocks formed from plagioclase crystal mush: Journal of Petrology, v. 31, p. 295–339.

Miller, J. D., Jr., Green, J. C., and Boerboom, T. B., 1989, Geology of the Illgen City Quadrangle, Lake County, Minnesota: Minnesota Geological Survey Miscellaneous Map Series, M-69, scale 1:24,000.

Miller, J. D, Jr., Schaap, B. D., and Chandler, V. W., 1990, The Sonju Lake intrusion and associated Keweenawan rocks: geochemical and geophysical evidence of petrogenetic relationships: Proceedings, Annual Institute on Lake Superior Geology, 36th, Part 1, Abstracts, p. 66–68.

Miller, J. D., Jr., Green, J. C., Boerboom, T. B., and Chandler, V. W., 1993, Geology of the Doyle Lake and Finland quadrangles, Lake County, Minnesota: Minnesota Geological Survey Miscellaneous Map Series M-73, scale 1:24,000.

Miller, J. D., Jr., Boerboom, T. B., and Jerde, E. A., 1994, Bedrock geologic map of the Cabin Lake and Cramer 7.5-minute quadrangles, Lake and Cook Counties, Minnesota: Minnesota Geological Survey Miscellaneous Map Series, M-82, scale 1:24,000.

Morey, G. B., 1972a, Mesabi Range, *in* Sims, P. K., and Morey, G. B., eds., Geology of Minnesota—A centennial volume: Minnesota Geological Survey, p. 204–217.

Morey, G. B., 1972b, Gunflint Range, *in* Sims, P. K., and Morey, G. B., eds., Geology of Minnesota—A centennial volume: Minnesota Geological Survey, p. 218–225.

Morrison, D. A., Ashwal, L. D., Phinney, W. C., Shih, C., and Wooden, J. L., 1983, Pre-Keweenawan anorthosite inclusions in the Keweenawan Beaver Bay and Duluth Complexes, northeastern Minnesota: Geological Society of America Bulletin, v. 94, p. 206–221.

Nicholson, S. W., and Shirey, S. B., 1990, Midcontinent Rift volcanism in the Lake Superior region: Sr, Nd, and Pb isotopic evidence for a mantle plume origin: Journal of Geophysical Research, v. 95, p. 10851–10868.

Paces, J. B., 1988, Magmatic processes, evolution and mantle source characteristics contributing to the petrogenesis of Midcontinent Rift basalts: Portage Lake Volcanics, Keweenaw Peninsula, Michigan [Ph.D. thesis]: Houghton, Michigan, Michigan Technological University, 413 p.

Paces, J. B., and Bell, K., 1989, Non-depleted subcontinental mantle beneath the Superior Province of the Canadian Shield: Nd-Sr isotopic and trace element evidence from Midcontinent Rift basalts: Geochemica et Cosmochemica Acta, v. 53, p. 2023–2035.

Paces, J. B., and Miller, J. D., Jr., 1993, Precise U-Pb ages of Duluth Complex and related mafic intrusions, northeastern Minnesota: geochonological insights to physical, petrogenetic, paleomagnetic and tectono-magmatic processes associated with the 1.1 Ga Midcontinent rift system: Journal of Geophysical Research, v. 98, no. B8, p. 13997–14013.

Shank, S. G., 1989, The petrology of the Beaver Bay Complex near Silver Bay, northeastern Minnesota [M.S. thesis]: Minneapolis, Minnesota, University of Minnesota, 127 p.

Shaw, H. R., 1980, The fracture mechanisms of magma transport from the mantle to the surface, *in* Hargraves, R. B., ed., Physics of magmatic processes: Princeton, New Jersey, Princeton University Press, p. 201–264.

Stevenson, R. J., 1974, A mafic layered intrusion of Keweenawan age near Finland, Minnesota [M.S. thesis]: Duluth, Minnesota, University of Minnesota, 160 p.

Sun, S. s., and McDonough, W. F., 1989, Chemical and isotopic systematics of ocean basalts: implications for mantle composition and processes, *in* Saunders, A. D., and Norry, M. J., eds., Magmatism in the ocean basins: Geological Society of London Special Publication 42, p. 313–345.

Thompson, M. D., McGinnis, L. D., Ervin, C. P., and Mudrey, M. G., Jr., 1990, Evolution of an aborted late Precambrian continental margin beneath Lake Superior: Eos (Transactions, American Geophysical Union), v. 71, p. 1563.

Wager, L. R., and Brown, G. M., 1968, Layered igneous rocks: San Francisco, W. H. Freeman, 588 p.

Weiblen, P. W., 1982, Keweenawan intrusive igneous rocks, *in* Wold, R. J., and Hinze, W. J., eds., Geology and tectonics of the Lake Superior basin: Geological Society of America Memoir, v. 156, p. 57–82.

Weiblen, P. W., and Morey, G. B., 1980, A summary of the stratigraphy, petrology and structure of the Duluth Complex: American Journal of Science, v. 280-A, p. 88–133.

White, W. S., 1966, Tectonics of the Keweenawan basin, western Lake Superior region: U.S. Geological Survey Professional Paper, v. 524-E, 23 p.

Winchell, N. H., 1899, The geology of Minnesota: Final Report of the Geological and Natural History Survey of Minnesota, v. 4.

MANUSCRIPT ACCEPTED BY THE SOCIETY JANUARY 16, 1996

Printed in U.S.A.

Geological Society of America
Special Paper 312
1997

A structural analysis of the Midcontinent Rift in Michigan and Minnesota

Kathleen M. Witthuhn-Rolf*
Department of Geology and Geophysics, University of Minnesota, Minneapolis, Minnesota 55455

ABSTRACT

The Midcontinent Rift is a 1.1-billion-year-old crustal structure that has been defined primarily by geophysical studies. Many studies have examined the rift from a petrologic and geophysical view, but few have examined the structural geology in any detail. This study examines the structural aspects of the rift in Minnesota and Michigan and elucidates the orientations of the principal stresses during the 10 m.y. of igneous activity by analyzing arrays of small-scale faults. The orientation of slickenlines relative to each respective fault surface provides an indication of the resultant stress field exerted in an area. Faults in Minnesota reflect three regimes: pure reverse faulting during northeast–southwest contraction, strike-slip motion during northwest–southeast contraction, and normal faulting during north-northwest to south-southeast extension. In Michigan, fault array analyses of data from the Keweenaw Peninsula reflect primarily north–south contraction, with secondary results indicating some north-northwest to south-southeast extension and east-southeast to west-northwest contraction. More than half the data from Isle Royale, Michigan, record east-southeast to west-northwest contraction, with minor fault populations supporting southeast–northwest contraction and extension. This paper evaluates existing models by applying geometrical and statistical methods to the data to determine the kinematics of the closing of the Midcontinent Rift System.

INTRODUCTION

Lake Superior was first recognized as a structural basin by Foster and Whitney (1850), an interpretation expanded upon by Irving (1883). Rich copper deposits were found in the brecciated flow tops and associated interflow conglomerates in Wisconsin, Minnesota, Ontario, Canada, and in both Isle Royale and the Upper Peninsula of Michigan, focusing early geological research on the Midcontinent Rift (Fig. 1).

The Midcontinent Rift is defined geophysically by large positive gravity (Klasner et al., 1982) and magnetic anomalies. These anomalies are thought to be caused by a thickness of basalt of at least 10 km (Cannon et al., 1989). Basalts are exposed on islands in the western half of Lake Superior and around its margins in south-

western Ontario, northeastern Minnesota, northern Michigan, and northwestern Wisconsin. Toward the southwest, the basalts are covered by Paleozoic sedimentary rocks; however, gravity and magnetic studies have determined that the basalt flows extend 800 km southwest into Kansas. From the central portion of Lake Superior, the basalts also extend southeast beneath the sedimentary strata of the Michigan Basin, terminating at the Grenville front.

Seismic profiles, cores, and outcrop patterns indicate the existence of a broad syncline (Fig. 1) associated with the rift (Craddock et al., 1970). The synclinal axis follows the gravity high, which becomes greatly diminished by the effect of sedimentary cover across the Michigan Basin. West and southwest of Lake Superior, the gravity high narrows to 30 km and is flanked by two gravity lows, resulting in an anomaly approaching 160 mgal.

*Present address: Minnesota Pollution Control Agency, 520 Lafayette Road, St. Paul, Minnesota 55155-4194.

Witthuhn-Rolf, K. M., 1997, A structural analysis of the Midcontinent Rift in Michigan and Minnesota, *in* Ojakangas, R. W., Dickas, A. B., and Green, J. C., eds., Middle Proterozoic to Cambrian Rifting, Central North America: Boulder, Colorado, Geological Society of America Special Paper 312.

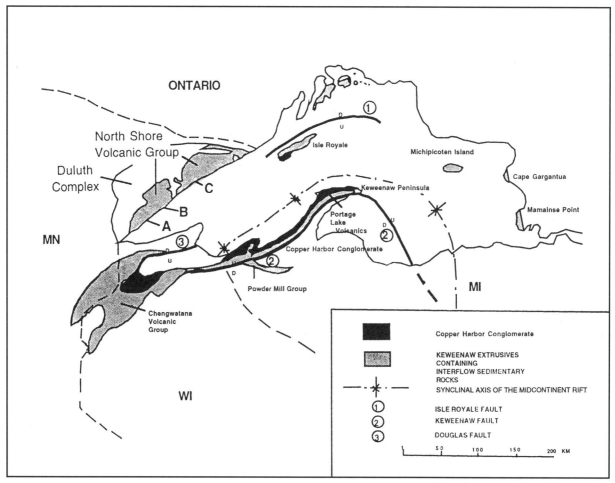

Figure 1. Areal distribution within the Lake Superior region of Keweenawan extrusives, extrusives containing interflow sedimentary rocks, and the Duluth Complex. The letters A, B, and C in along the Minnesota shoreline are the three main field areas: Two Harbors, Lafayette Bluff Tunnel, and Taconite Harbor, respectively. The figure also shows the general Minnesota and Michigan field areas, major faults, and the axis of the Lake Superior syncline.

Shortening of this originally extensional structure is indicated by large scale reverse faults bordering both limbs of the syncline (Cannon et al., 1989). In the axial zone of the syncline, the southern limb is bounded by the Keweenaw fault in Upper Michigan and the Lake Owen fault in northwestern Wisconsin, whereas the boundary of the northern limb is clearly defined geophysically by the Isle Royale fault in Lake Superior and the Douglas fault in Wisconsin (Cannon et al., 1989; White, 1966). Models offered to explain the opening and active closing of this rift structure fall into three categories: (1) a failed triple junction (Burke and Dewey, 1973); (2) convergence at the Grenville front causing extension in a north-northeast to south-southwest direction resulting in en-echelon pull-apart basins (Gordon and Hempton, 1986); and (3) the uneven docking of a plate initiating extension in a north–south direction, followed by contraction with full plate docking (Cambray and Fujita, 1991). In this chapter, small scale faults are analyzed on both limbs of the syncline, along a northwest-trending transect across Lake Superior from

the Keweenaw Peninsula to Isle Royale in Michigan and along the Minnesota north shore. A paleostress analysis is applied to the fault population (Angelier, 1979, 1984; Etchecopar et al., 1980), allowing principal stress directions to be determined, and the various tectonic models to be tested.

REGIONAL GEOLOGY

The Midcontinent Rift system cuts through older Archean metasediments, metavolcanics, and intrusives in Ontario, Canada, and Minnesota, Michigan, and Wisconsin in the United States (Dickas et al., 1989). Klasner et al. (1982) hypothesized that the orientation and presentation of the Midcontinent Rift is controlled by its presence along the boundary between the Archean granite-greenstone of the Superior Province to the north and the Penokean folded gneiss to the south. Sections of the Midcontinent Rift System in the Penokean rocks are narrower and more linear than those sections in the thinner crust of the Superior Province, possibly due to failure along multiple zones of weakness. This conclusion follows Sims (1980) definition of the Great Lakes Tectonic Zone.

STRATIGRAPHY

Geophysical studies have identified the edge of the Lake Superior basin, as defined by the thick basalt flows; these, however, are unexposed in the Southeastern portion of the basin (Cannon et al., 1989). In the Keweenaw Peninsula of Michigan, the stratigraphy (Fig. 2) consists of Middle Proterozoic basement rocks (Powder Mill Group, United States, and Osler Group, Canada) overlain by a sequence of middle Precambrian lava flows and interflow sediments (Portage Lake Volcanics, Michigan; Chengwatana Volcanics, Wisconsin; North Shore Volcanics, Minnesota), which are conformably overlain by sedimentary sequences (Oronto Group and Bayfield Groups). The Middle Keweenawan Powder Mill Group lavas, below the Portage Lake lava flows, exhibit reversed magnetic polarity in contrast to the overlying strata (Van Schmus et al., 1982).

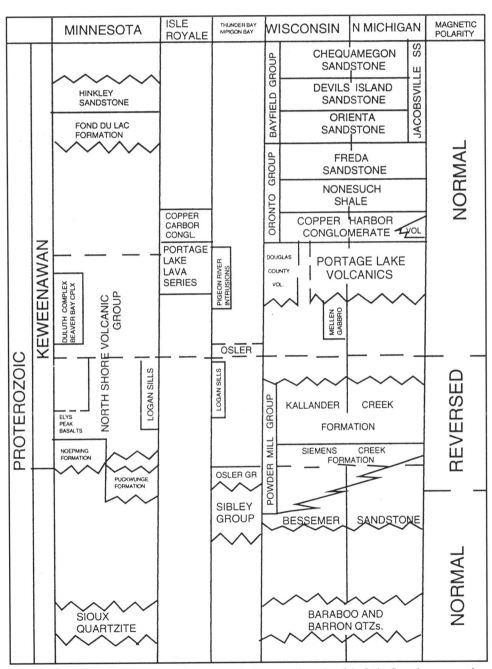

Figure 2. General stratigraphic column for Middle Proterozoic rocks of the Lake Superior area modified according to Miller (1992), showing the cogenesis of the Portage Lake Volcanics (MI) and the upper part of the North Shore Volcanics (MN). Adapted from Green, 1982.

The Portage Lake Volcanics (PLV), unconformably overlying the Powder Mills Group, are a series of tholeiitic flood basalts of middle Keweenawan age, reaching a thickness of about 10 km in the axis of the Lake Superior Basin. These volcanics are interlayered by relatively thin strata of interflow sediments (Brojanigo, 1984; Daniels, 1982) represented by immature fluvial sandstones and conglomerates, that vary in thickness from a few centimeters to a few tens of meters and extending laterally for up to several tens of kilometers. The 300 individual flows vary in thickness from a few tens of centimeters to 400 m, with an average of approximately 17 m. Several of the thicker lava flows in the Keweenaw Peninsula, such as the Greenstone flow, have a regional extent suggesting that they come from the same volcanic episode as flows on Isle Royale (Huber, 1973). The PLV contain normally polarized flows cogenetic with the uppermost lava flows seen in Minnesota. Recent precise zircon U/Pb dates (Davis and Sutcliffe, 1985; Davis and Paces, 1990; Paces and Miller, 1993; Davis et al., 1995) show that the entire magmatic history of the (exposed part of the) Midcontinent Rift System in the Lake Superior area spans the period from about 1,103 to 1,086 Ma, with two major pulses at 1,108 to 1,107 and 1,100 to 1,096 Ma. The magmatic polarity switch recorded in many of the sequences around the lake occurred at about 1,103 Ma during a period of quiescence.

PREVIOUS WORK

Geologic framework

Geologic studies in the Lake Superior region date back to the 1800s, with the most complete work, accompanied by geologic mapping of the region, accomplished by Irving (1883). A significant portion of the work in Michigan, following the discovery of copper in the early 1800s, was for the purpose of mineral exploration and evaluation of mineral resources. Logan (1863) published a discussion of Canadian geologic features in the Lake Superior region, especially the intrusive sill complexes extending from northern Minnesota through western Ontario, which are now known as the Logan sills. Grout (1918) described the Keweenawan Duluth Complex, located just northwest of the Lake Superior shore in Minnesota, as a "lopolith," introducing the term to the geologic literature. However, further studies found that the Duluth Complex did not match the definition of Grout (Wold and Hinz, 1982). The Michigan and U.S. Geologic Surveys geologically and topographically mapped the Upper Peninsula of Michigan, defining the boundaries of the series of volcanic rocks (Portage Lake Volcanics) and the sedimentary sequences (Cornwall, 1954a, 1955). Weiblen et al. (1985) gave a summary of the intrusive and extrusive igneous units in northeastern Minnesota.

Structural geology

Although the earliest recognition of Lake Superior as a structural basin was in 1850 (Foster and Whitney, 1850; Irving, 1883), it was not until 1911 that Van Hise and Leith first speculated that

rifting was at the origin of the Lake Superior Basin, and published an extensive summary of the geology of the region, accompanied by maps, which updated the work by Irving in the late 1800s (Van Hise and Leith, 1911; Leith et al., 1935). In the 1950s, workers of the Michigan and U. S. Geological Surveys conducted extensive mapping in the Keweenaw Peninsula, defining the various flows in the PLV along with their attitudes and major structures (Cornwall, 1954a, b, 1955; Cornwall and Wright, 1956a, b). Chase and Gilmer (1973) formulated a theory for the origin of the rift as due to the simple separation of Precambrian lithospheric plates (Wold and Hinze, 1982). This confirmed the application of the concept of plate tectonics to the region, originally proposed by White (1972). Huber (1973, 1983) mapped the geology of Isle Royale, including many of the gentle folds and larger faults. Mudrey (1976) conducted a petrologic and structural study of Pigeon Point, near the United States–Canadian border in Minnesota, determining the timing of faulting and folding events. Green (1982, 1983) has conducted numerous mapping and petrologic studies of the North Shore Volcanic Group in Minnesota. More recently, Miller (1992) completed more detailed work on the intrusive complexes in the northern Minnesota region, defining the boundaries between the units, and their petrologic evolution and timing of emplacement relative to the North Shore Volcanic complexes.

All of these studies are consistent with the model of the Midcontinent Rift defined by large thicknesses of basalt, cut by large reverse faults, indicating rifting, magmatism, and subsidence followed by contraction at the end of Keweenawan time.

Geophysical exploration

Geophysical studies of the Lake Superior region date back to the early 1900s, when iron-ore districts were delineated by measurements with the magnetic dip needle. Gravity techniques, implemented by Thiel (1956) and Bacon (1957), followed by magnetic studies by Thiel (1960), determined the existence of a fault trending north–south through the central portion of Lake Superior. This fault, called the Thiel fault, was later proposed to be an accommodation zone separating two major asymmetric basins within the Midcontinent Rift (Cannon et al., 1989). Early seismic studies of the area were conducted by Steinhart and Meyer (1961). Black (1955) postulated that the large positive gravity anomaly in the northern Midcontinent region represented an "en-echelon rift" pattern in the late Precambrian rocks related to the formation of the syncline. King and Zietz (1971) expanded this interpretation to describe the formation of the Midcontinent Rift System, while White (1972) proposed that a series of separate lava basins were responsible for the gravity anomaly along the rift zone. A joint effort between United States and Canadian researchers was initiated in 1962 to conduct a seismic-refraction study of the Precambrian Shield area. The complicated structure identified by this study prompted further work (Halls and West, 1971; Ocola and Meyer, 1973; Smith et al., 1966), which suggested rifting as a mechanism of crustal generation in the Lake Superior basin rather than contraction. Further geophysical work conducted over the Lake Superior region during the 1960s and 1970s was in the form of gravity studies (Canada Department of Energy, 1974; Craddock et al., 1970; Ervin and Hammer, 1974; Weber and Goodacre, 1966; Wold and Berkson, 1977), magnetic studies (Hinze et al., 1966; White, 1966; Wold and Ostenso, 1966), heat flow (Steinhart et al., 1968), and seismic reflection studies (Wold, 1979; Wold and Ostenso, 1966; Zumberge and Gast, 1961), all of which confirmed the synclinal structure of the Lake Superior Basin and approximated thicknesses of the volcanic and sedimentary rocks. Land gravity and magnetic studies of the Keweenawan rocks in the area were published by Bacon (1966), Meshref and Hinze (1970), Oray et al. (1973), and Klasner et al. (1979). Paleomagnetic studies during the past 20 years (Dubois, 1962; Palmer, 1970; Books, 1972; Green and Books, 1972; Halls, 1974) determined the polarity of the basalts, while geochronological studies were undertaken by Silver and Green (1972, 1963), Goldich (1968), and Van Schmus (1976) to obtain some of the ages for the Keweenawan Super Group. More recently, Cannon et al. (1989) worked on the Great Lakes International Multidisciplinary Program on Crustal Evolution (GLIMPCE), a seismic reflection profiling study that identified the extent of the rifted terrane and defined more clearly the thicknesses of the Keweenawan flows and sediments, particularly in the axis of the basin under Lake Superior. This study also constrained the amount of reverse motion on

the Isle Royale fault to be less than the 3-km throw estimated along the Keweenaw fault (Fig. 1).

FIELD DATA

Michigan

The study was concentrated in the Keweenaw Peninsula of Upper Michigan and in Isle Royale National Park (Figs. 1 and 3) where fault striae were both plentiful and of excellent quality. The two areas lie on opposite limbs of the Lake Superior syncline, resulting in a mirror image stratigraphy across the axis, as shown in Figure 3 (Cannon et al., 1989; Huber, 1983). Isle Royale exhibits excellent exposures along the southeast shores of the outer islands. The Keweenaw Peninsula and Isle Royale have the same stratigraphic sequence and the topography is similar in the two locations. It consists of a series of ridges, controlled by the PLV lava flows, separated by valleys of interflow sediments (Huber, 1983). The dramatic relief present on Isle Royale (230 m) and on the Keweenaw Peninsula (260 m) is due to the resistance of the massive Greenstone flow, the advancement direction of the Pleistocene glaciers scouring out the interflow sediments, and the geometry of the fractured lava flows. The exposures inland are rare and highly weathered, and the degree of fracturing makes it difficult to determine if blocks are in place. Along the Greenstone Ridge trail on Isle

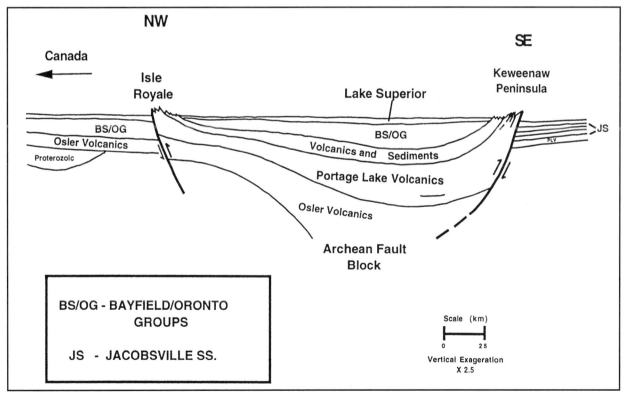

Figure 3. A northwest-southeast cross section of Lake Superior from the Keweenaw Peninsula, according to GLIMPCE profiles (see Cannon et al., 1989; Huber, 1983). PLV = Portage Lake Volcanics.

Royale, the exposures are frequent but have a rough to polished surface that, in some areas, is scoured with glacial striations. The best exposures of PLV are southeast-facing shorelines and to a lesser extent northwest-facing shorelines where, however, access proves difficult because of the high, sheer cliff faces.

The topography of the Keweenaw Peninsula is hilly but the ridge-valley sequence seen on Isle Royale is not as apparent here because the interflow sediment sequences are not as thick. Another factor influencing topography is the direction of glacial movement, which sculpted the Keweenaw Peninsula somewhat differently than Isle Royale.

Description of data. Examination of almost 600 small-scale fault structures on both of the synclinal limbs along a transect northwest across the Lake Superior syncline, from the Keweenaw Peninsula to Isle Royale in Michigan, yielded strikes with consistent bedding parallel trends (Fig. 4). The quality of fault sense criteria is excellent in most cases, with fault slickenfibers allowing for reliable macroscopic determination of fault motion.

Along the Keweenaw Peninsula on the southern limb of the Lake Superior syncline, the PLV form a broad anticline, with an axis plunging north-northwest. The lava flows in the northern one-third of the Keweenaw Peninsula trend northeast, while those in the southern two-thirds of the peninsula are east of the anticlinal

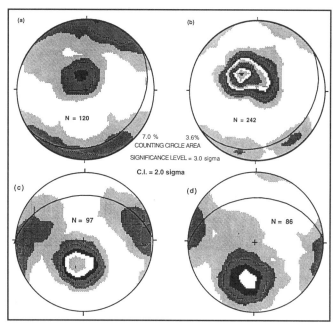

Figure 4. Top, Equal-area projection of poles-to-planes for normal (a) and reverse (b) faults on Isle Royale. Notice that although there are twice as many faults mapped as reverse than normal, both populations have similar trends. Bottom, Same type of diagram for the (c) normal and (d) reverse faults in the east-west trending Portage Lake Volcanics of the Keweenaw Peninsula. Notice that while the populations and trends of both types of faults are similar, the normal faults are more shallowly dipping. The great circle in all figures represents the average attitude of the lava flows: 055/20S (a and b) and 090/37N (c and d). N = number of faults.

hinge and trend mainly east-west. A majority of the small scale faults analyzed on the east-west trending flows of the Keweenaw Peninsula are east-west striking and contain steeply pitching striae, indicating dip-slip motion on the faults (Fig. 5a and b). However, there are two smaller fault populations trending slightly east and west of north, with gently pitching slickensides, suggesting strike-slip motion (Fig. 5a and b). Notice the large population of apparently normal faults represented in the data. These faults are present in the younger flows on the peninsula and were deposited after a significant amount of tilting by basin settling had already occurred. This is indicated by the 37° dip of the younger flows as compared to the 65° dip of the older flows. All of the faults measured from the northeast-trending PLV strike parallel to this northeast trend and contain gently plunging striae indicating strike-slip motion along these faults (Fig. 5c). Fault orientation data from Isle Royale, on the northern limb of the Lake Superior syncline, exhibit two gently dipping populations, one trending northeast and the other northwest, with striae pitching gently (Fig. 6a and b).

Minnesota

In contrast to the geology in Michigan, the North Shore of Minnesota has a more complex history. In Michigan, pulsations of volcanism alternated with periods of deposition. However, in Minnesota, these volcanic pulses were often concurrent with emplacement of intrusive complexes. If the Midcontinent region was initially in an extensional regime, these units were more easily emplaced, but not without some net stress imposed on the existing rock units. Since later contraction of the area is documented by the significant reverse motion along the Keweenaw, Isle Royale, Douglas, and Lake Owen faults (Cannon et al., 1989), it is likely this motion has been translated and modified by other activities along the North Shore of Minnesota. The undulating North Shore Volcanics (NSV) are well fractured, with quartz, calcite, and zeolite mineralized joints and faults. Slickenlines on faults in these flows are not as well developed or visible as those in the PLV in Michigan.

The orientations of 225 faults were measured from the well-exposed NSV at more than six locations along approximately 117 km (70 mi) of the northern shoreline of Lake Superior in Minnesota. These locations were from Two Harbors to Schroeder, with 71 of the faults occurring in the diabase unit, through which the Lafayette Bluff tunnel was built (Figs. 1 and 7). Since two-thirds of the data were collected from two locations separated by a distance of approximately 100 km (60 mi), the data were initially analyzed together then subsequently divided into three areas (Fig. 1), based upon geography and rock type.

Plots of the normal and reverse faults indicate a major northeast–southwest trend, with a spread to the northwest–southeast, that is, parallel and perpendicular to the shore. The ratio of normal to reverse faults is 1:1.52 (Fig. 7). This plot of the poles to fault planes also describes the east-northeast to west-southwest trending and shallow-dipping flow

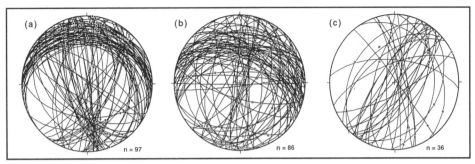

Figure 5. Equal-area plot of normal (a) and reverse (b) faults in the east-west trending Portage Lake Volcanics (PLV) and (c) the small population of faults in the northeast-southwest (045°)-trending PLV on the Keweenaw Peninsula. The pitch of the associated slickenside is indicated by the black dot (normal) or open square (reverse) on each great circle. Notice the shallowly pitching striae in (c) denoting strike-slip faulting. A significant proportion of the apparently normal faults in (a) are actually shallow thrust faults rotated by subsequent compression, as explained in the Discussion section of the text. N = number of faults.

stratification. Slickensides, or slickenlines, reflect mainly strike-slip motion along the steeply dipping faults, but dip-slip motion along the more gently dipping faults. The trend of a slickenside describes the resolved shear stress across a plane (Fig. 8). Therefore, if the slickenlines from all 225 faults are plotted up on an equal area diagram, their trends should indi-

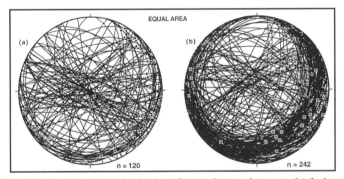

Figure 6. Lower hemispheric plot of normal (a) and reverse (b) faults from Isle Royale. Note the greater proportion of steeply dipping normal faults. The reverse faults tend to strike either northeast-southwest or northwest-southeast, as indicated by the dense lines. Black dots in (a) and white squares in (b) represent the orientation of the slickensides on each fault surface. N = number of faults.

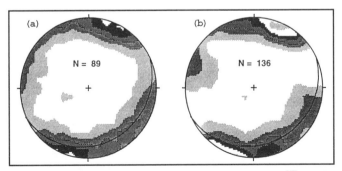

Figure 7. Poles-to-planes for (a) normal and (b) reverse faults in the North Shore Volcanics. Notice the faults are mainly steeply dipping unlike the attitude of the lava flows, which are represented by the great circle (065/17S). N = number of faults.

cate the general direction of motion during both normal and reverse faulting during Keweenawan time.

The Two Harbors field area (A, Fig. 1), composed of lava flows from Two Harbors, Split Rock River, and Beaver Bay, only yielded 33 faults; 17 are from the immediate Two Harbors area and are described in Figure 9 (bottom). The Lafayette Bluff Tunnel (B, Fig. 1) was built through a diabase unit, which would respond differently from the NSV to a set of stresses for compositional reasons. This justified the separate analysis of the 71 faults recorded there, even though the tunnel is close to the Two Harbor field area geographically (Fig. 9, center). The third and largest division (126 faults) is the Taconite Harbor area (C, Fig. 1), which is the area bounded by the Taconite Harbor railroad cut (108 faults) to the south and exposures in the Cross River, near Schroeder, to the north (Fig. 9, top).

These plots of the individual field areas show the uniformity in the strikes of faults and steep dips in field areas B (diabase) and A (NSV), but considerable scatter in the angles of dip in the Taconite Harbor railroad cut, even though the strikes of the fault population seem predominantly in the north-northeast to south-southwest and northwest–southeast directions.

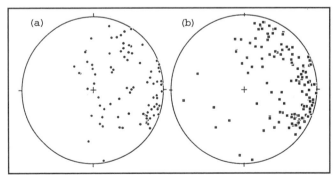

Figure 8. Equal-area plot of the orientation of slickenlines on normal (a) and reverse (b) faults in the North Shore Volcanics. Note the slickensides on normal and reverse faults have similar fields of orientation. A significant population of both normal and reverse faults trend northwest-southeast, but the normal faults have steeper dips (see Fig. 6).

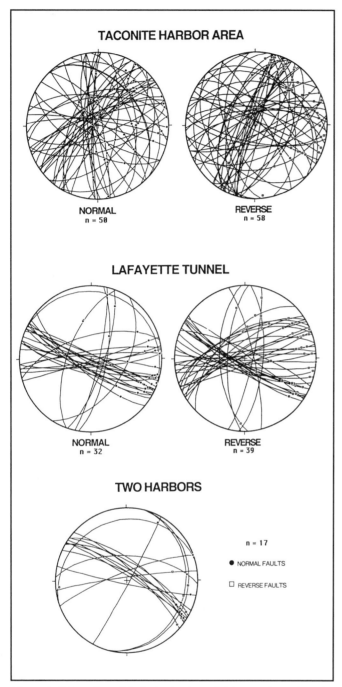

Figure 9. Orientation of faults in the three field areas along the Lake Superior shore in Minnesota. Circles and squares on the great circles denote the pitch of the striae on each fault surface. N = number of faults.

PALEOSTRESS ANALYSIS

The orientations of the principal stresses are usually determined by evaluating historical motions along major faults. In the absence of accessible major faults, it becomes necessary to use smaller scale faults to determine the paleostress orientation(s) (Petit and Green, 1987).

Etchecopar fault inversion analysis

Paleostress analysis of the data was accomplished by implementing the Etchecopar analysis program (Etchecopar et al., 1980) as adapted for the Macintosh computer. Resolving the forces (normal and shear) on the fault plane (Bott, 1959), the three principal components of the stress tensor (σ_1, σ_2, and σ_3 where $\sigma_1 \geq \sigma_2 \geq \sigma_3$) are calculated (Angelier, 1984; Lisle, 1987). This program uses a statistical analysis technique to satisfy a requested percentage of the fault population, chosen by the operator, and calculates a stress tensor consistent with that percentage of the faults, in the form of the orientations of the three principal stresses, σ_1, σ_2, and σ_3, along with the ratio R, where:

$$R = \frac{\sigma_2 - \sigma_3}{\sigma_1 - \sigma_3}$$

The program iterates through the data to find the specified percentage of data with minimal variance. The variance is defined by calculating the orientation (in a fault plane) of the maximum resolved shear stress consistent with the iterated state of stress (Angelier, 1984; Bott, 1959), and comparing the theoretical striation to that measured on the plane. This angular relationship, for all faults, is minimized until a stable stress tensor is found. Basic assumptions of paleostress inversion methods are the following: (1) the original stress regime is homogeneous on the scale of the studied area and does not allow for modifications of the stress state by the presence of preexisting faults (Angelier, 1979); (2) the direction of motion on a fault is in the direction of the maximum resolved shear stress (Bott, 1959); (3) the stress tensor is symmetric and without torque (Etchecopar, personal communication, 1981).

The initial input requirements of the program are (1) the data in a precise format, (2) the percentage of data the calculated tensor needs to satisfy.

Printed output of the Etchecopar program provides (1) trend and plunge of the three principal stresses; (2) ratio of the stresses, R, which defines the shape of the stress ellipsoid (Etchecopar et al., 1980); (3) a Mohr circle defining the precise positioning of each fault in the σ_n, σ_s space, where σ_n and σ_s are the normal and shear stresses, respectively; (4) a histogram of the variance of the measured slickenlines with the calculated ideal, listing each fault by number; (5) a stereogram with the three principal stresses plotted along with the scatter.

Populations of faults not selected by the program as a close match for the first calculated tensor can be reentered into the program and a second tensor calculated to match the remaining population. This process was continued until 95% of the data were found to be consistent with a tensor. If the deviation of each measured slickenline from the ideal, as calculated by the algorithm, is plotted as a histogram, an assessment of the fit of the data can be made. Data clustered at the left side of the histogram, where the deviation is minimal, indicates good compatibility between the computed tensor and observations.

Michigan

Isle Royale. Of the 362 faults measured on Isle Royale, only 299 could be analyzed by the Etchecopar program at one time. In order to achieve an unbiased assessment of the paleostress field, the data were divided by population at first, and then by latitude. The majority of the data from most of the subsets produce a stable east-southeast to west-morthwest maximum principal stress (σ_1) (Fig. 10, tensor #1). One hundred and ninety-six data, all with a variance of less than 0.4 to 0.5 radians (24° to 30°), were selected to be reanalyzed by the Etchecopar program, in order to "refine" the stress tensor. When these data were entered into the Etchecopar program, the isolated tensor was composed of principal stresses (σ_1, σ_2 and σ_3) with the following orientations, respectively: 101.9/10.4, 10.9/5.3, and 254.3/78.3 (Fig. 10). The R value, or ratio of the magnitudes of the three principal stresses listed above, is 0.020, indicating only a very small difference in the magnitudes of σ_2 and σ_3. When the magnitudes of the minimum and intermediate principal stresses (σ_2 and σ_3) are very close to each other, and σ_1 is subhorizontal, a transpressive stress regime results. In this regime, one would expect to see many oblique-reverse faults ranging from pure strike-slip to pure reverse included in the population of faults consistent with this stress tensor. By repeating this analysis process the change in the orientation of the far-field stresses, that is, the values of σ_1, σ_2 and σ_3, may be observed. This leads one to speculate the far-field stresses may have either changed direction, or the local expression of these stresses may have changed during the contraction of the Midcontinent Rift System.

Faults not included in the population of faults consistent with the first tensor were separated out and reanalyzed by the Etchecopar program to isolate a second tensor. Sixty-two faults with variances less than 36° were consistent with a stress tensor having principal stresses (σ_1, σ_2 and σ_3) of 144.4/05.5, 234.7/03.4, 356.2/83.6. The R-value is 0.013, again indicating that σ_2 and σ_3 are similar in magnitude (Fig. 10). With maximum principle compression in a southeast–northwest direction, a population of strike-slip to reverse faults determines this tensor.

Faults with orientations furthest from the ideal stress tensor, as determined by a large variance, were again analyzed by the same method. The results suggest that 51 data, with a variance less than 42, are consistent with an extensional regime where the minimum compressive stress (σ_3) has nearly the same orientation as the maximum principle compressive stress (σ_1) in tensor #2 (Fig. 10). The orientations of σ_1, σ_2 and σ_3 for the third stress tensor are: 303.3/66.6, 052.6/08.1, and 145.9/21.7, respectively, with an R-value of 0.470. An R-value close to 0.5 suggests that the magnitudes of the principal stresses are not similar, therefore, the stress regime here is consistent with a well-defined northwest–southeast direction of extension. Faults consistent with this tensor are mainly normal, although a significant number of faults are strike-slip.

Keweenaw Peninsula. Two-hundred nineteen faults were analyzed from the PLV on the Keweenaw Peninsula. Of these, 57 normal and 55 reverse faults were consistent with the following orientations of principal stresses (σ_1, σ_2 and σ_3): 356.7/30.5, 110.7/34.6, and 236.7/40.4, respectively. The ratio of the stresses, $R = 0.008$, indicates the magnitudes of σ_2 and σ_3 are very similar. Examination of geologic quadrangles indicate the older lava flows have steeper dips than those which are younger, indicating the strata was already tilted slightly due to subsidence of the rifted terrane (Cornwall, 1955). For this reason, all data were rotated to horizontal, with respect to the younger flows, in order to gain an accurate calculation of the stesses exerted on the terrane at that time (Fig. 11).

The remainder of the data unaccounted for by tensor #1, were reanalyzed to determine whether a second stress regime could be identified. Thirty-nine faults, 27 normal and 12 reverse-oblique (mainly strike-slip), were consistent (variance ≤24°) with principal stresses with the following orientations: 127.1/64.7, 248.4/13.8, and 343.7/20.8; and R-value of 0.904 (Fig. 11). An R-value of this dimension indicates that the magnitudes of σ_1 and σ_2 are similar, thereby defining a transtensional regime.

The third tensor was determined by reanalyzing all data (68) not included in the previous two tensors. Values for σ_1, σ_2 and σ_3 obtained for 40% (27) of these data are 283.3/11.9, 017.8/20.5, and 165.1/66, with an R-value of 0.293 (Fig. 11). The principal stresses of this tensor are also similar to tensor #1 from Isle Royale (Fig. 10).

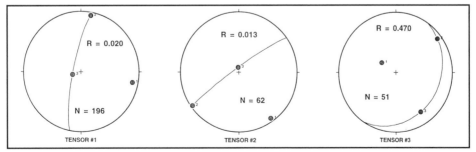

Figure 10. Calculated tensors from Isle Royale data. Note the first two tensors denote a transpressional regime. Extension in a southeast-northwest direction (the opposite of tensor #2) is represented by tensor #3. N = number of faults; R = ratio of the stresses.

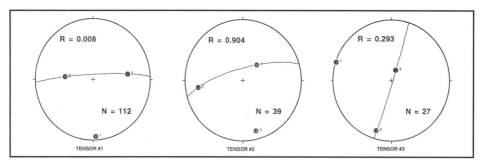

Figure 11. Equal-area, lower hemispheric projections of the stress tensors calculated from the Portage Lake Volcanics on the Keweenaw Peninsula. All values have been rotated to horizontal, with respect to the youngest lava flows. Notice tensor #1 describes north–south compression and reverse faulting, while tensor #3 closely matches tensor #1 calculated from Isle Royale data (Fig. 10). Tensor #2 is consistent with normal faulting. N = number of faults; R = ratio of the stresses.

Minnesota

Total field area. An analysis of all the data yields three major populations of faults, each of which is consistent with its own orientation of the three principal stresses. The deviation of the orientation of each slickenline from that calculated by the regressional analysis for the ideal slickenline is less than 36° for each tensor. Of the 225 faults analyzed, 98 (44%) were consistent with the following orientations of the three principal stresses (σ_1, σ_2 and σ_3): 62.4/0.4, 332.4/0.9, and 178.7/89, respectively (Fig. 12). The R-value of 0.032, indicating similar magnitudes of σ_2 and σ_3, suggests a transpressional regime and is consistent with the shore-parallel, dextral strike-slip faults consistent with tensor #1.

For tensor #2, 50 of the remaining 127 (39%) faults analyzed are consistent with a transtensional regime (R-value of 0.890) with σ_1, σ_2, and σ_3 oriented as follows: 332.9/3.6, 216.5/81.9, and 63.3/7.3 (Fig. 12). As expected, most of the faults are steeply dipping normal faults with gently pitching slickenlines, indicating strike-slip motion.

Of the remaining 77 faults, 33 (43%) are consistent with the following orientations of σ_1, σ_2, and σ_3: 112.5/78, 247.8/8.6, and 339/8.3 and an R-value of 0.204. This third tensor represent an extensional regime with σ_1 nearly vertical and similar magnitudes

of σ_2 and σ_3. These three tensors satisfy nearly two-thirds of the data for the combined field areas.

Individual field areas. In an attempt to refine the analysis, the effects of lithology and positioning of the rocks in the Midcontinent Rift region were then taken into account. For this reason, fault data from the two major field areas, the Lafayette Bluff Tunnel diabase and the Taconite Harbor railroad cut, which contain the majority (87%) of the fault data, have been analyzed separately. These data were used to extrapolate the orientations of the three principal stresses during fault formation. Three tensors were isolated from the 71 faults measured around the Lafayette Bluff Tunnel; two indicate compression in a northeast–southwest direction and extension in the northwest–southeast direction, and one describing reverse faulting by compression from the east-southeast to west-northwest. Figure 13 describes the three tensors isolated from the Lafayette Bluff Tunnel data. In tensor #1, the values of σ_1, σ_2 and σ_3 (027.9/5.9, 279.7/71.7, and 119.7/17.2) result in shore-parallel strike-slip faulting. With σ_3 positioned perpendicular to the shoreline and an R-value of 0.561, indicating unequal magnitudes of the σ_2 and σ_3 stresses, the faults are dynamically sinistral. The orientations of the three principal stresses in tensor #2 (294.5/9.5, 200.1/24.6, and 044/63) describe a transpressive regime. The R-value of 0.014 indicates the magnitudes of σ_2 and $\sigma3$ are similar. Another strike-

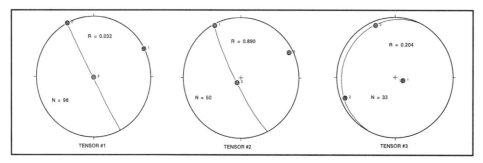

Figure 12. Calculated tensors from the combined field areas from Minnesota along the North Shore of Lake Superior. Note the first tensor denotes northeast-southwest compression, the second tensor indicates strike-slip motion along a north-northwest to south-southeast trend, while tensor #3 reflects extension in a north-northwest to south-southeast direction. N = number of faults; R = ratio of the stresses.

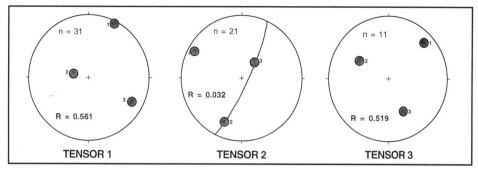

Figure 13. Equal-area plots of the three calculated tensors from the diabase through which the Lafayette Bluff Tunnel is constructed. Notice only in tensor #2 are the magnitudes of two of the principal stresses similar, as described by the great circle and R-value in the figure. Although tensors #1 and #3 have similar orientations of the three principal stresses, tensor #1 reflects more north-south maximum compressive stress (σ_1) and the intermediate compressive stress (σ_2) is more vertical, describing a more purely strike-slip regime with shore-perpendicular extension. N = number of faults; R = ratio of the stresses.

slip regime is described by the orientation of the principal stresses in tensor #3, where σ_1 is 027.5/5.8, σ_2 is 280.1/71.2, and σ_3 is 119.4/17.8. The R-value of 0.607 indicates the lack of similarity in magnitudes of the principle stresses. There is no indication of extension in the derived stress tensors for this area.

Data (118) from the NSV of the Taconite Harbor area, including the Cross River, describes three stress regimes, more closely approximating Tensor #1 in Figure 12 (Fig. 14), with the orientations of σ_1, σ_2, and σ_3. The first tensor is described by the following orientation of principle stresses (σ_1, σ_2 and σ_3): 243.0/0.8, 333.0/1.1, and 116.0/88.6, respectively. Notice that tensor #1 describes a compressional regime, but with an R-value of 0.022, the magnitudes of σ_2 and σ_3 are similar, therefore these data are describing a transpressional situation. This tensor is consistent with 50% of the analyzed data.

Tensor #2 has orientations of σ_1, σ_2, and σ_3 equaling 015.0/79.6, 244.3/6.8, and 153.4/7.8, respectively. It is consistent with 25% of the total data from this area and describes a northwest–southeast extensional setting; however, with similar magnitudes of σ_2 and σ_3, as denoted by the R-value of 0.044, the extension could also be northeast-southwest.

The third tensor, consistent with 12.5% of the total data, describes northwest–southeast compression. The orientations of σ_1, σ_2 and σ_3 are 143.2/13.8, 350.1/74.6, and 234.9/6.7, respectively, with an R-value of 0.673 again indicating the lack of similarity among the magnitudes of the principle stresses.

Since similar fault trends are seen in both field areas to some degree, each calculated stress tensor is consistent with some percentage of faults from these field areas. If this were not the case, a good argument could be made for changing orientations of the three principal stresses with changes in latitude or position along the Lake Superior shoreline in Minnesota. Since this is not seen, it is probable that the differences in lithology allow for varying response to the same stress tensors, applied at the same time but with nonsimilar orientation recorded as accommodation occurred.

DISCUSSION

Summary of results

The data in all field areas strongly suggest the Midcontinent Rift region was subjected to one or more significant con-

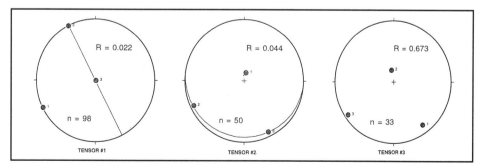

Figure 14. Equal-area plots of the three tensors calculated from the North Shore Volcanics (NSV) in the Taconite Harbor railroad cut. Notice that although extension and contraction in tensors #2 and #3, respectively, are both northwest-southeast, the minimal compressive stress (σ_3) in tensor #3 is northeast-southwest, resulting in a transpressional regime characterized by dextral strike-slip faulting. R = ratio of the stresses; N = number of faults.

traction periods. Data from the Keweenaw Peninsula and Isle Royale, Michigan, indicate compression of the central portion of the Midcontinent Rift in Lake Superior in north–south and east-southeast to west-northwest directions, respectively. Both Michigan field areas have recorded extension, but in slightly different orientations (Figs. 10, 11, and 15). The large population of apparently normal faults recorded in the PLV of the Keweenaw Peninsula were rotated thrust (reverse) faults and were rotated back to horizontal for the paleostress analysis (Fig. 16). The northeast trend and gentle dips of the flows on Isle Royale do not necessitate this same rotation because even after rotation, normal and reverse faults do not change their apparent direction of motion. The excellent quality of slickensides left indisputable macroscopic indicators of the motion along the fault surfaces. Plots of the orientations of slickenlines confirm the major directions of compression calculated by the fault inversion analyses.

Minnesota fault data are more diverse in orientation but have major trends in the northwest–southeast, north-northeast to south-southwest, and east-northeast to west-southwest directions. Data from field areas A and B (Fig. 1) are more consistent with the last two orientations (mentioned above), while data from field area C (Fig. 1) is represented by the first orientation. The large percentage of shore-parallel faults, along with

the gently pitching slickenlines, indicates that strike-slip faulting predominates.

Use of a fault inversion analysis (Etchecopar, personal communication, 1981) for data from the Keweenaw Peninsula resulted in a dominant stress tensor, consistent with north-south contraction, satisfying 51% of the data (Fig. 10). Analysis of Isle Royale fault measurements indicates two compressional tensors, one east-southeast to west-northwest and the other southeast–northwest, satisfying 55% and 17% of the data, respectively. One tensor, consistent with 14% of the data, was calculated supporting southeast–northwest extension.

Relative time sequence of events

Fault data suggest dominantly contractional events but the existence of minor extensional faulting requires a discussion of the timing of contractional versus extensional tectonism in the area. As discussed earlier, when bedding on the Keweenaw Peninsula is rotated back to horizontal, a significant proportion of normal faults become reverse faults and the steeper reverse faults also rotate to a shallower dip, but remain reverse faults. Even after the rotation of the fault population, normal faults still remain and need to be explained. In addition, three tensors, one from each of the Michigan localities and one from Minnesota, describe tensional regimes.

Unfortunately, no criteria could be used on the studied faults to determine whether extensional tectonism pre or postdated compression. In a rift setting, one expects to see extensional features documented in the rocks. If it is assumed the compression occurred prior to the extension, then it becomes necessary to postulate a mechanism for this extensional phase.

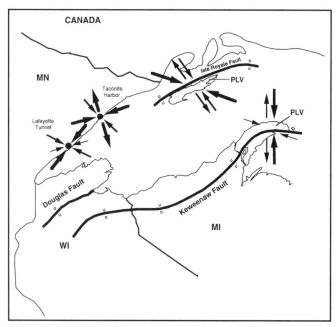

Figure 15. Summary diagram of the paleostress analyses for all the field areas. The size and thickness of the arrows indicates the maximum (σ_1), intermediate (σ_2), and minimum (σ_3) compressive stresses of a particular tensor in that area. The direction of the arrows indicate either compression (inward) or tension (outward). Note that while the Portage Lake Volcanics in Michigan record north-south and northwest-southeast compression, the North Shore Volcanics in Minnesota only record a minimal amount of northwest-southeast compression, while most of the data reflect northeast-southwest compression. Strata in areas identified by PLV are intercalated Portage Lake Volcanics and interflow sediments.

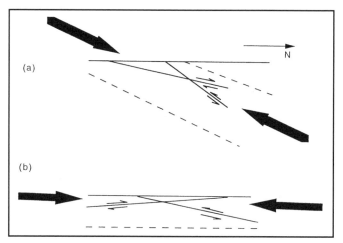

Figure 16. Representation of faults on the Keweenaw Peninsula. When the shallower Portage Lake Volcanic flows are rotated back to horizontal (b) according to Anderson (1955), the previously shallowly dipping normal faults (a) become thrust (reverse) faults. Before rotation, the normal faults are difficult to explain in a compressional setting, but after rotation, they become consistent with the stress field. Large arrows represent maximum compressive stress and dashed line represents "bedding."

It is also possible for an area that has been contracted to undergo "relaxation" after the contraction ceases, or it may extend due to gravitational collapse.

The second tensors for both the Keweenaw Peninsula and Taconite Harbor are consistent with extension in a north-northwest to south-southeast direction, σ_3 plunging 21° and 8°, respectively. If these tensors predated their respective tensor #1, then they should also be rotated to bring the younger strata to horizontal. When this is done it is noted that both σ_3 of tensor #2 are subhorizontal. It is reasonable to conclude on the basis of this analysis, that the extensional event could have predated the contraction. If extension postdated contraction, it would be more difficult to explain the rather high plunge value of σ_3. However, because of the lack of more direct evidence of overprinting relationships, it is impossible at this point to provide a more conclusive result, and future studies should concentrate on the relative timing of faulting.

There are geologic arguments, however, that support the contemporaneity of normal faulting and extrusion of the Portage Lake and North Shore Volcanics. The PLV flows near the Keweenaw and Isle Royale faults dip much more steeply (30) than at the opposite side of the peninsula and island, respectively. This would be expected if subsidence, by crustal sagging (Green, 1983), and extrusion of the flows were contemporaneous. Miller (1992) has determined that the North Shore Volcanic sequences were extruded in pulses and through geochronological correlation with the PLV extends this mode of pulsating extrusion to the rest of the rift. The older flows would then have time to cool sufficiently and record deformation by normal faulting in the vicinity of the well-documented "growth faults" (Dickas et al., 1989), such as the Keweenaw, Isle Royale, and Douglas faults. One other possible mode of formation of normal faults, as well as other types of faults, is accommodational motion of fractured lava flows during contraction of the region as explained below.

Accommodational faulting

When the plots of the faults consistent with each tensor are studied, one observes that many of the faults explained by a particular stress field appear to have formed under stresses incompatible with that type of faulting. In general, faults may be isolated but usually they occur in zones, defining blocks of varying sizes and shapes. Movement occurs along many of the faults when the area is subjected to deformation, and under these circumstances, motion along each fault may be different in order to accommodate the resulting motion of each block. Differences in competence, such as interflow sediments (as seen in the PLV), or varying lithology (NSV) within a lava flow, may provide zones of weakness along which slip may easily occur. In other situations a block may accommodate stress by deforming internally, by slip on a large number of smaller fractures.

Wallace (1951) suggested that the shape of the tectonic blocks can influence the distribution of shear stresses and the orientation of the rupture. Since the Midcontinent Rift is arcuate and synclinal in the Lake Superior cross section (Cannon et al., 1989; Huber, 1973), with an anticlinal overprint in the central portion of the Keweenaw Peninsula (perpendicular to the axis of the syncline; Dickas et al., 1989), it is most probable that tectonic shortening caused uneven stress distribution and fracturing along the major faults. There are data to support this phenomenon on the Keweenaw fault near the tip of the Keweenaw Peninsula, but the Isle Royale fault is under water and the only visualization has been via the GLIMPCE study (Cannon et al., 1989). While there is no clear break, as seen for the Keweenaw and Douglas faults in southern Lake Superior, the GLIMPCE data identifies the Isle Royale fault by the disturbance of the signal just northwest of Isle Royale (Cannon et al., 1989).

Experimental computer modeling

A small portion of the faults, on both limbs of the syncline, which are at high angles to and nearly parallel with σ_1, exhibit a sense of slip opposite to that expected from the calculated stress. Accommodation faults apparently violate the principle that fault slip takes place in the direction and sense of the maximum resolved shear stress. Such faults have been shown in two-dimensional computer experiments by Pershing (1990) to form a significant proportion of the artificial fault population used in the experiment where the stress field was known. Extrapolation of these two-dimensional findings to actual three-dimensional situations is not straightforward, but it is suggested that movement of faults which are nearly parallel or perpendicular to the maximum compression direction (σ_3), can be constrained to be opposite to that predicted from far-field stresses for compatibility reasons. These results may apply particularly well to the normal faults associated with tensor #2 (Keweenaw Peninsula), which show an orientation similar to faults associated with tensor #1, but are slightly more steeply dipping.

Variation of far field stresses?

If the Midcontinent Rift System was shortened in a north–south direction, we would expect to find strong evidence of these kinematics on both synclinal limbs, not just the southerly one. The fact that a large percentage of the data on the northern part of the Keweenaw Peninsula support north–south contraction, and the same percentage on Isle Royale support a more east–west compression, is strong evidence that either some reorientation of the stress field occurred locally, or that these contraction events are unrelated. Since the southeast–northwest σ_1 is recorded both at Isle Royale and the Keweenaw Peninsula, and the north–south σ_1 at the Keweenaw Peninsula can be associated with the northwest–southeast σ_1 at Isle Royale (both are nearly perpendicular to the strike of the strata), it is proposed that these two contraction events are, in fact, unrelated. Data from Minnesota more consistently supports contraction in a northeast–southwest direction and is assumed to reflect either the extension or be an accommodational motion as a result of local reorientation of stress.

Tectonic significance of compressional tensors

The expression of the stress tensors can be analyzed (Figs. 10 through 14) in all field areas by looking at different scales of deformation: (1) Plate scale, relation to Grenville Front; (2) Regional scale, dextral strike-slip motion along the Isle Royale fault; no supporting data; (3) Local scale, diffuse shear zone with accommodation occurring along faults.

Plate scale tectonism

Activity on the Grenville Front began between 1,100 and 1,200 Ma, with contraction in an east-southeast to west-northwest direction (Klasner et al., 1982), satisfying the major stress tensor calculated from Isle Royale and the second tensor from the North Shore of Minnesota (Fig. 12). These tensors, although not explaining the bulk of the data from the Keweenaw Peninsula, are nevertheless represented (tensor #3) by 10% of the total data. Considering the model of plate collision with uneven boundaries as proposed by Cambray and Fujita (1991), and the similar cooling dates of Grenville rocks along the northern portion of the Grenville front, the following hypothesis can be proposed. Collision along the Grenville front generated west-northwest to east-southeast contraction structures. This motion explains the data from Isle Royale and to a lesser extent that from the Keweenaw Peninsula and Minnesota. The north–south contraction recorded by a large percentage of the Keweenaw Peninsula data may be a result of the geometry of the rift itself, as is the northeast–southwest compression seen in data from Minnesota. When plates come into collisional contact with each other, the boundaries are usually not linear, but are of a more uneven to jagged shape. Assuming this plate boundary geometry between two plates converging in a north–south direction, the resulting local stresses will vary, reflecting the type of motion along the boundary at discrete increments. These increments can be as large as 30 km, or as small as 1 km, if the studied field areas are used as examples. The more oblique motion along the Grenville front will act as a restriction in movement direction allowing motion in only the north–south direction, for both opening and closing. In contrast, if the rift structure were passive, one would expect that motion along the Grenville Front alone would generate the east-southeast to west-northwest σ_1 direction as recorded at Isle Royale.

Failed arm of a triple junction

Burke and Dewey (1973) have speculated that a plume-generated triple junction was the formational mechanism for the Midcontinent Rift. The Trans-Superior Tectonic Zone, which is postulated to be the failed third arm (Klasner et al.,

1982; Green, 1983), has been dated slightly older than rocks in the other two arms. This fact casts doubt on the appropriateness of Burke and Dewey's mechanism.

Regional mechanism

Movement along the Isle Royale fault is reverse, with a throw of 3 km, as indicated by the GLIMPCE study. However, the motion along the small faults, as documented in this study on Isle Royale, is not strictly dip-slip but is oblique-slip. This lends support to a significant amount of dextral strike-slip movement that may have also occurred along the Isle Royale fault. It is possible the Isle Royale fault splayed out in an en echelon manner transferring motion along shore-parallel faults in Minnesota. Therefore two mechanisms may be possible. (1) The entire southern block, that is, all crust south of the Isle Royale fault, was rotated anticlockwise during contraction. There is no evidence to support this hypothesis at this time, (2) Some folding of flows in the deepest portions of Lake Superior, oblique to the axis of the syncline, provided an easterly component to the predominantly southerly direction of closing, accentuated by the anisotropies of the younger flows, which affected faulting on Isle Royale. Due to the large (and, therefore, detail-limiting) scale of the geophysical studies, there is also no evidence to support this hypothesis.

Local mechanisms

Since Isle Royale and the Minnesota North Shore are positioned on the southwest-trending limb of the Midcontinent Rift, the recorded motion might be expected to be more oblique if it (the limb) is a reflection north–south opening and closure. Motion along the Keweenaw fault on the Keweenaw Peninsula is mainly reverse dip-slip (>3 km), allowing it to be classified as a contraction-reactivated growth fault. This motion and classification is supported by the orientation of the structures (flows, fault, and slickensides), paleostress analysis, and the GLIMPCE data (Cannon et al., 1989).

The motion along the Isle Royale fault must have a significant amount of strike-slip motion for minor faults in the Isle Royale basalts to record east-southeast to west-northwest contraction. Perhaps there is a broad diffuse shear zone extending into Minnesota, encompassing the northeast islands, composed of many small faults, which allow for motion in this recorded direction, partly controlled by the attenuation and anisotropy of the lava flows. Since the Michigan and Minnesota field areas are essentially in different volcanic basins of Lake Superior and may be separated by an accommodation zone, it is reasonable to assume some degree of independent motion in each area (Cannon et al., 1989).

Let us assume north–south compression of the Midcontinent Rift occurs, leaving evidence of these kinematics behind. Because Isle Royale is located at the edge of the northern synclinal limb, very close to the arcuate shoreline of Minnesota and Canada, movement of the northern limb is confined. This situation is somewhat analogous to a basalt-filled pull-apart basin, which would be undergoing contraction. If the basin had an arcuate shape, the material could not form a simple syncline but would be constrained by the shape of its lateral boundaries. The varying shape of the basin causes a realignment of the local stress field around Isle Royale and Minnesota, resulting in extrusion of the PLV and NSV in an easterly and southerly direction, depending on the positioning along the margin. This behavior is reflected in the large number of faults (60%) on both Isle Royale and Minnesota recording the east-southeasterly motion. This could also be accomplished with "adjustments" along small faults in the PLV and NSV, such as the diffuse shear zone suggested in the previous paragraph.

In order to test this hypothesis, field measurements need to be taken around the eastern bend in the syncline, where northeast-southwest contraction should be observed.

CONCLUSIONS

The calculated stress tensors from each limb of the syncline were found to be similar, dominantly horizontal σ_1, but represented in very different proportion in the two studied areas. Isle Royale is characterized by an east-southeast to west-northwest σ_1 (and a subsidiary compression normal to strike), the Keweenaw Peninsula records a dominant, strike-normal, north–south compression (with a secondary east-southeast to west-northwest σ_1), and Minnesota is mainly characterized by northeast-southwest σ_1 and dominant strike-slip faulting. The relative timing of these contraction events is not constrained but their orientations suggest that the east-southeast to west-northwest σ_1 could be related to coeval collision in the Grenville orogen, and the north–south and northeast-southwest σ_1, arose because of plate motion constrained by the original, arcuate geometry of the rift.

In addition, both areas contain a small proportion of normal faults that are interpreted to have formed early, associated with the rifting event, although some of them can be accommodation fractures to an overall contraction event. Future studies should concentrate on the definition of criteria to decipher the absolute and relative timing of tectonic events in this part of the Midcontinent Rift System in addition to in-depth studies of faults in other exposed parts of the rift, such as the Osler volcanics and Mamainse Point in Ontario.

ACKNOWLEDGMENTS

Thanks to Dr. F. William Cambray, of Michigan State University, for his invaluable help and insight in obtaining data in the Upper Peninsula of Michigan. Structural mapping in Minnesota was supported in part by the Minnesota Geological Survey. Thanks, also, to the cooperative staff of Isle Royale National Park, who provided excellent facilities and, therefore, improved access to the smaller islands from which these data were collected.

REFERENCES CITED

Anderson, E. M., 1955, The dynamics of folding: Edinborough, Oliver and Boyd, 206 p.

Angelier, J., 1979, Determination of the mean principal directions of stresses for a given fault population: Tectonophysics, v. 56, T17–T26.

Angelier, J., 1984, Tectonic analysis of fault slip data sets: Journal of Geophysics Research, v. 89, no. B7, p. 5835–5848.

Bacon, L. O., 1957, Relationship of gravity to geological structure in Michigan's Upper Peninsula, *in* Snelgrove, A. K., ed., Geological Exploration: Proceedings and Abstracts, Institute on Lake Superior Geology, Houghton, Michigan, p. 54–58.

Bacon, L. O., 1966, Geologic structure east and south of the Keweenawan fault on the basis of geophysical evidence, *in* Smith, T. J., and Steinhart, J. S., eds., The Earth beneath the continents: American Geophysical Union Geophysical Monograph 10, p. 42–55.

Black, W. A., 1955, Study of the marked positive gravity anomaly in the northern Midcontinent region of the United States [abs.]: Geological Society of America Bulletin, *v.* 66, p. 1531.

Bott, M. H. P., 1959, The mechanics of oblique slip faulting: Geological Magazine, v. 96, p. 109–117.

Brojanigo, A., 1984, Keweenaw fault: Structures and sedimentology [M.S. thesis]: Houghton, Michigan Technological University, 195 p.

Books, K. G., 1972, Paleomagnetism of some Lake Superior Keweenawan rocks: U.S. Geological Survey Professional Paper 760, 42 p.

Burke, K., and Dewey, J. F., 1973, Plume generated triple junctions: Key indicators in applying plate tectonics to old rocks: Journal of Geology, v. 81, p. 406–433.

Cambray, F. W., and Fujita, K., 1991, Collision induced ripoffs, ancient and modern: The Midcontinent Rift System and the Red Sea–Gulf of Aden compared, *in* Chan, L. S., ed., Annual Institute on Lake Superior Geology, 37th, Program and Abstracts, Eau Claire, Wisconsin, p. 15–16.

Canada Department of Energy, Mines and Resources, 1974, Bouguer anomaly map of Canada: Earth Physics Branch Gravity Map Series, no. 74-1, scale 1:500,000.

Cannon, W. F., and 11 others, 1989, The North American Midcontinent Rift beneath Lake Superior from GLIMPCE seismic reflection profiling: Tectonics, v. 8, no. 2, p. 305–332.

Chase, C. G., and Gilmer, T. H., 1973, Precambrian plate tectonics: The Midcontinent Gravity High; Earth and Planetary Science Letters, v. 21, p. 70–78.

Cornwall, H. R., 1954a, Bedrock geology of the Lake Medora Quadrangle, Michigan: U.S. Geological Survey, scale 1:24,000.

Cornwall, H. R., 1954b, Bedrock geology of the Delaware Quadrangle, Michigan: U.S. Geological Survey, scale 1:24,000.

Cornwall, H. R., 1955, Bedrock geology of the Fort Wilkins Quadrangle, Michigan: U.S.Geological Survey, scale 1:24,000.

Cornwall, H. R., and Wright, J. C., 1956a, Geologic map of the Hancock Quadrangle, Michigan: U.S. Geological Survey, 1:24,000.

Cornwall, H. R., and Wright, J. C., 1956b, Geologic Map of the Laurium Quadrangle, Michigan: U.S. Geological Survey, scale 1:24,000.

Craddock, C., Mooney, H. M., and Kolehmainen, V., 1970, Simple Bouguer gravity map of Minnesota and northwestern Wisconsin: Minnesota Geological Survey Miscellaneous Map M-10, scale 1:1,000,000, 14 p.

Daniels, P. A., Jr., 1982, Upper Precambrian sedimentary rocks: Oronto Group, Michigan–Wisconsin, *in* Hinze, W. J., ed., Geology and tectonics of the Lake Superior Basin: Geological Society of America Memoirs, no. 156, p. 107–133.

Davis, D. W., Green, J. C., and Manson, M. L., 1995, Geochronology of the 1.1 Ga North American Mid-Continent Rift, *in* Institute on Lake Superior Geology, 41st, Marathon, Ontario, Program with Abstracts, p. 9–10.

Davis, D. W., and Paces, J. B., 1990, Time resolution of geologic events on the Keweenaw Peninsula and implications for development of the Midcontinent Rift System: Earth Planetary Science Letters, v. 97, p. 54–64.

Davis, D. W., and Sutcliffe, R. H., 1985, U-Pb ages from the Nipigon plate and northern Lake Superior: Geological Society of America Bulletin, v. 96, p. 1572–1579.

Dickas, A. B., Mudrey, M. G., Jr., Green, J. C., Ojakangas, R. W., Kalliokoski, J., Paces, J. B., and Bornhorst, R. H., 1989, Metamorphism and tectonics of eastern and central North America: Volume 1, Lake Superior Basin segment of the Midcontinent Rift System, in Field Trips for the International Geological Conference on Basement Tectonics: Washington, D. C., American Geophysical Union, 62 p.

Dubois, P. M., 1962, Paleomagnetism and correlation of Keweenawan rocks: Geological Survey of Canada Bulletin, v. 71, 75 p.

Ervin, C. P., and Hammer, S., 1974, Bouguer anomaly gravity map of Wisconsin: Wisconsin Geological and Natural History Survey, no. 14, scale 1:500,000.

Etchecopar, A., Vasseur, G., and Diagnieres, M., 1980, An inverse problem in microtectonics for the determination of stress tensors from fault striation analysis: Tectonophysics, v. 158, p. 51–65.

Foster, J. W., and Whitney, J. D., 1850, Report on the geology and topography of a portion of the Lake Superior Land District; Part I, Copper Lands: U.S. House of Representatives, Executive Documents, no. 69, 224 p.

Goldich, S. S., 1968, Geochronology in the Lake Superior region: Canadian Journal of Earth Science, v. 5, p. 715–724.

Gordon, M. B., and Hempton, M. R., 1986, Collision induced rifting: the Grenville orogeny and the Keweenawan Rift of North America: Tectonophysics, v. 127, p. 1–25.

Green, J. C., 1982, Geology of Keweenawan extrusive rocks, in Wold, W. A., and Hinze, W. J., eds., Geology and tectonics of the Lake Superior Basin: Geological Society of America Memoirs, v. 156, p. 47–55.

Green, J. C., 1983, Geologic and geochemical evidence for the nature and development of the Middle Proterozoic (Keweenawan) Midcontinent Rift of North America: Tectonophysics, v. 94, p. 413–437.

Green, J. C., and Books, K. G., 1972, Paleomagnetic evidence for the extent of lower Keweenawan lavas in Minnesota, Annual Institute on Lake Superior Geology, 18th, Program with Abstracts, Houghton, Michigan, 8th paper, 3 p.

Grout, F. F., 1918, The lopolith, an igneous form exemplified by the Duluth gabbro: American Journal of Science, series 4, v. 46, p. 516–522.

Halls, H. C., 1974, A paleomagnetic reversal in the Osler volcanic group, northern Lake Superior: Canadian Journal of Earth Science, v. 11, p. 1200–1207.

Halls, H. C., and West, G. F., 1971, A seismic refraction survey in Lake Superior: Canadian Journal of Earth Science, v. 8, p. 610–630.

Hinze, W. J., O'Hara, N. W., Trow, J. W., and Secor, G. B., 1966, Aeromagnetic studies of eastern Lake Superior, in Smith, T. J., and Steinhart, J. S., eds., The Earth beneath the continents: American Geophysical Union Geophysical Monograph, no.10, p. 95–110.

Huber, N. K., 1973, The Portage Lake Volcanics (middle Keweenawan) on Isle Royale, Michigan: U.S. Geological Survey Professional Paper 754-C, 30 p.

Huber, N. K., 1983, The geologic story of Isle Royale National Park: U.S. Geological Survey Bulletin, no. 1309, 66 p.

Irving, R. D., 1883, The copper-bearing rocks of Lake Superior: U.S. Geological Survey Monograph, no. 5, 464 p.

King, E. R., and Zeitz, I., 1971, Aeromagnetic study of the midcontinent gravity high of Central United States: Geological Society of America Bulletin 82, p. 2187–2208.

Klasner, J. S., Wold, R. J., Hinze, W. J., Bacon, L. O., O'Hara, N. W., and Berkson, J. M., 1979, Bouguer gravity anomaly map of the northern Michigan–Lake Superior region: U.S. Geological Survey Geophysical Investigations Map GP 930, scale 1:1,000,000.

Klasner, J. S., Cannon, W. F., and Van Schmus, W. R., 1982, Keweenawan tectonic history of southern Canadian Shield and its influence on formation of the Midcontinent Rift, in Wold, W. A., and Hinze, W. J., eds., Geology and tectonics of the Lake Superior Basin: Geological Society of America Memoirs, no. 156, p. 27–46.

Leith, C. K., Lund, R. J., and Leith, A., 1935, Precambrian rocks of the Lake Superior region: U.S. Geological Survey Professional Paper, no. 184, 34 p.

Lisle, R. J., 1987, Principle stress orientations from faults: an additional constraint: Annals of Tectonics, v. 1, no. 2, p. 155–158.

Logan, W. E., 1863, Report of progress from its commencement to 1863; Geological Survey of Canada, 983 p.

Meshref, W. M., and Hinze, W. J., 1970, Geological interpretation of aeromagnetic data in western Upper Peninsula of Michigan: Michigan Geological Survey Report of Investigations, v. 12, 25 p.

Miller, J. D., 1992, The need for a new paradigm regarding the petrogenesis of the Duluth Complex: Proceedings, Annual Institute on Lake Superior Geology, 38th, Hurley, Wisconsin, p. 65–67.

Mudrey, M. G., Jr., 1976, Late Precambrian structural evolution of Pigeon Point, Minnesota, and relations to the Lake Superior syncline: Canadian Journal of Earth Science, v. 13, no. 7, p. 877–888.

Ocola, L. C., and Meyer, R. P., 1973, Central North American rift system: 1. Structure of the axial zone from seismic and gravimetric data: Journal of Geophysical Research, v. 78,p. 5173–5194.

Oray, E., Hinze, W. J., and O'Hara, N. W., 1973, Gravity and magnetic evidence for the eastern termination of the Lake Superior syncline: Geological Society of America Bulletin, v. 84, p. 2763–2780.

Paces, J. B., and Miller, J. D., 1993, Precise U-Pb ages of Duluth Complex and related mafic intrusions, northeastern Minnesota: geochronological insights to physical, petrogenic, paleomagnetic and tectonomagmatic processes associated with the 1.1 Ga Midcontinent Rift System: Journal of Geophysical Research, v. 98, p. 13997–14013.

Palmer, H. C., 1970, Paleomagnetism and correlation of some Middle Keweenawan rocks, Lake Superior: Canada Journal of Earth Science, v. 7, p. 1410–1436.

Pershing, J. C., 1990, Paleostress analysis of fault populations: numerical test and application to the West Spitsbergen orogen, [Ph.D. thesis]: University of Minnesota, 273 p.

Petit, J. P., and Green, J. C., 1987, Criteria for the sense of movement on fault surfaces in brittle rocks: Journal of Structural Geology, v. 9, no. 5/6, p. 597–608.

Silver, L. T., and Green, J. C., 1963, Zircon ages for middle Keweenawan rocks of Lake Superior region: Transactions of the American Geophysical Union, Abstracts, v. 44, p 107.

Silver, L. T., and Green, J. C., 1972, Time constants for Keweenawan igneous activity: Geological Society of America Program with Abstracts, v. 4, p. 665–666.

Sims, p. K., 1980, The Great Lakes Tectonic Zone—A major crustal structure in Central North America: Geological Society of America Bulletin, v. 91, no. 1, p. 690–698.

Smith, T. J., Steinhart, J. S., and Aldrich, J. C., 1966, Lake Superior crustal structure: Journal of Geophysical Research, v. 71, p. 1141–1172.

Steinhart, J. S., and Meyer, R. P., 1961, Explosion studies of continental structure: Carnegie Institution of Washington Publication 662, 409 p.

Steinhart, J. S., Hart, S. R., and Smith, T. J., 1968, Heat flow: Carnegie Institution of Washington Yearbook 67, p. 360–367.

Thiel, E. C., 1956, Correlation of gravity anomalies with the Keweenawan geology of Wisconsin and Minnesota: Geological Society of America Bulletin, v. 67, p. 1079–1100.

Thiel, E. C., 1960, Geologic interpretation of airborne magnetometer profiles across Lake Superior: Annual Institute on Lake Superior Geology, 6th, Proceedings and Abstracts, Madison, Wisconsin, p. 32.

Van Hise, C. R., and Leith, C. K., 1911, The geology of the Lake Superior region: U.S. Geological Survey Monograph, no. 52, 641 p.

Van Schmus, W. R., 1976, Early and Middle Proterozoic history of the Great Lakes area, North America, in Global tectonics in Proterozoic times: Royal Society of London Philosophical Transactions, Series A., v. 280, p. 605–628.

Van Schmus, W. R., Green, J. C., and Halls, H. C., 1982, Geochronology of Keweenawan rocks of the Lake Superior region: A summary, in Wold,

W. A., and Hinze, W. J., eds., Geology and tectonics of the Lake Superior Basin: Geological Society of America Memoirs, no. 156, p. 165–171.

Wallace, R. E., 1951, Geometry of shearing stress and relation to faulting: Journal of Geology, v. 59, p. 118–130.

Weber, J. R., and Goodacre, A. K., 1966, A reconnaissance underwater gravity survey of Lake Superior, *in* Steinhart, J. S., and Smith, T. J., eds., The Earth beneath the continents: American Geophysical Union Monograph, v. 10, p. 56–65.

Weiblen, R. W., Saini-Eidukat, B., and Miller, J. D., 1985, Duluth Complex and associated rocks of the Midcontinent Rift System: International Geological Congress, 28th, Guidebook T-345, 43 p.

White, W. S., 1966, Geologic evidence for crustal structure in the western Lake Superior Basin, *in* Steinhart, J. S., and Smith, T. J., eds., The Earth beneath the continents: American Geophysical Union Geophysical Monograph, no. 10, p. 28–41.

White, W. S., 1972, The base of the upper Keweenawan, Michigan and Wisconsin: U.S. Geological Survey Bulletin 1354-F, 23 p.

Wold, R. J., 1979, Map showing bedrock topography of Lake Superior: U.S. Geological Survey Miscellaneous Field Studies Map MF 1174, scale 1:600,000.

Wold, R. J., and Berkson, J. M., 1977, Bouguer gravity anomaly map of Lake Superior: U.S. Geological Survey Miscellaneous Field Studies Map MF 884, scale 1:500,000.

Wold, R. J., and Ostenso, N. A., 1966, Aeromagnetic, gravity, and sub-bottom profiling studies in western Lake Superior, *in* Steinhart, J. S., and Smith, T. J., eds., The Earth beneath the continents: American Geophysical Union Geophysical Monograph, no. 10, p. 66–94.

Wold, R. J., and Hinze, W. J., 1982, Introduction, *in* Wold, W. A., and Hinze, W. J., eds., Geology and Tectonics of the Lake Superior Basin: Geological Society of America Memoirs, no. 156, 280 p.

Zumberge, J. H., and Gast, P., 1961, Geological investigations in Lake Superior: Geotimes, v. 6, p. 10–13.

MANUSCRIPT ACCEPTED BY THE SOCIETY JANUARY 16, 1996

Geological Society of America
Special Paper 312
1997

Post-extension shortening strains
preserved in calcites of the Midcontinent Rift

John P. Craddock, Alene Pearson*, Michelle McGovern, Elizabeth Kropf, and Andrew Moshoian*
Geology Department, Macalester College, St. Paul, Minnesota 55105
Katie Donnelly
Geology Department, Trinity University, San Antonio, Texas 78212

ABSTRACT

Analysis of mechanically twinned calcite throughout the Middle Proterozoic Midcontinent Rift (57 sites, 62 strain analyses) reveals the presence of two layer-parallel and subhorizontal shortening strains, one that is rift parallel and one that is rift normal. We interpret these two strain patterns to represent strike-slip and thrust motions, respectively, along the margins of the rift during closure between 1,060 Ma and <350 Ma. Calcite strains preserved in pre-rift, Penokean-aged rocks (5 sites) are unique compared to the rift calcite strains, and are included.

Within the rift there are six generations of calcite whose regional chronology has been determined by field cross-cutting relations: calcite amygdules (oldest; in basalt), Oronto Group cements, post–Oronto Group veins, Jacobsville Sandstone cements, post–Jacobsville Sandstone calcite-cemented clastic dikes, and deformed Ordovician-Devonian limestones (youngest). Not all of these calcite elements are present in the four structural subprovinces of the rift, the Douglas thrust (west margin, southeast dip), and Keweenaw thrust (east margin, northwest dip) foot and hanging walls, but there is a regional progression of horizontal shortening axes that are aligned parallel to the rift on the western margin, and become rift normal on the eastern margin through the time interval (1,060 to 350 Ma) in question.

INTRODUCTION

Recent studies of orogenic deformation in the Appalachian foreland have documented the presence of subtle, regional synorogenic chemical remagnetizations, magnetic anisotropy fabrics, and calcite twinning shortening strains (e.g., Craddock and van der Pluijm, 1989; Lu et al., 1990; Jackson et al., 1992; Halliday et al., 1992). Craddock et al. (1993) found that twinned calcite in Paleozoic limestones as

far as 2,200 km from the Appalachian-Ouachita thrust front preserve a subhorizontal, thrust transport-parallel shortening strain; this included Paleozoic limestones north and west of the Middle Proterozoic Midcontinent Rift. Calcite amygdules in the Keweenawan rift basalts are also mechanically twinned but contain a very different strain pattern than that found in the overlying Paleozoic rocks; this study is the first attempt at

**Present address: Department of Geology and Geophysics, University of California, Berkeley, California 94720.

Craddock, J. P., Pearson, A., McGovern, M., Kropf, E., Moshoian, A., and Donnelly, K., 1997, Post-extension shortening strains preserved in calcites of the Midcontinent Rift, *in* Ojakangas, R. W., Dickas, A. B., and Green, J. C., eds., Middle Proterozoic to Cambrian Rifting, Central North America: Boulder, Colorado, Geological Society of America Special Paper 312.

using calcite twinning in an igneous rock suite as a means of determining the deformational history of a rift margin that was once extensional (ca. 1.1 Ga; see Wold and Hinze, 1982) and later closed by thrust faulting along its margins (1,060 to <350 Ma; Cannon et al., 1993).

The Midcontinent Geophysical Anomaly (Keweenawan rift; see McSwiggen et al., 1987), is a failed triple junction system that underlies much of the Lake Superior and Michigan basin region (Wold and Hinze, 1982). Rocks of the southwestern arm are best exposed north of Twin Cities, Minnesota (Fig. 1); the rift contains a vast thickness of basalts (Green, 1982) that are overlain by clastic sediments of the Oronto Group (Ojakangas and Morey, 1982; Dickas, 1986). The Bayfield Group sandstones and the Jacobsville Sandstone (see Kalliokoski, 1982) overly the Oronto Group, and are offset along the bounding Douglas (western margin, southeast dip) and Keweenaw–Lake Owen–Hastings (eastern margin, northwest dip) thrust faults (Craddock, 1972).

The rift is host to substantial copper mineralization both in the basalts and the Oronto Group; calcite is an accessory mineral associated with copper mineralization and these fillings have been dated at 1,050 Ma, which is synchronous with the initiation of thrust closure (Bornhorst et al., 1988; Rb-Sr method). Within the four structural subprovinces of the rift, the Douglas hanging and footwalls and the Keweenaw hanging and footwalls, the regional crosscutting and field relations of the different calcite elements are as follows: (1) amygdules (oldest), (2) Oronto Group cements, (3) veins crosscutting the basalts and Oronto Group sediments, (4) Jacobsville Sandstone cement, (5) calcite-cemented clastic dikes in the Jacobsville Sandstone, and (6) adjacent, folded Paleozoic carbonates (youngest).

Calcite twin analysis methods

Calcite mechanically twins at low stresses (10 MPa) independent of temperature and normal stresses (Groshong,

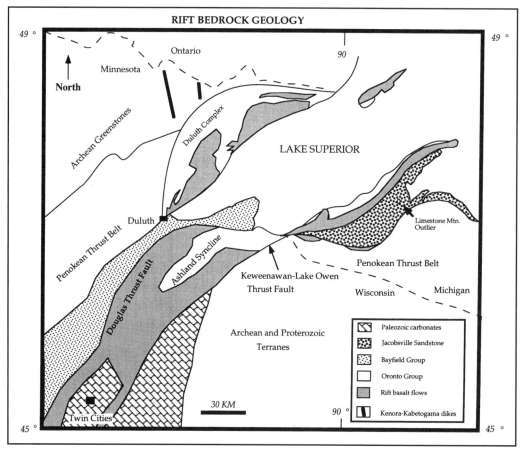

Figure 1. Generalized bedrock geology of the Midcontinent Rift and adjacent Archean and Proterozoic terranes. The Proterozoic Kenora-Kabetogama mafic dikes (thick black lines) and the Paleozoic Limestone Mountain outlier are shown.

1972, 1974). Twinning is possible along three glide planes and calcite strain hardens once twinned; subsequent twinning is possible only at higher stress levels if that stress is oriented >45° from the inital stress orientation (Teufel, 1980). Twinning causes the calcite lattice structure to strain harden such that subsequent mechanical twinning along another glide plane is difficult and twinning strain overprints (i.e., high percentages of negative expected values; NEVs) are difficult. A high percentage of negative expected values (>40%; NEVs) indicates that a second, non-coaxial twinning event occurred and these two twinning strains can be analyzed separately. Paleostresses responsible for twinning can be calculated in terms of their orientation (Turner, 1953) and magnitude (Jamison and Spang, 1976). Strain ellipsoids are computed using the Groshong method (1972, 1974) which is quite accurate (Groshong et al., 1984) for strains ranging from 1 to 17%.

The analysis of twinned calcite is usually applied best to sparry limestones and their inclusive calcite veins that are deformed but not severely metamorphosed. Utilizing calcite amygdules as strain indicators is a new approach at understanding stress transmission and the role of mechanical twinning in the deformational behavior of an igneous rock suite. The assumption we have made is that the basalt vesicles were filled with amygdule minerals shortly after crystallization of the melt. Local stress phenomena appear to be insignificant in twinning calcite as we have recorded a regionally consistent calcite amygdule strain pattern.

In general, extracting calcite-bearing components within igneous rocks proved a challenge, as compared to working with limestones. Many of the cemented clastic sediments, and most of the calcite amygdules, required multiple thin sections of the same orientation or two orthogonal sections to derive a meaningful strain result. In vein samples we cut one horizontal thin section parallel to the strike of the vein. Twenty to fifty thin twins were measured in each thin section.

PRE-RIFT
TWINNING STRAINS

As part of our regional rift strain study, we collected samples of pre-Keweenawan, calcite-rich rocks that are adjacent to, and older than, the rift as a means of documenting the uniqueness of the twinning strains preserved in the rift calcite elements. Four of these Archean and Early Proterozoic suites were sampled, and range in age from ca. 2.6 to 2.0 Ga.

North and west of the Penokean orogenic belt in Minnesota, the Kenora-Kabetogama mafic dike swarm is exposed throughout Archean country rock and forms a radiating pattern that spans ~110° of curvature (Southwick and Halls, 1987). Beck and Murthy (1982) have dated these dikes at 2.2 Ga using Rb-Sr techniques. The mafic dikes are everywhere parallel to a vertical fault-fracture array throughout the country rock of the region (Craddock and Moshoian, 1996); calcite fillings within this fault-fracture array or on the margins of the mafic dikes preserve a dike-parallel, horizontal shortening strain (Figs. 1, 2; sites 58, 59).

Along the Huron River at Big Eric's Crossing near L'Anse, Michigan, a thin (<1 m) Lower Proterozoic limestone rests unconformably on Archean basement (Kalliokoski and Lynott, 1987; site 61). This limestone has been correlated with the Michigamme Formation of the Baraga Group (Cannon and Gair, 1970), and contains a horizontal, north-south shortening strain, presumably of Penokean origin.

In the footwall of the Keweenaw thrust west of Mellen, Wisconsin, the Middle Proterozoic Bad River Dolomite is exposed in the Grandview quarry (Mudrey, 1979; site 60). These carbonates are equivalent to the Chocolay Group of the Marquette Supergroup to the east and contain abundant sparry limestones. Analysis of the twinned calcite revealed the presence of a horizontal shortening strain directed at N50°E.

Kalliokoski (1982) has described the outcrop associations of the Archean Presque Isle peridotite found along the shore of Lake Superior just north of Marquette, Michigan (Site 62). The peridotite is altered, unconformably overlain by a horizontal paleosol and the Jacobsville Sandstone, and contains sparry vein calcite. The calcite preserves a nearly vertical shortening strain directed toward 120°; this is nearly perpendicular to the north-south, horizontal shortening strain preserved by calcite cement in the overlying Jacobsville Sandstone (see below).

RIFT CLOSURE
TWINNING STRAINS

Within the rift there are four structural subprovinces, none of which contain all of the six calcite elements described below (Fig. 1 and 2; Table 1). The details of the calcite strains are presented in Table 2 and are shown chronologically in Figures 3 to 6.

Amygdule fillings

Calcite amygdules (16 samples, 17 strain analyses; sites 3–7, 17–20, 25, 26, 31–34, 35, 51) are present in all four structural subprovinces (Fig. 3) and preserve two distinctive subhorizontal shortening strains: one parallel to the rift on the

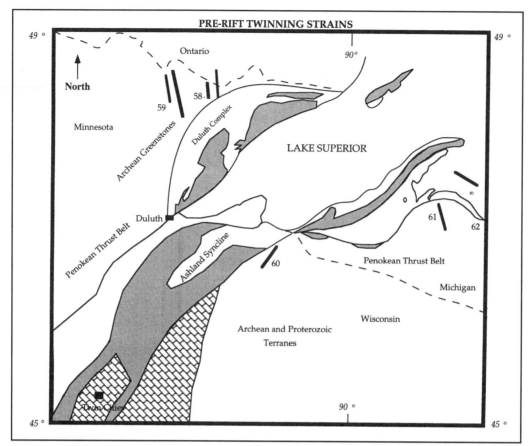

Figure 2. Keweenawan rift base map with calcite twinning strain shortening axes of pre-rift calcite elements plotted next to their adjacent sample number (Table 2). Sites 58 and 59 are Kenora-Kabetogama mafic dikes; the dikes are represented by the thicker lines. See Figure 1 for legend.

TABLE 1. SAMPLE DISTRIBUTION

	Douglas Thrust Fault		Keweenawan Thrust Fault	
	Footwall	Hanging wall	Hangingwall	Footwall
RIFT CALCITE				
Amygdules	5	4	6	2
Cements	0	0	6	7
Veins	8	5	11	1
Limestones	0	0	0	1
Clastic Dikes	0	0	0	1
Totals (Total 57)	13	9	23	12
NON-RIFT CALCITE				
Veins	2	0	0	1
Limestones	0	0	0	2
Totals (Total 5)	2	0	0	3

western margin, and one normal to the rift on the eastern margin (Fig. 4). Averages for the group are −5.79% (e_1%), 20.31% NEVs. Site 5 represents an anomalous result with a high percentage of NEVs requiring that NEVs and PEVs splits were analyzed separately. Both data sets preserve nearly vertical shortening strains in opposite directions. This is the only amygdule sample without a layer-parallel shortening strain fabric.

Oronto Group cements

The Oronto Group includes the basal Copper Harbor Conglomerate (CHC), the Nonesuch Shale, and the Freda Sandstone. The sands and conglomerates of the CHC contain abundant calcite cement (Fig. 5) of which we collected six samples (Fig. 6, Table 1; sites 21, 24, 27, 29, 37, 40, 52). All

TABLE 2. CALCITE TWINNING STRAIN DATA

Specimen	e₁ Trend† (°)	NEVs* (%)	e₁ Magnitude† (%)	Nominal Error§	Rock Type	Calcite Type	Location
1	27	38	-3.71	0.08	Basalt	Veins	Grand Portage
2	151	42	-5.72	0.41	Basalt	Veins	Grand Marais
3	38	22	-3.52	0.43	Basalt	Amygdules	Grand Marais
4	43	11	-1.43	0.74	Basalt	Amygdules	Grand Marais
5(PEVs)	167, 86	0	-8.68	0.03	Basalt	Amygdules	Cascade River
5(NEVs)	88, 57	100	-1.63	0.68	Basalt	Amygdules	Cascade River
6	52	37	-3.32	0.42	Basalt	Amygdules	Temperance River
7	12	0	-3.41	0.31	Basalt	Amygdules	Gooseberry Falls
8	36	30	-3.36	1.02	Granite	Veins	Silver Cliff
9	41	10	-2.31	0.54	Basalt	Veins	Lester River
10	10	21	-3.45	0.06	Basalt	Veins	Duluth
11(PEVs)	160	0	-4.32	0.76	Basalt Dike	Veins	Thomson Dam
11(NEVs)	2	100	-7.17	0.32	Basalt Dike	Veins	Thomson Dam
12(PEVs)	147	0	-8.71	0.71	Basalt Dike	Veins	Thomson Dam
12(NEVs)	330, 30	100	-2.63	1.05	Basalt Dike	Veins	Thomson Dam
13	7	26	-1.15	0.06	Basalt	Veins	Dresser
14(PEVs)	78, 38	0	-7.94	0.08	Bsaalt	Veins	Dresser
14(PEVs)	64	100	-1.16	0.13	Basalt	Veins	Dresser
15	30, 70	33	-7.82	0.21	Basalt	Veins	Dresser
16	113	0	-1.55	0.36	Basalt	Veins	Manitou Falls
17	78	0	-6.83	0.07	Basalt	Amygdules	South Amnicon
18	50	30	-9.88	0.03	Basalt	Amygdules	Amnicon
19	168	25	-1.54	0.37	Basalt	Amygdules	Amnicon
20	93	33	-5.8	0.27	Basalt	Amygdules	Grandview
21	155	10	-6.13	0.22	Conglomerate	Cements	Potato River Falls
22	150	26	-1.21	0.46	Conglomerate	Veins	Potato River Falls
23	96	25	-0.66	0.33	Conglomerate	Veins	Potato River Falls
24	63	0	-4.28	0.47	Conglomerate	Cements	Potato River Falls
25	125	14	-6.03	0.06	Basalt	Amygdules	Mellen
26	144	41	-9.62	0.07	Basalt	Amygdules	Mellen
27	17, 52	0	-4.67	0.29	Sandstone	Cements	Superior Falls
28	1, 51	33	-0.66	0.08	Sandstone	Veins	Superior Falls
29	27	5	-7.13	0.62	Conglomerate	Veins	Rainbow Falls
30	19	16	-6.16	0.05	Basalt	Veins	Wakefield
31	122	0	-8.43	0.16	Basalt	Amygdules	Wakefield
32	2	41	-9.53	0.32	Basalt	Amygdules	Wakefield
33	41	28	-3.44	0.51	Basalt	Amygdules	Bergland
34	157	18	-3.63	0.06	Basalt	Amygdules	Bergland
35	171	30	-6.73	0.08	Basalt	Veins	Bergland
36	27	25	-5.96	0.37	Basalt	Amygdules	Houghton
37	41	0	-6.72	0.08	Sandstone	Cements	Eagle River
38	162	14	-1.92	0.13	Basalt Sandstone	Veins	Eagle River
39	52	20	-5.23	1.01	Basalt Sandstone	Veins	Eagle River
40	77	16	-1.13	0.09	Conglomerate	Cements	Copper Harbor
41	35	13	-3.01	0.03	Conglomerate	Veins	Copper Harbor
42	122	33	-2.63	0.35	Conglomerate	Veins	Copper Harbor
43	1	41	-1.13	0.42	Conglomerate	Cements	Bette Grise Bay
44	108	18	-3.19	0.82	Limestone	Limestone	Limestone Mountain
45	177	28	-2.16	1.03	Sandstone	Cements	Huron Mountains
46	51	0	-1.54	0.05	Sandstone	Cements	Huron Mountains
47	138	40	-1.02	0.25	Sandstone	Cements	Huron Mountains
48	20	30	-3.65	0.04	Sandstone	Cements	Big Bay
49	47	20	-0.13	0.14	Sandstone	Cements	Marquette
50	3	25	-6.11	0.47	Sandstone	Cements	Huron Mountains
51	10	33	-2.34	0.63	Basalt	Amygdules	Scoville Point
52	43	14	-3.9	0.23	Conglomerate	Cements	Denzer, Michigan
53	38	10	-2.11	0.53	Basalt	Vains	Leif Erickson Park
54	33	0	-1.22	0.23	Basalt	Veins	Minong Ridge
55	11	22	-3.19	0.31	Basalt	Veins	Minong Ridge

TABLE 2. CALCITE TWINNING STRAIN DATA (continued)

Specimen	e_1 Trend[†] (°)	NEVs* (%)	e_1 Magnitude[†] (%)	Nominal Error[§]	Rock Type	Calcite Type	Location
56	38	12	-2.33	0.46	Basalt	Veins	Smithwick Island
57	21	0	-2.21	0.13	Basalt	Veins	Smithwick Island
58	10	6	-7.76	0.07	Basalt Dike	Veins	Tower
59	170	8	-5.7	0.18	Basalt Dike	Veins	Lake Kabetogama
60	36	28	-2.27	1.03	Limestone	Limestone	Mellen
61	164	12	-5.93	0.09	Limestone	Limestone	L'Anse
62	120, 81	40	-5.01	0.44	Peridotite	Veins	Marquette

*PEVs and NEVs are positive and negative expected values, respectively.
[†]e_1 directions are azimuthal; plunges indicate a shortening axis that is not subhorizontal.
[§]For nominal error calculation, see Groshong et al., 1984.

Figure 3. Photomicrograph of a twinned calcite amygdule that preserves two sets of twin lamellae in a basalt. Amygdule diameter is 300 microns.

six samples contain layer-parallel shortening strains, three of which are rift parallel (sites 24, 37, 40) and three which are rift normal (sites 21, 27, 29). Bedding at site 27 is quite steep and the shortening strain is layer parallel. Strain magnitudes average −5.01% and the percentage of NEVs for the group was low (5.16%).

Calcite veins

Calcite vein fillings predominate throughout the basalts and Oronto Group sediments in the rift but are rare in adjacent Bayfield Group and Jacobsville Sandstone (Fig. 7, Table 2; sites 1, 2, 8, 9–16, 22, 23, 28, 30, 35, 38, 39, 41, 42, 53–57). Within the rift a variety of vein orientation subsets were observed, but the dominant vein orientations are vertical fillings that are normal to the rift axis (e.g., sites 37, 40); some calcite veins are 2 m wide and extend laterally for hundreds of meters. The calcite veins as a group contain high strains (-3.91%) and the most noise (NEVs 30.50%). Regionally, both rift-parallel and rift-normal horizontal shortening strains are represented, with the former better developed on the western rift margin. In the Keweenawan thrust footwall near Wakefield (site 30), the Powder Mill Group basalts are displaced along the Pelton Creek thrust (Hedgman, 1992) and the calcite preserves a clear north-south, thrust-transport shortening strain pattern. Three vein sets are present in the Chengwatana basalts at sites 13 to 15 and a wide variety of strain results is observed, one of which required a NEVs split (site 14); shallow, east-dipping thrust faults are present in these west-dipping basalt flows

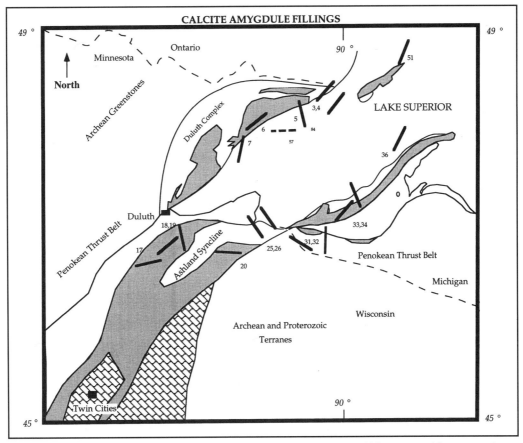

Figure 4. Shortening strains preserved by twinned amygdule calcite. Solid lines are positive expected values (PEVs) data splits; thick dashed lines are negative expected values (NEVs) splits (site 5). Medium numbers refer to specimen locations described in Table 2. Smallest numbers represent the plunge (°) of e_1. See Figure 1 for legend.

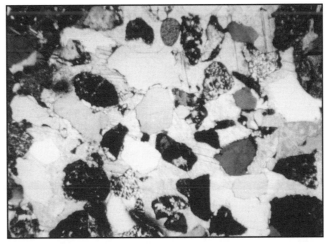

Figure 5. Mechanical twins preserved in the calcite cement of the Jacobsville Sandstone, Huron Mountains, Michigan. Photomicrograph width is 3 mm.

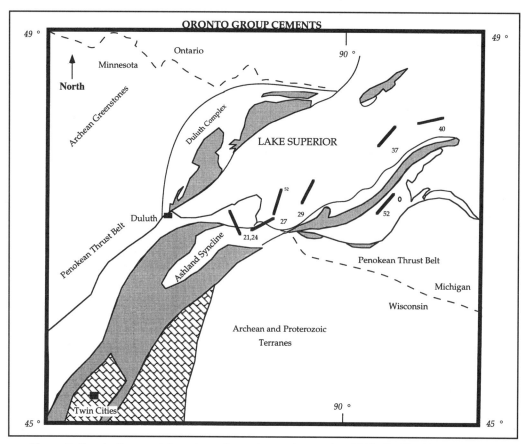

Figure 6. Oronto Group calcite cement shortening strain axes (see Table 2). Site 27 has a steep plunge that is bedding-parallel. Medium numbers refer to specimen locations described in Table 2. The smaller number "52" represents plunge (°) of e_1. Figure 1 gives legend.

and may be related to the complex strains. Calcite is interlayered with Keweenawan-aged basalt dikes at Thomson Dam; samples 11 and 12 are from the same dike but the margin (north-south strike) and interior (east-west strike), respectively. The complex deformation history probably records local strains associated with dike emplacement.

Jacobsville Sandstone cements

Calcite cements in the Jacobsville Sandstone have been described by Hamblin (1958) and are best preserved either along the Keweenawan thrust contact or between the Huron Mountains and Marquette, Michigan, along Lake Superior. We analyzed the twinning strains of six samples (sites 44–50) from the Keweenawan footwall and found low strain values (–1.60%), relatively low NEVs (26.5%) and a consistent north-south, horizontal shortening strain pattern (average e1 trend: N13°E; Fig. 8).

Clastic dikes in Jacobsville Sandstone

Post–Jacobsville Sandstone clastic dikes within the Keweenawan footwall have been identified at three locations (see Kalliokoski, 1982, p. 151), and more recently by Wilson (1990). The clastic dikes in the Huron Mountains crosscut the Jacobsville Sandstone, are unconformably overlain by Cambrian Munising Sandstone, and are cemented by calcite. The dikes strike north-south with vertical dips and along their margins have north-south, horizontal fault striations indicating minor dextral strike-slip motion has occurred (Fig. 8). Site 50 is the location of one clastic dike that has a calcite cement which preserves a horizontal shortening twinning strain that is parallel to the dike.

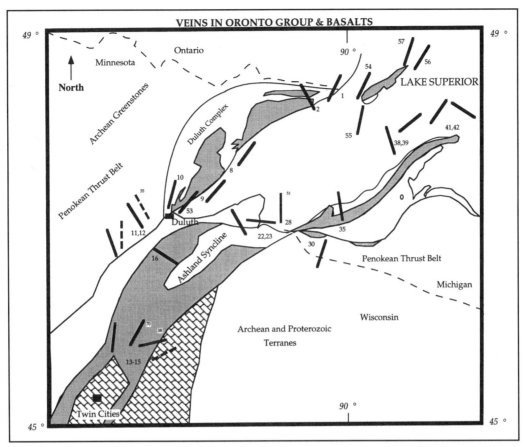

Figure 7. Vein calcite shortening strains. Positive expected values (PEVs) and negative expected values (NEVs; dashed lines) data splits are present at sites 11 and 12 and 13 to 15 (see Table 2). Site 28 has a steep dip to the northwest, and the shortening is bedding parallel. Medium numbers refer to specimen locations de-scribed in Table 2. Figure 1 gives legend.

Paleozoic carbonates

Milstein (1987) has reviewed the setting of the Limestone Mountain outliers (site 44) of folded Ordovician-Devonian (see Coleville, 1983) carbonates and commented on their origins. The nearest rocks of similar age are 242 km (~150 mi) to the southeast on the erosional margin of the Michigan basin. We have found a layer-parallel shortening twinning strain that is parallel to both the Keweenawan and Appalachian thrust transport directions (Fig. 9; see Craddock et al., 1993). The underlying Jacobsville Sandstone is also folded harmonically with the Paleozoic carbonates in two locations.

DISCUSSION

In this study we document the spatial and temporal changes in the stress and strain fields, as preserved by twinned calcite of the different rift calcite elements, formed during the closure of the Midcontinent Rift. We also document the uniqueness of twinning strains in pre-rift, Archean and Early Proterozoic calcite and limestones as compared to the younger rifting strains.

The pre-rift calcite twinning reflects a Penokean orogen strain, with the exception of the Archean peridotite (site 55). None of these strain analyses indicate a noncoaxial overprint from deformation associated with the younger, crosscutting Midcontinent Rift as evidenced by the low NEVs percentages, although block rotations are a possibility to explain the shortening axis of site 53.

The Douglas thrust fault seems to be a discontinuity along which layer-parallel, subhorizontal, and rift-parallel shortening (e.g., strike-slip; see Lomando and Engelder, 1984) strains are separated from the remainder of the rift where the twinning strains record a mixture of layer-parallel and subhorizontal rift-parallel and rift-normal (e.g., thrusting)

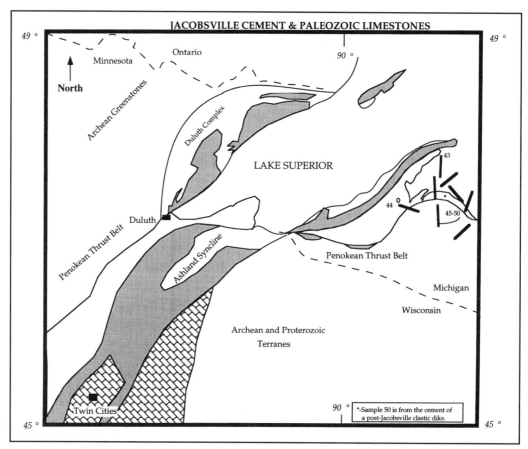

Figure 8. Jacobsville Sandstone calcite cement shortening strains, and strain results from a post-Jacobsville clastic dike (site 50), and calcite from the Limestone Mountain outlier (site 44; see Table 2). Figure 1 gives legend.

shortening. East of the rift in the Keweenaw thrust footwall, the Jacobsville Sandstone and younger clastic dikes preserve a north-south, horizontal shortening strain. This fabric is overlain by a horizontal, northwest-southeast shortening strain in Paleozoic limestones related to reactivation of the Keweenaw thrust. Consequently, thrusting was the most recent tectonic event and was likely preceeded by strike-slip motion (rift-parallel horizontal shortening strains) that was preceeded by rift-forming normal faulting.

The Keweenaw rift had a thrust closure history that began at 1,060 Ma and, along the Keweenawan thrust, sediments as young as Devonian are deformed. The twinning strains indicate that rift-parallel shortening (strike-slip motion?) predominated along the western margin of the rift and a combination of rift-parallel and rift-normal (e.g., thrusting) shortening dominated the eastern rift margin. Northerly, horizontal shortening preserved in the Jacobsville Sandstone is also found in the younger, calcite-cemented clastic dikes. At Limestone Mountain, the folded

Ordovician-Devonian carbonates preserve a layer-parallel shortening strain that is normal to both the Keweenawan thrust and the Appalachian-Ouachita front (see Craddock et al., 1993). The rift-normal horizontal shortening strains probably reflect both the coincident Grenville orogen (ca. 1.0 Ga) to the east (Gordon and Hempton, 1986) and, if the folds at Limestone Mountain are related to reactivation of the Keweenawan thrust, a Permian displacement on this fault (see Roberts, 1940).

ACKNOWLEDGMENTS

This project was funded by National Science Foundation grant EAR-9004181, the Pew Charitable Trust, the Blandin Foundation, the Huron Mountain Wildlife Foundation, and the Beltman Trust. Reviews of the manuscript by Bill Cannon and Steve Marshak greatly improved the manuscript. Field assistance was cheerfully provided by Andrea Troolin, Mark Schmitz, Karl Wirth, and Annie Craddock.

Figure 9. Clastic dike, which is cemented by twinned calcite, Lake Superior shoreline, Huron Mountains, Michigan. Pen for scale in center of photo.

REFERENCES CITED

Beck, W., and Murthy, V. R., 1982, Rb-Sr and Sm-Nd isotopic studies of Proterozoic mafic dikes in northeastern Minnesota: Institute on Lake Superior Geology, 28th, International Falls, Minnesota, p. 5.

Bornhorst, P. J., Paces, J. B., Grant, N. K., Obradovich, J. D., and Huber, N. K., 1988, Age of native copper mineralization, Keweenaw Peninsula, Michigan: Economic Geology, v. 83, p. 619–625.

Cannon, W. F. and Gair, J. E., 1970, A revision of stratigraphic nomenclature for middle Precambrian rocks in northern Michigan: Geological Society of America Bulletin, v. 81, p. 2843–2846.

Cannon, W. F., Peterman, Z. E., and Sims, P. K., 1993, Crustal-scale thrusting and origin of the Montreal River monocline—a 35 km thick cross section of the Midcontinent Rift in northern Michigan and Wisconsin: Tectonics, v. 12, p. 728–744.

Coleville, V. R., 1983, Ordovician-Silurian eustatic sea level changes at Limestone Mountain, Houghton County, Michigan: [M.S. thesis]: Milwaukee, University of Wisconsin, 143 p.

Craddock, J. C., 1972, Late Precambrian regional geological setting, in Sims, P.K., and Morey, G.B., eds., Geology of Minnesota: A centennial volume: Minneapolis, Minnesota Geological Survey, p. 281–291.

Craddock, J. P., Jackson, M., van der Pluijm, B. A., and Versical, R. T., 1993, Regional shortening fabrics in eastern North America: Far-field stress transmission from the Appalachian-Ouachita orogenic belt: Tectonics, v. 12, p. 257–264.

Craddock, J. P. and Moshoian, A., 1996, Continuous Proterozoic strike-slip fault–en echelon fracture arrays in deformed Archean rocks: Implications for fault propagation mechanics and dike emplacement: Proceedings, International Basement Tectonics Conference, 10th, Duluth, Minnesota (August, 1992), p. 379–407.

Craddock, J. P. and van der Pluijm, B. A., 1989, Late Paleozoic deformation of the cratonic carbonate cover of eastern North America: Geology, v. 17, p. 416–419.

Dickas, A., 1986, Comparative Precambrian stratigraphy and structure along the Mid-continent Rift, American Association of Petroleum Geologists Bulletin, v. 70, p. 225–238.

Gordon, M. B., and Hempton, M. R., 1986, Collision-induced rifting: The Grenville orogeny and the Keweenawan rift of North America: Tectonophysics, v. 127, p. 1–25.

Green, J. C., 1982, Geology of Keweenawan extrusive rocks, in Wold, R. J., and Hinze, W. J., eds., The geology and tectonics of the Lake Superior basin: Geological Society of America Memoir 156, p. 47–55.

Groshong, R. H., Jr., 1972, Strain calculated from twinning in calcite: Geological Society of America Bulletin, v. 83, p. 2025–2038.

Groshong, R. H., Jr., 1974, Experimental test of the least squares strain gauge technique using twinned calcite: Geological Society of America Bulletin, v. 85, p. 1855–1864.

Groshong, R. H., Jr., Teufel, L. W., and Gasteiger, C. M., 1984, Precision and accuracy of the calcite strain-gage technique: Geological Society of America Bulletin, v. 95, p. 357–363.

Halliday, A. N., Ohr, M., Mezger, K., Chesley, J. T., Nakai, S. and DeWolf, C. P., 1992, Recent developments in dating ancient crustal fluid flow: Reviews of Geophysics, v. 29, p. 577–584.

Hamblin, W. K., 1958, The Cambrian sandstones of northern Michigan: Michigan Geological Survey Publication 51, 231 p.

Hedgman, C. A., 1992, Petrology and provenance of a conglomerate facies of the Jacobsville Sandstone: Ironwood to Bergland, Michigan: Proceedings, Institute on Lake Superior Geology, 38th, p. 33–34.

Jackson, M., Sun, W. and Craddock, J. P., 1992, The rock magnetic fingerprint of chemical remagnetization in Midcontinental Paleozoic carbonates: Geophysical Research Letters, v. 19, p. 781–784.

Jamison, W. R. and Spang, J. H., 1976, Use of calcite twin lamellae to infer differential stress: Geological Society of America Bulletin, v. 87, p. 868–872.

Kalliokoski, J., 1982, Jacobsville Sandstone, in Wold, R. J., and Hinze, W. J., eds., The geology and tectonics of the Lake Superior basin: Geological Society of America Memoir 156, p. 147–155.

Kalliokoski, J., and Lynott, J. S., 1987, Huron River: Precambrian unconformities and alteration at and near Big Eric's Crossing, Michigan, in D. L. Biggs, ed., North Central Section of the Geological Society of America: Centennial Field Guide Volume 3: Boulder, Colorado, Geological Society of America, p. 273–276.

Lomando, A. J., and Engelder, T., 1984, Strain indicated by calcite twinning: implications for deformation of the early Mesozoic northern Newark Basin, New York: Northeastern Geology, v. 6, p. 192–195.

Lu, G., Marshak, S., and Kent, D. V., 1990, Characteristics of magnetic carriers responsible for late Paleozoic remagnetization in the carbonate strata in the Midcontinent, USA: Earth and Planetary Science Letters, v. 99, p. 351–361.

McSwiggen, P. L., Morey, G. B., and Chandler, V. W., 1987, New model of the Midcontinent Rift in eastern Minnesota and western Wisconsin: Tectonics, v. 6, p. 677–685.

Milstein, R. L., 1987, Anomalous Paleozoic outliers near Limestone Mountain, Michigan in D. L. Biggs, ed., North Central Section of the Geological Society of America: Centennial Field Guide Volume 3: Boulder, Colorado, Geological Society of America, p. 263–268.

Mudrey, M. G., Jr., 1979, Middle Precambrian geology of northern Wisconsin: Institute on Lake Superior Geology, 25th, Field Trip Guidebook 4, 44 p.

Ojakangas, R. W., and Morey, G. B., 1982, Keweenawan sedimentary rocks of the Lake Superior region: a summary, in Wold, R. J. and Hinze, W. J., eds., Geology and tectonics of the Lake Superior basin: Geological Society of America Memoir 156, p. 157–164.

Roberts, E., 1940, Geology of the Alston district, Houghton and Baraga Counties, Michigan [M.S. thesis]: Pasadena, California Institute of Technology, 48 p.

Southwick, D. L., and Halls, H. C., 1987, Compositional characteristics of the Kenora-Kabetogama dike swarm (Early Proterozoic), Minnesota and Ontario: Canadian Journal of Earth Sciences, v. 24, p. 2197–2205.

Teufel, L. W., 1980, Strain analysis of experimental superposed deformation using calcite twin lamellae: Tectonophysics, v. 65, p. 291–309.

Turner, F. J., 1953, Nature and dynamic interpretation of deformation lamellae in calcite of three marbles: American Journal of Science, v. 251, p. 276–298.

Wilson, G. C., 1990, Late Archean structural evolution and Keweenawan and Cambrian structure and sedimentation in the Huron Mountains, Marquette County, Michigan [Ph.D thesis]: University of Wisconsin, 317 p.

Wold, R. J. and Hinze, W. J., eds., 1982, The geology and tectonics of the Lake Superior basin: Geological Society of America Memoir 156, 280 p.

MANUSCRIPT ACCEPTED BY THE SOCIETY JANUARY 16, 1996

Geological Society of America
Special Paper 312
1997

Tectonic context of native copper deposits
of the North American Midcontinent Rift System

Theodore J. Bornhorst
Department of Geological Engineering and Sciences, Michigan Technological University, 1400 Townsend Drive, Houghton, Michigan 49931

ABSTRACT

Economic deposits of native copper are an important feature of the Precambrian North American Midcontinent Rift System. The rift was progressively filled by a thick succession of basalt and then red clastic sedimentary rocks from about 1,109 to 1,060 Ma. Regional contraction of the rift at about 1,060 Ma produced faulting, fracturing, folding, and uplift of rocks on the edges of the rift, including transformation of original graben-bounding faults into high-angle reverse faults. Native copper ore deposits are contemporaneous with these faults/fractures that integrated the paleo-hydrologic system and provided pathways for upward movement and focussing of ore fluids into permeable and porous tops of basalt lava flows and interflow sedimentary rocks. Potential ore fluids were likely generated throughout the tectonic history of the rift by burial of strata and elevated basal heat flow. It is postulated that the coincidence of available "burial" metamorphic fluids, generated via the thermal pulse related to the rifting event, and faults/fractures, generated via regional Grenvillian contraction, provided the critical component in the genetic model of the native copper deposits.

INTRODUCTION

The North American Midcontinent Rift System is host to the world's largest native-copper mining district. Cumulative production between 1845 and 1968 was about 5 billion kg of refined copper (Weege and Pollack, 1971). Although about 96% of the production came from a 45-km-long belt in vicinity of Houghton and Calumet (Fig. 1), many tops of basalt flows in the Keweenaw Peninsula contain small uneconomic amounts of native copper (Broderick, 1931). Similar uneconomic amounts of native copper occur in basalt flows throughout the exposed portions of the rift, including Isle Royale in Michigan, Wisconsin, Minnesota, and Ontario, Canada, (Cannon and McGervey, 1991), providing evidence that hydrothermal fluids capable of precipitating native copper extensively permeated rift-filling volcanic rocks. The Keweenaw native copper deposits are a unique feature of the Earth's geological history. Why is the North American Midcontinent Rift System host to large native copper deposits when other rift systems are not; and why are economic deposits only known from the Keweenaw Peninsula?

The Keweenaw Peninsula native copper district has been subjected to a wide variety of geologic studies over the past 100 years. These studies provide cumulative evidence regarding the relative timing and localization of the native copper ore deposits (e.g., Butler and Burbank, 1929; White, 1968; Bornhorst et al., 1988). Integration of geologic constraints with absolute age constraints from recent geochronologic studies makes it possible to place the deposition of native copper into the context of the tectonic evolution of the rift.

Bornhorst, T. J., 1997, Tectonic context of native copper deposits of the North American Midcontinent Rift System, *in* Ojakangas, R. W., Dickas, A. B., and Green, J. C., eds., Middle Proterozoic to Cambrian Rifting, Central North America: Boulder, Colorado, Geological Society of America Special Paper 312.

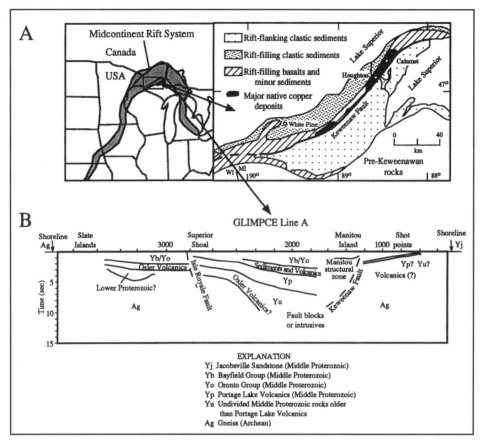

Figure 1. A, Location of the Keweenawan Peninsula native copper district within the North American Midcontinent Rift System. Large black area contains all major deposits, two smaller areas contain minor deposits. B, Interpretative cross section across the Lake Superior segment of the Midcontinent Rift System by Cannon et al. (1989) from seismic-reflection profile Line A. Depths are estimates of Cannon et al. (1989).

GEOLOGIC SETTING OF THE NATIVE COPPER DEPOSITS

Rift-filling volcanic and sedimentary rocks of the Keweenaw Peninsula dip moderately toward Lake Superior and include the Portage Lake Volcanics, Copper Harbor Conglomerate, Nonesuch Shale, and Freda Sandstone (Fig. 2). The Portage Lake Volcanics (PLV) are the host rock unit for the native copper ore deposits. It is composed of over 200 individual subaerial high-alumina, olivine tholeiitic flood basalt lava flows with occasional interflow sediments. A total thickness of about 5 km of the PLV is exposed on the Keweenaw Peninsula. Recently acquired seismic profiles across Lake Superior show that near the center of the rift the total volcanic section is more than 25 km thick, including about 10 km of PLV (Fig. 1; Cannon et al., 1989; Hinze et al., 1990; Cannon, 1992). In the currently accepted tectonomagmatic model for the Midcontinent Rift System large volumes of magma were extruded in response to a period of rifting over an asthenospheric mantle plume (Hutchinson et al., 1990). Rift magmatism extended from 1,109 to 1,087 Ma while the PLV was erupted during a 2-

to 3-m.y. span of time at about 1,095 Ma (Fig. 3; Davis and Paces, 1990; Paces and Miller, 1993).

A typical subaerial lava flow has an average total thickness of about 10 to 20 m and consists of a massive (vesicle-free) interior capped by a vesicular flow top (Paces, 1988; White, 1960). About one-fourth of the vesicular flow tops are also brecciated. A few of the thicker flows and interflow sedimentary rocks can be traced laterally along strike for as much as 90 km, although many flows have much less continuity in strike direction. Interflow conglomerates and sandstones are as much as 40 m thick and make up less than 5 volume percent (Merk and Jirsa, 1982) of the PLV. They increase in abundance upward in the PLV section and are dominated by well-lithified pebble to boulder conglomerate with lesser interbedded sandstone and occasional significant thickness of siltstone and shale. Both tops of lava flows, especially brecciated varieties, and interflow conglomerates host native copper ore deposits in the Keweenaw Peninsula and in general, secondary minerals. Massive interiors of lava flows are barren and usually do not host large amounts of secondary minerals except along fractures and faults. The hydrothermal fluids were more readily

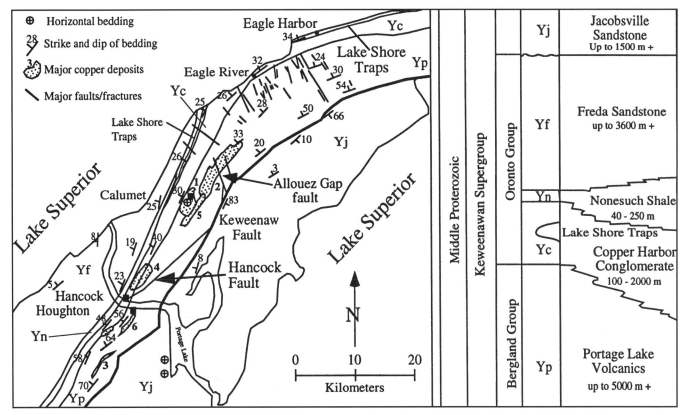

Figure 2. Geologic map and stratigraphic column of the main segment of the Keweenaw Peninsula native copper district showing attitude of bedding and major native copper deposits (modified from White, 1968). Stratigraphic nomenclature of Cannon and Nicholson (1992). Major deposits (total production of refined copper) are marked by numerals: 1, Calumet and Hecla conglomerate (1,922 million kg); 2, Kearsarge flow top (1,029 million kg); 3, Baltic flow top (839 million kg); 4, Pewabic (Quincy) flow tops (490 million kg); 5, Osceola flow top (263 million kg); and 6, Isle Royale flow top (155 million kg). Total district production equals 5,013 million kg of refined copper. The numerals 1–6 indicate the locations of the mines in Figure 4B.

available in the more permeable and porous tops of lava flows and interflow conglomerates.

A thick succession of rift-filling clastic sedimentary rocks (Oronto Group) overlies the PLV (Fig. 2) and represents a change from a period of tectonic development of the rift dominated by volcanism to one dominated by sedimentation. The sedimentary fill represents a waning of magmatic activity but continued subsidence of the rift basin as the thermal anomaly of the asthenoshperic plume decayed (Cannon and Hinze, 1992; Hutchinson et al., 1990). A maximum of about 6 km of rift-filling clastic sedimentary rocks are exposed in western Upper Michigan with as much as 8 km existing beneath the center of Lake Superior (Fig. 1; Cannon, 1992; Cannon et al., 1989). These clastic sedimentary rocks are dominated by red-colored conglomerates (Copper Harbor Conglomerate) and red-colored sandstones (Freda Sandstone) with a relatively thin intervening interval of gray to black shale (Nonesuch Shale). Initiation of deposition of mature red-colored sandstones (Jacobsville Sandstone) across the entire basin late in the thermal subsidence phase of the rift is proposed by Cannon et al. (1989) and Hinze

et al. (1990). The age of overlying rift-filling strata is poorly constrained but is likely older than about 1,060 Ma (approximate age of regional contraction) as the Freda Sandstone and at least some of the Jacobsville Sandstone was deposited prior to initiation of contractional faulting (Fig. 3).

The Keweenawan rift-filling strata dip from 5 to 70° north to northwest toward the center of the rift (Lake Superior). Dip angles increase toward the base of the section such that volcanic strata tend to dip more than 30° and overlying sedimentary strata dip less than 30° (Fig. 2). White (1960) demonstrated that a relatively horizontal datum level was maintained throughout the filling of the rift, implying significant syndepositional downwarpage of rift-filling strata. White (1968) estimated that syndepositional downwarpage could explain about half of the present-day dip of the PLV.

The last phase of the Midcontinent Rift System is characterized by transformation of the original graben bounding normal faults into reverse faults (Fig. 3). The Keweenaw fault was originally a major graben bounding growth fault based on interpretation by Cannon et al. (1989) of a seismic line (line A)

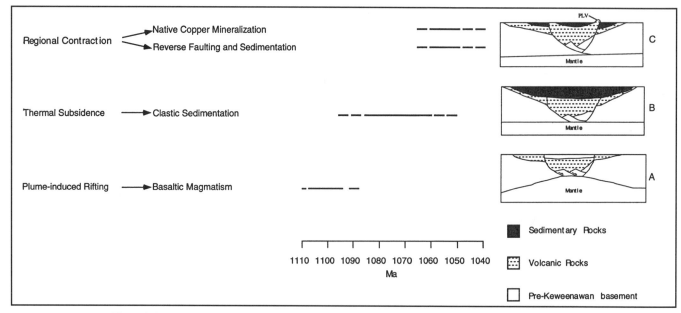

Figure 3. Temporal progression of major geologic events of the North American Midcontinent Rift System. Schematic cross sections of development of the rift from Cannon et al. (1989). A, An initial broad crustal sag filled with volcanic rocks is followed by extension that results in normal growth faults and eruption of large volumes of plume-induced basalt into the central graben. B, After volcanism wanes, thermal subsidence continues with the basin progressively filled with clastic sediments. C, The last phase of development of the rift is regional Grenvillian contraction, which inverts original graben bounding faults into reverse faults. This results in the uplift of buried rift strata, erosion, and exposure of the Portage Lake Volcanics (PLV) in the Keweenaw Peninsula. Contraction generated faults/fractures provided for upward movement and focussing of ore fluids into permeable and porous tops of basalt lava flows and interflow sedimentary rocks within the Portage Lake Volcanics.

crossing the tip of the Keweenaw Peninsula, but is now a low- to high-angle reverse fault (Figs. 1 and 2). The Keweenaw fault has several kilometers of reverse displacement that caused steepening of already tilted strata (due to syndepositional downwarpage) to their present attitudes. Within rift-filling strata of the Keweenaw Peninsula, faults, fractures, and broad open folds developed in response to this contractional event (White, 1968; Butler and Burbank, 1929). The shallow-dipping Jacobsville Sandstone, more than 3 km thick, was deposited during and after active reverse movement along the Keweenaw fault in a rift-flanking basin (Figs. 2 and 3; Hedgman, 1992). Near the Michigan-Wisconsin border, Cannon et al. (1993) have determined that high-angle reverse faulting occurred about 1,060 ± 20 Ma based on reset Rb-Sr biotite ages. The timing and probable cause of this contractional event is continental collision along the Grenville front (Cannon, 1994; Cannon and Hinze, 1992; Hoffman, 1989). Although the duration of regional contraction is unknown, thermal models (Price and McDowell, 1993; Price et al., 1996) suggest that completion by about 1,040 Ma is reasonable.

NATIVE COPPER DEPOSITS

Native copper represents over 99% of the metallic minerals in the mined orebodies of the Keweenaw Peninsula and is localized as lens and blanket-like orebodies along certain stratigraphic horizons of brecciated and vesicular flow tops (58.5% of total production) and interflow conglomerate beds (39.5% of production), and as veins in fracture systems (2% of production; Butler and Burbank, 1929; White, 1968). Most of the 5 billion kg of refined native copper was produced from 13 major stratabound deposits with the 4 largest deposits producing 85% of the copper at a grade of about 2% (Weege and Pollack, 1971). Small persistent quantities of native silver (less than 0.1%; White, 1968) accompany the native copper. Most of the native copper carries a small amount of arsenic in solid solution (less than 5% arsenic in total copper + silver + arsenic, typically less than 0.2%; Broderick, 1929). Copper-nickel arsendies occur in veins that are paragenetically late (Butler and Burbank, 1929). Within native copper deposits chalcocite, also paragenetically late, occurs as small veins cutting flow-top deposits and as coatings on joints with calcite in conglomerate deposits (White, 1968). The relationship between several copper sulfide deposits that occur in flow tops near the base of the PLV within the Keweenaw Peninsula (Broderick et al., 1946; Robertson, 1975) is not well documented.

Native copper mineralization is closely associated in time and space with secondary minerals formed during alteration of the PLV, although individual deposits may not exactly follow the districtwide paragenesis (Fig. 4). Metamorphic zoning based on distribution of amygdule-filling minerals, equivalent to low temperature and pressure burial metamorphism to zeolite

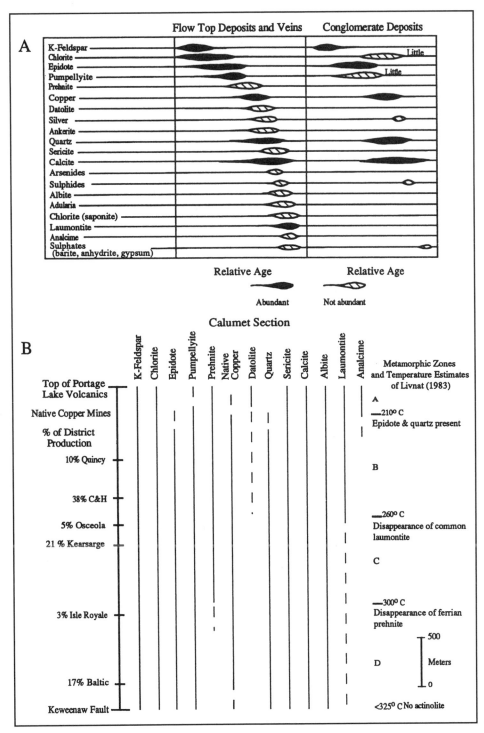

Figure 4. A, Paragenesis of secondary minerals in flow top deposits and veins and conglomerate deposits of the Keweenaw Peninsula native copper district (modified from Butler and Burbank, 1929). B, Distribution of amygdule- and vein-filling minerals in the Calumet cross section of the Portage Lake Volcanics (compiled from Livnat, 1983; Stoiber and Davidson, 1959). Vertical distribution of secondary minerals is similar throughout the area of the major native copper mines listed on the vertical stratigraphic column (see Fig. 2 for mine locations).

and prehnite-pumpellyite facies (Jolly and Smith, 1972), varies vertically within the PLV with significant deposits of native copper occurring throughout the stratigraphic/metamorphic section. Lower metamorphic grades extend deeper into the PLV northeast of the major native copper deposits (Cornwall, 1955; Cornwall and White, 1955) indicating that metamorphic zones laterally cross stratigraphy. Broderick (1929) and Livnat (1983) showed that the metamorphic mineral zones (isograds) within the PLV dip more shallowly towards Lake Superior compared to the volcanic strata, which dip moderately, implying that the volcanic units were tilted prior to metamorphism and associated native copper mineralization. While flow tops are intensively altered, massive interiors of lava flows are much less altered and for thicker interiors hydrothermal alteration is confined to the vicinity of fractures. Secondary minerals exist in all flows regardless of their thickness, but original igneous minerals are present in the massive interiors of flows (Paces, 1988). This and the unaltered geochemical composition of many flow interiors are consistent with an impermeable character of the interiors of lava flows. Interflow conglomerate beds make up only a small volume of the PLV (<5%) but, host comparatively large amount of economic native copper (~40%). Many stratabound deposits are closely associated with veins that contain the same gangue minerals as in the stratabound deposits (Butler and Burbank, 1929; Broderick, 1931; White, 1968) implying they are synchronous. Veins are indicative of the overall epigenetic timing of mineralization.

STRUCTURE

Numerous faults/fractures occur in the PLV of the Keweenaw Peninsula. Some, perhaps many, were originally formed during the extensional phase of the Midcontinent Rift System, most notably the Keweenaw fault, and were reactivated during the contractional phase. Most faults and folds are likely related to reverse movement along the Keweenaw fault (White, 1968). The Keweenaw fault is the major fault in the Keweenaw Peninsula and is presently a low- to high-angle reverse fault with a minimum of 3 km of reverse motion (originally a graben-bounding normal fault; Cannon et al., 1989). The Keweenaw fault strikes more or less parallel to the bedding of the PLV (Fig. 2) and measured dips of the fault plane are generally subparallel to dips of the adjacent PLV (Butler and Burbank, 1929). Broderick et al. (1946) noted that the Keweenaw fault would make an ideal conduit for ore fluids. Although no deposits are located along the main fault itself, some fluid movement along the Keweenaw fault and adjacent rocks is indicated by highly altered rocks and by several small native copper deposits along nearby subparallel branch faults. The St. Louis deposit, target of recent evaluation, contains 8 million tons of ore grading 0.8% copper within a fault zone subparallel and within about 50 m of the Keweenaw fault (Hong, 1992; Northern Miner, 1990). Like the Keweenaw Fault, in other districts the main faults show little or no mineralization (Sibson, 1987).

Several reverse faults transverse the PLV at higher angles to the strike of bedding than the Keweenaw fault, including the Hancock and Isle Royale faults. High-angle faults striking north to northwest are common in the area of the major native copper deposits near the crest of a regional anticlinal structure (White, 1968). Displacement along these faults varies from none (fractures) to 200 m (Butler and Burbank, 1929). Native copper deposits are spatially associated with some of these faults/fractures.

In the Baltic and Isle Royale flow-top deposits, which produced 20% of district copper, faults were the principal pathway for the upward movement and focusing of ore fluids into the stratabound ore horizon (Broderick, 1931). A similar situation exists for the Caledonia, Mass, and Adventure mines (Fig. 1 in PLV east of White Pine). There native copper is hosted in flow tops in close association with abundant faults and fractures (Butler and Burbank, 1929; Bornhorst and Whiteman, 1992). Faults host native copper, some with quite high grades. Native copper occurs in the massive interiors of lava flows where cut by faults. Fault brecciated and recemented alteration minerals indicate that faulting occurred before, during, and after deposition of native copper. For these mines it is likely that faults were a major pathway for transport of ore fluids to the flow tops.

The Allouez Gap fault (Fig. 2) cuts the PLV at a high angle. Almost every permeable horizon near the Allouez Gap fault contains above average amounts of native copper; nowhere else in the district are there so many mineralized beds. R. J. Weege, former chief geologist of C & H Mining Company, suggested in an unpublished company report that the Allouez Gap fault was the single most important fluid pathway in the district, perhaps linking 60% of the district production. The largest flow-top deposit in the district, the Kearsarge deposit (1,026 million kg of copper, 21% of district production) occurs where the Allouez Gap fault bisects the thickest segment of the Kearsarge flow along its 55-km-strike length (Fig. 2). The small Kingston deposit (9 million kg of copper, <1% of district production) is also bisected by the Allouez Gap fault. Distribution of ore and alteration led Weege et al. (1972) to conclude that fluids moved outward from the fault. The Houghton and Allouez conglomerate deposits (50 million kg of copper, ~1% of district production) are also near the Allouez Gap fault with local faults and the conglomerates themselves connected directly to the fault. The giant Calumet and Hecla Conglomerate deposit (1,900 million kg of copper, 38% of district production) is about 1.8 km west of the Allouez Gap fault. The axis of the orebody trends obliquely at depth toward the fault (Butler and Burbank, 1929). Genetic connection between the Calumet and Hecla Conglomerate deposit and the Allouez Gap fault is speculative with a possible model being one where ore fluids move up the Allouez Gap fault and then updip along the permeable conglomerate with ore localized by laterally thinned conglomerate.

Northeast of the major native copper district small vein dominated deposits (Fig. 2) are localized just beneath the thickest basalt flow in the district, the Greenstone flow. These were the first mines in the district. A reasonable model is one in which hydrothermal fluids moved up along cross fractures until stopped by the impermeable massive interior of the Greenstone flow.

The coincidence of abundant faults and fractures with thickening of permeable strata along strike in the Calumet area is well documented for the two largest deposits in the district (60% of district copper production; Kearsarge and Calumet and Hecla deposits). Documentation of the lateral variation in thickness of most stratigraphic horizons is lacking. Perhaps primary depositional thickening represents an "accommodation" zone during rifting such as discussed by Dickas and Mudrey (1989). This zone could have also been the loci of later contractional related structures.

Broad open synclines and anticlines with wavelengths of around 10 km and various orientations are superimposed on the regional dip and are likely related to reverse movement along the Keweenaw fault (White, 1968). Faults with displacement and mineralized tension breaks are common near the crests of anticlines (Butler and Burbank, 1929). Dips of specific horizons tend to steepen 20 to 30° along strike from northeast to southwest of the major native copper deposits, yielding a gently twisted surface (Fig. 2). The strike of bedding changes to a more east-west orientation at the northern end of this twist.

Butler and Burbank (1929), Broderick (1929, 1931), Broderick et al. (1946), White (1968), Weege et al. (1972), and other geologists have pointed out that a close connection exists between native copper mineralization and faults/fractures related to contraction of the rift. Did the faults act as pathways for the movement of mineralizing fluids into stratabound permeable horizons? Sufficient evidence exists to support the general hypothesis that secondary permeability provided by the network of faults/fractures produced during the contractional phase of the rift integrated the plumbing system allowing for upward movement and focussing of ore fluids. At least some faults were synchronous with mineralization. The genetic connection between some deposits and faults is still equivocal.

AGE OF MINERALIZATION

Stratabound and vein deposits of native copper are contemporaneous and epigenetic, thus younger than the host PLV. Native copper mineralization is also younger than the Copper Harbor Conglomerate (Copper Harbor Conglomerate is about 1,086 Ma based on U-Pb dating of interformational volcanic rocks; Davis and Paces, 1990) because it hosts rare veins of calcite and native copper. White (1968) interpreted the age of native copper mineralization as after deposition of much or all of the Freda Sandstone. Minor native copper has been reported in the base of a drill hole within the Jacobsville Sandstone near Rice Lake (Weege, personal communication, 1986), suggesting that some mineralization may postdate deposition of at least some of the Jacobsville Sandstone. Since the Jacobsville Sandstone was deposited contemporaneous with reverse movement

along the Keweenaw Fault, the age of mineralization could be either syn- or post contraction.

Bornhorst et al. (1988) used the Rb-Sr method on amygdule-filling microcline, calcite, epidote, and chlorite to determine the absolute age of native copper mineralization in the Keweenaw Peninsula to be between 1060 and 1047 Ma (± ~20 Ma). This age is consistent with the approximate age of the reverse movement along Keweenaw fault of 1060 ± 20 Ma (Cannon et al., 1993).

At the White Pine mine (Fig. 1), Mauk et al. (1992a, b) show evidence for two distinct episodes of copper mineralization: main stage copper sulfides and subordinate native copper formed during diagenesis of the Nonesuch Shale and second stage native copper and subordinate copper sulfide synchronous with thrust faulting. The thrust faults are interpreted as contemporaneous with the Keweenaw fault leading to the inference by Mauk et al. (1992a, b) that second stage copper at White Pine is related to the native copper deposits of the Keweenaw Peninsula. Ruiz et al. (1984) reported an age of 1047 ± 35 Ma for calcite from White Pine that, if second stage, is the same as both the age of the reverse movement along the Keweenaw fault of around 1060 ± 20 Ma (Cannon et al., 1993) and the age of native copper mineralization in the Keweenaw Peninsula of 1060 to 1047 ± ~20 Ma (Bornhorst et al., 1988).

Both field relationships and radiometric dating indicate an age of native copper mineralization in the Keweenaw Peninsula of about 1,060 to 1,050 Ma (Fig. 3). The age of native copper mineralization is the same as the approximate age of the Keweenaw Fault, 1060 ± 20 Ma as determined by Cannon et al. (1993). This is consistent with the hypothesis that faults/fractures generated during the contractional phase of the rift were important pathways for movement of ore fluids.

RELATIONSHIP WITH RIFT DEVELOPMENT

Why should the genesis of native copper ore deposits be related to the contractional phase of the Midcontinent Rift System? In order to understand this relationship one must consider the temporal evolution of hydrothermal fluids in the rift, the effect of contraction on expulsion of fluids, and the channelling of ore fluids by faults.

The thermal history of the rift can be predicted by one-dimensional thermal models (McDowell et al., 1992; Price and McDowell, 1993; Price et al., 1996). These models, using techniques described by Furlong and Edman (1989) and Huntoon and Furlong (1993), suggest that temperatures of a given rift strata increased to a broad temperature peak within about 20 m.y. after deposition of the PLV (about 1,075 Ma) due to burial and elevated basal heat flow. The temperature then began to decrease, but had not substantially declined, before the onset of contraction at about 1,060 Ma. The decay of the thermal anomaly increased at the onset of contraction due to uplift by reverse faulting. Although not included in model calculations, the upward expulsion of fluids during contraction would also increase decay of the thermal anomaly. While model calculations cannot produce unique results, these results are similar to those reported by Cannon et al. (1992). The occurrence of native copper throughout exposed rift-filling basalts and the 30-m.y. time lag between the emplacement of native copper and the end of significant magmatic activity suggests that ore-bearing fluids are related to the regional thermal event rather than a local body of magma. A genetic model relying on regional burial metamorphic-type ore fluids has been advocated by numerous authors, with copper leached from the rift-filling basalt strata themselves (Jolly, 1974; White, 1968; Stoiber and Davidson, 1959).

Two end-member scenarios are reasonable for the source of the ore fluids. The temperature history models predict that the PLV in the center of the rift was dehydrated by as much as 10 to 15 m.y. prior to ore deposition. These fluids could have slowly migrated towards the edge of the filled rift where millions of years after their generation they were then available to be tapped by contraction-related faults and focused into ore-bearing horizons. Alternatively, thermal models suggest that temperature of buried basalt strata on the flanks of the rift was sufficiently high to have been the source of ore-bearing fluids. Thus, the ore fluids need to be tapped from only some few (<10) kilometers beneath the present ore horizons, in the basalt strata on the edge of the filled rift directly downdip from the ore-bearing rocks after burial prior to or during tapping by faults, and not from very thick sections of basalt in the center of the basin.

The minimum amount of rock necessary to provide the 5 billion kg of copper mined in the Keweenaw Peninsula is about 1,400 km³, based on a preleached basalt copper concentration of 60 ppm (Paces, 1988) with the top, porous 15% of each flow (White, 1968) available for leaching at 50% efficiency and density of 2.85 gm/cm³. Seismic data indicate a thickness of 10 km or more of basalts exist within the rift proper immediately downdip from the 45-km-strike length of ore-rich belt in the Keweenaw Peninsula (Hinze et al., 1990; Cannon et al., 1989). A downdip width of only a little more than 3 km of rock is necessary to provide sufficient copper in the ore fluids suggesting that more than enough copper source rock exists within the edge of the rift fill. While migration of burial metamorphic fluids from center to the edge of the rift fill is not precluded, it is not necessary. A column of fluids likely existed prior to onset of the contractional phase with a continuum of fluid types from hotter "burial" metamorphic fluids at depth upward to cooler connate fluids and near the surface meteoric fluids.

The paleohydrologic section on the edge of the rift consisted of relatively thin permeable and porous lava flow tops and interflow sedimentary rocks separated by impermeable massive interiors of lava flows. At the time of ore deposition these stratified permeable and impermeable rocks were tilted (25 to 50°), since ore mineralization was emplaced after both syndepositional downwarpage and at least some of the contractional phase. Variable dip and strike extent and thickness of lava flows and interflow unconformities likely created a spectrum of situations for stratified "paleoaquifers" from isolated to par-

tially connected to continuous (Bornhorst, 1992). It seems reasonable to speculate that simple updip movement of ore fluids was relatively slow and dispersed in these dipping, thin, separated, and sometimes isolated, "paleoaquifers." Wholesale simple updip movement seems an unlikely mechanism to provide large volumes of fluids localized/focussed into the horizons of the native copper ore deposits.

The onset of the contractional phase late in the history of the rift provided a network of faults/fractures that integrated the plumbing system and allowed for easier and more rapid upward movement of fluids. Major faults were principal pathways for focussing of ore fluids where they intersect locally thick permeable strata. Faults may have behaved as valves and become highly permeable pathways immediately postfailure (Sibson et al., 1988; Sibson, 1990) and resulted in periodic upward movement of ore fluids. For faults to act as valves they must cut a vertical pressure gradient with overpressure related to low permeability barriers (Sibson, 1990). The PLV does contain numerous localized and more regionally extensive low permeability barriers in the form of interiors of lava flows. Faults such as high angle reverse faults resulting from reactivation of normal faults, are especially likely to have valve behavior (Sibson, 1990).

Fluid flow in small fractures cutting massive interior of lava flows (e.g., Deloule and Turcotte, 1989) may have also been an important mechanism for the upward transport of fluids between aquifers separated by aquicludes in the paleohydrologic system. White (1968) suggested that during ore mineralization permeability due to fracturing may have been more important than primary permeability through flow tops and interflow sedimentary units. At the ore mineralized horizon, fluid pressures were likely greater than hydrostatic but less than lithostatic (e.g., Powley, 1990), with the pressure dependent on the height of rocks above the ore horizon ("mountains") being maintained by contractional uplift rather than stratigraphic depth (see schematic section C in Fig. 3).

Mixing of ore fluids channeled upward through faults with cooler, more dilute resident fluids may have been an important mechanism for precipitation of native copper. Oxygen isotope data for calcite shows about a 10 per mil spread at a given stratigraphic depth and for quartz, based on less data, the spread is about 5 per mil (Livnat, 1983). This spread could be interpreted as due to fluid mixing. The needed reducing conditions for precipitation of native copper seems to rule out mixing of ore fluids with oxidized groundwaters. Reduction of ore fluids during deposition of native copper would have yielded copper sulfides if sulfur was present in the fluids, supporting low sulfur content of the ore fluids. The oxidation of magnetite to hematite and the prograde metamorphic reaction of pumpellyite to epidote occurred along with native copper deposition and also could have provided reducing conditions needed for the deposition of native copper (Jolly, 1974). Overall, a combination of fluid mixing, fluid-rock interaction and cooling may have caused precipitation of native copper.

If reversed faults were important conduits of mineralizing

fluids with pressure valve type of behavior, then why is the Keweenaw fault relatively unmineralized? Northwest-southeast–oriented horizontal maximum contraction could have caused the generally northeast–trending Keweenaw fault (Fig. 1) to be under high contractional stress, whereas perpendicular structures could result by tension stress and thereby provide more open channelways for movement of ore fluids. Curvature of the Keweenaw fault could have enhanced this effect. The Keweenaw Peninsula is notable for the bending of the strike of rift-filling rocks and the twisting of dips in the highly mineralized Calumet area (Fig. 2).

The Midcontinent Rift System does not seem unusual in either igneous activity or in geothermal gradient as compared to other rifts (Hutchinson et al., 1990). Low grade burial metamorphism/hydrothermal alteration of mafic volcanics is observed throughout the world, yet native copper ore deposits of the Keweenaw Peninsula are unique. Major contractional faulting late in the history of the Midcontinent Rift System, related to Grenvillian deformation (Cannon, 1994), distinguishes it from other flood-basalt provinces. It is postulated that the superposition of this deformation event on temporally available "burial" metamorphic fluids being generated via the thermal pulse related to rifting provided the critical component in the genetic model of the native copper deposits by providing the fluid pathways regardless of whether or not fluids were actually expelled by contraction or simply by thermal/density buoyancy. The thick section of basalts in the rift and a high geothermal gradient may have also played a role (Nicholson et al., 1992), albeit secondary, in the genesis of the native copper deposits.

The localization of large native copper deposits to the Keweenaw Peninsula may be controlled by several factors. White (1968) observed that the major native copper deposits occur in an area of the rift where the strikes of rift-filling rocks bend and where there are many faults and fractures and broad open folds. The volcanic strata dip quite steeply in this segment of the rift. In addition, several stratigraphic horizons are thickest in the Calumet area of the Keweenaw Peninsula. The coincidence of abundant potential fluid pathways along faults and fractures and thickened permeable flow tops and conglomerates could explain the favorable conditions for mineralization in the Keweenaw Peninsula. A favorable orientation of the Keweenaw Peninsula within the regional Grenvillian contractional deformation may have also been important. However, since much of the Midcontinent Rift System is buried beneath younger cover, perhaps another area of native copper deposits remains hidden.

CONCLUSIONS

The Lake Superior segment of the North American Midcontinent Rift System is host to the most significant native copper ore district in the world. Known economic deposits of native copper are confined to the Keweenaw Peninsula of Michigan's Upper Peninsula. Potential ore fluids were likely generated throughout the tectonic history of this continental rifting in response to burial of strata and elevated basal heat flow. However, the formation of economic deposits needed the happy coincidence of active faulting during the still near peak, but slowly dying, thermal pulse to provide pathways for focussing of fluids. These faults were generated by a regional contractional event due to plate collision hundreds of kilometers away along the Grenville front. The contraction event caused a perturbation in the thermal history of the rift by uplift of strata and allowed upward movement of fluids, without influx of additional heat.

ACKNOWLEDGMENTS

The Donors of The Petroleum Research Fund administered by the American Chemical Society provided partial support of this research. S. D. McDowell provided comments on an early draft and on a revised version. Reviews by J. D. Miller and L. G. Woodruff improved the quality of this manuscript.

REFERENCES CITED

Bornhorst, T. J., 1992, An overview of the Keweenaw Peninsula native copper district, Michigan: Society of Economic Geologists Guidebook Series, v. 13, p. 33–62.

Bornhorst, T. J., and Whiteman, R. C., 1992, The Caledonia native copper mine, Michigan: Society of Economic Geologists Guidebook Series, v. 13, p. 139–144.

Bornhorst, T. J., Paces, J. B., Grant, N. K., Obradovich, J. D., and Huber, N. K., 1988, Age of native copper mineralization, Keweenaw Peninsula, Michigan: Economic Geology, v. 83, p. 619-625.

Broderick, T. M., 1929, Zoning in Michigan copper deposits and its significance: Economic Geology, v. 24, p. 149–162, 311–326.

Broderick, T. M., 1931, Fissure vein and lode relations in Michigan copper deposits: Economic Geology, v. 26, p. 840–856.

Broderick, T. M., Hohl, C. D., and Eidemiller, H. N., 1946, Recent contributions to the geology of the Michigan copper district: Economic Geology, v. 41, p. 675–725.

Butler, B. S., and Burbank, W. S., 1929, The copper deposits of Michigan: U.S. Geological Survey Professional Paper 144, 238 p.

Cannon, W. F., 1992, The Midcontinent rift in the Lake Superior region with emphasis on its geodynamic evolution: Tectonophysics, v. 213, p. 41–48.

Cannon, W.F., 1994, Closing of the Midcontinent Rift—A far-field effect of Grenvillian contraction: Geology, v. 22, p. 155–158.

Cannon, W. F., and Hinze, W. J., 1992, Speculations on the origin of the North American Midcontinent Rift: Tectonophysics, v. 213, p. 49–55.

Cannon, W. F., and McGervey, T. A., 1991, Map showing mineral deposits of the Midcontinent Rift, Lake Superior region, United States and Canada: U.S. Geological Survey Miscellaneous Field Studies Map MF-2153, scale 1:500,000.

Cannon, W. F., and Nicholson, S. W., 1992, Revisions of stratigraphic nomenclature within the Keweenawan Supergroup of northern Michigan: U.S. Geological Survey Bulletin 1970-A, 8 p.

Cannon, W. F., and 11 others, 1989, The North American Midcontinent Rift beneath Lake Superior from GLIMPCE seismic reflection profiling: Tectonics, v. 8, p. 305–332.

Cannon, W. F., Woodruff, L. G., and Daines, M. J., 1992, A thermal lag model for native copper mineralization in the Midcontinent Rift System of Lake Superior [abs.]: Geological Society of America Program with Abstracts, v. 24, no. 7, p. A61.

Cannon, W. F., Peterman, Z. E., and Sims, P. K., 1993, Crustal-scale thrusting and origin of the Montreal River monocline—A 35-km-thick cross section of the Midcontinent Rift in northern Michigan and Wisconsin: Tectonics, v. 12, p. 728–744.

Cornwall, H. R., 1955, Geologic Map of the Fort Wilkins Quadrangle, Michigan: U.S. Geological Survey Geologic Quadrangle Maps of the United States, Map GQ-74, scale 1:24,000.

Cornwall, H. R., and White, W. S., 1955, Bedrock geology of the Manitou Island Quadrangle, Michigan: U.S. Geological Survey Geologic Quadrangle Map of the United States, Map GQ-73, scale 1:24,000.

Davis, D. W., and Paces, J. B., 1990, Time resolution of geologic events on the Keweenaw Peninsula and implications for development of the Midcontinent Rift System: Earth and Planetary Science Letters, v. 97, p. 54–64.

Deloule, E., and Turcotte, D. L., 1989, The flow of hot brines in cracks and the formation of ore deposits: Economic Geology, v. 84, p. 2217–2225.

Dickas, A. B., and Mudrey, M. G., Jr., 1989, Central North American case for segmented rift development [abs.]: International Geological Congress Abstracts, 28th, Washington, D.C., v. 1, p. 396–397.

Furlong, K. P., and Edman, J. D., 1989, Hydrocarbon Maturation in thrust belts: thermal considerations, *in* Price, R. A., ed., Origin and evolution of sedimentary basins and their energy and mineral resources: American Geophysical Union Geophysical Monograph 48, p. 137–144.

Hedgman, C. A., 1992, Provenance and tectonic setting of the Jacobsville Sandstone, from Ironwood to Keweenaw Bay, Michigan [M.S. thesis]: Cincinnati, University of Cincinnati, 158 p.

Hinze, W. J., Braile, L. W., and Chandler, V. W., 1990, A geophysical profile of the southern Margin of the Midcontinent Rift System in western Lake Superior: Tectonics, v. 9, p. 303-310.

Hoffman, P. F., 1989, Precambrian geology and tectonic history of North America, *in* Bally, A. W., and Palmer, A. R., eds., The geology of North America—An overview: Geological Society of America, The Geology of North America, v. A, p. 447–512.

Hong, S. M., 1992, Alteration of the Portage Lake Volcanics in the St. Louis mine area near Calumet, Michigan [M.S. thesis]: Houghton, Michigan Technological University, 135 p.

Huntoon, J. E., and Furlong, K. P., 1993, Thermal evolution of the Newark basin: Journal of Geology, v. 100, p. 579–591.

Hutchinson, D. R., White, R. S., Cannon, W. F., and Schulz, K. J., 1990, Keweenaw hot spot: Geophysical evidence for a 1.1 Ga Mantle plume beneath the Midcontinent Rift System: Journal of Geophysical Research, v. 95, p. 10869–10884.

Jolly, W. T., 1974, Behavior of Cu, Zn, and Ni during prehnite-pumpellyite rank metamorphism of the Keweenawan basalts, northern Michigan: Economic Geology, v. 69, p. 1118–1125.

Jolly, W. T., and Smith, R. E., 1972, Degradation and metamorphic differentiation of the Keweenawan lavas of northern Michigan, U.S.A.: Journal of Petrology, v. 13, p. 273–309.

Livnat, A., 1983, Metamorphism and copper mineralization of the Portage Lake Lava Series, northern Michigan [Ph.D. thesis]: Ann Arbor, University of Michigan, 292 p.

Mauk, J. L., Kelly, W. C., van der Pluijm, B. A., and Seasor, R. W., 1992a, Relations between deformation and sediment-hosted copper mineralization: Evidence from the White Pine portion of the Midcontinent Rift System: Geology, v. 20, p. 427–430.

Mauk, J. L., Brown, A. C., Seasor, R. W., and Eldridge, C. S., 1992b, Geology and stable isotope and organic geochemistry of the White Pine sediment-hosted stratiform copper deposit: Society of Economic Geologists Guidebook Series, v. 13, p. 63–98.

McDowell, S. D., Price, K. L., Huntoon, J., and Bornhorst, T. J.,1992, Thermal modeling and illite/smectite geothermometry of the Precambrian Oronto Group, Wisconsin and Michigan [abs.]: Annual Institute on Lake Superior Geology, 38th, Program and Abstracts,v. 38, p. 59–60.

Merk, G. P., and Jirsa, M. A., 1982, Provenance and tectonic significance of the Keweenawan interflow sedimentary rocks: Geological Society of America Memoir 156, p. 97–105.

Nicholson, S. W., Cannon, W. F., and Schulz, K. J., 1992, Metallogeny of the Midcontinent Rift System of North America: Precambrian Research, v. 58, p. 355–386.

Northern Miner, 1990, Great Lakes hits high grade zones with infill drilling: Northern Miner, v. 76, no. 32, p. 3.

Paces, J. B., 1988, Magmatic processes, evolution and mantle source characteristics contributing to the petrogenesis of Midcontinent Rift basalts: Portage Lake Volcanics, Keweenaw Peninsula, Michigan [Ph.D. thesis]: Houghton, Michigan Technological University, 413 p.

Paces, J. B., and Miller, J. D., Jr., 1993, Precise U-Pb ages of Duluth Complex and related mafic intrusions, northeastern Minnesota: Geochronological insights to physical, petrogenetic, paleomagnetic, and tectonomagmatic processes associated with the 1.1 Ga Midcontinent Rift System: Journal of Geophysical Research, v. 98, p. 13997–14013.

Powley, D. E., 1990, Pressures and hydrogeology in petroleum basins: Earth Science Reviews, v. 29, p. 215–226.

Price, K. L., and McDowell, S. D., 1993, Illite/smectite geothermometry of the Proterozoic Oronto Group, Midcontinent Rift System: Clays and Clay Minerals, v. 41, p. 134–147.

Price, K. L., Huntoon, J. E., and McDowell, S. D., 1996, Thermal history of the 1.1 Ga Nonesuch Formation, North American Mid-continent Rift at White Pine, Michigan: American Association of Petroleum Geologists Bulletin, v. 80, p. 1–15.

Robertson, J. M., 1975, Geology and mineralogy of some copper sulfide deposits near Mount Bohemia, Keweenaw County, Michigan: Economic Geology, v. 70, p. 1202–1224.

Ruiz, J., Jones, L. M., and Kelly, W. C., 1984, Rubidium-strontium dating of ore deposits hosted by Rb-rich rocks using calcite and other common Sr-bearing minerals: Geology, v. 12, p. 259–262.

Sibson, R. H., 1987, Earthquake rupturing as a mineralizing agent in hydrothermal systems: Geology, v. 15, p. 701–704.

Sibson, R. H., 1990, Conditions for fault-valve behaviour: Geological Society of London Special Publication, no. 54, p. 14–28.

Sibson, R. H., Robert, F., and Poulsen, K. H., 1988, High-angle reverse faults, fluid-pressure cycling, and mesothermal gold-quartz deposits: Geology, v. 16, p. 551–555.

Stoiber, R. E., and Davidson, E. S., 1959, Amygdule mineral zoning in the Portage Lake Lava Series, Michigan copper district: Economic Geology, v. 54, p. 1250–1277, 1444–1460.

Weege, R. J., and Pollack, J. O., 1971, Recent developments in native-copper district of Michigan: Society of Economic Geologists Field Conference, Michigan Copper District, September 30–October 2, 1971, p. 18–43.

Weege, R. J., Pollock, J. P., and the Calumet Division Geological Staff, 1972, The geology of two new mines in the native copper district: Economic Geology, v. 67, p. 622–633.

White, W. S., 1960, The Keweenawan lavas of Lake Superior, an example of flood basalts: American Journal of Science, v. 258A, p. 367–374.

White, W. S., 1968, The native-copper deposits of northern Michigan, *in* Ridge, J. D., ed., Ore deposits of the United States, 1933–1967 (the Graton Sales volume): American Institute of Mining, Metallurgical and Petroleum Engineering, New York, p. 303–325.

MANUSCRIPT ACCEPTED BY THE SOCIETY JANUARY 16, 1996

Geological Society of America
Special Paper 312
1997

An overview of the geology and oxide, sulfide, and platinum-group element mineralization along the western and northern contacts of the Duluth Complex

S. A. Hauck, M. J. Severson, L. Zanko
Natural Resources Research Institute, University of Minnesota, Duluth, Minnesota 55811
S.-J. Barnes
Module des Sciences de la Terre, Université du Québec à Chicoutimi, Chicoutimi, Québec PQ G7H 2B1, Canada
P. Morton
Department of Geology, University of Minnesota, Duluth, Minnesota 55812
H. Alminas and E. E. Foord
U.S. Geological Survey, Box 25046, MS 973 and 905, Denver Federal Center, Denver, Colorado 80225-0046
E. H. Dahlberg
Department of Natural Resources, Division of Minerals, 1525 East 3rd Avenue, Hibbing, Minnesota 55746

ABSTRACT

Sulfide and oxide mineralization occurs along the western and northern footwall contacts of the 1.1-Ga Duluth Complex (Complex) in northeastern Minnesota, which was emplaced during the formation of the Midcontinent Rift System. Platinum-group element mineralization is known to occur only along the western contact. The Duluth Complex is composed of troctolitic, gabbroic, and anorthositic rocks that form a series of individual intrusions that make up the Complex. These were emplaced into footwall rocks of Archean granite-greenstone terranes and Lower Proterozoic metasedimentary rocks (Biwabik and Gunflint Iron-Formations; Rove, Virginia, and Thomson Formations) and into the penecontemporaneous North Shore Volcanic Group basalts, which form the hanging wall, and, in some areas, the footwall. Development of rift-related structures, both extensional and contractional, affected development of mineralization. Structural discontinuities provided conduits for emplacement of late-stage granophyric and pegmatitic iron-rich ultramafic rocks. Faults and folds were responsible for localizing massive sulfide mineralization, as well as, for providing conduits for later syn- to post-magmatic hydrothermal fluids.

On the northern margin of the Complex, 81.6 million tons of low-grade titaniferous magnetite ore (approximately 12 to 14% TiO_2) occur in 14 bodies that range in size from 1 to 19 million tons. Oxide mineralization along the western contact contains 245 million tons of $\geq$10% TiO_2. These include the Wyman Creek area, Longnose, Longear, and Section 17 bodies of the Partridge River Intrusion, and the Water Hen Intrusion of the southern Duluth Complex. Overall three types of oxide mineralization occur in the Duluth Complex: (1) oxide-rich metasedimentary inclusions in mafic or ultramafic rocks that exhibit metasedimentary textures and/or can be traced laterally into footwall iron-formation; (2) banded or layered oxide segregations that include cumulus oxide-rich horizons; and (3) late discordant oxide-bearing ultra-

Hauck, S. A., Severson, M. J., Zanko, L., Barnes, S.-J., Morton, P., Alminas, H., Foord, E. E., Dahlberg, E. H., 1997, An overview of the geology and oxide, sulfide, and platinum-group element mineralization along the western and northern contacts of the Duluth Complex, *in* Ojakangas, R. W., Dickas, A. B., and Green, J. C., eds., Middle Proterozoic to Cambrian Rifting, Central North America: Boulder, Colorado, Geological Society of America Special Paper 312.

mafic pipes and bodies (OUI). The latter type is often associated with inclusions of iron-formation or appears to be related to windows of iron-formation subcrop within the Complex. These bodies are remarkably similar to iron-rich, ultramafic pegmatites and pipes located within the Bushveld and Stillwater Complexes that have been interpreted to have a metasomatic origin.

Along the western margin of the Complex, the copper-nickel sulfide resources are estimated at 4.4 billion tons of 0.66% copper with a Cu:Ni ratio of 3.3:1. There are five types of mineralization: (1) sedimentary (Cu-Fe-Zn) sulfides in the footwall, (2) disseminated Cu-Ni sulfides within basal troctolitic rocks, (3) semimassive to massive sulfides, (4) sulfides in OUI, and (5) late-stage secondary sulfides. The primary sulfide minerals in types 2 to 5 include pyrrhotite, chalcopyrite, cubanite, and pentlandite. Only types 2 and 5 are known to occur along the northern contact.

Platinum-group minerals (PGMs) are only known to occur along the western margin of the Complex. PGMs found include: ferro-alloys and sulfides to arsenides, bismuthinides, stannides, tellurides, and antimony-based minerals. Platinum-group elements (PGEs) are found associated with disseminated, massive, and secondary sulfides, and with massive oxides. Anomalous PGE-enriched zones are also associated with structural discontinuities and alteration; however, remobilization appears to be locally confined. The PGE/base metal ratios in their igneous host rocks suggest that most of the PGEs were removed from the troctolitic magmas before emplacement into their present positions. Exceptions to this interpretation include the Birch Lake area of the South Kawishiwi Intrusion and stratabound Cu-Ni sulfides within the Dunka Road deposit.

Sulfur isotope data indicate that sulfur contamination and homogenization in the magma occurred prior to emplacement. Carbon, oxygen, and hydrogen isotope data demonstrate that devolatization of the footwall sedimentary rocks contributed to syn- to post-magmatic fluids that altered the Complex rocks and remobilized the sulfide and PGE. Stable isotope data also show the presence of meteoric and magmatic components to these fluids. Limited fluid inclusion data indicate that as many as four fluid events affected the rocks along the western contact. Secondary chloride precipitates on drill core and in footwall rocks show that chlorine-rich fluids affected the entire western margin of the Complex. This event may also have been related to remobilization of Cu-Ni ± PGE mineralization.

INTRODUCTION

The Duluth Complex (Complex) is a large, composite tholeiitic mafic intrusion that was emplaced into comagmatic flood basalts along a portion of the Middle Proterozoic (1.1 Ga, Keweenawan) Midcontinent Rift System (Fig. 1). The Complex is exposed along the northwest portion of the Midcontinent Rift System. It is the largest of several known rift-related intrusive complexes in the Lake Superior region (Weiblen, 1982). The Duluth Complex, like many other rift-related intrusions, contains numerous oxide, sulfide, and platinum-group element (PGE) occurrences (Nicholson et al., 1992). Located in northeastern Minnesota, the Complex is an arcuate body composed of numerous mafic intrusions, which stretches from Duluth north toward Ely and east to Lake Superior (Fig. 2).

Recent papers on the Midcontinent Rift System suggest that it developed in response to a mantle plume that was active over a 23-m.y. time span (Cannon, 1992; Cannon and Hinze, 1992; Hutchinson et al., 1990; Klewin and Shirey, 1992; Nicholson and Shirey, 1990). Cannon and Hinze (1992) suggest that the mantle plume began to form while the area was under contraction in response to the Grenville collision to the east, but this contractural regime prevented full opening of the continent. When the plume energy subsided and extension ceased, the area returned to a contractural tectonic regime. Evidence of extensional and contractional tectonism is found on both sides of Lake Superior (Cannon et al., 1989; Mauk et al., 1992a, c; Witthuhn, 1992). Weiblen and Morey (1980) propose that a step and riser system of extensional faults developed in the Complex parallel to the rift axis in response to the opening of the rift.

GENERAL GEOLOGY

Footwall rocks

Footwall rocks of the Complex consist of contact metamorphosed (1) supracrustal Archean granite (Giants Range, Snowbank, and Kekekabic) and greenstone (Vermilion District) on the west and north (Figs. 3, 4), (2) Lower Proterozoic (1.7 Ga)

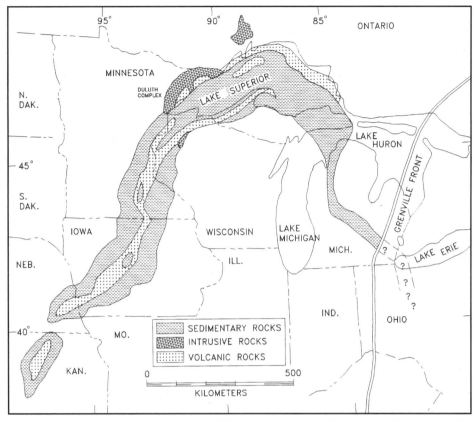

Figure 1. Location map of the Midcontinent Rift System (after Chandler, 1990; Miller et al., 1993).

Animikian sediments (Pokegama Quartzite, Biwabik/Gunflint Iron-Formations, and Thomson/Virginia/Rove Formations), and (3) Upper Proterozoic sedimentary and volcanic rocks (sandstone/quartzite of the Nopeming and Puckwunge Formations and mafic to felsic volcanics of the North Shore Volcanic Group) on the west and north (Figs. 3, 4). All but the Pokegama Quartzite are known to occur at or in close proximity to the footwall contact of the Complex. Contact metamorphism dehydration of the footwall rocks is recognizable from a few hundred meters to as far as 5 km from the basal contact of the Complex along both the northern and western contacts (French, 1968; Floran and Papike, 1975, 1978; Labotka et al., 1981). Of these footwall rocks, the Biwabik Iron-Formation (BIF) and the Virginia Formation (VF) have had the greatest influence on mineralization within the Complex, and discussion is primarily limited to the rocks of these formations.

Biwabik Iron-Formation. The Biwabik Iron-Formation (BIF) consists of alternating sequences of ferruginous chert and slate that are subdivided (Wolff, 1919) into four lithostratigraphic members (from top to bottom): (1) Upper Slaty, (2) Upper Cherty, (3) Lower Slaty, and (4) Lower Cherty. Gundersen and Schwartz (1962) further subdivide these four members into 22 submembers—A to V (Fig. 5). Iron mining is confined to the Upper Cherty, primarily to submembers K to M, at the eastern end of the Mesabi Range near the footwall con-

tact of the Complex, where the iron-formation has undergone pyroxene hornfels contact metamorphism (Bonnichsen, 1975). In the western Complex, the footwall is age transgressive from the Lower Proterozoic Virginia Formation on the southwest to Biwabik Iron-Formation farther east to Archean Giants Range Batholith on the northeast (Fig. 3). However, large inclusions of Biwabik Iron-Formation occur along strike to the northeast well into the Complex (Green et al., 1966). Severson (1994) believes he recognizes large rafts of metasomatized Biwabik Iron-Formation (up to 30 m thick) within an igneous stratigraphic zone of the South Kawishiwi Intrusion.

Farther to the northeast along the northern contact of the Complex, the footwall rocks along the basal intrusive contact are age transgressive, but in the reverse order compared to the western footwall rocks, that is, from Archean rocks on the west to the Gunflint Iron-Formation (equivalent to the Biwabik Iron-Formation) to the Lower Proterozoic Rove Formation (equivalent to the Virginia Formation) on the east (Fig. 4; Nathan, 1969; Morey, 1972; Morey and Nathan, 1977, 1978; Morey et al., 1978).

Virginia Formation. Overlying the Biwabik Iron-Formation is the Virginia Formation (Lucente and Morey, 1983), which also occurs at the basal contact along the western margin of the Complex and, like the Biwabik Iron-Formation, "pinches out" against the basal contact (Fig. 3). The Virginia Formation, like its south-

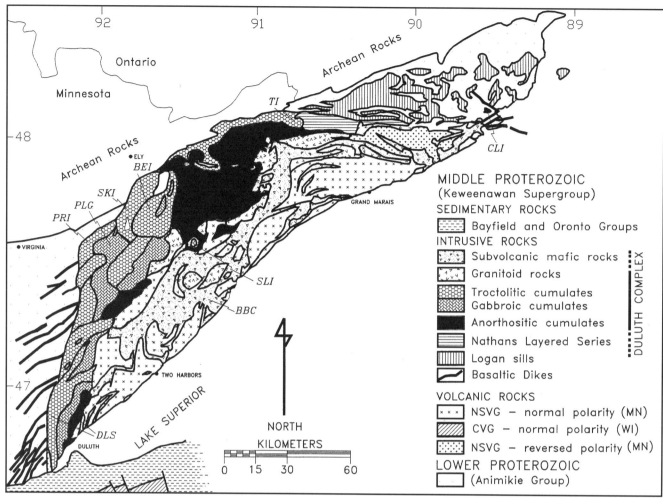

Figure 2. Generalized geologic map of the Duluth Complex showing various intrusions and rock types (Miller et al., 1993). Intrusion designators are: *DLS*, Duluth Layered Series; *BBC*, Beaver Bay Complex; *SLI*, Sonju Lake Intrusion; *CLI*, Crystal Lake Intrusion; *PRI*, Partridge River Intrusion; *PLG*, Powerline Gabbro; *SKI*, South Kawishiwi Intrusion; *BEI*, Bald Eagle Intrusion; and *TI*, Tuscarora Intrusion.

ern equivalent—the Thomson Formation (Morey and Ojakangas, 1970)—and its northern equivalent—the Rove Formation (Morey, 1965, 1967, 1969)—consists primarily of argillite and graywacke, with minor interbeds of siltstone, graphitic argillite, chert, and carbonate/calc-silicate. Numerous large inclusions of Virginia Formation (up to 50 m thick and continuous over 1 square kilometer) also occur in the basal troctolites of the Complex. A unique sulfide-rich layer, informally called the Bedded Pyrrhotite Unit (up to 10 m thick), is found locally in inclusions and within the footwall Virginia Formation in the Babbitt, Serpentine, and Dunka Pit areas (Hauck, 1993; Martineau, 1989; Severson, 1994; Zanko et al., 1994). It contains regular, 1- to 3-mm pyrrhotite beds separated by 1- to 5-cm layers of graywacke.

North Shore Volcanic Group. Penecontemporaneous volcanic rocks of the North Shore Volcanic Group (NSVG), and associated sedimentary rocks, are also locally present at the

basal contact of the Complex. Near the city of Duluth, the basal unit of the NSVG, the Ely's Peak basalt, is in sharp contact with the Duluth Layered Series. The basalt has been metamorphosed to pyroxene hornfels facies (Kilburg, 1972). Metabasalt is interpreted to occur in the footwall of the Water Hen deposit (Strommer et al., 1990), but Severson (1995) reinterprets this unit to be a chilled fine-grained troctolite. Numerous hanging-wall inclusions of NSVG occur in the Complex along the western margin (Severson and Hauck, 1990, and references therein). Some of the metabasalt inclusions are associated with minor quartzite in the Babbitt-Serpentine area and in the footwall of the Water Hen Intrusion (Severson, 1994; Zanko et al., 1994). These quartzite inclusions may be related to the Nopeming/Puckwunge sandstone/quartzite (Mattis, 1972; Ojakangas, 1992) that locally underlies the Complex.

Pre-Complex sills. Numerous pre-Complex sills are related to the development of the Midcontinent Rift System (Fig. 2). At

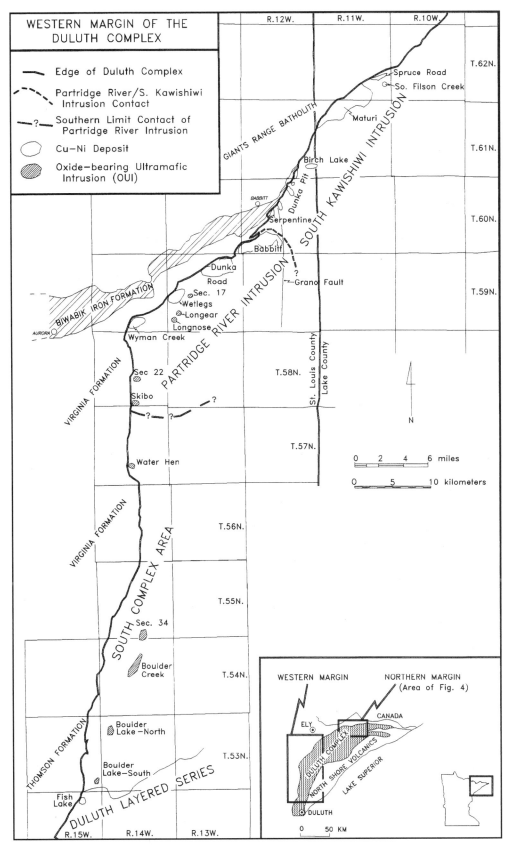

Figure 3. Location map of Cu-Ni sulfide and Fe-Ti oxide deposits along the western margin of the Duluth Complex.

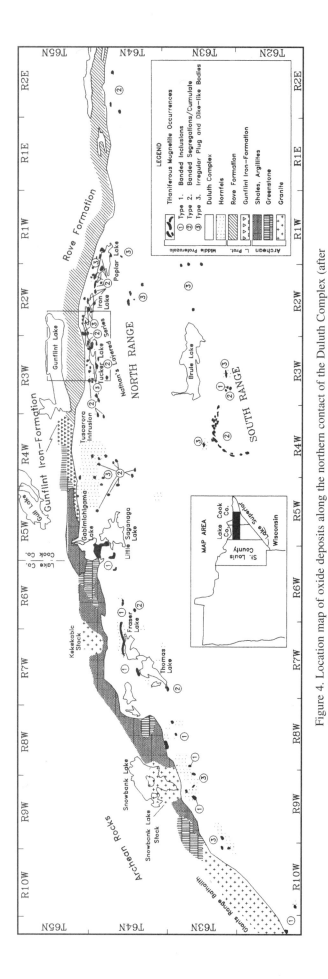

Figure 4. Location map of oxide deposits along the northern contact of the Duluth Complex (after Grout, 1950).

the northern margin of the Complex are two sets of diabase sills, known as the Logan Sills or Intrusions (Lawson, 1893; Grout and Schwartz, 1933) and the Pigeon River Intrusions (Geul, 1970; Mudrey, 1973). They intrude the Rove Formation and the Gunflint Iron-Formation (Jones, 1963, 1984; Mathez, 1971; Mathez et al., 1977; Weiblen et al., 1972b). The Logan Sills (quartz-normative, tholeiitic) are the oldest rocks associated with the rift system (1,108.8 Ma; Davis and Sutcliffe, 1985) and are reversely magnetized (Robertson and Fahrig, 1971; Jones, 1984), whereas the Pigeon River Intrusions (Geul, 1970; Mudrey, 1973) have a normal polarity and are composed of olivine diabase (high-alumina tholeiitic basalt). Mudrey (1973,

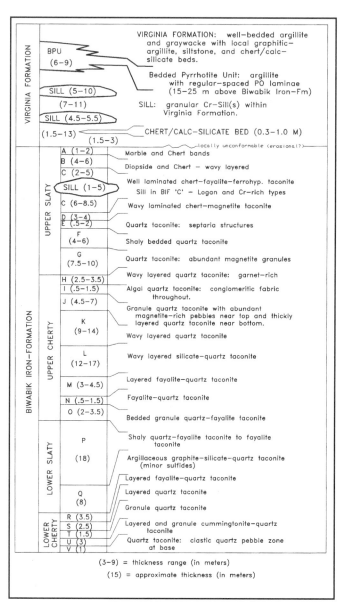

Figure 5. Detailed stratigraphy of the Virginia Formation and Biwabik Iron-Formation (modified after Gundersen and Schwartz, 1962; Hauck et al., 1996; Severson, 1994).

1976) relates the emplacement of the Pigeon River Intrusions and Logan Sills to development of northeast and east-west faults, respectively. Antonellini and Cambray (1992) suggest that the Logan Sills were emplaced (southwest to northeast) along discontinuities that formed during extension in the development of the Midcontinent Rift System.

Gundersen and Schwartz (1962) identified sills intruding the Biwabik Iron-Formation along the western contact. Severson (1991, 1994) subsequently described sills that intruded both the Virginia Formation and the Biwabik Iron-Formation. These sills were contact metamorphosed by the Complex, which produced a fine-grained granular textured rock with hornblende. Geochemical investigation (Hauck et al., 1996) of these sills indicated that they can be subdivided into two types: Logan-type sills, and Cr-rich sills (>600 ppm Cr).

The Logan-type sills can have plagioclase phenocrysts, and occasionally clinopyroxene phenocrysts, and are geochemically similar to the Logan Sills described by Jones (1984). Geochemical data of Severson (1994) show that the Cr-rich sills (>600 ppm Cr) are geochemically similar to basalt inclusions in the South Kawishiwi Intrusion. Hauck et al. (1996) suggest the Cr-rich sills may be related to an early feeder system for the basalts.

The Cr-rich sills are found in both the Virginia Formation and the Biwabik Iron-Formation (Fig. 5). In contrast, Logan-type sills occur only in the Biwabik Iron-Formation along the western contact, but along the northern contact, they intrude both the Rove Formation and the Gunflint Iron-Formation. The significance of the identification of Logan-type sills only in the Biwabik Iron-Formation along the western contact is unknown; however, this feature may be due to limited sampling and exposure.

Duluth Complex units

Rock types of the Complex are varied and include troctolitic, gabbroic, anorthositic, granodioritic, and granitic intrusive bodies. Within this general classification, the Complex consists of a series of both layered and unlayered mafic intrusions whose relations to each other, and to the overall emplacement history of the Complex, is poorly understood. Previous workers (Taylor, 1964; Davidson, 1972) divide the rocks into an older Anorthoitic Series and a younger Troctolitic Series. However, recent age dating (Paces and Miller, 1993) and other studies (Hauck, 1993; Zanko et al., 1994; Severson, 1994) suggest that there are penecontemporaneous and younger anorthositic rocks.

The individual intrusions (Table 1) were emplaced over an 11-m.y. period from 1,107 Ma to 1,096 Ma (Paces and Miller, 1993). The oldest dated rocks of the Complex belong to Nathan's Layered Series (1,106 Ma; Paces and Miller, 1993), which is exposed along the northern margin of the Duluth Complex (Fig. 2; Nathan, 1969). The youngest dated rocks of the Complex (1,096 Ma) are associated with the Silver Bay Gabbro and Sonju Lake Intrusion of the Beaver Bay Complex (Fig. 2; Paces and Miller, 1993).

Igneous stratigraphy of the western margin. Recent relogging of over 550 drill holes (>175 km of drill core) in the western portion of the Complex has generated an igneous stratigraphy for both the Partridge River and South Kawishiwi Intrusions (Figs. 3, 6, 7; Severson and Hauck, 1990; Severson, 1991, 1994). Both intrusions consist dominantly of troctolitic rocks (plagioclase-olivine cumulates with variable lesser amounts of intercumulus augite); however, a few subhorizontal horizons of ultramafic rock (olivine-plagioclase cumulates) are also present and exhibit excellent correlations in drill holes.

Partridge River Intrusion. The Partridge River Intrusion consists of at least seven subhorizontal igneous units (Fig. 6) that are present in the basal 900 m (3,000 ft) along a 24-km (15-mi) strike length. From the base of the Complex up, these units are characterized by: Unit I, sulfide-bearing, heterogeneous troctolites with variable sedimentary hornfels inclusions; Unit II, homogeneous troctolite with a persistent ultramafic base; Unit III, mottled-textured troctolites with distinctive olivine oikocrysts ("marker bed"); and Units IV through VII, each consists of homogeneous troctolites with a basal ultramafic member. The ultramafic members of the upper units (II through VII) probably represent the inception of episodic magma injection that crystallized more primitive ultramafic layers before mixing with the resident magma. Laterally extensive ultramafic members are not as common in Unit I; however, some members directly overlie PGE-enriched troctolitic rocks.

South Kawishiwi Intrusion. An entirely different igneous stratigraphy is present within the South Kawishiwi Intrusion. At least twelve different units (Fig. 7) are present; however, not all units are equally present along a 30-km (19-mi) strike length. Unlike the Partridge River Intrusion, the South Kawishiwi Intrusion contains thick correlative ultramafic packages (units U1, U2, and U3) in the lowermost troctolitic units rather than in the upper units (Fig. 7). Massive oxide pods (magnetite dominant) are associated with the U3 ultramafic package. Spatial relations suggest that the massive oxides were derived from intruded and assimilated Biwabik Iron-Formation (Severson, 1994). In addition, PGE-enriched and Cr-enriched rocks are associated with the U3 Unit.

South Complex area. Rock types to the south of the Partridge River Intrusion, in the South Complex area (Fig. 3), also consist of variable troctolitic rocks with rare, subhorizontal horizons of ultramafic rock. In addition, layered oxide-rich gabbroic rock, late oxide-bearing ultramafic intrusions (OUIs), and fine-grained hornfelsed troctolitic rock are locally present. Detailed relogging of all available drill holes indicates that there is no consistent igneous stratigraphy present over the entire South Complex area (Severson, 1995). Rather, correlative units in several drill holes for one small area do not extend into an adjacent drilled area that is located a few miles distant. This lack of large-scale continuity in igneous units suggests that the South Complex constitutes an area that actually includes several smaller intrusive bodies. Cu-Ni mineralization occurs sporadically in only a few areas of the South Complex. It is generally associated with late Fe-Ti-enriched ultramafic intrusions (OUIs).

TABLE 1. GENERALIZED AREAS OF THE DULUTH COMPLEX WITH RELATED INTRUSIONS, CU-NI±PGE DEPOSITS/OCCURRENCES, AND FE-TI(±CU-NI) DEPOSITS/OCCURRENCES

Area	Formal and Informal Names of Intrusions	Cu-Ni ± PGE Deposits/Occurrences	Fe-Ti(±Cu-Ni) Deposits/Occurrences
Northern Margin	Anorthositic Series Felsic Series/Granophyre Logan Sills Pigeon River Intrusions and Sills Nathan's Layered Series		
			North Range South Range
	Tuscarora Intrusion Crystal Lake Intrusion	Great Lake Nickel	
Western Margin	Partridge River Intrusion	Wyman Cr., Wetlegs, Dunka Road, Babbitt	Longnose, Longear, Sec. 17, Wyman Cr., Sec. 22, Skibo
	Powerline Gabbro		
	South Kawishiwi Intrusion	Spruce Road, South Filson Cr., Maturi, Birch Lake, Dunka Pit, Serpentine	Birch Lake
	Bald Eagle Intrusion		
	South Complex (Area)		Water Hen, Sec. 34, Boulder Cr., N. and S. Boulder Lakes
	Duluth Layered Series	Fish Lake	Fish Lake?, Bardon Peak
	Anorthositic Series		
Eastern Margin (Beaver Bay Complex)	Felsic Series/Granophyre Lax Lake Gabbro Silver Bay Gabbro Sonju Lake Intrusion Beaver River Diabase Other sills within the North Shore Volcanic Group		

Duluth Layered Series. South of the South Complex area, in the vicinity of Duluth, is the Duluth Layered Series of Miller et al. (1993). The Duluth Layered Series consists of layered gabbroic and troctolitic cumulates, which are bounded at the base by olivine gabbro and overlain by a complex zone of ferrogabbro, ferromonzodiorite, and quartz ferromonzodiorite (Miller et al., 1993). Anorthositic Series rocks form a cap to the Duluth Layered Series rocks and are composed of coarse-grained, plagioclase cumulates (Miller et al., 1993). The relationship between the two rocks is unclear. The Duluth Layered Series rocks have an apparent mafic "chill" against the Anorthositic Series rocks (Miller et al., 1993), but U-Pb zircon age dates on these two units give essentially identical dates of 1,099 Ma (Paces and Miller, 1993). These dates imply that the two series may have been rapidly emplaced one after the other (Miller et al., 1993).

Fe-Ti mineralization occurs in crosscutting OUI bodies at Bardon Peak (Ross, 1985). Ross (1985) proposed a metasomatic origin for this mineralized occurrence based on similar rocks in the Bushveld Complex. Cu-Ni and minor Fe-Ti mineralization

also occur in the Fish Lake occurrence near the northern contact with the South Complex area. The disseminated Cu-Ni sulfide, PGE, and Fe-Ti mineralization has been described by Sassani and Pasteris (1988, 1992) and Sassani (1992).

Igneous stratigraphy of the northern margin. The igneous stratigraphy of the northern margin is defined by the work of Babcock (1959), Grout et al. (1959), Nathan (1969), Phinney (1972); the reconnaissance mapping of Davidson (1969a, b, 1972, 1977a through h), and Davidson and Burnell (1977); and the detailed mapping of Mathez (1971), Mathez et al. (1977), Morey and Nathan (1977, 1978), and Morey et al. (1978). Four main rock series are defined in the region: Nathan's Layered Series or Early Mafic Series, Anorthositic Series, Troctolitic Series, and Felsic Series.

The Crystal Lake Intrusion will not be discussed further since it is a separate intrusion north of the northern contact. However, additional information on the geology and the Cu-Ni-Cr-PGE mineralization in the intrusion can be found in Cogulu (1984, 1985, 1990, 1993a, b), Eckstrand and Cogulu (1986,

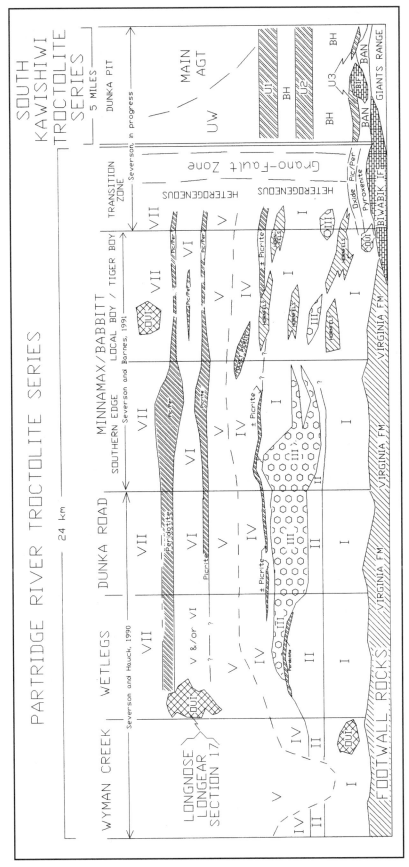

Figure 6. Schematic cross section of the igneous stratigraphy in the Partridge River Intrusion (from Severson, 1994). Stratigraphic units from bottom to top are: Unit I, Sulfide-bearing heterogeneous mixture of augite troctolite and troctolite containing minor discontinuous ultramafic horizons. Unit I is the major sulfide-bearing horizon in all the Cu-Ni deposits; Unit II, Homogeneous-textured troctolite, with abundant ultramafic horizons in the Wetlegs area. Minor sulfide-bearing horizons are scattered throughout Unit II; Unit III, Mottled-textured, fine-grained, anorthositic troctolite exhibiting a very irregular distribution of olivine oikocrysts that are unique to this unit and make it a major "marker horizon"; Unit IV, Augite troctolite (homogeneous) with a moderately continuous basal ultramafic horizon that is termed "±picrite" in the Babbitt area; Unit V, Coarse-grained, homogeneous anorthositic troctolite that grades downward into Unit IV (contact is shown by dashed line); Units VI and VII, Homogeneous troctolite to anorthositic troctolite. Each unit is marked by a continuous ultramafic base.

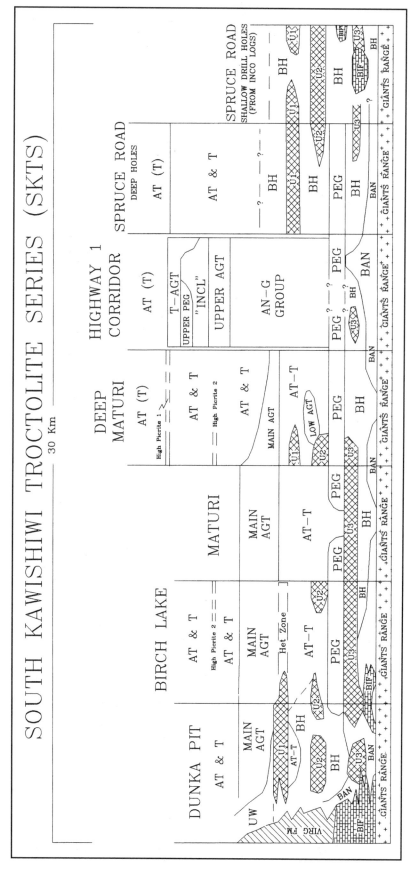

Figure 7. Schematic cross section of the igneous stratigraphy in the South Kawishiwi Intrusion (from Severson, 1994). Stratigraphic units from bottom to top are: BAN, Bottom Augite Troctolite and Norite zone: Heterogeneous mixture of augite troctolite and norite (both sulfide-bearing), containing local inclusions of Biwabik Iron-Formation (BIF) and Giants Range Batholith (GRAN); BH, Basal Heterogeneous zone: The BH is the main sulfide-bearing ore zone in each of the Cu-Ni deposits; U3, Ultramafic 3 Unit: A package of alternating layers of ultramafic and troctolitic rocks. U3 typically contains massive oxide horizons and/or BIF inclusions and is generally sulfide-bearing; PEG, Pegmatitic unit of Foose (1984): A poorly defined unit in which pegmatoids and gradational pegmatite zones are sporadically present; U2, Ultramafic 2 Unit: A package of alternating layers of ultramafic and troctolitic rocks; massive oxide layers are only rarely present; AT-T, Anorthositic Troctolite to Troctolite zone: Homogeneous package of both rock types that are gradational into each other; U1, Ultramafic 1 Unit: Package of alternating layers of ultramafic and troctolitic rocks, massive oxides are totally lacking; UW, Updip Wedge: Package of heterogeneous sulfide-bearing troctolite that occurs in the western portion of the Dunka Pit area and extends southwestward to the Serpentine deposit. The UW Unit only occurs where the Virginia Formation is present at the basal contact; MAIN AGT, Zone where augite troctolite is the dominant rock type. In the Birch Lake area there is a heterogeneous (HET) zone of mixed rock types and grain sizes at the base of the unit; AT&T, Thick, monotonous zone of anorthositic troctolite and troctolite with minor augite troctolite and troctolitic anorthosite. The AT&T unit also contains two ultramafic horizons tentatively called High Picrite #1 and #2; AT(T), Thick, monotonous zone of anorthositic troctolite containing minor troctolite and rare augite troctolite zones; Highway One Corridor Rocks, A thick stratigraphic sequence (large raft of Anorthositic Series) delineated in five deep drill holes. The most definitive rock types of the sequence are: AN-G GROUP, intermixed anorthositic and gabbroic rocks; and the "Incl" unit, a fine-grained, granular inclusion of magnetite-rich basalt; BIF, Biwabik Iron-Formation.

1989), Eckstrand et al. (1989), Geul (1970), Heslop (1968), Lightfoot and Lavigne (1995), MacRae and Reeve (1968), Mainwaring (1968a, b), Mainwaring and Watkinson (1981), Nicholson et al. (1992), Reeve (1969), Smith and Sutcliffe (1987), Sutcliffe (1991), Tanton (1935), Wanless et al. (1966), and Whittaker (1986).

Nathan's Layered Series or Early Mafic Series. Nathan (1969) defined 27 stratigraphic cumulate igneous units, which have concordant and discordant iron-titanium oxides, along the northern contact. Age dating indicates emplacement of these layered gabbros, troctolites, and ferrogabbros at 1,106.9 ± 0.5 Ma (Paces and Miller, 1993). These rocks are also similar in magnetic polarity and composition to the Logan Sills (Weiblen et al., 1972b), which have been dated at 1,108.8 Ma (Davis and Sutcliffe, 1985). Davidson (1977a to f, h) and Davidson and Burnell (1977) recognized during reconnaissance mapping that these units were much more widespread and renamed the unit the Early Mafic Series.

Anorthositic Series. According to Davidson (1972), almost half of the bedrock along the northern contact is anorthosite. Recent work by Miller and Weiblen (1990) indicates the series is "dominated by laminated, but unlayered gabbroic to troctolitic anorthosite." Miller and Weiblen (1990) concluded that the anorthosites crystallized from a plagioclase crystal mush and that the liquid composition of this plagioclase-rich magma ranged from moderately to highly evolved tholeiitic basalt that was saturated (or nearly so) in both olivine and plagioclase. These mushes originated by suspension and segregation of plagioclase in lower crustal magma chambers and were emplaced (by ponding) beneath a "comagmatic volcanic edifice as multiple intrusions of irregular shape and generally small (kilometer size) size" (Miller and Weiblen, 1990, p. 335). Miller and Weiblen (1990) suggested that the anorthosites were emplaced during the intermediate stages of rifting, but formed in the lower crustal magma chambers during the initial stages of rifting and crustal attenuation.

Troctolitic Series. The Troctolite Series consists primarily of moderately layered olivine gabbro and troctolite. Miller and Weiblen (1990) estimate that the series is >1,500 m thick in the Snowbank Lake Quadrangle. The contacts between layers are abrupt to gradational over several meters (Davidson, 1972; Miller and Weiblen, 1990). Davidson (1972) indicates that the series is primarily troctolite near the contact and overlain by olivine gabbro, with minor troctolite layers. Davidson (1972) also notes that the olivine gabbro unit can be distinguished from the troctolitic unit by the abundance of oxide minerals.

Within the northern area is a distinct troctolitic intrusion, the Tuscarora Intrusion (Fig. 4; Johnson, 1970; Mogessie, 1976; Morey et al., 1978; Nathan, 1969; Weiblen et al., 1972a). Mogessie (1976) suggests that the Tuscarora Intrusion may be the northeastern terminus of the troctolitic intrusions along the western contact. The Tuscarora Intrusion is primarily a medium-grained troctolite, with gabbro and troctolite near the top and a finer-grained troctolitic basal unit. Disseminated copper-nickel sulfide mineralization is present at the base of this intrusion (Johnson, 1970; Mogessie, 1976).

Felsic Series. Davidson (1972) describes the Felsic Series as two types: (1) small, local plutons associated with Anorthositic Series rocks; and (2) large sheets of mostly flat-lying, homogeneous granophyric granite that overlies the Troctolitic Series rocks. The granophyre in both cases is massive fine- to medium-grained and red, due to hematite-stained feldspars. These late-stage, red, granophyric bodies are prevalent near the hanging wall contact of the Complex (Davidson, 1972; Nelson, 1991). An intermediate composition rock exists between the gabbro and the granophyre, suggesting a gradual separation of the gabbro and granophyre (Babcock, 1959; Phinney, 1972). Davidson (1972) suggests that the large granophyric sheets are late differentiates of the gabbro. He also notes that the smaller plutons occur over possible buried projections of the Giants Range Batholith beneath the Duluth Complex, suggesting that the plutons may be a mobilized partial melt of the Giants Range Batholith.

Hanging wall and post-Complex units

Hanging wall rocks of the Complex consist primarily of mafic volcanic rocks of the North Shore Volcanic Group (Fig. 2). North- to northwest-trending, post-Complex diabasic dikes crosscut the Complex and the footwall rocks along the western contact. They are present in the drill holes of the Babbitt deposit (Fig. 3) and in the iron mines north of the basal contact. The geological map of Morey and Cooper (1977) also shows scattered post-Complex dikes within both Partridge River and South Kawishiwi Intrusions. Green et al. (1987) describe several basaltic dike and sill swarms related to the Midcontinent Rift System, some of which are peripheral to or intrude the Complex (Fig. 2).

STRUCTURE

Structure within the Duluth Complex is related to the opening of the Midcontinent Rift System. GLIMPCE seismic reflection data in Lake Superior illustrates that normal faulting (extension) occurred during opening of the rift and reverse faulting occurred during collapse (compression) of the rift (Cannon et al., 1989). Rift-related normal faults and reactivated/pre-Complex, northwest-striking, (transform) scissor faults are present along the western margin (Chandler and Ferderer, 1989; Cooper et al., 1977; Holst et al., 1986; Severson, 1988, 1994; Severson and Hauck, 1990; Weiblen et al., 1977).

Several northeast-trending normal faults parallel to the Midcontinent Rift System occur near the margin of and within the Complex; these faults support the half-graben model of Weiblen and Morey (1980), who proposed a step and riser model for the configuration of the basal contact of the Complex. They envisioned the model as steep, southeast-dipping, northeast-trending normal faults that formed during the opening of the Midcontinent Rift. These faults then controlled the emplacement of the troctolitic and later intrusions. For example, in the Wetlegs area, a pre-Complex fault occurs where an inferred window of Biwabik Iron-Formation subcrops at the base of the Complex (Fig. 8). Positioned above this fault/subcrop zone are three (Longnose, Longear, and Section 17) Oxide-bearing Ultramafic Intrusions (OUIs).

In the transition zone between the Partridge River Intrusion and South Kawishiwi Intrusion is the Grano Fault (Severson, 1994; Fig. 3). This north-trending, scissors fault, which shows greater movement to the south toward the interior of the Complex, is characterized by the presence of numerous, small granophyric intrusions. Along this and other faults in the Babbitt

area, both granophyres and OUIs occur within fault zones, and are excellent indicators of structural discontinuities.

Pre- to syn-Complex subhorizontal folds are present in the footwall Virginia Formation. Structure contours on the top of the Biwabik Iron-Formation and at the base of the Complex define anticlinal and synclinal structures in the Bathtub and Local Boy ore zones of the Babbitt deposit (Figs. 9, 10; Holst et al., 1986; Severson, 1988; Severson and Barnes, 1991; Hauck, 1993; Severson et al., 1994). These folds may be related to syn-Complex folding and faulting (Hauck, 1993).

Many workers have suggested that the faulting and other structures controlled the distribution of the basal copper-nickel

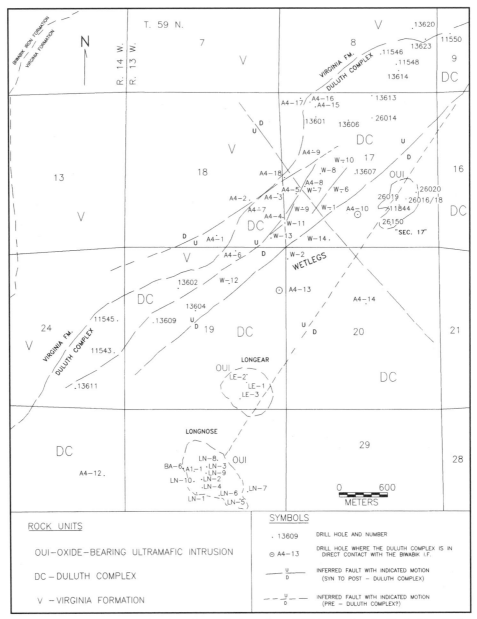

Figure 8. Location of Oxide-bearing Ultramafic Intrusions (OUIs) associated with faulting in the Wetlegs area, western margin of the Duluth Complex (after Severson, 1988).

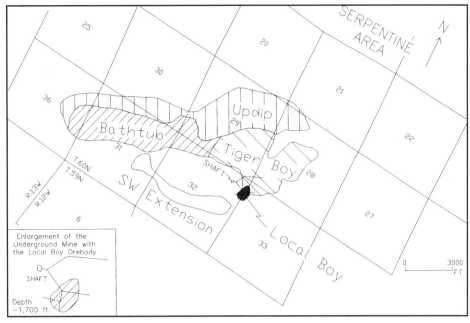

Figure 9. Location of ore zones in the Babbitt Cu-Ni deposit.

mineralization (Chandler and Ferderer, 1989; Severson and Barnes, 1991; Weiblen and Morey, 1975, 1976; Weiblen et al., 1977). Also, late-stage fractures filled with serpentine, biotite, secondary sulfides, and secondary chloride-bearing mineral phases occur along the entire western margin of the Complex, indicating that one or more regional, late-stage, tectonic events occurred, with accompanying fluid activity. For example, in the Bathtub ore zone in the Babbitt deposit, the inferred sequence of intrusion, structural, and alteration events illustrates how multiple faulting, folding, and fracturing events (Fig. 11) affected the emplacement of the intrusive rocks as well as the localization of subsequent fluid/alteration processes.

MINERALIZATION

The majority of known oxide, sulfide, and platinum-group element (PGE) mineralization was discovered within the Partridge River and South Kawishiwi Intrusions within 1.5 to 2.5 km of the western contact of the Duluth Complex (Fig. 3). Several Cu-Ni deposits were defined along the western margin of the Complex by drilling from 1950 through 1980 (Fig. 3; Table 1). Also, Fe-Ti oxide deposits have been outlined along the western and northern margins of the Complex (Figs. 3, 4; Table 1). The Cu-Ni sulfide mineralized areas identified were Spruce Road, South Filson Creek, Maturi, Birch Lake, Dunka Pit, Serpentine, Babbitt, Dunka Road, Wetlegs, and Wyman Creek (Fig. 3; Table 1). Fe-Ti mineralization identified within the Partridge River Intrusion included the Longnose, Longear, Section 17, Section 22, and Skibo bodies. Water Hen and other smaller Fe-Ti deposits were discovered within the southern Duluth Complex (Table 1).

Along the western margin, most sulfide occurs as dissemi-

nations in the basal units of both the Partridge River and South Kawishiwi Intrusions. Semimassive to massive sulfide bodies are related to structural discontinuities in the Complex rocks and/or proximity to sulfide-rich facies within the Virginia Formation. PGE-enriched zones within individual intrusions are associated with massive sulfide (e.g., Local Boy ore zone, Babbitt deposit), stratabound disseminated sulfide zones (e.g., Dunka Road deposit), massive oxides (e.g., Birch Lake), or steeply inclined fault zones (e.g., South Filson Creek occurrence).

Disseminated sulfides and oxide mineralization, similar in mineralogy to sulfides and oxides along the western contact, occur sporadically in Nathan's Layered Series and at the base of the Tuscarora Intrusion to the west (Fig. 4; Grout, 1950; Johnson, 1968, 1970; Mogessie, 1976; Nathan, 1969). Sulfides are also related to the Logan Intrusions that intrude the Rove Formation and Gunflint Iron-Formation (Johnson, 1968, 1970). No known large Cu-Ni resources or PGE enrichments are reported in the literature for the northern area.

The basal units of the Duluth Complex assimilated and devolatilized the footwall rocks (Virginia Formation and Biwabik Iron-Formation) during contact metamorphism. As a result, the basal units are highly heterogenous in composition, texture, and sulfide content. In addition, all igneous units have been subjected to varying degrees of pervasive alteration, including saussuritization, uralitization, serpentinization, and chloritization. Miarolitic cavities also formed in plagioclase-rich rocks in areas with more intense alteration. In Nathan's Layered Series, sulfide mineralization is rare, except in areas of deuteric alteration where secondary veins of copper-rich and nickel-poor minerals are present (Nathan, 1969); however, there are large tonnages of Fe-Ti oxides (Grout, 1950; Nathan, 1969).

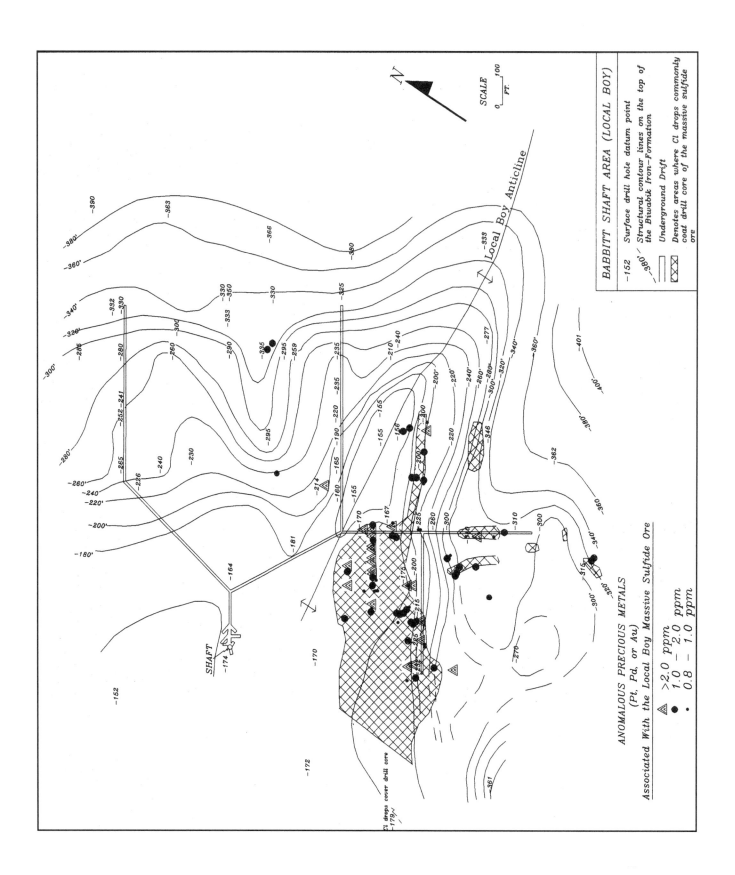

BABBITT SHAFT AREA (LOCAL BOY)

-152   Surface drill hole datum point

380'   Structural contour lines on the top of
        the Biwabik Iron-Formation

        Underground Drift

        Denotes areas where CI drops commonly
        coat drill core of the massive sulfide
        ore

ANOMALOUS PRECIOUS METALS
(Pt, Pd, or Au)
Associated With the Local Boy Massive Sulfide Ore

        >2.0 ppm
        1.0 – 2.0 ppm
        0.8 – 1.0 ppm

Figure 10. Location of the Local Boy Anticline in the Local Boy ore zone (defined by contours on top of Biwabik Iron-Formation), Babbitt Cu-Ni deposit. Note locations of elevated platinum-group elements and of chloride-rich droplets on drill core (after Severson and Barnes, 1991).

In this paper, we discuss oxide mineralization first, then sulfide and PGE mineralization.

Oxide mineralization

According to Winchell (1897), the titaniferous iron ores of the Duluth Complex were discovered about 1867, about the same time as the Mesabi Range iron ores. However, the iron ores were not subdivided into titaniferous and nontitaniferous iron ores by the Minnesota Geological and Natural History Survey (now the Minnesota Geological Survey) until 1887. Grout (1950) estimated that along the northern margin of the Complex 81.6 million tons of low-grade titaniferous magnetite ore (about 12 to 14% TiO_2) occurred in 14 bodies, ranging in size from 1 to 19 million tons. Oxide mineralization along the western contact was discovered during drilling of aeromagnetic highs during copper-nickel exploration in the 1960s (Bonnichsen, 1972a). Listerud and Meineke (1977) estimated that 220 million tons of

oxide $\geq 10\%$ TiO_2 occurred in three areas: Wyman Creek and Section 17 of the Partridge River Intrusion, and the Water Hen Intrusion of the southern Duluth Complex. Additional reserves (>15% TiO_2) of 25 million tons are found in the Longnose deposit within the Partridge River Intrusion (W. Ulland, personal communications, 1989).

Disseminated, interstitial/intercumulate to included oxides in cumulate mineral phases are a ubiquitous constituent of most Complex rocks. The primary oxide minerals are ilmenite and magnetite (generally ilmenite > magnetite), along with minor amounts of other spinels (e.g., chromite, ulvöspinel, hercynite, and pleonaste; Severson, 1991; Mogessie and Stumpfl, 1992; Sassani, 1992). Whereas most Complex rocks average 1 to 10% oxide, some intrusions can contain much more. For example, the Powerline Gabbro of Severson and Hauck (1990) contains 5 to 20% oxide, primarily in the form of magnetite.

Pasteris (1984, 1985) recognized five varieties of oxides in the Babbitt and Spruce Road Cu-Ni deposits: (1) spinel and ilmenite rods in plagioclase, (2) individual ilmenite grains, (3) individual magnetite grains, (4) granular (composite?) ilmenite-magnetite, and (5) lamellar ilmenite in magnetite. Severson and Hauck (1990) and Severson (1994) reported hypersthene-oxide symplectite coronas around olivine in the Partridge River and South Kawishiwi Intrusions. Geerts (1991) recognized two types of ilmenite in the Dunka Road deposit:

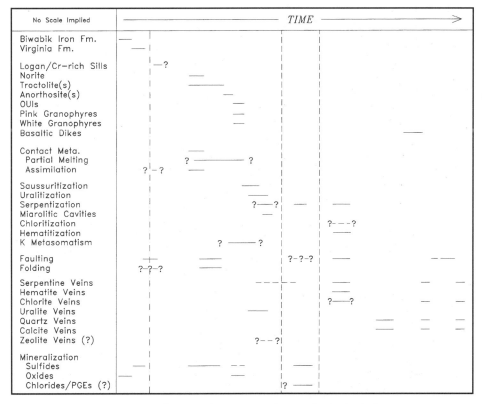

Figure 11. Paragenetic sequence of events in the Bathtub ore zone, Babbitt Cu-Ni deposit (Hauck, 1993).

(1) interstitial ilmenite and ilmenite blebs in sulfides, which were compositionally similar (Morton and Geerts, 1992); and (2) rods of spinel in plagioclase.

Broderick (1917) was the first to recognize four distinct types of oxide mineralization in the northern part of the Complex based on field relationships and mineralogy. He classified them as: (A) inclusions of Gunflint Iron-Formation; (B) gabbroic banded segregations; (C) irregular late bodies of titaniferous magnetite; and (D) dike-like bodies of titaniferous magnetite. The mineralogy of Broderick's various oxide types consists of individual grains of ilmenite and magnetite, ilmenite intergrowths in magnetite, and ilmenite rims on magnetite (Broderick, 1917; Schwartz, 1930; Grout, 1950). While working in the western and southern parts of the Complex, Severson (1988) and Severson and Hauck (1990) recognized the last two varieties of Broderick's classification (Fig. 4) and described them as a single type of oxide mineralization. They called them Oxide-bearing Ultramafic Intrusions (OUIs), based on their ultramafic composition, high oxide content, and crosscutting relationships.

Overall, we have classified ilmenite and magnetite mineralization in the Complex into three general types: Type 1, banded or layered, oxide-rich metasedimentary inclusions in mafic and ultramafic rocks; Type 2, banded or layered oxide segregations (cumulates) in mafic rocks; and Type 3, discordant Oxide-bearing Ultramafic Intrusions (OUIs) with semimassive to massive oxide zones.

Type 1 oxides. Type 1 oxides, Fe-rich metasedimentary inclusions, are difficult to define in the Complex because they represent a spectrum from recognizable metasedimentary inclusions grading to totally recrystallized/metasomatized zones that are void of recognizable sedimentary features. Spatial relationships suggest that the totally recrystallized Type 1 oxides are related to magma contamination by assimilation of portions of the Biwabik and Gunflint Iron-Formations.

In the northern Complex, Broderick (1917), Grout (1950), and Grout et al. (1959) described Fe-rich inclusions as inclusions of Gunflint Iron-Formation, on the basis of mineralogy and field relationships. The magnetite, pyroxene, and olivine compositions of these inclusions were similar to those of metamorphosed Gunflint Iron-Formation, except the former had no quartz. In the field, the inclusions and titaniferous magnetite occurrences could be traced east into the Complex from their footwall contact (Broderick, 1917; Grout, 1950).

Broderick (1917) reports, based on field and mineralogical evidence, that there was large-scale addition of titanium to the Gunflint Iron-Formation inclusions from the surrounding gabbro. However, clear-cut recognition of the inclusions is complicated by alteration of the Gunflint Iron-Formation to coarsely crystalline aggregates of olivine, pyroxene, and Ti-rich (?) magnetite. Schwartz (1930) and Grout et al. (1959) note that the inclusions were also associated with hornfelsed Rove Formation (Fig. 4), and that the titanium content in the inclusions varied from 1 to 25%. Like Broderick (1917), Grout et al. (1959) believed the titanium was added to the inclusions from the gabbro, because

the content of titanium in unaltered Gunflint Iron-Formation was insignificant. A similar mechanism of Ti-migration from the Complex into the Biwabik Iron-Formation has been documented by Muhich (1993) in the Dunka Pit (Fig. 3).

At the base of the Partridge River Intrusion, Severson (1994) identifies a Basal Ultramafic Unit (BU Unit, Fig. 7) that occurs above the Biwabik Iron-Formation footwall contact. The unit consists of an upper picrite-peridotite zone that grades down into a lower pyroxenite zone. Massive oxide zones, consisting of titanomagnetite with minor ilmenite and green pleonaste, are common in the upper zone. The lower zone consists of coarse-grained orthopyroxenite that grades down into recrystallized pyroxenitic Biwabik Iron-Formation, with occasional beds of magnetite and chert (Severson, 1994). Because the orthopyroxenite and recrystallized iron-formation are similar in appearance, the contact between BU Unit and Biwabik Iron-Formation is arbitrarily chosen in places.

Massive oxide horizons are also found associated with ultramafic rocks (U3 Unit; Fig. 7) near the base of the South Kawishiwi Intrusion. Severson (1994) notes that the U3 Unit contains recognizable Biwabik Iron-Formation inclusions, and suggests that the massive oxides are related to assimilated iron-formation, which was also postulated by Dahlberg (1989). These spatial relationships suggest that massive oxide zones in the U3 Unit at Birch Lake area of the South Kawishiwi Intrusion are highly altered or metasomatized Type 1 oxides. However, recent isotope work (Hauck et al., 1996) and the PGE-Cr mineralization associated with the Birch Lake massive oxides suggest that some of the oxides may be magmatic (Type 2 oxides). Alapieti (1991) described "atoll textures" in oxides from massive oxide horizons in the Birch Lake area in which the oxides were poikilitically enclosed by optically continuous plagioclase. The "atoll texture" suggests sintering of Biwabik Iron-Formation magnetite, in which case these oxide horizons could be related to assimilation of Biwabik Iron-Formation.

Type 2 oxides. Type 2 oxides display an igneous texture and occur as banded segregations of magnetite/ilmenite ± Cr-spinels and plagioclase, up to 3 m thick or as oxide-rich ultramafic horizons. The first variety are best exposed at the northern contact of the Complex. They consist of semimassive to massive oxide with aligned plagioclase laths. Their origin is questionable because in some cases they are associated with assimilated iron-formation. Broderick (1917) notes that these segregations can be traced for as much as 1.6 km (1 mi) along strike (Fig. 4), and they can be correlated in the subsurface. The segregations grade from dominantly gabbro, with minor oxide, to nearly pure oxide (titanomagnetite). The segregations are generally of higher grade (more titanomagnetite) than the disseminated oxides in metasedimentary inclusions. Broderick (1917) infers that the iron in the "banded segregations" came from inclusions of Gunflint Iron-Formation. However, Lister (1966) believes these types of oxide bodies could be formed by segregation of an immiscible liquid in conjunction with convection currents and crystal sedimentation (partial assimilation of inclusions and extraction of water from the inclusions); the resulting oxide-rich liquid collected in disc-shaped depressions on the floor of the chamber. Nathan (1969) considers the banded segregations to be the result of cumulate processes, and magnetite and ilmenite are cumulate minerals.

Along the western contact Type 2 oxides occur as either oxide-rich horizons or modally layered oxide-rich gabbros. Type 2 oxides are known to be present (1) north of the Water Hen Fe-Ti deposit in drill hole SL-19A associated with ultramafic layers (Severson, 1995), (2) scattered throughout the South Complex area (Severson, 1995), and (3) at Fish Lake, in the Duluth Layered Series (Fig. 3; Sassani, 1992; Sassani and Pasteris, 1992). However, at Birch Lake, in the South Kawishiwi Intrusion, massive oxide horizons (primarily magnetite>>ilmenite, associated with other Cr-spinels: Sabelin, 1987; Sabelin and Iwasaki, 1985; Severson, 1994; Hauck et al., 1996) are associated with a layered ultramafic/troctolite package (U3 Unit). Because the U3 Unit of the South Kawishiwi Intrusion is spatially situated downdip of the Biwabik Iron-Formation, the massive oxides at Birch Lake may be related to intruded and metasomatized iron-formation (Severson, 1994). Recent isotope work by Hauck et al. (1996) suggests that magmatic massive oxides may also be present at Birch Lake. In addition, massive oxides and ultramafic layers of the U3 Unit are often Cr-rich, suggesting a strong magmatic component. For these reasons, the massive oxide zones at Birch Lake are included under both Type 1 and Type 2 oxides.

Type 3 oxides. The troctolitic rocks of the Complex are intruded by Oxide-bearing Ultramafic Intrusions (OUIs). The OUIs are characterized by coarse-grained to pegmatitic pyroxenite, peridotite, and dunite, most of which contain lenses of massive oxide. The OUIs generally occur as sill or dikelike bodies (Fig. 12), or as large irregular to regular "pipelike" bodies (Figs. 8, 13). The Water Hen OUI appears to be rootless as defined by detailed drilling (Fig. 14); however, detailed three-dimensional data on the other OUIs are lacking. In the Complex, OUIs are generally related to faults or are positioned in the immediate vicinity of fault zones (Fig. 8). Biwabik Iron-Formation inclusions or footwall subcrops are also present. The ultramafic bodies may be zoned, as at Longnose (Fig. 13), and may have numerous apophyses around a main ultramafic body (Water Hen, Fig. 14). Oxide mineralization occurs as disseminated (15 to 30%), semimassive (30 to 70%), and massive (>70%) and is composed of ilmenite and/or magnetite; in some cases, the magnetite along the western contact can be enriched in vanadium (3,000 to 8,000 ppm V). OUIs have many similarities to the discordant, iron-rich, pegmatitic ultramafic bodies of the Bushveld Complex (Wagner, 1929; Cameron and Desborough, 1964; Schiffries, 1982; Scoon, 1987; Stumpfl and Rucklidge, 1982; Viljoen and Scoon, 1985; Schiffries and Skinner, 1987; Rudashevsky et al., 1992), the Stillwater Complex (Volborth and Housley, 1984; Raedeke and McCallum, 1984, 1985; Nicholson and Lipin, 1985), and the Rio Jacare Intrusion of Brazil (Sá et al., 1993).

The origin of the OUIs is unclear. Along the western margin of the Complex, Severson (1988, 1991, 1994), and Severson and Hauck (1990) propose a spatial/empirical relationship between OUIs and the Biwabik Iron-Formation. Severson (1988) shows that the Longnose, Longear, and Section 17 OUIs are associated with a fault-related subcrop of the Biwabik Iron-Formation (Fig. 8), and he suggests a genetic relationship between the OUIs and the incorporation of the Biwabik Iron-Formation at the basal contact. In the South Kawishiwi Intrusion, massive oxide and ultramafic layers of the U3 Unit (Fig. 7; Severson, 1994) occur at the same stratigraphic horizon as identifiable inclusions of Biwabik Iron-Formation and suggest the same spatial relationship. However, the high Ti content of the OUIs and massive oxide layers is problematic if they were directly derived from Biwabik Iron-Formation that was intruded by and subsequently assimilated by Complex magmas. Muhich (1993) finds that local titanium enrichment in the Biwabik Iron-Formation at Dunka Pit was due to migration of titanium across the contact from the Complex into the iron-formation. Oxygen isotopes further indicate fluid transport across this contact. These observations suggest that the Fe-Ti-enriched OUIs in fact may be genetically related to inclusions and/or subcrops of iron-formation.

Summary of oxide types. Previous work regarding the genesis of Type 2 and 3 oxides along the northern contact defined a spatial and genetic relationship to variably assimilated sedimentary iron-formations. Broderick (1917) suggested that iron-rich bodies near the northern margin formed from inclusions of Gunflint Iron-Formation, and that the titanium in these bodies was derived from the Complex. Grout et al. (1959) suggested that OUI-like bodies formed by fusion of inclusions and by injections of the resultant liquid into the adjacent rocks. Lister (1966) proposed that partial assimilation of iron-rich inclusions contributed to the formation/segregation of an iron-rich immiscible liquid that subsequently settled in ponds within a magma chamber that was co-precipitating plagioclase and olivine crystals. However, Ross (1985) theorized that iron-rich peridotites at Bardon Peak near Duluth were formed by infiltration metasomatism, a mechanism invoked for the Bushveld iron-rich ultramafic bodies. A similar metasomatic origin was considered by Severson (1988, 1994) and Severson and Hauck (1990) for the OUIs in both the Partridge River and South Kawishiwi Intrusions. On the other hand, Bonnichsen (1972a), and later Mainwaring and Naldrett (1977), who proposed a similar origin for the Water Hen Intrusion, proposed that the OUIs may have formed from a ferrogabbro magma that had abundant suspended plagioclase crystals, and due to the high specific gravity of the liquid, the plagioclase separated (rose to the top) leaving an iron-rich titaniferous ultramafic magma.

Sulfide mineralization

Listerud and Meineke (1977) estimate copper-nickel resources of the Complex at 4.4 billion tons of 0.66% copper with a copper-to-nickel ratio of 3.3:1. Included within this reserve esti-

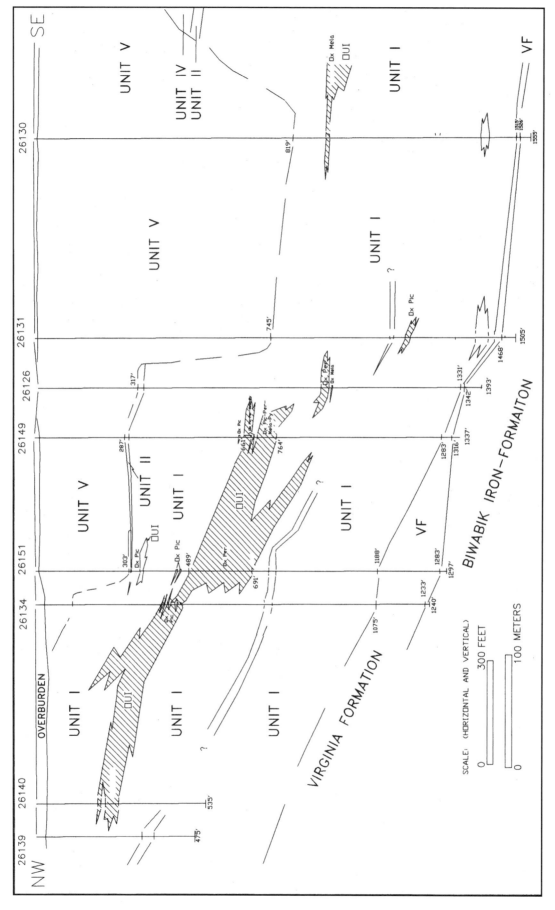

Figure 12. Representative cross section through the Wyman Creek area, T.59N., R.14W. (after Severson and Hauck, 1990). See Figure 3 for location.

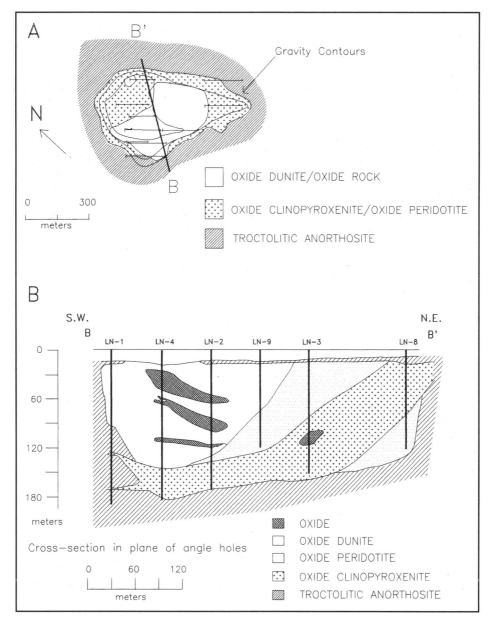

Figure 13. A, Plan view (with 0.5 mgal gravity contours). Horizontal straight lines represent the projection of drill hole on to the plane; B, cross section of the Longnose Oxide-bearing Ultramafic Intrusion (after Linscheid, 1991; W. Ulland, personal communication, 1995). Vertical lines represent the drill holes along B–B′, which are in the plane of the cross section (inclination of 45°). See Figures 3 and 8 for location.

mate are five types of sulfide mineralization known to occur along the western margin: (1) sedimentary sulfide in the footwall, (2) disseminated sulfide, (3) semimassive to massive sulfide, (4) sulfide in OUIs, and (5) late-stage secondary sulfide. The sulfide characteristics of these types are described in Bonnichsen (1972b) and elsewhere (see Severson, 1991, 1994; Severson and Hauck, 1990; Weiblen and Morey, 1976). Only Types 2 and 5 are known to occur along the northern contact.

Sedimentary sulfide. Sedimentary sulfide mineralization (syngenetic with deposition ?) occurs in the Bedded Pyrrhotite Unit of the Virginia Formation (Fig. 5). At the Babbitt, Serpentine (Fig. 15), and Dunka Pit deposits, the Bedded Pyrrhotite Unit is characterized by thin (≤2 mm) pyrrhotite/pyrite (95 to 100%) ± chalcopyrite ± sphalerite ± galena layers (0.5 to 5 cm apart) that are parallel to bedding. The Bedded Pyrrhotite Unit generally is about 10 m thick, but it is >50 meters thick in the footwall of the Bathtub ore zone. It occurs in both the footwall and as hornfels inclusions, and its lateral extent away from the Complex is not known. Sulfur isotope data from samples of the Bedded Pyrrhotite Unit in the Serpentine area ($\delta^{34}S$ 15.3 to

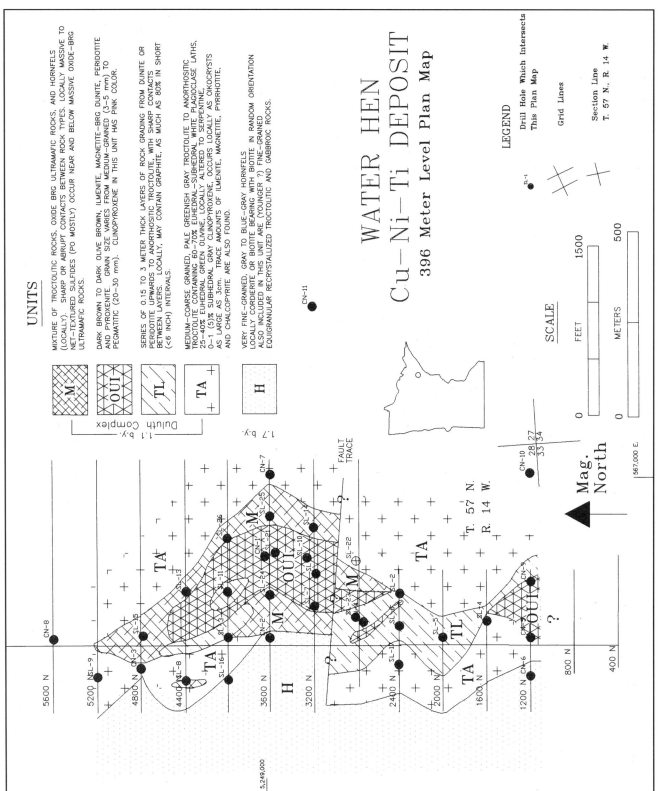

Figure 14. Plan view of the Water Hen Oxide-bearing Ultramafic Intrusion (after Strommer et al., 1990). See Figure 3 for location.

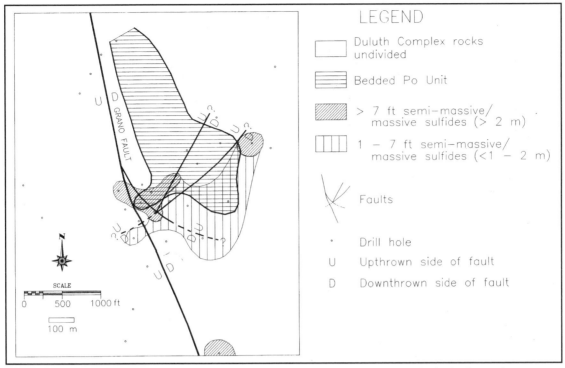

Figure 15. Distribution of semimassive and massive sulfide zones in drill holes in the Serpentine Cu-Ni deposit (T.60N., R.12W.) relative to the Bedded Pyrrhotite Unit of the Virginia Formation (after Zanko et al., 1994). See Figure 3 for location.

29.1‰) suggest that the Bedded Pyrrhotite Unit formed in a closed basin (Zanko et al., 1994). Severson (1994) reports similar sulfur isotope values (14.0 to 21.1‰) for this unit.

In addition to the Bedded Pyrrhotite Unit, sulfide-enriched (syngenetic ?) rocks are locally present throughout the Virginia Formation. At Dunka Road, disseminated sphalerite occurs in the footwall Virginia Formation (0.76 m of 1.08% Zn; Geerts, 1991); other sedimentary rocks of the Virginia Formation also contain sphalerite and chalcopyrite in the Babbitt area (Hauck, 1993).

Some quartzite inclusions associated with metabasalt in the Serpentine–Dunka Pit area have disseminated chalcopyrite and occasionally pyrrhotite and pentlandite mineralization (Bonnichsen, 1972b; Severson, 1994; Zanko et al., 1994). These inclusions may be related to the Puckwunge/Nopeming Formations. The origin of the mineralization is unknown.

Disseminated igneous sulfide. Disseminated interstitial sulfide includes both irregular and sporadic zones and stratabound, PGE-enriched sulfide zones (Bonnichsen, 1972b; Geerts, 1991, 1994; Geerts et al., 1990a; Morton and Geerts, 1992). The sulfides in both types are predominantly composed of pyrrhotite, chalcopyrite, pentlandite, and cubanite, with a Cu:Ni ratio of 3.3:1. They are primarily confined to the basal igneous units in the South Kawishiwi and Partridge River Intrusions. Other disseminated sulfides occur locally above the basal units of both intrusions. Sulfur isotope data (ranges from 8 to 12‰) indicate that sulfur in the magma was contaminated by

sulfur derived from assimilation and devolatization of footwall rocks ($\delta^{34}S$ ranges from 0 to 25‰ in the Babbitt area) prior to emplacement (Ripley, 1981; Ripley and Al-Jassar, 1987). Ripley (1990a) also reports $\delta^{34}S$ values as low as –1.4‰ that may represent magmatic sulfur at Babbitt associated with PGEs near to the Local Boy ore zone.

In the Dunka Road deposit, elevated Cu, Ni, and PGEs occur in stratabound zones beneath serpentinized ultramafic units (Geerts, 1991, 1994; Geerts et al., 1990a, b; Morton and Geerts, 1992). Within the augite troctolites and troctolites of the basal unit (Unit I) at Dunka Road, three stratabound, PGE-enriched sulfide layers occur beneath stratigraphically continuous to discontinuous ultramafic units (Fig. 16). A fourth zone occurs in a similar situation in Unit VI (Fig. 16). These zones are enriched in Pt + Pd (>800 to 2,400 ppb) and Cu (>0.5 wt.%) with fine fractures and minor alteration (i.e., saussuritization, uralitization, and serpentinization; Geerts et al., 1990a, b; Geerts, 1994). Geerts (1994) considers the stratabound PGE-mineralization at Dunka Road to be from a primary magmatic mantle source that was not contaminated by crustal material.

Semimassive to massive sulfide. Semimassive to massive sulfide ore occurs in highly localized and structurally controlled zones near the basal contact of both the South Kawishiwi and Partridge River Intrusions. At the Babbitt deposit (Local Boy ore zone; Figs. 9, 10), high-grade Cu-Ni massive sulfide ore is situated dominantly within the footwall Virginia Formation

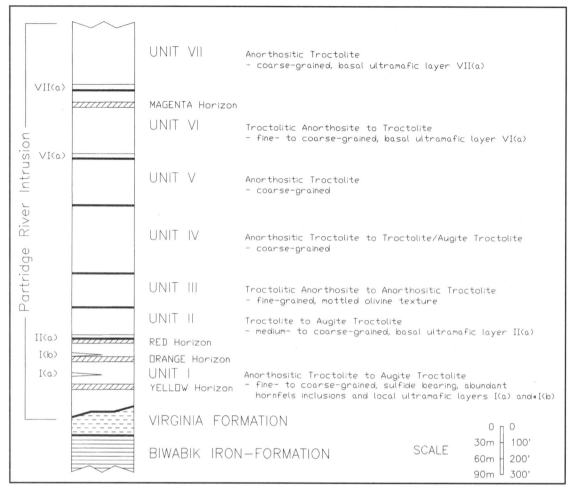

Figure 16. Generalized stratigraphic column for the Dunka Road Cu-Ni deposit (T.59N., R.13W.) showing stratabound, disseminated, platinum-group element enriched copper-nickel zones—red, orange, yellow, and magenta horizons beneath ultramafic horizons I(a), I(b), II(a), and VII(a) (after Geerts, 1991, 1994). See Figure 3 for location.

along the axis and flanks of a major east-west–trending anticline (Fig. 10). Ripley (1986) and Severson and Barnes (1991) propose that an immiscible sulfide melt, formed in an auxiliary magma chamber at depth, was injected into structurally prepared zones along the anticline to form the Local Boy ores. Low-grade Cu-Ni massive sulfide (pyrrhotite dominant) ores at the nearby Serpentine area were formed in a different manner. There the ore-forming process was spatially related to (1) partial to complete melting of nearby sulfide-rich footwall units (Bedded Pyrrhotite Unit; Fig. 15), and (2) proximity to a north-trending fault zone, the Grano Fault (Zanko et al., 1994). Footwall rocks within the Serpentine area were intensely fractured along the Grano Fault zone prior to emplacement of the Complex. During subsequent later intrusion, the highly fractured footwall units were easily assimilated, and the Bedded Pyrrhotite Unit provided a local source of sulfur for the massive sulfide ore. As a result, the semimassive to massive ore (pyrrhotite >> chalcopyrite) at Serpentine was distributed

around the periphery of the Bedded Pyrrhotite Unit in the area of the fault zone (Fig. 15; Zanko et al., 1994).

Sulfide in oxide-bearing ultramafic intrusions. Most OUIs contain disseminated sulfide (trace to 3%), primarily pyrrhotite, chalcopyrite, cubanite, and pentlandite. In the Water Hen, Skibo, and Boulder Lake South bodies, net-textured, pyrrhotite-rich sulfide occurs within the OUI (Bonnichsen, 1972b; Mainwaring and Naldrett, 1977; Seitz and Pasteris, 1988; Strommer et al., 1990). Also at Water Hen, late orthopyroxenite dikes cut the chilled fine-grained troctolite and contain Cu-rich sulfide and anomalous PGE values (Morton and Hauck, 1987; Severson, 1995).

Secondary sulfide. Late-stage secondary, fracture-controlled sulfides occur all along the western margin of the Complex from Fish Lake to Spruce Road (Weiblen and Morey, 1976; Foose and Weiblen, 1986; Strommer et al., 1990; Kuhns et al., 1990; Severson and Barnes, 1991; Mogessie and Stumpfl, 1992; Sassani, 1992; Hauck et al., 1996). The sulfide-filled fractures appear to be associated with syn- to late-Complex deformation. Mineralization consists of bornite, digenite, chalcocite, covellite, native copper and silver, pyrite, mackinawite, valleriite, and violarite. The nickel arsenide, maucherite, is a prevalent, minor metallic mineral. Silicates and oxides filling the fractures are biotite, serpentine, chlorite, sericite, and magnetite. Preliminary geochemical analysis of these secondary mineralized zones indicate that they can be anomalously enriched in Sb, As, Bi, Cl, Pb, Se, Te, Sn, and Zn (Morton and Hauck, 1987; Geerts, 1991, data in Appendix B).

Secondary copper-rich veins and copper-rich disseminated mineralization occur in the footwall Giants Range Batholith of the South Kawishiwi Intrusion (Bonnichsen et al., 1980; Severson, 1994; Hauck et al., 1996). Sulfur isotope data on these sulfides (from 8.7 to 10‰) are similar to values (10 to 11‰) found in sulfide in the overlying South Kawishiwi Intrusion (Sever-

son, 1994). Both types of sulfides show similar sulfur isotopic values to those found in contaminated Partridge River Intrusion sulfides. Severson (1994; Fig. 17A, B) shows that these sulfide zones in the Giants Range Batholith, along with abundant late-stage granitoid and pyroxenitic veins in the overlying troctolitic rocks, form linear (northeast-southwest) trends, and he suggests that they define areas of faulting. The Giants Range Batholith and the overlying South Kawishiwi Intrusion rocks are moderately to strongly altered in these zones, and the ultramafic and troctolitic rocks above these zones also have the highest PGE values (Severson, 1994; Hauck et al., 1996).

Platinum-group elements

Several PGE-enriched zones have been defined within the margin of the western Complex. These include (1) stratabound

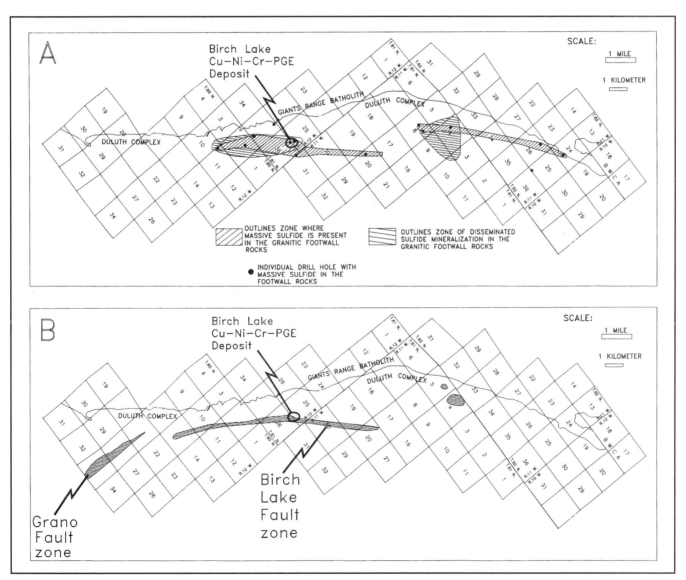

Figure 17. A, Linear distribution of massive sulfides, disseminated sulfides, and copper-rich veins in the Giants Range Batholith footwall (after Severson, 1994). B, Linear distribution of late-stage granitoid and pyroxenitic rocks in the Duluth Complex (after Severson, 1994). PGE, platinum-group element.

zones with sulfides at Dunka Road, (2) structurally controlled zones within massive sulfide ore at the Babbitt deposit, (3) stratabound/structurally controlled oxide/sulfide zones at Birch Lake, (4) structurally controlled zones with sulfides at South Filson Creek, and (5) a stratabound zone with semimassive oxides at Fish Lake.

Platinum-group minerals. Platinum-group minerals (PGM) are only known to occur along the western margin of the Complex. The exclusion of mineral exploration from the northern margin (the Boundary Waters Canoe Area) has been accompanied by minimal scientific mineral research. The first PGM identified in the Complex was sperrylite from the Minnamax shaft in the Babbitt deposit (Ryan and Weiblen, 1984). Sabelin (1985) and Sabelin and Iwasaki (1986) described Pt-Fe alloys and platinum-group element sulfides, sulfarsenides, and arsenides associated with disseminated sulfides and massive oxides from the Birch Lake area (Fig. 18). Several other platinum-group minerals have since been identified (Table 2) in both the Partridge River and South Kawishiwi Intrusions. These PGM are associated with As, Bi, Sb, Sn, and Te (the same anomalous metals as in secondary sulfide zones). However, PGM in most areas are rare to absent even when geochemical analyses indicate significant amounts of PGEs (>800 ppb). For example, Ripley (1990a) reported that some PGE-enriched samples that were analyzed from the Babbitt deposit contained no optically visible PGM; however, Ripley and Chryssoulis (1994) determined with an ion microprobe that parts per million quantities of Pd were present within the crystal structure of cubanite, maucherite, and secondary mackinawite.

Copper, nickel, and noble element distributions. The following discussion is based on 350 partial and full analyses for PGE, Au, Re, Cu, and Ni analyses; samples containing >0.1% copper are referred to as mineralized. About 250 samples are from published work (Geerts, 1991, 1994; Geerts et al., 1990a; Kuhns et al., 1990; Mogessie and Stumpfl, 1992; Naldrett and Duke, 1980; Ripley, 1990a; Sabelin and Iwasaki, 1986; Saini-Eidukat, 1991; Saini-Eidukat et al., 1990; Sassani, 1992; Severson and Barnes, 1991; Strommer et al., 1990). An additional 100 samples were subsequently analyzed at the Université du Québec à Chicoutimi (Severson, 1994; Hauck et al., 1996; Zanko et al., 1994).

Naldrett (1981) has divided PGE deposits into two types: (1) PGE-dominated deposits that are mined for their PGE content, and (2) Ni-Cu-sulfide dominated deposits with the PGE produced as a by-product. The classic examples of PGE-dominated deposits (or reefs) are the Merensky Reef of the Bushveld Complex and the JM-reef of the Stillwater Intrusion. The reefs contain only minor amounts of sulfides. The classic examples of Ni-Cu sulfide deposits are the Sudbury and Noril'sk deposits.

Broadly speaking, there are two schools of thought on the origin of PGE-dominated deposits. The igneous model proposed by workers such as Naldrett (1981) suggests that a mafic magma may become saturated with droplets of iron sulfide liquid. The sulfide droplets collect the metals from the silicate magma. They then settle to form a layer of disseminated sulfides in the intrusion. The hydrothermal model proposed by workers such as Boudreau and McCallum (1992) suggests that mafic magma becomes saturated with a Cl-rich hydrous phase, and the metals partition into this hydrous phase. This hydrous phase percolates through the magma until a suitable substrate, such as an oxide layer, is encountered, and the metals may be precipitated.

There is a greater consensus on the origin of the Ni-Cu sul-

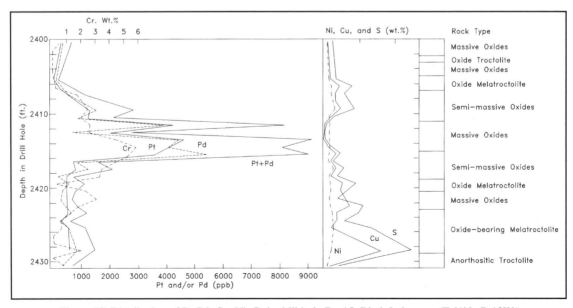

Figure 18. Distribution of Pt, Pd, Cu, Ni, Cr in drill hole Du-15, Birch Lake area (T.61N., R.12W.) South Kawishiwi Intrusion (after Sabelin and Iwasaki, 1986; geology from Severson, 1994). See Figures 3 and 17 for location.

TABLE 2. PLATINUM-GROUP MINERALS IDENTIFIED IN THE DULUTH COMPLEX

	Composition	Location	Reference*		Composition	Location	Reference*
Isoferroplatinum	Pt_3Fe	Birch Lake	2, 4	?	PdSnSb	Babbitt	5
Tetraferroplaninum	PtFe	Birch Lake	2, 4	Sperrylite	$PtAs_2$	S. Filson Creek	4
?	Pd-alloys	Birch Lake	2			Babbitt	1, 6, 8
Laurite	RuS_2	Birch Lake	2			Birch Lake	8
?	PtPdSbAs	Birch Lake	5	Michenerite	PtBiTe	Dunka Road	7
?	$PtFe_x$	Birch Lake	4			Wetlegs	8, 9
?	PtIrOsFe	Birch Lake	5	Froodite	$PbBi_2$	Dunka Road	4, 7
?	IrPtRh	Birch Lake	5			Wetlegs	8, 9
?	IrOs	Birch Lake	5	Moncheite	$PtTe_2$	Birch Lake	9
?	OsRuAsS	Birch Lake	5			Wetlegs	8, 9
Platinum	Pt	Birch Lake	5	Taimyrite	$(Pd,Cu)_3Sn$	Babbitt	8
?	PdPt	Birch Lake	5, 8	?	$Pt(Bi,Sb)_2$	Babbitt	4
?	PtIrOsFe	Birch Lake	5	?	PdBiAs	Babbitt	4
?	PtCuFe	Birch Lake	5	Urvantsevite	$Pd(Bi,Pb)_2$	Babbitt	4
?	PtPdCuFe	Birch Lake	5	Sudburyite	PdSb	Babbitt	4
?	PtAsAg	Babbitt	5	Stannopalladinite	$Pd_3Sn_2(?)$	Babbitt	4
Palladium	Pd	Birch Lake	5	?	PbBiTe	Birch Lake	8
?	PdSn	Babbitt	5			Wetlegs	8
?	PtPdSn	Babbitt	5	?	$Pd_7(Sb,Bi)_8$	Dunka Road	7
?	BiPdS	Babbitt	5	Cabriite	Pd_2SnCu	Babbitt	4
?	PdSbPb	Babbitt	5	?	Pt/Ni Arsenide	Babbitt	1
?	PtPdSnCu	Babbitt	5	?	PdBi	Fish Lake	3, 10
?	PdSbPbBi	Babbitt	5	?	$Pd_7(Sb,Bi)_8$	Dunka Road	7

*Table from Dahlberg, 1992 — Sources: 1 = Ryan and Weiblen, 1984; 2 = Sabelin et al., 1986; 3 = Sassani and Pasteris, 1988; 4 = B. Cannon, written communication, 1990; 5 = Jedwab, written communication, 1991, 1993; 6 = Kuhns et al., 1990; 7 = Morton and Hauck, 1987; 8 = Mogessie et al., 1991; 9 = Holst et al., 1986; 10 = Sassani, 1992.

fide deposits. Most workers believe that a mafic magma may become saturated with iron sulfide droplets. The droplets collect the metals and settle to the margins of the intrusion where they form a pool of sulfide liquid (Naldrett, 1981). The sulfide pool crystallizes to form massive sulfide ore. The sulfides may undergo fractional crystallization of monosulfide solid solution (MSS) to produce a zoned ore body with an Fe-rich sulfide cumulate and a Cu-rich sulfide liquid (Distler et al., 1977). Recent experimental work on the partition coefficients of PGE into MSS (Barnes et al., 1994) shows that the MSS is enriched in Ir and Rh and depleted in Pt, Pd, and Cu; thus, fractional crystallization of MSS could produce zoned ore bodies. After the solidification of the sulfides, metals might be remobilized by metamorphic fluids or late Cl-rich magmatic fluids. However, there is some debate as to whether this is possible (Wood et al., 1992).

As outlined by Barnes et al. (1993), a simple plot of Cu/Pd versus Pd may be used to classify Ni-Cu sulfide deposits in terms of their PGE potential. Most of the Complex samples contain less than 1 ppm Pd and have Cu/Pd ratios greater than mantle values (Fig. 19A, B). This type of deposit is usually described as depleted, meaning depleted in PGE. However, not all of the Complex samples are depleted. Samples at Dunka Road and some samples from the U3 Unit at Birch Lake contain 1 to 10 ppm Pd and have Cu/Pd ratios close to mantle values. These samples may be described as undepleted. Finally,

there are a group of samples from the oxide units and pegmatites of Birch Lake containing 1 to 10 ppm Pd and Cu/Pd ratios very much lower than mantle. In terms of Cu/Pd and Pd content, these samples resemble PGE-dominated or reef deposits and may be described as enriched.

To establish whether sulfides are the main factor effecting the distribution of the metals, Cu, Ni, Au, and the PGE have been plotted against S (Fig. 20a to h). There is a strong correlation between Cu and S and between Re and S (Fig. 20a, b), indicating that sulfide minerals are the main hosts for these metals. Above 0.2% S, nickel shows a positive correlation with S, but wide scatter in data is evident below 0.2% S (Fig. 20c). This situation arises because some of the silicate and oxide phases in the rocks (i.e., olivine and magnetite) contain considerable Ni, whereas others (i.e., plagioclase) contain very little. At low sulfide contents, two mixing lines are required, one for the mixing of sulfides with ferromagnesium phases, and one for mixing with Ni-poor phases. Above 0.2% S, the Ni content of the rock is largely controlled by the sulfide minerals, and only one mixing line is needed.

Iridium and Rh show positive correlations with S (Fig. 20d, e), similar to that seen for Cu, suggesting that sulfides are the main host for these metals. However, for a given S content, samples from Dunka Road and Birch Lake (stars and crosses) appear to be richer in Ir and Rh than samples from other areas in the Complex. This situation suggests that the sulfides at these

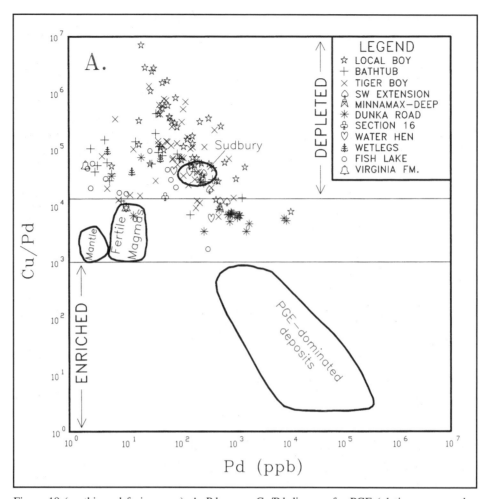

Figure 19 (on this and facing page). A, Pd versus Cu/Pd diagram for PGE (platinum-group element) samples in the Partridge River Intrusion and its footwall rocks (diagram after Barnes et al., 1993). In the legend, Local Boy, Bathtub, Tiger Boy, SW Extension, Minnamax Deep all refer to Cu-Ni orebodies or areas in the Babbitt deposit (T.60N., R.12W). Dunka Road, Wetlegs, and Section 16 are in T.59N., R.13W. (see Fig. 3). Fish Lake is located in the Duluth Layered Series (see Fig. 3; T.52N., R.15W.). Water Hen (T.57N., R.14W.) is in the South Complex area (see Fig. 3). B, Pd versus Cu/Pd diagram for PGE samples in specific stratigraphic units and mineral deposits of the South Kawishiwi Intrusion (SKI) and its footwall rocks (diagram after Barnes et al., 1993). IF, iron-formation.

two localities are richer in Ir and Rh than sulfides from other localities in the Duluth Complex.

For most samples, Pd, Pt, and Au show a positive correlation with S from 0.01 to 10% S (Fig. 20f to h). This suggests that the metal content is controlled by sulfide content (between 0.01 and 10% S). As in the case of Ir and Rh for a given S content, the Birch Lake and Dunka Road samples are richer in Pd, Pt, and Au, indicating that the sulfides at Birch Lake and Dunka Road are richer in PGE than sulfides from elsewhere in the Complex. There are five samples from Birch Lake that are enriched in Pd and Pt relative to the general trend, and sulfide control of the metals in these samples is not obvious. These are the same samples that plot in the PGE-enriched field on Figure 19B. (There are no S values for some

of the samples from Birch Lake so more samples are shown in the PGE-enriched field on Fig. 19 than on Fig. 20f to h). Above 10% S, there is no direct correlation between S content of the rock and the Pd, Pt, and Au contents. This point has previously been noted by Ripley (1990a) for samples from the Babbitt deposit.

On the basis of the S versus metal plots, it may be concluded that in rocks containing disseminated sulfides (apart from the five Pt-Pd enriched Birch Lake samples) the distribution of all of the metals is largely controlled by sulfides. In rocks containing more than 10% S, the Cu, Ni, Ir, Rh, and Re contents still seem to be controlled by sulfides. The lack of correlation between Pt, Pd, and Au for sulfur-rich samples could be interpreted to indicate that Pt, Pd, and Au have been remobi-

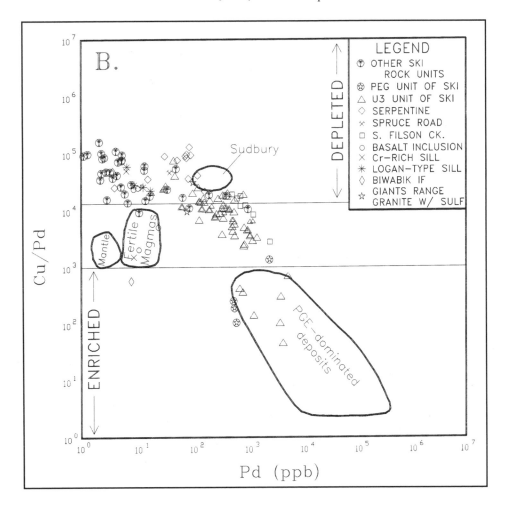

lized in these samples. However, it could also be a result of crystal fractionation of a sulfide liquid.

A third approach of looking at the Complex data is to use mantle normalized metal plots as described by Barnes et al. (1988). The Pt-Pd enriched samples from Birch Lake (not recalculated to 100% sulfides; Fig. 21a) have arched-shaped mantle-normalized metal patterns that are characteristic of PGE-reefs in the Stillwater and Bushveld Complexes. On Figure 21a, showing high-grade PGE rocks from Birch Lake, both the patterns and concentrations are similar to those of the Merensky Reef of the Bushveld Complex. The average for the Birch Lake mineralized samples is listed in Table 3.

The concentration of the metals in the disseminated ores at all localities has been recalculated to 100% sulfides to allow comparison between rocks with different sulfide contents. As mentioned earlier, samples from Dunka Road and the remaining mineralized samples from Birch Lake are richer in PGE than samples from other areas in the Complex. The Dunka Road and Birch Lake samples have fairly smooth metal patterns. The noble metals are not enriched relative to Cu (Fig. 21b), and there is a steady increase from Ir to Pd where the patterns flatten out between Pd and Cu. The Cu content of the sulfides is high (22%; Table 3) compared with the Cu-content of disseminated ores

associated with other flood basalts (Insizwa, Muskox, and Noril'sk; Table 3). However, Cu-rich ores do occur at these localities, and compared to these sulfides, the Dunka Road and Birch Lake sulfides have Pt, Pd, Au, and Cu contents similar to those in the Oktyabr'sky orebody of the Talnakh Intrusion in the Noril'sk area, Russia and the Muskox Intrusion in the Northwest Territories of Canada; however, the Duluth Complex sulfides are richer in Ni, Os, Ir, and Ru (Fig. 21b; Table 3).

Most of the disseminated ores from the Complex are depleted in PGE relative to Ni and Cu (Fig. 21b; Table 3). On the metal patterns, Ir is depleted relative to Ni, and Pt is depleted relative to Pd. The concentrations of the metals in the sulfides are similar to those found in the Insizwa intrusion. The reason for the depletion is probably that the magma has previously segregated sulfides, and these sulfides have scavenged the PGE from the magma leaving it depleted in PGE. Subsequently, when a second generation of sulfides form, they are depleted in PGE (Barnes et al., 1988).

The massive ore samples (samples containing >10% S) come from the Babbitt area. The concentration of the metals in these ores has also been recalculated to 100% sulfides, although as mentioned above, this may not be the correct procedure for Pt, Pd, and Au if these elements have been remobilized. The

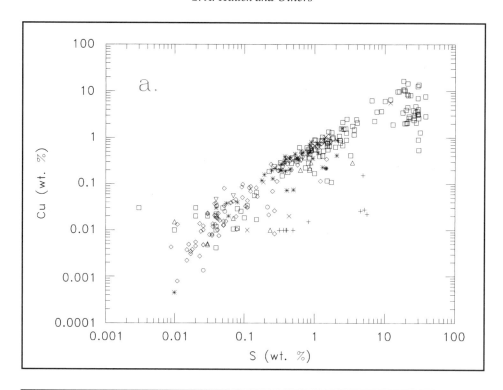

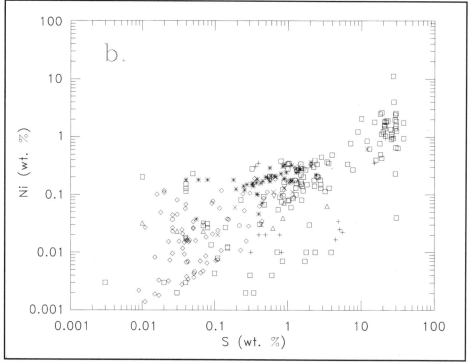

Figure 20 (on this and following three pages). Scatter plots of various metals versus S (wt. %): a, Cu versus S; b, Ni versus S; c, Pt versus S; d, Pd versus S; e Au versus S; f, Ir versus S; g, Rh versus S; and h, Re versus S. Different mineralized areas are represented by the following symbols: × = Dunka Road; * = Birch Lake; + = Virginia Formation and Biwabik Iron-Formation sediments; □ = Babbitt, all ore zones; ● = Spruce Road; ◇ = South Kawishiwi Intrusion units; △ = Fish Lake; ▽ = Wetlegs; ○ = pre-Complex sills (Logan-type and Cr-rich). Data from: open-file private company data at the Department of Natural Resources, Division of Minerals, Hibbing, Minnesota; Geerts et al., 1990a; Hauck, 1993; Hauck et al., 1996; Kuhns et al., 1990; Mogessie and Stumpfl, 1992; Morton and Hauck, 1987; Ripley, 1990a, b; Sabelin and Iwasaki, 1985, 1986; Saini-Eidukat et al., 1990; Sassani, 1992; Severson, 1994; Strommer et al., 1990; Zanko et al., 1994.

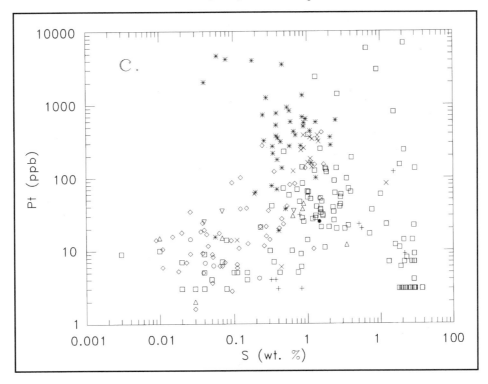

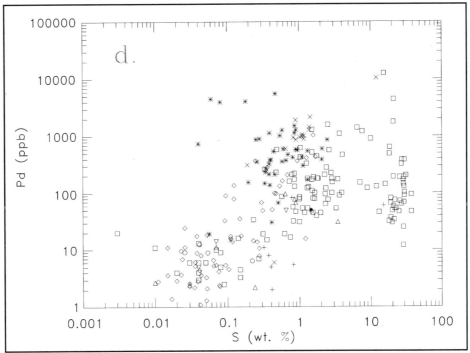

recalculation does not effect the shape of the metal pattern, only the level, so comments based on the shape of the pattern remain valid. To allow for the possible effects of fractional crystallization of a sulfide liquid, two averages have been calculated, one for Cu-rich samples (presumably the fractionated liquid), and one for Cu-poor samples (presumably the cumulate). Both Cu-rich and Cu-poor samples are strongly depleted in PGE with

concentrations well below those found in other flood basalt ores and in disseminated ore at the Complex (Table 3; Fig. 21b). Further, both Cu-rich and Cu-poor samples show large negative Pt anomalies. (Barnes and Naldrett, 1986, defined Pt anomalies to be analogous to Eu anomalies, and a formula for calculating them can be found in Barnes and Francis, 1995.) The Cu-rich ore is enriched in Pt, Pd, and Au relative to the Cu-

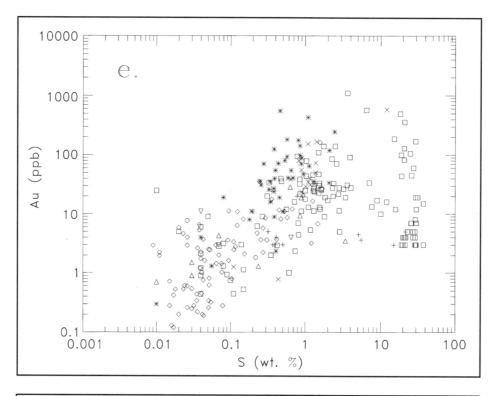

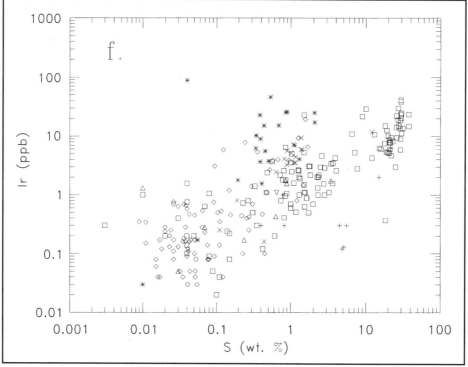

poor ore and slightly depleted in Ir and Rh. This is consistent with crystal fractionation of MSS from a sulfide liquid to produce the zonation, and could also explain the lack of correlation between these elements and S.

Origin of platinum-group element mineralization. An underlying assumption in much of the data presentation is that the metal distribution in the Duluth Complex is essentially the product of magmatic processes. The overall model is that a large magma chamber was periodically filled with mafic magma. Most magma pulses were depleted in PGE on arrival, possibly due to

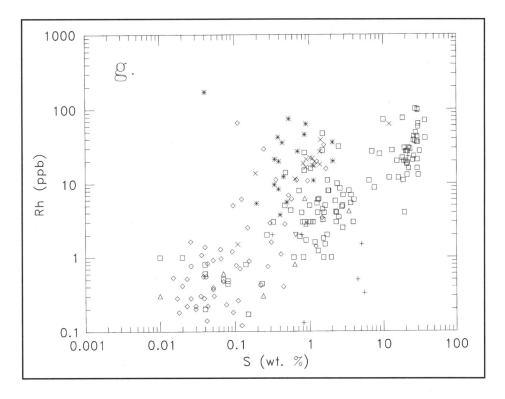

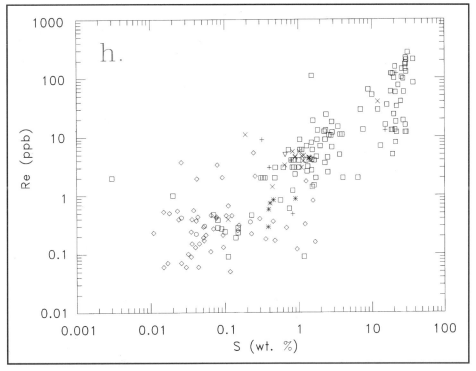

earlier scavenging of the PGE by sulfides at depth. On emplacement in the chamber, the magma once again became sulfide saturated and a second generation of PGE depleted sulfides formed. In much of the Complex, the sulfide droplets never collected together and disseminated sulfides formed. In the Babbitt area, the droplets coalesced to form a sulfide pool that subsequently underwent crystal fractionation of MSS to produce Cu-rich and Fe-rich sulfides. In the Dunka Road and Birch Lake areas, the magma had not previously experienced sulfide segregation, and therefore, the sulfides that formed were not depleted in PGE.

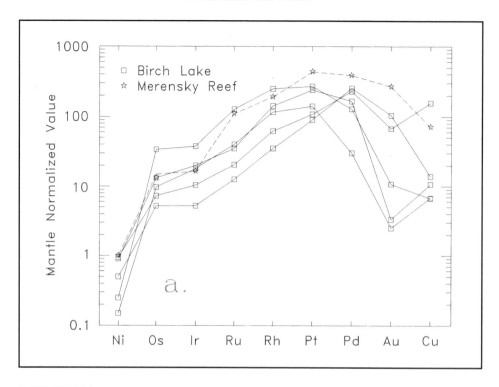

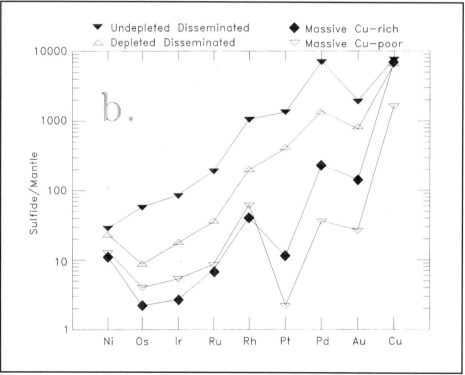

Figure 21. Comparison of PGEs in sulfide ores in the Duluth Complex with those in other flood basalts (mantle normalized): a, comparison of Merensky Reef and high-grade Birch Lake mantle normalized PGE signatures, not normalized to 100% sulfides; b, comparison of undepleted disseminated Cu-Ni ores (Dunka Road and Birch Lake samples), depleted disseminated Cu-Ni ores (other Complex samples), massive Cu-rich ores (Babbitt), and massive Cu-poor ores (Babbitt).

TABLE 3. METAL CONTENT OF MINERALIZED ROCKS FROM THE DULUTH COMPLEX

	Locality	n	S (wt. %)	Ni (wt. %)	Cu (wt. %)	Os (ppb)	Ir (ppb)	Ru (ppb)	Rh (ppb)	Pt (ppb)	Pd (ppb)	Au (ppb)
PGE-dominated ore	Birch Lake	9	0.16	0.13	0.08	57	84	305	155	2,762	2,510	81
Undepleted disseminated ore	Dunka Road and Birch Lake	52	0.83	0.18	0.49	5	9	29	21	438	706	71
Depleted disseminated ore	Various localities	123	1.81	0.24	0.96	2	4	10	10	219	290	61
Massive ore	Various localities	44	24.00	1.59	5.69	9	12	31	33	201	551	47
Recalculated to 100% sulfide												
Undepleted disseminated ore	Dunka Road and Birch Lake	30	35.90	5.77	22.27	258	387	1,260	1,057	11,628	32,024	2,332
Depleted disseminated ore	Various localities	89	36.96	4.70	18.13	38	77	225	199	3,476	6,100	940
Massive ore Cu-rich	Various loaclities	11	37.15	2.23	20.32	10	12	48	41	97	1,032	177
Massive ore Cu-poor	Various localities	29	40.12	2.65	4.76	18	23	56	62	19	164	32
Sulfides associated with other flood basalts												
Disseminated ore	Insizwa*		38.00	6.32	6.26	237.0	413.0	523	325	7,238	2,851	2,235
Massive ore Cu-rich	Insizwa		38.00	7.71	18.20	81.0	134.0	595	325	2,552	1,985	446
Massive ore Cu-poor	Insizwa		38.00	8.62	8.62	90.0	284.0	450	408	5,965	2,122	345
Disseminated ore	Oktyabr'sky†	8	37.52	3.15	8.71		35	134	367	4,383	17,063	90
Massive ore Cu-rich	Oktyabr'sky	13	33.91	2.53	25.50	3	8	17	57	25,103	63,624	3,650
Massive ore Cu-poor	Oktyabr'sky	24	37.80	4.01	6.53	12	29	77	327	2,670	12,873	260
Disseminated ore	Muskox ol. gabbro§	15	37.54	4.88	6.30	39.1	15.9		225	1,470	12,999	926
Massive ore Cu-rich	Muskox	1	35.15	1.56	22.87	85.8	23.5	121	474	17,536	185,686	5,998

*Lightfoot and Naldrett, 1984.

†Zientek et al., 1994.

§Barnes and Francis, 1995.

This model may be overly simplistic as many authors have found evidence of hydrothermal action in the Duluth Complex, and some authors suggest that the hydrothermal fluids redistributed the PGE. In the massive ores in the Local Boy ore zone at Babbitt, Severson and Barnes (1991) suggest that the PGE have also been locally remobilized and concentrated within structurally controlled zones. Many other authors (e.g., Mogessie and Stumpfl, 1992) have argued that Pt has been hydrothermally mobilized, and thus, the Pt anomalies may in fact reflect the mobility of Pt. A remobilization/hydrothermal event is further supported by stable isotopic data (presented later), alteration effects, mineral compositions, and the presence of structural zones associated with PGE-enriched zones. The presence of late hydrothermal fluids that could have remobilized PGEs is documented by mineral compositions and fluid inclusion data. Microprobe data on recrystallized plagioclase in one of the stratabound zones at Dunka Road are consistent with hydrothermal alteration produced by 700 °C saline solutions (Morton and Geerts, 1992; Geerts, 1994). Also, the presence of serpentine-filled fractures in the same zone suggests that temperatures either decreased to a range of 350 to 500 °C, or that later lower temperature fluids were introduced. Microprobe data on four different spinel-rhombohedral oxide pairs in a drill hole at Fish Lake (Sassani, 1992) show that the rocks were affected by at least four episodes of fracture-controlled hydrothermal alteration in T-fO_2 space, beginning at 750 °C (approx. FMQ-2) and decreasing to 350 °C (approx. FMQ-8). Within the Fish Lake drill hole is a PGE-Cr-bearing (572 ppb Pt+Pd, 48,000 ppm Cr; Dahlberg et al., 1987), semimassive oxide horizon beneath an ultramafic horizon. Sassani (1992) interprets the PGE enrichment beneath the impermeable oxide layer as due to magmatic hydrothermal fluids (four episodes) that locally increased the pH, decreased the oxygen fugacity, and decreased the temperature of the rocks they infiltrated. According to Sassani (1992), these oxide data are similar to those from oxides in a PGE-enriched zone in the Birch Lake area in the South Kawishiwi Intrusion (Du-15 drill hole), except that the Birch Lake area experienced higher oxygen fugacity. Sassani (1992) attributes the difference in the oxygen fugacities of the hydrothermal fluids to the difference in the lithologies of the footwall rocks from which the fluids were generated or interacted (i.e., organic-rich Virginia Formation in the Fish Lake area versus oxidizing Giants Range Batholith in the Birch Lake area).

Mogessie and Stumpfl (1992) report an association of PGM with hydrous silicates (rich in Cl), graphite, and zoned spinel in several drill cores taken from the western margin of the Complex. They suggest that chloride complexes, associated with late magmatic fluids, were responsible for transporting

S. A. Hauck and Others

and depositing PGEs. At Fish Lake, Pasteris et al. (1995), analyzing fluid inclusions, have shown that high-chloride, high-temperature brines passed through the rocks.

Structural zones are also important in concentrating late hydrothermal solutions capable of remobilizing PGEs. In the Local Boy ore zone, elevated values of PGEs occur along the crest of the Local Boy Anticline and suggest possible redistribution of PGEs by later fluids (Fig. 10). In the Birch Lake area, Severson (1994) proposes that PGE enrichment is due to circulating hydrothermal fluids along faults. The faults are delineated by linear zones of sulfide-enrichment in the footwall and intrusion of late-stage granitoid and pyroxenitic veins (Fig. 17A, B) into the basal troctolites. Extensive hydrothermal alteration in drill core in the area supports this hypothesis.

While all of the above evidence indicates extensive hydrothermal activity has taken place, most of the variations in the distribution of the metals can be explained by a simple magmatic model.

Stable isotope data

Sulfur, carbon, oxygen, and hydrogen isotopic data (Figs. 22 to 25) exist only for igneous and footwall rocks along the western margin of the Complex, except for some carbon data from the Rove Formation (Fig. 23c) and limited sulfur data for sulfides from the northern margin (Fig. 22a; Jensen, 1959). Whole rock and mineral sulfur isotope data (Figs. 22a to c) are tightly distributed around a mean of about $\delta^{34}S$ 7 to 8‰. There is virtually no difference between intrusive rocks and sedimentary footwall rocks, except for the Bedded Pyrrhotite Unit. The latter shows markedly heavier sulfur isotopic values (14 to 29‰) than all other footwall rocks. Carbon data for sedimentary and igneous rocks are very similar in their range of values, $\delta^{13}C$ –37 to –19‰; however, carbon values in sedimentary rocks are highly positively skewed, whereas those from the igneous rocks are more uniformly distributed (compare Fig. 23c with 23a, b). Oxygen and hydrogen isotopic signatures from igneous and sedimentary rocks are clearly different (Figs. 24 to 25). For example, the mean $\delta^{18}O$ is about 11 to 12‰ for the sedimentary rocks and between 6 and 7‰ for the igneous rocks. Hydrogen isotopic values are generally heavier in the sedimentary rocks, δD –80 versus –95‰ for igneous rocks.

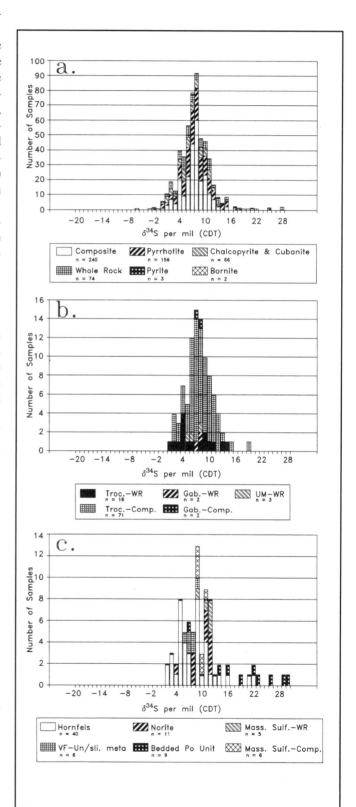

Figure 22. Distribution of sulfur isotopes from the western portion of the Duluth Complex: a, mineral and all whole rock (WR) data (whole rock data from analysis of crushed rock samples; composite (Comp.) data is from sulfide concentrates only); b, troctolite (Troc.), gabbro (Gab.), and ultramafic (UM) rock data; c, unmetamorphosed and slightly metamorphosed Virginia Formation (VF–Un/sli. meta.), i.e., hornfels, norite, and massive sulfide (Mass. Sulf.) data (data from Butler, 1989; Hauck et al., 1996; Jensen, 1959; Mainwaring and Naldrett, 1977; Mogessie and Stumpfl, 1992; Rao, 1981; Ripley, 1981, 1990a; Ripley and Al-Jassar, 1987; Ripley and Taib, 1989; Ryan, 1984; Severson, 1994; Taib, 1989; Zanko et al., 1994).

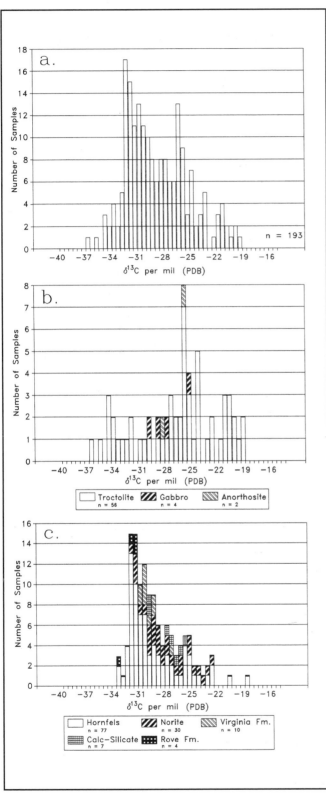

Figure 23. Distribution of carbon isotopes from the western and northern portion of the Duluth Complex: a, all carbon isotope data; b, data from troctolite, gabbro, and anorthosite; c, data from Virginia and Rove Formations, hornfels, and norite (data from Andrews, 1987; Andrews and Ripley, 1989; Butler, 1989; Ripley and Al-Jassar, 1987; Taib, 1989).

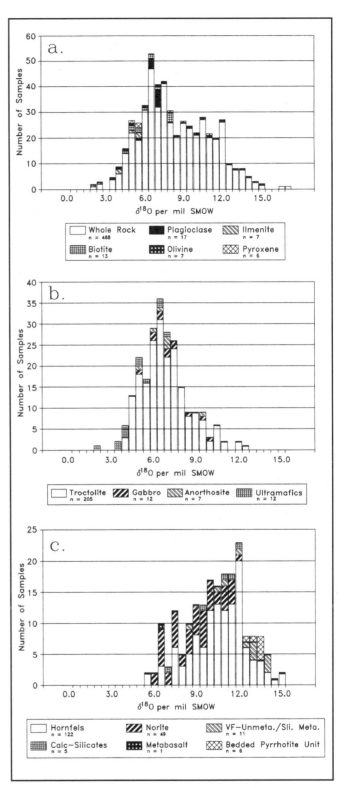

Figure 24. Distribution of oxygen isotopes only from the western portion of the Duluth Complex: a, all whole rock and mineral oxygen isotope data; b, igneous rock data; c, unmetamorphosed and slightly metamorphosed Virginia Formation (VF-Unmeta./sli. meta.), i.e., hornfels, norite, and metabasalt (data from Andrews, 1987; Andrews and Ripley, 1989; Churchill, 1978; Hauck et al., 1996; Mogessie and Stumpfl, 1992; Mudrey, 1973; Muhich, 1993; Rao, 1981; Ripley and Al-Jassar, 1987; Ripley and Taib, 1989; Ripley et al., 1992, 1993; Ryan, 1984; Taib, 1989; Zanko et al., 1994).

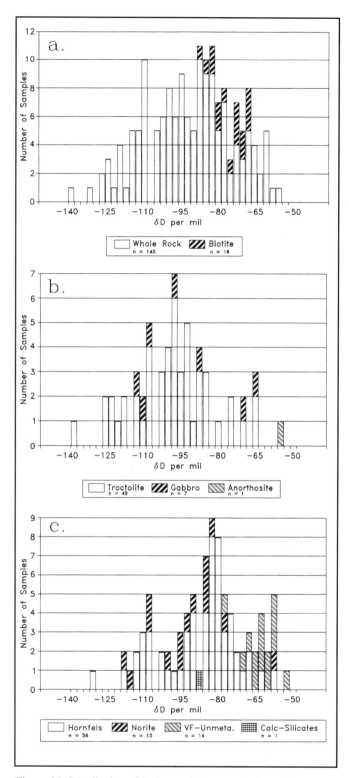

Figure 25. Distribution of hydrogen isotopes only from the western portion of the Duluth Complex: a, all whole rock and biotite hydrogen isotope data; b, igneous rock data; c, Virginia Formation (VF), hornfels, and norite data (data from Andrews, 1987; Andrews and Ripley, 1989; Butler, 1989; Hauck et al., 1996; Ripley and Al-Jassar, 1987; Ripley and Taib, 1989; Ripley et al., 1992, 1993; Taib, 1989). Unmeta., unmetamorphosed.

The lack of difference between the sulfur data in the igneous rocks and their footwall sediments suggests that much of the sulfur now residing in the Complex rocks was incorporated in a very large volume of magma and homogenized prior to emplacement (E. Ripley, personal communication, 1993). In contrast, localized, heavy sulfur isotopic values in the Bedded Pyrrhotite Unit xenoliths and at the footwall contact (Fig. 26), appears to have only minimal effect on the overall sulfur isotopic composition of the more voluminous intruding magmas (Zanko et al., 1994). The large-scale homogenization effect is not evident in the hydrogen and oxygen data; rather, these data document the local effects of contamination due to wall-rock assimilation. For example, noritic rocks are most commonly found along the basal contact of the Complex and around xenoliths where the effects of contamination to the magma should have been highest. Not surprisingly, carbon, oxygen, and hydrogen isotopic values for norites fall between and overlay the isotopic values of the sedimentary and igneous rocks (Figs. 23 to 25).

Sulfur isotopic data and mineralogical studies in Unit I (Fig. 6) of the Babbitt deposit (Ripley and Alawi, 1986; Ripley and Al-Jassar, 1987; Andrews and Ripley, 1989; Taib and Ripley, 1993) support derivation of sulfur from the footwall rocks. However, the data suggest that sulfur was not derived in situ from the immediately surrounding footwall, but rather in shallow crustal staging chambers at depth. Sulfur isotopic values of $0 \pm 2‰$ in the upper units at Babbitt are evidence that these units were undersaturated with respect to sulfur upon emplacement. The latter suggests these rocks had a shorter chamber residence time and were uncontaminated by the country rocks (Taib and Ripley, 1993).

Sulfur appears to have been homogenized in the magma prior to emplacement, except in one localized setting. Severson (1994) and Zanko et al. (1994) both recognize within the South Kawishiwi Intrusion that basal massive sulfide locally occurs at the periphery of Bedded Pyrrhotite Unit inclusions, which suggests a genetic relationship (Fig. 15). Sulfur and oxygen isotopic data of samples collected around a Bedded Pyrrhotite Unit inclusion in the Serpentine deposit demonstrate localized isotopic effects due to contamination (Fig. 26). Based upon mineralogy of the xenolith, Zanko et al. (1994) have suggested that the top of the xenolith was hotter than the base. The magmatic sulfur isotopic value at the base of the xenolith may be the result of no contamination by the heavier sulfur isotopic values of the xenolith. Because of the volumetric dominance of lower $\delta^{18}O$ of the magma, the higher $\delta^{18}O$ content of the xenolith has a very localized effect (over 1 to 3 m) on the intruding melt (Fig. 26).

Ripley et al. (1993) infer that there were three isotopically distinct fluids responsible for hydrothermal alteration in the Complex, derived from (1) the devolatization of the footwall rocks and inclusions, (2) magmatic fluids, and (3) meteoric waters (Fig. 27a, b) that produced a loss of about 2.5% H_2O from footwall rocks in addition to a decrease in whole rock δD to values from approximately -80 to approximately $-100‰$ (Ripley et al., 1992; Hauck et al., 1996). This interpretation is

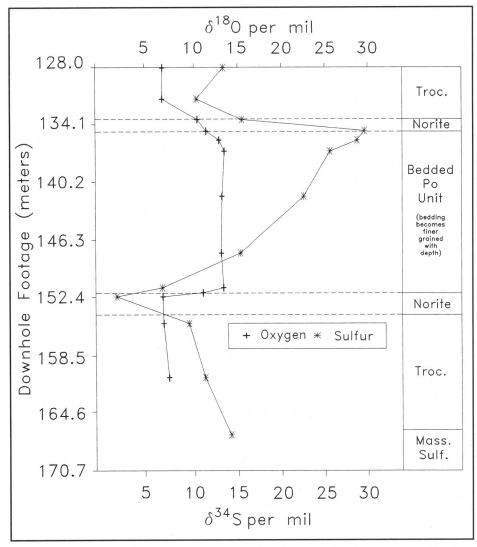

Figure 26. Distribution of sulfur and oxygen isotopic values around and within a xenolith of the Bedded Pyrrhotite (Po) Unit within the Serpentine deposit (after Zanko et al., 1994).

based on hydrogen isotope data (δD of –56 to –138‰), H_2O content of igneous rocks (Fig. 27a, b), and alteration mineralogy (serpentine, chlorite, amphibole, epidote, calcite, prehnite, and greenalite). The similarity between Figure 27a and b indicates that the alteration process is widely variable, but that the effects are similar on both the footwall and Duluth Complex rocks. The meteoric water component is evident in the Birch Lake samples shown on Figure 28. Ripley et al. (1993) explain that extreme serpentinization of olivine-rich rocks has produced anomalously low $\delta^{18}O$ values of 2 to 4‰ in the Babbitt area. Based on hydrogen isotopes, Lee and Ripley (1993) also interpret the alteration at Spruce Road as the product of low-temperature fluids derived from country rocks. In Figure 28, oxygen and hydrogen isotope data illustrate the evolution of a hydrothermal fluid from unaltered Virginia Formation (upper

right hand corner of the diagram) to progressively more evolved fluids (lower left hand corner) that are a mixture of igneous and footwall fluids. Ripley et al. (1992) indicate these data are consistent with a fluid that can contain up to ~15% methane.

Limited fluid inclusion data from quartz and olivine from Pasteris (1989, 1990) and Pasteris et al. (1992, 1995) also support devolatization of footwall material as a source of fluids associated with hydrous alteration. These data, and data from Pasteris (1984, 1988), also show that methane is an important component of fluid inclusions along with chlorine, nitrogen, graphite, and carbon dioxide. Pasteris and et al. (1992, 1995) suggest that the Complex rocks have been subjected to multiple injections of fluid, derived from both the footwall and the melt, over a wide range of pressure, temperature, and time. In large part, the complexity of the fluid story is due to episodic overprinting of

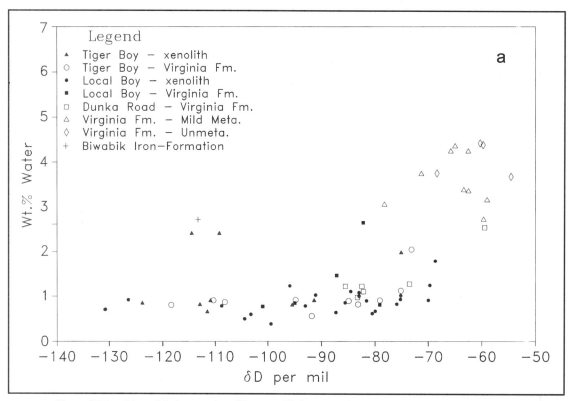

Figure 27 (on this and facing page). a, δD versus H₂O content for unmetamorphosed and metamorphosed pelitic rocks of the Virginia Formation (after Ripley et al., 1992, 1993; Hauck et al., 1996). b, δD versus H₂O content for igneous rocks of the Partridge River and South Kawishiwi Intrusions (after Ripley et al., 1992, 1993; Hauck et al., 1996).

hydrothermal cells that developed around each new intrusion in the Complex (Pasteris et al., 1995). This conclusion is based on fluid compositions and the heterogeneity of the inclusions.

Relation of chlorine-rich samples and mineralization

Drill cores (Fig. 29) of olivine-bearing and olivine-, oxide-, and sulfide-rich rocks commonly display a variety of white, green, blue, and rusty brown encrustations (or efflorescences) and liquid drops; both encrustations and drops are Cl-enriched. These drops and encrustations have been reported from a number of locations in the Complex from South Boulder Lake to Spruce Road; none, as yet, have been reported from the northern area. Brown liquid droplets, containing up to 3,000 ppm Cl (Dahlberg and Saini-Eidukat, 1991), generally occur on drill core that are (1) very weakly to strongly serpentinized mafic to ultramafic rocks, (2) semimassive to massive oxides, (3) semimassive to massive sulfides (Local Boy), (4) disseminated sulfide and oxide, and (5) occasionally on pyroxenes (Severson and Barnes, 1991; Hauck, 1993; Hauck et al., 1996). SEM analysis of the precipitates or encrustations that form from droplets show that they are dominantly iron and magnesium chlorides, chlorocarbonates, and/or carbonates with lesser amounts of Cu ± Ni Cl-CO₃ phases. The Fe-O-Cl precipitates are probably dehydra-

tion products of hibbingite, γ-Fe₂Cl(OH)₃, a recently identified mineral (Saini-Eidukat et al., 1994). These authors believe that hibbingite is a relict of late- to post-magmatic Cl-enriched fluids that produced serpentinization.

Nickel-rich drops and encrustations also may occur in rocks with sulfide mineralization. Associated with these drops and encrustations are (1) Sn oxy/hydroxy chlorides, (2) Sn, Bi, Pd, Zn, Cu, and Ag chlorides and sulfides (these same cations are also anomalously enriched in PGE-enriched zones), (3) native or oxide forms of Pb, Zn, Ag, Cu, and Bi, and barite, and (4) native gold. For the most part, the Cl drops and encrustations are confined to rocks of the Complex. However, Severson (1994) and Severson et al. (1994) recently identified iron chloride drops and precipitates along fractures of the Biwabik Iron-Formation and in a pre-Complex sill where they are footwall to the Complex. Alminas et al. (1993) report chloride precipitates in Archean outcrops north and west of the Complex.

During scanning electron microscope (SEM) examination of 12 samples from various drill holes in the Bathtub ore zone of the Babbitt deposit and elsewhere in the Duluth Complex (Fig. 29) efflorescence developed on the surfaces of the samples examined, through dehydration in the SEM vacuum. These samples had no efflorescences or mineral coatings when first sampled. The appearance and mineralogy of the efflorescences, and

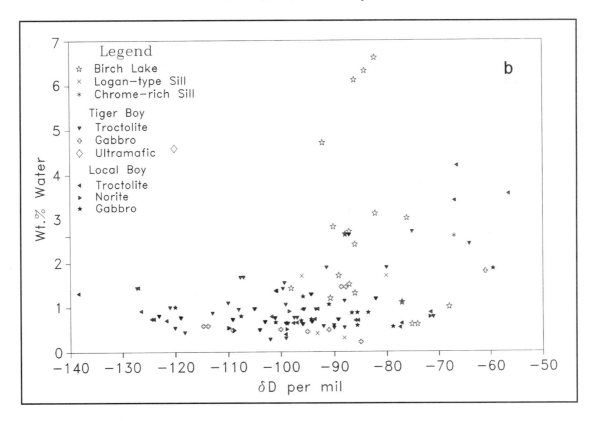

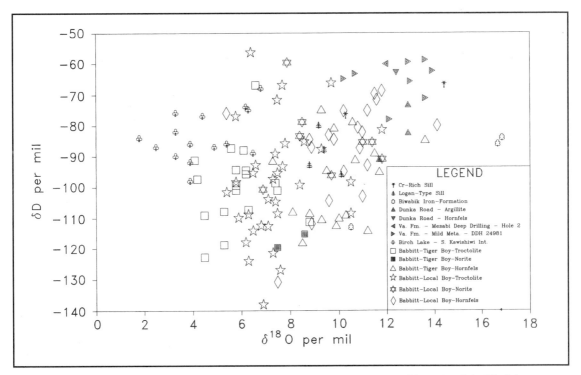

Figure 28. δD versus δ¹⁸O for unmetamorphosed and metamorphosed pelitic rocks of the Virginia Formation and for igneous rocks of the Partridge River and South Kawishiwi Intrusions (after Hauck et al., 1996).

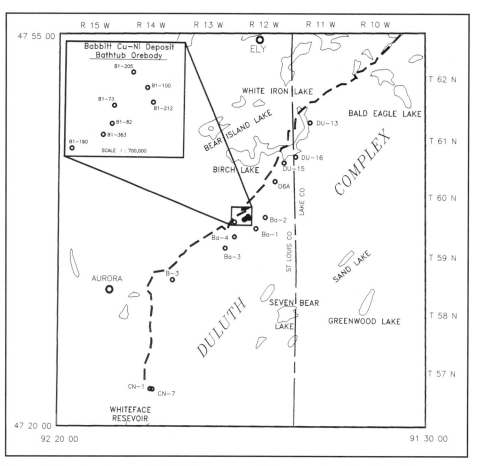

Figure 29. Location of some drill holes that were sampled for secondary chloride precipitates along the western margin. Note wide occurrence of the secondary chlorides.

subsequent encrustations, was strongly dependent upon the local mineralogy of the core. For example, copper and nickel (blue, blue-green, and green), Cl-CO$_3$ efflorescences were present on sulfide-bearing or sulfide-rich samples in direct proportion to the amount of sulfides. Samples with no visible sulfides had white (Mg-Ca CO$_3$) efflorescences, and red-brown to orange-brown to yellow-brown (Fe-Cl-oxides-hydroxides) drops and blistered drops. White and red-orange-yellow brown efflorescences and drops also occurred on sulfide-bearing drill core, but were much less abundant.

Table 4 lists the minerals and compounds that were identified by SEM-EDS, X-ray diffraction, and optical techniques from the cores. Two samples have microcrystalline green efflorescences; X-ray diffraction patterns indicate the mineral is CuCl$_2$ · 3Ca(OH) (only known as a synthetic compound). Other samples have less Ca and more Cu than required, but have the same X-ray pattern (the Cu analogue: CuCl$_2$ · 3Cu(OH)$_2$?). The extremely small grain size, several microns or less, and the paucity of material prevent complete characterization as a new mineral species.

No hibbingite is identified with these efflorescences. Oxida-

tion of hibbingite could probably produce some or all of the FeO(OH) polymorphs: akaganeite, lepidocrocite, and goethite, depending on the Eh and pH conditions. However, only akaganeite and lepidocrocite are identified. The akaganeite and lepidocrocite contain significant Cl in place of OH, possibly reflecting their parentage from hibbingite.

Also coeval with the Fe-O-Cl and Cu-Cl-CO$_3$ phases are white prismatic to platy calcite; white, microcrystalline, powdery giorgiosite, Mg$_5$(CO$_3$)$_4$(OH)$_2$ · 5H$_2$O (third known occurrence in the world); a mineral that appears to be the Ca-analogue of northupite; and chalconatronite, Na$_2$Cu(CO$_3$)$_2$ · 3H$_2$O.

Encrustation samples obtained by brushing selected drill core consist predominantly of assorted chlorides/oxychlorides of iron, magnesium, and sodium. SEM scanning of the material indicates a disproportionately high content of Au, Ag, Sn, and Pb relative to the metal contents of typical rock types occurring within the Complex. The gold primarily occurs in native form with a generally low silver content and a high copper content. Silver originates predominantly in native/oxide form, but silver chloride/oxychloride and sulfate/sulfide are also present. Tin occurs in native form, as well as tin oxides (romarchite, hydro-

TABLE 4. SECONDARY SURFICIAL MINERALS AND COMPOUNDS IDENTIFIED FROM SELECTED DRILL CORE SAMPLES FROM THE DULUTH COMPLEX, MINNESOTA

Drill Hole No. and Interval (ft)	Mineral/Compound	Formula	Description
B1-363 1480.2 Sulfide-bearing	New species	$CuCl_2 \cdot 3Ca(OH)_2$	Green efflorescence micro-crystals
B1-363 1480.2 Sulfide-bearing	Xonotlite	$Ca_6Si_6O_{17}(OH)_2$	White, matted fibers
B1-116 1671 Sulfide-rich	New species	$CuCl_2 \cdot 3Ca(OH)_2$	Green efflorescence micro-crystals
B1-116 1671 Sulfide-rich	Akaganeite	$\beta\text{-}FeO(OH,Cl)$	Earthy, brown spheres extremely fine-grained
B1-205 385.6 No sulfides present	Giorgiosite(?) + at least one other mineral	$Mg_5(CO_3)_4(OH)_2 \cdot 5H_2O(?)$	Milky-white vug-filling
B1-205 385.6 No sulfides present	Northupite(?)	Ca-analogue of northupite(?)	White vug-filling
B1-363 1450.3 Sulfide-bearing	Chalconatronite	$Na_2Cu(CO_3)_2 \cdot 3H_2O$	Blue-green spheres
B1-73 936.0-936.3 Sulfide-bearing	Calcite	$CaCO_3$	White, lacy effloresence and fibrous habit
B1-100 1318.8 No sulfides present	Cummingtonite	$(Mg,Fe)_7Si_8O_{22}(OH)_2$	White to pale green matted; fibers in vug
B1-212 1521.7 Sulfide-bearing	Akaganeite	$\beta\text{-}FeO(OH,Cl)$	Brown efflorescence
B1-212 1521.7 Sulfide-bearing	Chalconatronite	$Na_2Cu(CO_3)_2 \cdot 3H_2O$	Blue-green efflorescence
B1-212 1521.7 Sulfide-bearing	Unknown	Cu-Na-Carbonate	Green efflorescence
B1-212 1521.7 Sulfide-bearing	Unknown	Na-Carbonate	White efflorescence
B1-205 390.0-390.2 No sulfides present	Giorgiosite(?)	Na-Carbonate	White efflorescence
B1-205 390.0-390.2 No sulfides present	Lepidocrocite	$FeO(OH,Cl)$	Red-brown efflorescence

romarchite, cassiterite) and as tin and tin-lead chloride/oxychlorides. Lead, like tin, is widespread, occurring predominantly in native or sulfide form, but commonly also as a chloride/oxychloride.

Uncovered thin sections exposed to a water-saturated atmosphere, and then dried, produce extensive chloride crusts on the surface of the thin sections, demonstrating the abundance and water soluble nature of chlorides filling pore spaces in the core. These encrustations consist primarily of chlorine-bearing iron, magnesium, and sodium minerals. The dominant minerals are chlorides, and possibly oxy-chlorides. Carbonates are not found. This treatment of thin sections also induces the growth of extensive amounts of sulfates over and near sulfide grains as well as over and within microfractures in the thin sections.

Occurrence of other Cl-bearing minerals. Alapieti et al.

(1992) show 2.41% Cl in chlorite within a veinlet within a PGE-enriched magnetite-ilmenite adcumulate (South Kawishiwi Intrusion). Chlorine occurs in chlorite-hematite fractures in PGE-enriched Local Boy massive sulfide ore (Severson and Barnes, 1991). High Cl contents of biotite, serpentine, and apatite and changes in plagioclase composition suggest potential transport of PGEs by Cl complexing (Boudreau, 1993; Mogessie and Stumpfl, 1992; Mogessie et al., 1991; Ripley, 1990a; Saini-Eidukat and Kucha, 1991; Saini-Eidukat et al., 1990). In addition, Pasteris et al. (1995) give extensive data on fluid inclusions that contain up to 48 wt.% total dissolved salts.

Possible origins for chlorides and related elements. The large amount of chlorides associated with rocks within and adjacent to the Complex suggests that Cl-rich fluids were intimately linked to the evolution of the Duluth Complex and possibly to the Midcontinent Rift System. These Cl-rich fluids may have been capable of transporting and concentrating PGEs and other metals within the Duluth Complex.

In some instances, the distribution of chloride-encrusted drill core demonstrates the importance of structural control (Dahlberg and Saini-Eidukat, 1991). For example, chloride-encrusted core is spatially situated along an east-west–trending anticline in the Local Boy area of the Babbitt deposit (Fig. 10; Severson and Barnes, 1991). In the Bathtub ore zone at Babbitt, various chlorides and carbonate precipitates (drops and encrustations) occur on drill core from holes spatially situated along the north limb of the Bathtub Syncline and along the axis and limbs of the Local Boy Anticline. This relationship suggests that Cl-rich fluid flow (hydrothermal?) was confined to an east-west-trending zone. This zone of increased porosity may be related to syn-Complex fracturing in both the Complex and footwall rocks. In addition, the most pervasive saussuritization-uralitization in the Bathtub ore zone also produced miarolitic cavities along the north limb of the same syncline; this alteration event appears to precede serpentinization. Structural and alteration events that postdate the Cl-rich fluid flow are north-trending basaltic dikes. Troctolitic rocks adjacent to the dikes are brecciated and altered; however, neither dike nor brecciated troctolite show evidence of secondary Cl precipitates.

Severson's (1994) data on the Duluth Complex support a model like that proposed by Boudreau and McCallum (1992) for generation of the PGE-bearing, chlorine-rich, late-stage magmatic hydrothermal fluids. According to this model, chlorine-rich fluids are produced from the intercumulus melt during the latter stages of crystallization. These Cl-rich fluids are then expelled upwards by the weight of the overlying magma. The fluids were either initially enriched in PGEs and/or scavenged PGEs during migration. Deposition of the PGEs would then occur at stratigraphic or geochemical barriers (e.g., ultramafic or oxide layers) where there would be a change in oxygen fugacity of the fluid.

In the McCreedy East and Deep Victor Ni-Cu deposits in the Sudbury Igneous Complex, Jago et al. (1994) report alkali-, metal-, and halogen-rich (chlorine and fluorine) precipitates in microfractures or "bleeding fractures" and interstitial chlorine-bearing grunerite, fluorite, and iron-manganese chloride. Mineralogical evidence from Jago et al. (1994) indicates that fluid-filled microfractures occurred in wide zones that also are enriched in Zn, Pb, Bi, Te, Sn, and minor Sb. The PGEs in these zones correlate with the presence of these trace metals and the occurrence of halogen phases. Similar chlorine-bearing silicates and nonsilicates (e.g., iron chlorides), as well as high-salinity fluid inclusions, have been recognized in other Sudbury Ni-Cu deposits (Springer, 1989; Li and Naldrett, 1993; Farrow and Watkinson, 1992). Jago et al. (1994) support the suggestions of Li and Naldrett (1993) that halogen-rich fluids were exsolved from a highly fractionated sulfide melt as it approached the solidus to produce these late-stage hydrothermal minerals and precipitates in the microfractures. Similar late-stage hydrothermal origins have been proposed for the origin of PGEs in the

Duluth Complex (Mogessie et al., 1991; Mogessie and Stumpfl, 1992; Mogessie and Saini-Eidukat, 1992).

Another explanation for Cl-rich fluids occurs in recent work by Arehart et al. (1990) and Mauk et al. (1989, 1992a, b, c) in the White Pine, Michigan, area. These authors suggest that cupriferous brines were expelled along thrust and tear faults on the east side of the Midcontinent Rift System (Fig. 1) during the closing (compression) of the basin at 1,060 Ma. Stable isotope data indicate that these late-stage fluids included both meteoric and igneous sulfur (and copper) components. Mauk and Hieshima (1992) also indicate that petroleum and pyrobitumen inclusions are associated with late-stage compressional faults in the White Pine region, which illustrates migration of petroleum along with the brines from the hotter axial regions of the collapsing Midcontinent Rift System basin. These observations lead us, and Pasteris et al. (1995), to believe that hydrothermal brines/fluids could also have been expelled along the western margin of the basin in the Duluth Complex area during the compressional phase of the rift, where compressive structures have been documented (Witthuhn, 1992). Morton and Ameel (1985) also document the presence of highly saline wells within the North Shore Volcanic Group and within the Duluth Complex. Existing structural discontinuities in the Duluth Complex could have concentrated the fluid movement and be responsible for localized concentrations of (1) PGEs, (2) chloride precipitates(?), (3) secondary sulfides, and (4) alteration phases. The movement of the fluids probably took place prior to the emplacement of the north- to northeast-trending dikes that lack any indications of chlorides, or there was more than one chlorine event.

In summary, numerous Cl-precipitates have been documented throughout, and adjacent to, the Complex. However, the presence of Cl-precipitates is only an indication for the former presence of Cl-fluids. While some of the reported PGE occurrences may be related to transport and concentration of PGE by Cl-complexing, not all Cl-enriched rocks are PGE-enriched. Still in question is the ability of Cl-complexes to transport PGEs over other complexes in transport and deposition (Wood et al., 1992). The PGE concentration mechanism(s), initial PGE sources, and origin of the Cl-rich fluids are poorly understood and require additional study.

SUMMARY AND CONCLUSIONS

The Duluth Complex, formed during activation of the Midcontinent Rift System at approximately 1.1 Ga, contains significant amounts of both oxide and sulfide mineralization. Three major types of oxide-rich (Fe-Ti ± Cu-Ni) deposits have been delineated along both the northern and western margins of the Complex. All three types appear to be related in varying degrees to intrusion into and assimilation of footwall iron-formation. A model involving metasomatism of iron-formation has been proposed by Severson and Hauck (1990) and Severson (1994) as a means for producing these bodies. Type 1 oxides are banded to layered oxide-rich metasedimentary inclusions in mafic and ultramafic rocks that exhibit metasedimentary textures and/or can be traced laterally into iron-formation. Type 2 oxides are banded or layered oxide segregations that include cumulus oxide-rich horizons. Many of the latter oxides are also spatially related to rafts of iron-formation. Type 3 oxides, late, discordant oxide-bearing ultramafic pipes and bodies, are often in contact with inclusions of iron-formation, or appear to be related to windows of iron-formation subcrop within the Complex. However, Type 3 oxides in the South Complex area are not related to iron-formation; rather they are associated with oxide-rich layered gabbroic rocks (Severson, 1995). Overall, these oxide-rich bodies are very similar in mineralogy and texture to late ultramafic pipelike bodies in the Bushveld, Stillwater, and Rio Jacare Complexes.

Numerous Cu-Ni (±PGE) deposits have also been outlined along the western contact of the Complex. Disseminated copper-nickel sulfides formed at the basal contact of the Complex, and they originated by sulfide segregation from the intruding magmas. Isotope data indicate that initial sulfur contamination of the magmas by metasedimentary country rock may have occurred in shallow crustal staging chambers before emplacement into their present positions. Localized geochemical changes in the magma due to assimilation of footwall material and/or fluids produced by devolatization of the footwall produced the heterogeneity of the basal troctolitic rocks and disseminated sulfides. Some massive sulfide resulted from in situ incorporation of local sulfides from the footwall rocks into the intruding magma. Other massive sulfides were formed from the generation of an immiscible sulfide melt in auxiliary magma chambers, followed by injection of the melts into structurally prepared zones. In all cases, the ultimate source of sulfur appears to have been the sedimentary footwall rocks. Late-stage hydrothermal fluids remobilized and precipitated secondary sulfides within structurally prepared zones. Sassani (1992) and Pasteris et al. (1992, 1995), using petrographic and fluid inclusion data, propose that rocks from Fish Lake were subjected to at least four different events of fluid movement, and that some of the fluid had a strong footwall component. These events varied in temperature, pressure, and timing; fluid movement was concentrated along structural discontinuities. Secondary effects of these fluids include hydrous alteration, secondary copper sulfides, and enriched zones of PGEs.

Stable isotope data from the basal zone of the Complex show that at least three different fluid sources (magmatic, devolatization of footwall rocks, and meteoric) affected the Complex rocks, either singularly or in combination. Upper portions of the Partridge River and South Kawishiwi Intrusions are less contaminated and retain their more primitive composition.

Cu/Pd versus Pd plots and mantle-normalized PGE plots indicate that prior to emplacement, the South Kawishiwi and Partridge River magmas were already depleted in PGEs, probably due to some earlier sulfide segregation. However, localized stratabound sulfide zones at Dunka Road and massive oxide and disseminated sulfide zones at Birch Lake are not depleted in PGE, and exhibit mantle signatures. In these cases, sulfide segregation

probably occurred after the magma was emplaced. Thus, a magmatic enrichment process is indicated for the PGEs. Elsewhere within the Complex, Sassani (1992), Mogessie and Saini-Eidukat (1992), and Mogessie and Stumpfl (1992) suggest that late magmatic, chlorine-rich hydrothermal fluids (C-H-O-S-Cl) were responsible for transport of PGEs, primarily in the Partridge River and South Kawishiwi Intrusions, and the magmatic hydrothermal fluid event was related to the introduction of volatiles from the footwall metasediments. Even at Dunka Road, Morton and Geerts (1992) recognize that patchy plagioclase recrystallization, associated with PGE-enriched sulfide horizons, is consistent with alteration by saline solutions at 700 °C.

Based on the PGE data, however, both magmatic and hydrothermal events seem to have affected the PGE distribution. PGEs at Dunka Road and Birch Lake underwent segregation on site, but with minimal redistribution in some zones. Other rocks in the Complex were depleted in PGEs prior to emplacement, which indicates an earlier sulfide segregation event. Isotopic data, fluid inclusion compositions, and hydrous silicate paragenesis support the occurrence of later hydrothermal events that redistributed copper and some precious metals. The oxide, sulfide, and precious metal mineralization reflects the complexities related to the emplacement and crystallization of the numerous intrusions within the Duluth Complex.

ACKNOWLEDGMENTS

Portions of this work were supported by funding from the Minerals Diversification Plan of the Minnesota Legislature, the Minnesota Mining and Minerals Resources Research Institute, Minnesota Technology, Inc., and private industry for which we are greatly appreciative. We greatly appreciate the patience and diligence of the editors in seeing the manuscript to completion. The manuscript is much improved due to comments from Jill Dill Pasteris and Bernhardt Saini-Eidukat, but the final content of the paper is the responsibility of the authors.

REFERENCES CITED

Alapieti, T. T., 1991, Preliminary report on the microscopic study of drill hole Du-15: Minnesota Department of Natural Resources, Division of Minerals, Report 291, 56 p.

Alapieti, T. T., Dahlberg, E. H., and Nelson, S., 1992, Preliminary results of electron microprobe studies of heavy mineral concentrates of the central region Minnesota and chlorine-bearing minerals of the Duluth Complex, Minnesota [abs.]: Minnesota Department of Natural Resources, Division of Minerals, Drill Core Library Newsletter, v. 2, p. 6.

Alminas, H. V., Foord, E. E., Cartwright, D. F., and Dahlberg, E. H., 1993, Geochemical studies in the Vermilion District and the Duluth Complex, northeastern Minnesota—A progress report [abs.]: Annual Institute of Lake Superior Geology, 39th, Eveleth, Minnesota, v. 39, p. 4–5.

Andrews, M. S., 1987, Contact metamorphism of the Virginia Formation at Dunka Road, Minnesota: Chemical modifications and implications for ore genesis in the Duluth Complex [M.S. thesis]: Bloomington, Indiana University, 93 p.

Andrews, M. S., and Ripley, E. M., 1989, Mass transfer and sulfur fixation in the contact aureole of the Duluth Complex, Dunka Road Cu-Ni

Deposit, Minnesota: Canadian Mineralogist, v. 27, p. 293–310.

Antonellini, M. A., and Cambray, F. W., 1992, Relations between sill intrusions and bedding-parallel extensional shear zones in the Mid-continent Rift System of the Lake Superior Region: Tectonophysics, v. 212, p. 331–349.

Arehart, G. B., O'Neil, J. R., and Mauk, J. L., 1990, Stable isotope compositions of fluid inclusions in native copper from White Pine, Michigan [abs.]: Geological Society of America Abstracts with Programs, v. 22, no. 7, p. A250.

Babcock, R. C., Jr., 1959, Petrology of the granophyre and intermediate rock in the Duluth Gabbro of northern Cook County, Minnesota [M.S. thesis]: Madison, University of Wisconsin, 47 p.

Barnes, S-J., and Francis, D., 1995, The distribution of platinum-group elements, nickel, copper and gold in the Muskox layered intrusion, Northwest Territories, Canada: Economic Geology, v. 90, p. 135–155.

Barnes, S-J., and Naldrett, A. J., 1986, Variations in platinum group element concentrations in the Alexo mine komatiite, Abitibi Greenstone Belt, northern Ontario: Geological Magazine, v. 123, p. 515–524.

Barnes, S-J., Boyd, R., Kornelliussen, A., Nilssen, L-P., Often, M., Pedersen, R-B., and Robins, B., 1988, The use of mantle normalization and metal ratios in discriminating the effects of partial melting, crystal fractionation and sulphide segregation on platinum group elements, gold, nickel and copper: Examples from Norway, *in* Prichard, H. M., Potts, P. J., Bowles, J. F. W., and Cribb, S. J., eds., Geo-platinum 87: London, Elsevier, p. 113–143.

Barnes, S-J., Couture, J-F., Sawyer, E. W., and Bouchaid, C., 1993, Nickel-copper occurrences in the Belleterre-Angliers Belt of the Pontiac Subprovince and the use of Cu-Pd ratios in interpreting platinum-group element distributions: Economic Geology, v. 88, p. 1402–1418.

Barnes, S-J., Makovicky, E., Karup-Moller, S., Makovicky, M., and Rose-Hansen, J., 1994, Partition coefficients for Ni, Cu, Pd, Pt, Rh and Ir between monosulphide solid solution and sulphide liquid and the implications for the formation of compositionally zoned Ni-Cu sulphide bodies by fractional crystallization of sulphide liquid: Mineralogical Magazine, Goldschmidt Conference, Edinburgh, Aug. 28–Sept. 2.

Bonnichsen, B., 1972a, Southern part of the Duluth Complex, *in* Sims, P. K., and Morey, G. B., eds., Geology of Minnesota: A centennial Volume: St. Paul, Minnesota Geological Survey, p. 361-387.

Bonnichsen, B., 1972b, Sulfide minerals in the Duluth Complex, *in* Sims, P. K., and Morey, G. B., eds., Geology of Minnesota: A centennial volume: St. Paul, Minnesota Geological Survey, p. 388–393.

Bonnichsen, B., 1975, Geology of the Biwabik Iron Formation, Dunka River area, Minnesota: Economic Geology, v. 70, p. 319–340.

Bonnichsen, B., Fukui, L. M., and Chang, L. L. Y., 1980, Geologic setting, mineralogy, and geochemistry of magmatic sulfides, South Kawishiwi Intrusion, Duluth Complex, Minnesota: Proceedings, IAGOD Symposium, 5th, Stuttgart, Germany, p. 545–565.

Boudreau, A. E., 1993, Chlorine as an exploration guide for the platinum-group elements in layered intrusions: Journal of Geochemical Exploration, v. 48, p. 21–37.

Boudreau, A. E., and McCallum, I. S., 1992, Concentration of platinum-group elements by magmatic fluids in layered intrusions: Economic Geology, v. 87, p. 1830–1848.

Broderick, T. M., 1917, The relation of the titaniferous magnetites of northeastern Minnesota to the Duluth Gabbro: Economic Geology, v. 12, p. 663–696.

Butler, B. K., 1989, Hydrogen isotope study of the Babbitt Cu-Ni deposit, Duluth Complex, Minnesota [M.S. thesis]: Bloomington, Indiana University, 180 p.

Cameron, E. N., and Desborough, G. A., 1964, Origin of certain magnetite-bearing pegmatites in the eastern part of the Bushveld Complex, South Africa: Economic Geology, v. 59, p. 197–225.

Cannon, W. F., 1992, The Midcontinent Rift in the Lake Superior region with emphasis on its geodynamic evolution: Tectonophysics, v. 213, p. 41–48.

Cannon, W. F., and Hinze, W. J., 1992, Speculations on the origin of the North American Midcontinent Rift: Tectonophysics, v. 213, p. 49–55.

Cannon, W. F., and 11 others, 1989, The North American Midcontinent Rift beneath Lake Superior from GLIMPCE seismic reflection profiling: Tectonics, v. 8, p. 305–332.

Chandler, V. W., 1990, Geologic interpretation of gravity and magnetic data over the central part of the Duluth Complex, northeastern Minnesota: Economic Geology, v. 85, p. 816–829.

Chandler, V. W., and Ferderer, R. J., 1989, Copper-nickel mineralization of the Duluth Complex, Minnesota—A gravity and magnetic perspective: Economic Geology, v. 84, p. 1690–1696.

Churchill, R. K., 1978, A geochemical and petrological investigation of the Cu-Ni sulfide genesis in the Duluth Complex, Minnesota [M.S. thesis]: Minneapolis, University of Minnesota, 101 p.

Cogulu, E. H., 1984, Magmatic cycles in the Crystal Lake gabbros, Thunder Bay, Ontario [abs.]: Geological Association of Canada Program with Abstracts, p. 53.

Cogulu, E., 1985, Platinum-group elements and variations in chromium spinel in the Crystal Lake Gabbro, Thunder Bay, Ontario [abs.]: Canadian Mineralogist, v. 23, p. 299–300.

Cogulu, E. H., 1990, Mineralogical and petrological studies of the Crystal Lake Intrusion, Thunder Bay, Ontario: Geological Survey of Canada, Open File Report 2277, 67 p.

Cogulu, E. H., 1993a, Factors controlling postcumulus compositional changes of chrome-spinels in the Crystal Lake Intrusion, Thunder Bay, Ontario: Geological Survey of Canada, Open-File 2748, 79 p.

Cogulu, E. H., 1993b, Mineralogy and chemical variations of sulphides from the Crystal Lake Intrusion, Thunder Bay, Ontario: Geological Survey of Canada, Open-File 2749, 51 p.

Cooper, R. W., Morey, G. B., and Weiblen, P. W., 1977, Lineament and structural analysis of Precambrian rocks in northeastern Minnesota [abs.]: Geological Society of America Abstracts with Program, v. 9, no.7, p. 934–935.

Dahlberg, E. H., 1989, Lithochemistry of the platiniferous oxide-ultramafic association of the basal zone, Duluth Complex, Birch Lake–Duval area [abs.]: Minnesota Geological Survey, Information Circular 30, p. 49–50.

Dahlberg, E. H., ed., 1992, Minnesota Department of Natural Resources, Division of Minerals, Drill Core Library Newsletter, v. 2, p. 14–16.

Dahlberg, E. H., and Saini-Eidukat, B., 1991, A chlorine-bearing phase in drill core of serpentinized troctolitic rocks of the Duluth Complex, Minnesota: Canadian Mineralogist, v. 29, p. 239–244.

Dahlberg, E. H., Frey, B. A., Gladen, L. W., Lawler, T. L., Malmquist, K. L., and McKenna, M. P., 1987, 1986–1987 Geodrilling report: Minnesota Department of Natural Resources, Division of Minerals, Report 251, 179 p.

Davidson, D. M., Jr., 1969a, Kawishiwi Lake Quadrangle, Lake and Cook Counties, Minnesota: Minnesota Geological Survey Miscellaneous Map Series (with text) Map M-7, scale 1:24,000.

Davidson, D. M., Jr., 1969b, Perent Lake Quadrangle, Lake County, Minnesota: Minnesota Geological Survey Miscellaneous Map Series Map M-8, scale 1:24,000.

Davidson, D. M., Jr., 1972, Eastern part of the Duluth Complex, *in* Sims, P. K., and Morey, G. B., eds., Geology of Minnesota: A centennial volume: St. Paul, Minnesota Geological Survey, p. 354–360.

Davidson, D. M., Jr., 1977a, Alice Lake Quadrangle, Lake County, Minnesota: Minnesota Geological Survey Miscellaneous Map Series Map M-33, scale 1:24,000.

Davidson, D. M., Jr., 1977b, Beth Lake Quadrangle, Cook County, Minnesota: Minnesota Geological Survey Miscellaneous Map Series Map M-26, scale 1:24,000.

Davidson, D. M., Jr., 1977c, Cherokee Lake Quadrangle, Cook County, Minnesota: Minnesota Geological Survey Miscellaneous Map Series Map M-30, scale 1:24,000.

Davidson, D. M., Jr., 1977d, Eagle Mountain Quadrangle, Cook County, Minnesota: Minnesota Geological Survey Miscellaneous Map Series Map M-28, scale 1:24,000.

Davidson, D. M., Jr., 1977e, Kelso Mountain Quadrangle, Cook County, Minnesota: Minnesota Geological Survey Miscellaneous Map Series Map

M-27, scale 1:24,000.

Davidson, D. M., Jr., 1977f, Lake Polly Quadrangle, Lake and Cook Counties, Minnesota: Minnesota Geological Survey Miscellaneous Map Series Map M-34, scale 1:24,000.

Davidson, D. M., Jr., 1977g, Lima Mountain Quadrangle, Cook County, Minnesota: Minnesota Geological Survey Miscellaneous Map Series Map M-32, scale 1:24,000.

Davidson, D. M., Jr., 1977h, Pine Mountain Quadrangle, Cook County, Minnesota: Minnesota Geological Survey Miscellaneous Map Series Map M-31, scale 1:24,000.

Davidson, D. M., Jr., and Burnell, J. R., 1977, Brule Lake Quadrangle, Cook County, Minnesota: Minnesota Geological Survey Miscellaneous Map Series Map M-29, scale 1:24,000.

Davis, D. W., and Sutcliffe, R. H., 1985, U-Pb ages from the Nipigon plate and north Lake Superior: Geological Society of America Bulletin, v. 96, p. 1572–1579.

Distler, V. V., Malevsky, A. Yu., and Laputina, I. P., 1977, The distribution of PGE between pyrrhotite and pentlandite during the crystallization of the sulfide melts: Geochimia, v. 11, p. 1646–1658.

Eckstrand, O. R., and Cogulu, E. H., 1986, Se/S evidence relating to genesis of sulphides in the Crystal Lake Gabbro, Thunder Bay, Ontario [abs.]: Geological Association of Canada Program with Abstracts, v. 11, p. 66.

Eckstrand, O. R., and Cogulu, E. H., 1989, The role of sulphur contamination in magmatic Cu-Ni-PGE mineralization in the Crystal Lake intrusion, Thunder Bay area, Ontario [abs.]: Geological Society of Finland Bulletin, v. 61, p. 6–7.

Eckstrand, O. R., Cogulu, E. H., and Scoates, R. F. J., 1989, Magmatic Ni-Cu-Pge mineralization in the Crystal Lake layered intrusion, Ontario and the Fox River sill, Manitoba [abs.]: Minnesota Geological Survey, Information Circular 30, p. 45–46.

Farrow, C. E. G., and Watkinson, D. H., 1992, Alteration and the role of fluids in Ni, Cu, and platinum-group element deposition, Sudbury Igneous Complex contact, Onaping-Levack area, Ontario: Mineralogy and Petrology, v. 46, p. 67–83.

Floran, R. J., and Papike, J. J., 1975, Petrology of the low-grade rocks of the Gunflint Iron-Formation, Ontario-Minnesota: Geological Society of America Bulletin, v. 86, p. 1169–1190.

Floran, R. J., and Papike, J. J., 1978, Mineralogy and petrology of the Gunflint Iron-formation, Minnesota-Ontario: Correlation of compositional and assemblage variations at low to moderate grade: Journal of Petrology, v. 19, p. 215–288.

Foose, M. P., 1984, Logs and correlation of drill holes within the South Kawishiwi Intrusion, Duluth Complex, northeastern Minnesota: U. S. Geological Survey, Open-File Report 84-14, 230 p.

Foose, M. P., and Weiblen, P. W., 1986, The physical and chemical setting and textural and compositional characteristics of sulfide ores from the South Kawishiwi Intrusion, Duluth Complex, Minnesota, U.S.A., *in* International Geological Congress (Moscow), 27th, Special Copper Symposium: New York, Springer-Verlag, p. 8–24.

French, B. M., 1968, Progressive contact metamorphism of the Biwabik Iron-Formation, Mesabi Range, Minnesota: Minnesota Geological Survey Bulletin 45, 103 p.

Geerts, S. D., 1991, Geology, stratigraphy, and mineralization of the Dunka Road Cu-Ni prospect, northeastern Minnesota: Duluth, University of Minnesota, Natural Resources Research Institute Technical Report, NRRI/TR-91/14, 59 p.

Geerts, S. D., 1994, Petrography and geochemistry of a platinum group element-bearing horizon in the Dunka Road prospect, (Keweenawan) Duluth Complex, northeastern Minnesota [M.S. thesis]: Duluth, University of Minnesota, 155 p.

Geerts, S., Barnes, R. J., and Hauck, S. A., 1990a, Geology and mineralization in the Dunka Road copper-nickel mineral deposit, St. Louis County, Minnesota: Duluth, University of Minnesota, Natural Resources Research Institute Technical Report, NRRI/TR-89-16, 69 p.

Geerts, S. D., Hauck, S., and Morton, P., 1990b, Palladium mineralization within the Dunka Road copper-nickel mineral deposit, Duluth Complex, NE Minnesota [abs.]: Geological Society America Abstracts with Programs, v. 22, no. 7, p. A137.

Geul, J. J. C., 1970, Geology of the Devon and Pardee Townships and the Stuart location: Ontario Department of Mines and Geology Report 87, 52 p.

Green, J. C., Phinney, W. C., and Weiblen, P. W., 1966, Gabbro Lake Quadrangle, Lake County, Minnesota: Minnesota Geological Survey Miscellaneous Map Series Map M-2, scale 1:31,600.

Green, J. C., Bornhorst, T. J., Chandler, V. W., Mudrey, M. G., Jr., Myers, P. E., Pesonen, L. J., and Wilband, J. T., 1987, Keweenawan dykes of the Lake Superior region: Evidence for evolution of the Middle Proterozoic Midcontinent Rift of North America, *in* Halls, H. C., and Fahrig, W. F., eds., Mafic dyke swarms: Geological Association of Canada Special Paper 34, p. 289–302.

Grout, F. F., 1950, The titaniferous magnetites of Minnesota: St. Paul, Minnesota, Office of the Commissioner of the Iron Range Resources and Rehabilitation, 117 p.

Grout, F. F., and Schwartz, G. M., 1933, The geology of the Rove Formation and associated intrusives of northeastern Minnesota: Minnesota Geological Survey Bulletin 24, 103 p.

Grout, F. F., Sharp, R. P., and Schwartz, G. M., 1959, The geology of Cook County: Minnesota Geological Survey Bulletin 39, 163 p.

Gundersen, J. N., and Schwartz, G. M., 1962, The geology of the metamorphosed Biwabik Iron-Formation, eastern Mesabi district, Minnesota: Minnesota Geological Survey Bulletin 43, 139 p.

Hauck, S. A., 1993, Geology and mineralization of the Bathtub Cu-Ni area, Minnamax Cu-Ni deposit: Duluth, University of Minnesota, Natural Resources Research Institute, Unpublished Draft Report, NRRI/TR-93/48, 59 p.

Hauck, S. A., Alapieti, T., Severson, M. J., Ripley, E. M., and Goldberg, S., 1996, Geology and origin of PGE-Cr mineralization in the Birch Lake area, South Kawishiwi intrusion, Duluth Complex, Minnesota: Duluth, University of Minnesota, Natural Resources Research Institute Technical Report (in press).

Heslop, J. B., 1968, Mineralogy and textural relationships of the Mount Mollie sulphides, Pine Bay area, Thunder Bay District, Ontario [B.Sc. thesis]: London, Ontario, Canada, University of Western Ontario, 68 p.

Holst, T. B., Mullenmeister, E. E., Chandler, V. W., Green, J. C., and Weiblen, P. W., 1986, Relationship of structural geology of the Duluth Complex to economic mineralization: Minnesota Department of Natural Resources, Division of Minerals, Report 241-2, 165 p.

Hutchinson, D. R., White, R. S., Cannon, W. F., and Schulz, K. J., 1990, Keweenaw hot spot: Geophysical evidence for a 1.1-Ga mantle plume beneath the Midcontinent Rift System: Journal of Geophysical Research, v. 95, p. 10869–10884.

Jago, B. C., Morrison, G. G., and Little, T. L., 1994, Metal zonation patterns and microtextural and micromineralogical evidence for alkali- and halogen-rich fluids in the genesis of the Victor Deep and McCreedy East footwall copper orebodies, Sudbury Igneous Complex, *in* Lightfoot, P. C., and Naldrett, A. J., eds., Proceedings, Sudbury-Noril'sk Symposium: Ontario Geological Survey Special Volume 5, p. 65–75.

Jensen, M. L., 1959, Sulfur isotopes and hydrothermal mineral deposits: Economic Geology, v. 54, p. 374–394.

Johnson, R. G., 1968, Copper-nickel mineralization in the basal Duluth Gabbro Complex, northeastern Minnesota: A case study [M.S. thesis]: Iowa City, University of Iowa, 91 p.

Johnson, R. G., 1970, Economic geology of a portion of the basal Duluth Complex, northeastern Minnesota [Ph.D. thesis]: Iowa City, University of Iowa, 136 p.

Jones, N. W., 1963, The relationship between the Duluth gabbro and the dikes and sills in the vicinity of Hovland, Minnesota [M.S. thesis]: Minneapolis, University of Minnesota, 90 p.

Jones, N. W., 1984, Petrology of the some Logan diabase sills, Cook County, Minnesota: Minnesota Geological Survey, Report of Investigations 29, 40 p.

Kilburg, J. A., 1972, Petrology, structure, and correlation of the Ely's Peak basalt [M.S. thesis]: Minneapolis, University of Minnesota, 97 p.

Klewin, K. W., and Shirey, S. B., 1992, The igneous petrology and magmatic evolution of the Midcontinent Rift System: Tectonophysics, v. 213, p. 33–40.

Kuhns, M. J. P., Hauck, S. A., and Barnes, R. J., 1990, Origin and occurrence of platinum group elements, gold and silver in the South Filson Creek copper-nickel mineral deposit, Lake County, Minnesota: Duluth, University of Minnesota, Natural Resources Research Institute, Technical Report, NRRI/TR-89-15, 67 p.

Labotka, T. C., Papike, J. J., Vaniman, D. T., and Morey, G. B., 1981, Petrology of contact metamorphosed argillite from the Rove Formation, Gunflint Trail, Minnesota: American Mineralogist, v. 66, p. 79–86.

Lawson, A. C., 1893, The laccolithic sills of the northwest coast of Lake Superior: Geological and Natural History Survey of Minnesota, Bulletin No. 8, p. 24–48.

Lee, I. S., and Ripley, E. M., 1993, Stable isotopes as indicators of melt-country rock interactions at the Spruce Road Cu-Ni deposit, Duluth Complex, Minnesota [abs.]: Geological Society of America Abstracts with Programs, v. 25, no. 6, p. A400.

Li, C., and Naldrett, A. J., 1993, Platinum-group minerals from the Deep Copper Zone of the Strathcona deposit, Sudbury, Ontario: Canadian Mineralogist, v. 31, p. 31–44.

Lightfoot, P. C., and Lavigne, M. J. J., 1995, Nickel, copper, and platinum group element mineralization in Keweenawan intrusive rocks: New targets in Keweenawan of the Thunder Bay region, northwestern Ontario: Ontario Geological Survey, Open-File Report 5928, 32 p.

Lightfoot, P. C., and Naldrett, A. J., 1984, Chemical variation in the Insizwa Complex, Transkei, and the nature of the parent magma: The Canadian Mineralogist, v. 22, p. 111–123.

Linscheid, E. K., 1991, The petrography of the Longnose peridotite deposit and its relationship to the Duluth Complex [M.S. thesis]: Duluth, University of Minnesota, 121 p.

Lister, G. F., 1966, The composition and origin of selected iron-titanium deposits: Economic Geology, v. 61, p. 275–310.

Listerud, W. H., and Meineke, D. G., 1977, Mineral resources of a portion of the Duluth Complex and adjacent rocks in St. Louis and Lake Counties, northeastern Minnesota: Minnesota Department of Natural Resources, Division of Minerals, Report 93, 74 p.

Lucente, M. E., and Morey, G. B., 1983, Stratigraphy and sedimentology of the Lower Proterozoic Virginia Formation, northern Minnesota: Minnesota Geological Survey, Report of Investigations 28, 28 p.

MacRae, N. D., and Reeve, E. J. 1968, Differentiation sequence of the Great Lakes Nickel intrusion [abs.]: 14th Annual Institute on Lake Superior Geology, Superior, Wisconsin, v. 14, p. 28.

Mainwaring, P. R., 1968a, Sulfide assemblages of the Great Lakes Nickel intrusion [abs.]: 14th Annual Institute on Lake Superior Geology, Superior Wisconsin, v. 14, p. 29.

Mainwaring, P. R. 1968b, The sulphide assemblage of the Great Lakes nickel intrusion [B.Sc. thesis]: London, Ontario, Canada, University of Western Ontario, 63 p.

Mainwaring, P. R., and Naldrett, A. J., 1977, Country rock assimilation and genesis of Cu-Ni sulfides in the Water Hen Intrusion, Duluth Complex, Minnesota: Economic Geology, v. 72, p. 1269–1284.

Mainwaring, P. R., and Watkinson, D. H., 1981, Origin of chromium spinel in the Crystal Lake Intrusion, Pardee Township, Ontario: Ontario Geological Survey, Miscellaneous Paper 98, p. 180–186.

Martineau, M. P., 1989, Empirically derived controls on Cu-Ni mineralization: A comparison between fertile and barren gabbros in the Duluth Complex, Minnesota, U.S.A., in Prendergast, M. D., and Jones, M. J., eds., Magmatic sulphides—The Zimbabwe volume: London, England, Institute of Mining and Metallurgy, p. 117–137.

Mathez, E. A., 1971, Geology and petrology of the Logan intrusives of the Hungry Jack Lake Quadrangle, Cook County, Minnesota [M.S. thesis]: Tucson, University of Arizona, 111 p.

Mathez, E. A., Nathan, H. D., and Morey, G. B., 1977, Hungry Jack Lake

Quadrangle, Cook County, Minnesota: Minnesota Geological Survey Miscellaneous Map Series Map M-39, scale 1:24,000.

Mattis, A. F., 1972, Puckwunge Formation of northeastern Minnesota, in Sims, P. K., and Morey, G. B., eds., Geology of Minnesota: A centennial volume: St. Paul, Minnesota, Minnesota Geological Survey, p. 412–415.

Mauk, J. L., and Hieshima, G. B., 1992, Organic matter and copper mineralization at White Pine, Michigan, U.S.A.: Chemical Geology, v. 99, p. 189–211.

Mauk, J. L., Seasor, R. O., Kelly, W. C., and van der Pluijm, B. A., 1989, The relationship between structure and second-stage copper mineralization in the White Pine District of the Midcontinent Rift, northern Michigan [abs.]: Geological Society of America Abstracts with Programs, v. 21, no. 6, p. A130.

Mauk, J. L., Brown, A. C., Seasor, R. W., and Eldridge, C. S., 1992a, Geology and stable isotope and organic geochemistry of the White Pine sediment-hosted stratiform copper deposit, in Bornhorst, T. J., ed., Keweenawan copper deposits of western Upper Michigan: Society of Economic Geologists, Guidebook Series, v. 13, p. 63–98.

Mauk, J. L., Eldridge, C. S., Hieshima, G. B., Woodruff, L. G., and Seasor, R. W., 1992b, An igneous contribution to sediment-hosted stratiform copper mineralization at White Pine, Michigan: A working hypothesis [abs.]: Geological Society of America Abstracts with Programs, v. 24, no. 7, p. A23.

Mauk, J. L., Kelly, W. C., van der Pluijm, B. A., and Seasor, R. W., 1992c, Relations between deformation and sediment-hosted copper mineralization: Evidence from the White Pine part of the Midcontinent rift system: Geology, v. 20, p. 427–430.

Miller, J. D., Jr., and Weiblen, P. W., 1990, Anorthositic rocks of the Duluth Complex: Examples of rocks formed from plagioclase crystal mush: Journal of Geology, v. 31, pt. 2, p. 295–339.

Miller, J. D., Jr., Green, J. C., and Chandler, V. W., 1993, The geology of the Duluth Complex at Duluth: Institute on Lake Superior Geology, 39th, Eveleth, Minnesota, Part 2, Field Trips, p. 131–157.

Mogessie, A., 1976, Petrologic study of copper-nickel mineralization in the Tuscarora Intrusion, Duluth Complex, northeastern Minnesota [M.S. thesis]: Minneapolis, University of Minnesota, 137 p.

Mogessie, A., and Saini-Eidukat, B., 1992, A review of the occurrence of platinum group elements in the Duluth Complex, Minnesota, U.S.A.: Trends in Mineralogy, v. 1, p. 65–85.

Mogessie, A., and Stumpfl, E. F., 1992, Platinum-group element and stable isotope geochemistry of PGM-bearing troctolitic rocks of the Duluth Complex, Minnesota: Australian Journal of Earth Science, v. 39, p. 315–325.

Mogessie, A., Stumpfl, E. F., and Weiblen, P. W., 1991, The role of fluids in the formation of platinum-group minerals, Duluth Complex, Minnesota: Mineralogic, textural, and chemical evidence: Economic Geology, v. 86, p. 1506–1518.

Morey, G. B., 1965, The sedimentology of the Precambrian Rove Formation in northeastern Minnesota [Ph.D. thesis]: Minneapolis, University of Minnesota, 292 p.

Morey, G. B., 1967, Stratigraphy and sedimentology of the middle Precambrian Rove Formation in northeastern Minnesota: Journal of Sedimentary Petrology, v. 37, p. 1154–1162.

Morey, G. B., 1969, The geology of the middle Precambrian Rove Formation in northeastern Minnesota: Minnesota Geological Survey, Special Publication 7, 62 p.

Morey, G. B., 1972, Gunflint Range, in Sims, P. K., and Morey, G. B., eds., Geology of Minnesota: A centennial volume: Minnesota Geological Survey, p. 218–226.

Morey, G. B., and Cooper, R. W., 1977, Bedrock geology of the Hoyt Lakes–Kawishiwi area, St. Louis and Lake Counties, Minnesota: Minnesota Geological Survey, Open-File Map, scale 1:48,000.

Morey, G. B., and Nathan, H. D., compilers, 1977, South Lake Quadrangle, Cook County, Minnesota: Minnesota Geological Survey, Miscellaneous Map M-38, scale 1:24,000.

Morey, G. B., and Nathan, H. D., compilers, 1978, Gunflint Lake Quadrangle,

Cook County, Minnesota: Minnesota Geological Survey, Miscellaneous Map M-42, scale 1:24,000.

Morey, G. B., and Ojakangas, R. W., 1970, Sedimentology of the middle Precambrian Thomson Formation, east-central Minnesota: Minnesota Geological Survey, Report of Investigations 13, 32 p.

Morey, G. B., Weiblen, P. W., Papike, J. J., and Anderson, D. H., compilers, 1978, Long Island Lake Quadrangle, Cook County, Minnesota: Minnesota Geological Survey, Miscellaneous Map Series, Map M-46, scale 1:24,000.

Morton, P., and Ameel, J., 1985, Saline waters as indicators of economic mineralization, Minnesota: Minnesota Department of Natural Resources, Division of Minerals, Report 241-1, 39 p.

Morton, P., and Geerts, S. D., 1992, High temperature hydrothermal alteration associated with PGE enrichment of Cu-Ni sulfides, Dunka Road, Duluth Complex, NE Minnesota [abs.]: Annual Institute on Lake Superior Geology, 38th, Hurley, Wisconsin, v. 38, p. 72–74.

Morton, P., and Hauck, S. A., 1987, PGE, Au and Ag contents of Cu-Ni sulfides found at the base of the Duluth Complex, northeastern Minnesota: Duluth, University of Minnesota, Natural Resources Research Institute, Technical Report, NRRI/GMIN-TR-87-04, 85 p.

Muhich, T. G., 1993, Movement of titanium across the Duluth Complex— Biwabik Iron Formation contact at Dunka Pit, Mesabi Iron Range, northeastern Minnesota [M.S. thesis]: Duluth, University of Minnesota, 154 p.

Mudrey, M. G., Jr., 1973, Structure and petrology of the sill on Pigeon Point, Minnesota [Ph.D. thesis]: Minneapolis, University of Minnesota, 310 p.

Mudrey, M. G., Jr., 1976, Late structural evolution of Pigeon Point, Cook County, Minnesota, and relations to the Lake Superior syncline: Canadian Journal of Earth Science, v. 13, p. 877–888.

Naldrett, A. J., 1981, Platinum-group element deposits: Canadian Institute of Mining and Metallurgy, Special Volume 23, p. 197–232.

Naldrett, A. J., and Duke, J. M., 1980, Platinum metals in magmatic sulfide ores: Science, v. 208, p. 1417–1424.

Nathan, H. D., 1969, The geology of a portion of the Duluth Complex, Cook County [Ph.D. thesis]: Minneapolis, University of Minnesota, 198 p.

Nelson, N. S., 1991, A petrologic and geochemical study of granophyric granite in the roof of the Keweenawan Duluth Complex, Eagle Mountain, Cook County, Minnesota [M.S. thesis]: Duluth, University of Minnesota, 151 p.

Nicholson, S. W., and Lipin, B. R., 1985, Guide to the Gish mine area, *in* Czamanske, G. K., and Zientek, M. L., eds., The Stillwater Complex, Montana: Geology and guide: Montana Bureau Mines and Geology, Special Publication 92, p. 358–367.

Nicholson, S. W., and Shirey, S. B., 1990, Midcontinent Rift volcanism in the Lake Superior region: Sr, Nd, and Pb isotopic evidence for a mantle plume origin: Journal of Geophysical Research, v. 95, p. 10851–10868.

Nicholson, S. W., Cannon, W. F., and Schulz, K. J., 1992, Metallogeny of the Midcontinent Rift System of North America: Precambrian Research, v. 58, p. 355–386.

Ojakangas, R. W., 1992, Sedimentary fill of the 1100 Ma (Keweenawan) Midcontinent Rift System in the Lake Superior region [abs.]: International Basement Tectonics Conference, 10th, Programs with Abstracts and Mid-conference Field Trip Guide, August 1–11, 1992, Duluth, Minnesota, p. 159–162.

Paces, J. B., and Miller, J. D., Jr., 1993, Precise U-Pb ages of Duluth Complex and related mafic intrusions, northeastern Minnesota: New insights for physical, petrogenetic, paleomagnetic and tectono-magmatic processes associated with 1.1 Ga Midcontinent Rift system: Journal of Geophysical Research, v. 98, p. 13997–14013.

Pasteris, J. D., 1984, Further interpretation of the Cu-Fe-Ni sulfide mineralization in the Duluth Complex, northeastern Minnesota: Canadian Mineralogist, v. 22, p. 39–53.

Pasteris, J. D., 1985, Relationships between temperature and oxygen fugacity among Fe-Ti oxides in two regions of the Duluth Complex: Canadian Mineralogist, v. 23, p. 111–127.

Pasteris, J. D., 1988, Secondary graphitization in mantle-derived rocks: Geology, v. 16, p. 804–807.

Pasteris, J. D., 1989, Methane-nitrogen fluid inclusions in igneous rocks from the Duluth Complex, Minnesota [abs.]: PACROFI II, Program and Abstracts, v. 2, p. 51.

Pasteris, J. D., 1990, Hydrous alteration zones along the SE contact of the Duluth Complex, MN: Fluid-inclusion evidence for igneous and metamorphic processes [abs.]: Geological Society of America Abstracts with Programs, v. 22, p. A347.

Pasteris, J. D., Harris, T. N., and Sassani, D. C., 1992, Involvement of aqueous fluids in the rift environment of the Duluth Complex, Minnesota [abs.]: Geological Society of America Abstracts with Programs, v. 24, p. A61.

Pasteris, J. D., Harris, T. N., and Sassani, D. C., 1995, Interactions of mixed volatile-brine fluids in rocks of the southwestern footwall of the Duluth Complex, Minnesota: Evidence from aqueous fluid inclusions: American Journal of Science, v. 295, p. 125–172.

Phinney, W. C., 1972, Northern prong, Duluth Complex, *in* Sims, P. K., and Morey, G. B., eds., Geology of Minnesota: A centennial volume: Minnesota Geological Survey, p. 346–353.

Raedeke, L. D., and McCallum, I. S., 1984, Investigations in the Stillwater Complex: Part II. Petrology and petrogenesis of the Ultramafic Series: Journal of Petrology, v. 25, p. 395–420.

Raedeke, L. D., and McCallum, I. S., 1985, Guide to the Chrome Mountain area, *in* Czamanske, G. K., and Zientek, M. L., eds., The Stillwater Complex, Montana: Geology and guide: Montana Bureau Mines and Geology, Special Publication 92, p. 277–285.

Rao, B. V., 1981, Petrogenesis of sulfides in the Dunka Road Cu-Ni deposit, Duluth Complex, Minnesota, with special reference to the role of contamination by county rock [Ph.D. thesis]: Bloomington, Indiana University, 372 p.

Reeve, E. J., 1969, Petrology and mineralogy of a gabbroic intrusion in Pardee Township near Port Arthur, Ontario [M.S. thesis]: Milwaukee, University of Wisconsin, 79 p.

Ripley, E. M., 1981, Sulfur isotope studies of the Dunka Road Cu-Ni deposit, Duluth Complex, Minnesota: Economic Geology, v. 76, p. 610–620.

Ripley, E. M., 1986, Origin and concentration mechanisms of copper and nickel in Duluth Complex sulfide zones—a dilemma: Economic Geology, v. 81, p. 974–978.

Ripley, E. M., 1990a, Platinum group element geochemistry of Cu-Ni mineralization in the basal zone of the Babbitt deposit, Duluth Complex, Minnesota: Economic Geology, v. 85, p. 830–841.

Ripley, E. M., 1990b, Se/S ratios of the Virginia Formation and Cu-Ni sulfide mineralization in the Babbitt area, Duluth Complex, Minnesota: Economic Geology, v. 85, p. 1935–1940.

Ripley, E. M., and Alawi, J. A., 1986, Sulfide mineralogy and chemical evolution of the Babbitt Cu-Ni deposit, Duluth Complex, Minnesota: Canadian Mineralogist, v. 24, p. 347–368.

Ripley, E. M., and Al-Jassar, T. J., 1987, Sulfur and oxygen isotope studies of melt-country rock interaction, Babbitt Cu-Ni deposit, Duluth Complex, Minnesota: Economic Geology, v. 82, p. 87–107.

Ripley, E. M., and Chryssoulis, S. L., 1994, Ion microprobe analysis of platinum-group elements in sulfide and arsenide minerals from the Babbitt Cu-Ni deposit, Duluth Complex, Minnesota: Economic Geology, v. 89, p. 201–210.

Ripley, E. M., and Taib, N. I., 1989, Carbon isotopic studies of metasedimentary and igneous rocks at the Babbitt Cu-Ni deposit, Duluth Complex, Minnesota: Isotope Geoscience, v. 73, p. 319–342.

Ripley, E. M., Butler, B. K., and Taib, N. I., 1992, Effects of devolatilization on the hydrogen isotopic composition of pelitic rocks in the contact aureole of the Duluth Complex, northeastern Minnesota, U.S.A.: Chemical Geology, v. 102, p. 185–197.

Ripley, E. M., Butler, B. K., Taib, N. I., and Lee, I., 1993, Hydrothermal alteration in the Babbitt Cu-Ni deposit, Duluth Complex: Mineralogy and hydrogen isotope systematics: Economic Geology, v. 88, p. 679–696.

Robertson, W. A., and Fahrig, W. F., 1971, The great paleomagnetic loop—the polar wandering path from Canadian Shield rocks during the Neohelikian Era: Canadian Journal of Earth Science, v. 8, p. 1355–1372.

Ross, B. A., 1985, A petrologic study of the Bardon Peak peridotite, Duluth Complex [M.S. thesis]: Duluth, University of Minnesota, 140 p.

Rudashevsky, N. S., Avdontsev, S. N., and Dneprovskaya, M. B., 1992, Evolution of PGE mineralization in hortonolitic dunites of the Mooihoek and Onverwacht pipes, Bushveld Complex: Mineralogy and Petrology, v. 47, p. 37–54.

Ryan, R. J., and Weiblen, P. W., 1984, Pt and Ni arsenide minerals in the Duluth Complex [abs.]: 20th Annual Institute on Lake Superior Geology, Wausau, Wisconsin, v. 20, p. 58–60.

Ryan, R. M., 1984, Chemical, isotopic, and petrographic study of the sulfides in the Duluth Complex Cloud zone [M.S. thesis]: Bloomington, Indiana University, 88 p.

Sá, J. H. S., Barnes, S-J., Prichard, H. M., and Bowles, J. F. W., 1993, Platinum-palladium in iron-rich rocks of Rio Jacare Complex, Bahia, Brazil [abs.]: Encontro Brasileiro Sobre Elementos do Grupo da Platina, Brasilia, April 12–14.

Sabelin, T., 1985, Platinum group element minerals in the Duluth Complex [abs.]: 31st Annual Institute of Lake Superior Geology, Kenora, Ontario, v. 31, p. 83–84.

Sabelin, T., 1987, Association of platinum deposits with chromium occurrences: An overview with implications for the Duluth Complex: Skillings Mining Review, v. 76, no. 47, p. 4–7.

Sabelin, T., and Iwasaki, I., 1985, Metallurgical evaluation of chromium-bearing drill core samples from the Duluth Complex: University of Minnesota Minerals Resources Research Center, Contract Report to the Minnesota Department of Natural Resources, Division of Minerals, 58 p.

Sabelin, T., and Iwasaki, I., 1986, Evaluation of platinum group metal occurrence in Duval 15 drill core from the Duluth Complex: Minneapolis, University of Minnesota, Minerals Resources Research Center, Internal Report, 23 p.

Sabelin, T., Iwasaki, I., and Reid, K. J., 1986, Platinum group minerals in the Duluth Complex and their beneficiation behaviors: Skillings Mining Review, v. 75, no. 34, p. 4–7.

Saini-Eidukat, B., 1991, Platinum group elements in anorthositic rocks of the Duluth Complex, Minnesota [Ph.D. thesis]: Minneapolis, University of Minnesota, 146 p.

Saini-Eidukat, B., and Kucha, H., 1991, An iron chloride hydroxide from the Duluth Complex, Minnesota with implications for metal mobility in hydrothermal systems, *in* Pagel, M., and Leroy, J. L., eds., Source, transport and deposition of metals: Rotterdam, Balkema Publishing Company, p. 127–130.

Saini-Eidukat, B., Weiblen, P., Bitsianes, G., and Glascock, M., 1990, Contrasts between platinum group element contents and biotite compositions of Duluth Complex troctolitic and anorthositic series rocks: Mineralogy and Petrology, v. 42, p. 121–140.

Saini-Eidukat, B., Kucha, H. K., and Keppler, H., 1994, Hibbingite, γ-Fe$_2$(OH)$_3$Cl, a new mineral from the Duluth Complex, Minnesota, with implications for oxidation of iron-bearing compounds and transport of metals: American Mineralogist, v. 79, no. 5–6, p. 555–561.

Sassani, D. C., 1992, Petrologic and thermodynamic investigation of the aqueous transport of platinum-group elements during alteration of mafic intrusive rocks [Ph.D. thesis]: St. Louis, Washington University, 2 vols., 952 p.

Sassani, D. C., and Pasteris, J. D., 1988, Preliminary investigation of alteration in a basal section of the southern Duluth Complex, Minnesota, and the effects on sulfide and oxide mineralization, *in* Kisvarsanyi, G., and Grant, S. K., eds., North American conference on tectonic control of ore deposits and the vertical and horizontal extent of ore systems: Proceedings of the Volume, University Missouri-Rolla, p. 280–291.

Sassani, D. C., and Pasteris, J. D., 1992, Three-phase intergrowths in an oxide horizon of the basal zone of the Duluth Complex [abs.]: Minnesota Department of Natural Resources, Division of Minerals, Drill Core Library Newsletter, v. 2, p. 5.

Schiffries, C. M., 1982, The petrogenesis of a platiniferous dunite pipe in the Bushveld Complex: Infiltration metasomatism by a chloride solution: Economic Geology, v. 77, p. 1439–1453.

Schiffries, C. M., and Skinner, B. J., 1987, The Bushveld hydrothermal system: Field and petrologic evidence: American Journal of Science, v. 287, p. 566–595.

Schwartz, G. M., 1930, The relation of magnetite and ilmenite in the magnetite deposits of the Duluth Gabbro: American Mineralogist, v. 15, p. 243–252.

Scoon, R. N., 1987, Metasomatism of cumulus magnesian olivine by iron-rich postcumulus liquids in the upper Critical Zone of the Bushveld Complex: Mineralogical Magazine, v. 51, p. 389–396.

Seitz, J. C., and Pasteris, J., 1988, Oxide and sulfide mineralization at the base of the Duluth Complex—Skibo Tower, Minnesota [extended abs.], *in* Annual Current Activities Forum, 5th, Chisholm, Minnesota: Hibbing, Minnesota, Minnesota Department of Natural Resources, Division of Minerals, 17 p.

Severson, M. J., 1988, Geology and structure of a portion of the Partridge River Intrusion: A progress report: Duluth, University of Minnesota, Natural Resources Research Institute, Technical Report, NRRI/GMIN-TR-88-08, 78 p.

Severson, M. J., 1991, Geology, mineralization, and geostatistics of the Minnamax/Babbitt Cu-Ni deposit (Local Boy area), Minnesota, Part I: Geology: Duluth, University of Minnesota, Natural Resources Research Institute, Technical Report, NRRI/TR-91/13a, 96 p.

Severson, M. J., 1994, Igneous stratigraphy of the South Kawishiwi Intrusion, Duluth Complex, northeastern Minnesota: Duluth, University of Minnesota, Natural Resources Research Institute, Technical Report, NRRI/TR-93/34, 210 p.

Severson, M. J., 1995, Geology of the southern portion of the Duluth Complex: Duluth, University of Minnesota, Natural Resources Research Institute, Technical Report, NRRI/TR-95/26, 185 p.

Severson, M. J., and Barnes, R. J., 1991, Geology, mineralization, and geostatistics of the Minnamax/Babbitt Cu-Ni deposit (Local Boy area), Minnesota, Part II: Mineralization and geostatistics: Duluth, University of Minnesota, Natural Resources Research Institute, Technical Report, NRRI/TR-91/13b, 221 p.

Severson, M. J., and Hauck, S., 1990, Geology, geochemistry and stratigraphy of a portion of the Partridge River Intrusion: Duluth, University of Minnesota, Natural Resources Research Institute, Technical Report, NRRI/GMIN-TR-89-11, 230 p.

Severson, M. J., Patelke, R. L., Hauck, S. A., and Zanko, L. M., 1994, The Babbitt copper-nickel deposit, Part B: Structural datums: Duluth, University of Minnesota, Natural Resources Research Institute, Technical Report, NRRI/TR-94/21B, 48 p.

Smith, A. R., and Sutcliffe, R. H., 1987, Keweenawan intrusive rocks of the Thunder Bay area, *in* Summary of field work and other activities 1987: Ontario Geological Survey, p. 248–255.

Springer, G., 1989, Chlorine-bearing and other uncommon minerals in the Strathcona Deep Copper Zone, Sudbury District, Ontario: Canadian Mineralogist, v. 27, p. 311–313.

Strommer, J., Morton, P., Hauck, S. A., and Barnes, R. J., 1990, Geology and mineralization of a cyclic layered series, Water Hen Intrusion, St. Louis County, Minnesota: Duluth, University of Minnesota, Natural Resources Research Institute, Technical Report, NRRI/GMIN-TR-89-17, 29 p.

Stumpfl, E. F., and Rucklidge, J. C., 1982, The platiniferous dunite pipes of the Eastern Bushveld: Economic Geology, v. 77, p. 1419–1431.

Sutcliffe, R. H., 1991, Proterozoic geology of the Lake Superior area, *in* Thurston, P. C., ed., Geology of Ontario: Ontario Geological Survey, p. 627–660.

Taib, N. I., 1989, A carbon isotope study of the Babbitt Cu-Ni deposit, Duluth Complex, northeastern Minnesota: The devolatization of graphic-carbon bearing metasediments and the resulting contamination of the magma [M.S. thesis]: Bloomington, Indiana University, 86 p.

Taib, N. I., and Ripley, E. M., 1993, Distribution and genesis of Cu-Ni sulfides in a multiple intrusive sequence, Babbitt area, Duluth Complex, MN [abs.]: Geological Society of America Abstracts with Programs, v. 25, no. 6, p. A400–A401.

Tanton, T. L., 1935, Copper-nickel occurrences in Pigeon River area, Ontario: Canada Geological Survey, Paper 35-D, 11 p.

Taylor, R. B., 1964, Geology of the Duluth gabbro complex near Duluth, Minnesota: Minnesota Geological Survey, Bulletin 44, p. 1–63.

Viljoen, M. J., and Scoon, R. N., 1985, Distribution and main geologic features of discordant bodies of iron-rich ultramafic pegmatite in the Bushveld Complex: Economic Geology, v. 80, p. 1109–1128.

Volborth, A., and Housley, R. M., 1984, A preliminary description of complex graphite, sulphide, arsenide, and platinum group element mineralization in a pegmatoidal pyroxenite of the Stillwater Complex, Montana, U.S.A.: TMPM Tschermaks Mineralogische und Petrographische Mitteilungen, v. 33, p. 213–230.

Wagner, P. A., 1929, Platinum deposits and mines of South Africa: Edinburgh, Oliver and Boyd, 326 p.

Wanless, R. K., Stevens, R. D., Lachance, G. R., and Rimsaite, J. Y. H., 1966, Age determinations and geological studies: Canada Geological Survey, Paper 65-17, 79 p.

Weiblen, P. W., 1982, Keweenawan intrusive rocks, *in* Wold, R. J., and Hinze, W. J., eds., Geology and tectonics of the Lake Superior basin: Geological Society of American Memoir 156, p. 57–82.

Weiblen, P. W., and Morey, G. B., 1975, The Duluth Complex: A petrologic and tectonic summary: Proceeding, American Institute of Mining Engineers, Minnesota Section, Annual Meeting, 48th, and Annual Mining Symposium, 36th, University of Minnesota, Minneapolis, p. 72–95.

Weiblen, P. W., and Morey, G. B., 1976, Textural and compositional characteristics of sulfide areas from the basal contact zone of the South Kawishiwi Intrusion, Duluth Complex, northeastern Minnesota: Proceedings, American Institute Mining Engineering, Minnesota Section, Annual Meeting, 49th, and Annual Mining Symposium, 37th, University of Minnesota, Minneapolis, 24 p.

Weiblen, P. W., and Morey, G. B., 1980, A summary of the stratigraphy, petrology and structure of the Duluth Complex: American Journal of Science, v. 280, p. 88–133.

Weiblen, P. W., Davidson, D. M., Jr., Morey, G. B., and Mudrey, M. G., Jr., 1972a, Geology of Cook County, *in* Field trip guidebook for Precambrian geology of northeastern Cook County, Minnesota: Minnesota Geological Survey Guidebook Series No. 6, p. 1–61.

Weiblen, P. W., Mathez, E. A., and Morey, G. B., 1972b, Logan intrusions, *in* Sims, P. K., and Morey, G. B., eds., Geology of Minnesota: A centennial volume: Minnesota Geological Survey, p. 361–388.

Weiblen, P. W., Cooper, R. W., and Churchill, R. K., 1977, Relationship between mineralization and structure in the Duluth Complex [abs.]: Geological Society of America Abstracts with Program, v. 9, no. 7, p. 1220.

Whittaker, P. J., 1986, Chromite deposits in Ontario: Ontario Geological Survey, Study 55, 97 p.

Winchell, H. V., 1897, Reputed nickel mines in Minnesota: Engineering Mining Journal, v. 64, p. 578.

Witthuhn, K. H., 1992, The response to rifting by North Shore Volcanics in Minnesota [abs.]: International Basement Tectonics Conference, 10th, Programs with Abstracts and Mid-conference Field Trip Guide, August 1–11, 1992, Duluth, MN, p. 171–173.

Wolff, J. F., 1919, Recent geologic developments on the Mesabi range, Minnesota: American Institute Mining Metallurgy Engineering Transactions, v. 56, p. 142–169.

Wood, S. A., Mountain, B. W., and Pan, P., 1992, The aqueous geochemistry of platinum, palladium, and gold: Recent experimental constraints and a re-evaluation of theoretical predictions: Canadian Mineralogist, v. 30, p. 955–982.

Zanko, L., Severson, M., and Ripley, E. M., 1994, Geology and mineralization of the Serpentine Cu-Ni deposit: Duluth, University of Minnesota, Natural Resources Research Institute, Technical Report, NRRI/TR-93-52, 90 p.

Zientek, M. L., and 8 others, 1994, Cumulus processes and the composition of magmatic ore deposits: Examples from the Talnakh District, Russia, *in* Lightfoot, P. C., and Naldrett, A. J., eds., Proceedings, Sudbury-Noril'sk Symposium: Ontario Geological Survey Special Volume 5, p. 373–392.

MANUSCRIPT ACCEPTED BY THE SOCIETY JANUARY 16, 1996

Geological Society of America
Special Paper 312
1997

The Port Coldwell veins, northern Ontario: Pb-Zn-Ag deposits in a rift setting

T. Campbell McCuaig* and Stephen A. Kissin
Department of Geology, Lakehead University, Thunder Bay, Ontario P7B 5E1, Canada

ABSTRACT

The Port Coldwell Pb-Zn-Ag veins occur mainly in Archean supracrustal rocks in a linear belt on the north shore of Lake Superior between Schreiber and Marathon, Ontario. Although tonnages are small and only very minor production has come from the deposits, ore grades are high. The deposits contain galena, sphalerite, pyrite, chalcopyrite, covellite, digenite, and freibergite, the silver carrier, in calcite and quartz gangue. The veins are typical open-space fillings with generally well-developed, coarse-grained crustification textures.

Fluid inclusions in quartz, calcite, and sphalerite are H_2O dominated and contain no detectable daughter minerals or second fluid phases (e.g., liquid CO_2) at room temperature. Typically, a pattern of declining homogenization temperatures from early quartz and calcite to sphalerite and late calcite is observed. Eutectic temperatures (clustering at −21 and −28°C) and final melting temperatures (ranging from 0 to −12.5°C) are consistent with freezing point depression due to low to moderate salinity (0 to 16 wt% NaCl equiv.) in the undersaturated NaCl-H_2O system. Evidence for boiling is sparse, although in the Deadhorse Creek South deposit, homogenization to vapor at 171.5°C is associated with inclusions homogenizing to liquid at the same temperature, both with salinities of 11.7 equiv. wt% NaCl. The calculated pressure of 15 bars corresponds to a depth of 56 m, under lithostatic or 156 m under hydrostatic conditions in agreement with the shallow depth of deposition suggested by mineral textures.

Sulfur isotopic compositions of paired galena-sphalerite samples lie in the range +1.0 to +6.0 per mil. Isotopic partitioning temperatures are not in agreement with fluid inclusion temperatures, and a lack of isotopic equilibrium is suggested. The source of sulfur is unclear.

The deposits are demonstrably younger than the Coldwell Alkaline Complex (1,108 Ma), and lead isotopic data indicate that they are likely cogenetic with silver deposits of the Thunder Bay area, 250 km to the west. Their temporal and spatial relationship to the Keweenawan (Midcontinent) Rift suggests a genetic connection to extension. The Port Coldwell veins resemble Ag-Pb-Zn veins in clastic metasedimentary terranes, as defined by Beaudoin and Sangster (1992), a class of deposits related to crustal-scale faults commonly in extensional environments.

*Present address: Etheridge Henley Williams Geoscience Consultants, Suite 34A, 25 Walters Drive, Osborne Park, Western Australia, 6017.

McCuaig, T. C., and Kissin, S. A., 1997, The Port Coldwell veins, northern Ontario: Pb-Zn-Ag deposits in a rift setting, *in* Ojakangas, R. W., Dickas, A. B., and Green, J. C., eds., Middle Proterozoic to Cambrian Rifting, Central North America: Boulder, Colorado, Geological Society of America Special Paper 312.

INTRODUCTION

The Port Coldwell veins are a group of Pb-Zn-Ag occurrences scattered within a 40-km-long portion of the Terrace Bay Greenstone Belt on the north shore of Lake Superior approximately 250 km east of Thunder Bay, Ontario, Canada (Fig. 1). The veins were described briefly by Walker (1967) and more recently by Patterson et al. (1984, 1985). They include the Deadhorse Creek North, Deadhorse Creek South, and McKellar Creek occurrences, and possibly others including the Morley High Grade occurrence (Fig. 1). The latter lies some 40 km west of the Deadhorse Creek area, and Patterson et al. (1984) originally designated it as a gold occurrence, based on reported low gold values and proximity to the Terrace Bay Batholith.

All of these occurrences contain low tonnages but high ore grades. The Deadhorse Creek North deposit contains 36,000 t of ore grading 7.28% Zn, 1.45% Pb, and 8.27 oz. Ag/t (Patterson et al., 1984). The Deadhorse Creek South deposit similarly contains 35,000 t of ore grading 9.08% Zn, 19.87% Pb, and 27.65 oz. Ag/t (Patterson et al., 1985). Based on 30 ore samples, the Morley High Grade deposit has a grade of 25.85% Zn, 7.06% Pb, 0.185% Cu, and 41.57 oz. Ag/t with 0.196 oz. Au/t, as well. In spite of the high silver values, previous investigations have not identified the silver-carrying mineral in any of these occurrences.

For reasons of accessibility of the veins, most of the work reported in this paper is based on the Deadhorse Creek North, Deadhorse Creek South, and Morley High Grade occurrences. Other occurrences shown in Figure 1 are believed on the basis of mineralogy and vein form to be cogenetic with the first three. Other

Pb-Zn-Ag deposits are known, in the light of recent prospecting, to occur farther to the west (e.g., the Burning Rock occurrence 40 km north-northwest of Schreiber; Schnieders et al., 1992).

The suggestion has been made that the Port Coldwell veins are genetically related to silver-bearing vein deposits of the Thunder Bay area, approximately 250 km to the west (Franklin et al., 1986). The Thunder Bay area veins are of the five-element (Ni-Co-As-Ag-Bi) type, a deposit type distributed worldwide and of generally controversial origin. Kissin (1992) has reviewed these deposits, pointing out that as many as seven genetic hypotheses with variants have been proposed including hydrothermal/magmatic models, a hydrothermal/metamorphic model, a syngenetic model, and various nonmagmatic models. Among the latter, Kissin pointed out the strong temporal and spatial relation to rifting in a number of districts including Kongsberg-Modum (Oslo Graben), Cobalt-Gowganda (Temiskaming Rift), and especially the Thunder Bay district (Keweenawan Rift). The elevated heat flow associated with rifting in conjunction with the circulation system afforded by deep crustal fractures are proposed as the underlying mechanism for the formation of these deposits. The purpose of this study is to document the Port Coldwell Pb-Zn-Ag occurrences and to suggest a genetic model, considering the relationship to the silver deposits of the Thunder Bay area and Keweenawan rifting.

ANALYTICAL METHODS

Fluid inclusion studies were carried out on doubly polished 50 μm thick sections using an SGE Fluid Inclusion Research System mounted on a specially adapted Leitz SM-POL polariz-

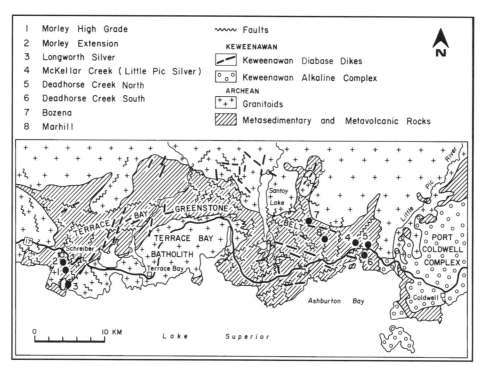

Figure 1. Location map and generalized geology of the Schreiber-Terrace Bay area, Ontario.

ing microscope. The system was calibrated using standards of known melting and freezing points. As all calibration errors were ±1°C or less, no corrections were applied. Pressures for some homogenization data were calculated by means of the FLINCOR program of Brown (1989).

Analyses of selected sulfide minerals were carried out using a Hitachi 570 scanning electron microscope equipped with a Tracor Northern TN2000 energy dispersive analyzer (SEM-EDA). The standardless software routine SQ from Tracor Northern Program 5022 was utilized in obtaining analyses, which were automatically normalized to near 100%.

Sulfur isotopic analyses were conducted on 10 galena and sphalerite separates by standard techniques by Geochron Laboratories Division, Kreuger Enterprises Incorporated, Cambridge, Massachusetts.

GEOLOGICAL SETTING

The Pb-Zn-Ag vein deposits occur in the Archean Terrace Bay Greenstone Belt, also known as the Schreiber lithotectonic assemblage (Williams et al., 1991). Based on the descriptions of Harcourt and Bartley (1938), Walker (1967) Marmont (1984), and Williams et al. (1991), the geology of the area is summarized below (Fig. 1).

The oldest rocks of the area are Archean metavolcanic and associated metasedimentary rocks of the Wawa Subprovince of the Superior Province, which constitute the Terrace Bay Greenstone Belt. Iron-rich tholeiites and calc-alkaline felsic metavolcanic rocks are interbedded with metagraywackes and meta-argillites, with minor iron formation and metaconglomerate. Minor dikes and sills of gabbro and diorite intrude the metavolcanic and metasedimentary rocks. Metamorphic grades range from greenschist to near granulite facies, increasing to the north toward the contact with granitic terrane, which has intrusive relations with the greenstone lithologies. Intruding all of the above is the Late Archean Terrace Bay Batholith, which in outcrop forms an ellipse with axes 25 by 8 km, near the center of the study area. The batholith consists predominantly of granodiorite with minor diorite, quartz monzonite, tonalite, and biotite-hornblende granite. A metamorphic aureole extending approximately 0.5 km into the country rock surrounds the batholith.

Proterozoic rocks associated with Keweenawan events include the Port Coldwell Alkaline Complex, which is located on the eastern edge of the study area, and diabase dikes and sills. A well-defined contact aureole extends approximately 1 km beyond the margin of the Port Coldwell Complex. Scattered diabasic hypabyssal intrusions are minor features, commonly occupying east- to northeast-striking faults. Although some older structures are present, most vertical displacement appears to have been related to Keweenawan events.

VEIN MORPHOLOGY AND PARAGENESIS

The Port Coldwell veins cut Archean greenstone lithologies of the Terrace Bay Greenstone Belt, as well as diabase dikes of both Archean and Proterozoic age. The Deadhorse

Creek veins strike perpendicular to the well-defined contact aureole of the Coldwell Complex. The veins are clearly younger than the contact aureole and, thus, have a maximum age of 1,108 Ma, the zircon U-Pb age of the complex (Heaman and Machado, 1992).

As noted previously, three occurrences were selected for detailed study (Fig. 1): Deadhorse Creek South (DHS), Deadhorse Creek North (DHN), and Morley High Grade (MHG). The DHN and DHS occurrences are similar in terms of vein geometry, paragenesis, and texture, consisting of growth-banded calcite, sphalerite, and galena with minor chalcopyrite, pyrite, and quartz. Both are narrow veins with low-grade alteration zones about 1 to 2 m wide on either side. Alteration is visible megascopically as bleaching, but is limited to propylitic assemblages of sericite-calcite-(epidote), which leave much relict texture and mineralogy unaltered. The DHS vein has a strike of N70°W and dips 60°N with a width of 1 to 10 cm, as compared to N80°W, 80°N, and a width of <5 cm to 1.5 m for DHN. The DHS vein does not appear to be brecciated and appears to have been deposited in a single pulse, an observation supported by fluid inclusion studies. The DHN vein has clearly experienced at least one episode of brecciation in which fragments of early vein material were cemented by later material. The DHS vein is hosted in metasediments, whereas the DHN vein is in metavolcanics near the contact of an Archean metagabbro intrusion. Both veins cut earlier structures and fabrics including the contact aureole of the Port Coldwell Alkaline Complex. A north-striking fault following Deadhorse Creek offsets both veins laterally for some meters.

No silver-carrying mineral is evident in hand specimens from the veins; however, in polished section a mineral resembling tetrahedrite is seen to occur in considerable amounts in association with galena. SEM-EDA analysis yields the following weight percent compositions: Cu = 17.92, Ag = 27.30, Fe = 5.70, Zn = 2.10, Sb = 24.66, S = 22.27; total = 99.95 wt%. Based on 29 atoms total, the analysis corresponds to the formula $(Cu_{5.22}Ag_{4.68})_{\Sigma 9.90}(Fe_{1.89}Zn_{0.61})_{\Sigma 2.50}Sb_{3.75}S_{12.85}$, which is freibergite, the silver-rich member of the tetrahedrite series. Based on its abundance and tenor of silver, this mineral clearly accounts for the silver values in the Deadhorse Creek veins.

In the DHS vein the paragenetic sequence is simple, and the vein is symmetrically banded. From the vein wall to the center of the vein, the sequence is as follows: (a) hydrothermal pyrite in wall rock adjacent to the vein; (b) calcite and pyrite adjacent to the wall rock; (c) growth-banded sphalerite with minor amounts of galena, calcite, and chalcopyrite; (d) large, euhedral galena crystals associated with sphalerite, chalcopyrite, and freibergite; and (e) barren, coarse-grained calcite in the vein center (Fig. 2).

The paragenetic sequence of the DHN vein is more difficult to decipher owing to one or more episodes of brecciation followed by additional mineralization (Fig. 3). However, an early phase of calcite was followed by cockade-textured galena and sphalerite. The vein contains noticeably more pyrite and

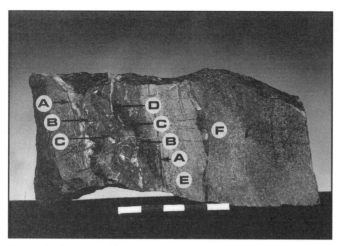

Figure 2. The paragenetic sequence in the Deadhorse Creek South vein, from the wall inward: A, early quartz and calcite; B, growth-banded sphalerite; C, coarse-grained galena; and D, late calcite. Bleached, altered metasedimentary wall rock (E) and a periferal quartz-calcite-pyrite (F) are shown separately. Scale in centimeters.

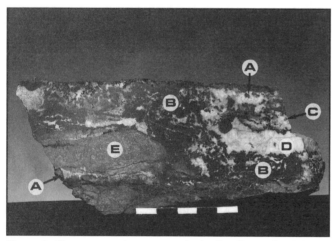

Figure 3. The paragenetic sequence in the Deadhorse Creek North vein, where the upper surface is a vein wall. From the vein walls inward are: A, early calcite with minor sulfides; B, calcite with sphalerite-galena-freibergite; C, brecciated early vein deposits; and D, late calcite. (E) is an altered wall-rock inclusion. Scale in centimeters.

chalcopyrite than does the DHS vein, as well as minor quartz. Late, coarse-grained calcite occupies the center of the veins. Freibergite is associated with galena and chalcopyrite, and a small amount of a chalcopyrite-digenite intergrowth is present. As the latter represents the breakdown products of a high-temperature digenite-chalcocite solid solution (Barton, 1973), hypogene origin of the intergrowth is likely.

The Morley High Grade (MHG) vein strikes N5°W with a dip of 60°W, and in its 7 m of exposure its width is approximately 20 cm. The paragenetic sequence is fairly well preserved, although an episode of brecciation and recementation has occurred. Although the felsic metavolcanic host rocks,

lying near the thermal aureole of the Terrace Bay Batholith, are of higher metamorphic grade than those hosting the DHS and DHN veins, the MHG vein is clearly not metamorphosed. The mineralogy consists of fine-grained sphalerite + calcite ± pyrite near the vein walls and wall rock inclusions grading to fine-grained sphalerite + galena + chalcopyrite + freibergite. The carbonate content is low, and no gold is visible, although gold values have been reported (Patterson et al., 1987).

In summary, the three occurrences DHS, DHN, and MHG are similar in terms of their mineralogy. All are Pb-Zn-Ag veins with freibergite as the silver carrier. The paragenetic sequence is broadly similar with early calcite + sphalerite ± pyrite followed by sphalerite + galena ± chalcopyrite ± freibergite, with freibergite and chalcopyrite later than galena. Late, barren calcite, sometimes with quartz, is the final stage of mineralization.

FLUID INCLUSION STUDIES

Fluid inclusions were found in sphalerite, calcite, and quartz. Inclusions were first frozen in order to determine initial melting (eutectic) temperatures (Te) and final melting temperatures (Tm), which yield information on the solution composition and salinity, respectively. Inclusions were then heated in order to obtain homogenization temperatures (Th). The latter are uncorrected for pressure and therefore represent minimum trapping temperatures for the inclusions.

The Tm and Th data are summarized in the histograms of Figures 4 and 5; complete data are given by Kissin and McCuaig (1988). All fluid inclusions studied are liquid-vapor, two-phase inclusions, lacking daughter minerals and second liquid phases (e.g., liquid CO_2) at room temperature. DHS and DHN display a similar pattern of homogenization temperatures, beginning with early, high-temperature calcite, as well as some quartz at DHN, which began to crystallize in the 360 to 340°C range. Temperatures of main stage sulfide (+ calcite) deposition are recorded by fluid inclusions in sphalerite that homogenize in the 200 to 120°C range at DHS and 210 to 150°C range at DHN. Late calcite at DHS crystallized at 105 to 95°C and, with quartz, at 160 to 130°C at DHN.

One inclusion in sphalerite at DHS homogenized to vapor at 171.5°C amid inclusions that homogenized to liquid at similar temperatures. This is considered to be evidence of boiling. As frozen inclusions from this specimen melted consistently at −8°C, it was possible to calculate a salinity of 11.7 equiv. wt% NaCl, using the equation of Potter et al. (1978). As the pressure and temperature of trapping is equal to the pressure and temperature of homogenization in a boiling solution (cf. Roedder, 1984), it is possible to calculate the pressure of trapping by the FLINCOR program and, further, to make a calculation of the depth at which the inclusion was trapped. The calculated pressure is 15 bars (1.5 MPa); if the rock density is estimated at 2,700 kg · m⁻³, the depth at which the inclusion formed is 56 m under lithostatic conditions or 156 m under hydrostatic conditions. Such shallow depths are consistent with the open-space filling texture of the vein.

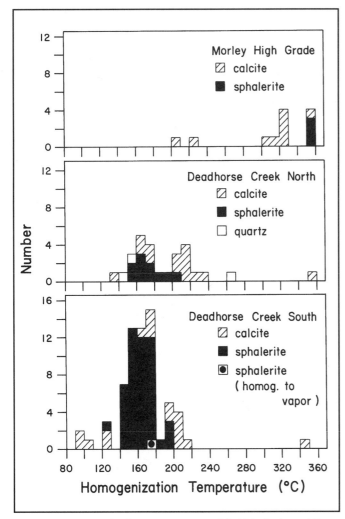

Figure 4. Homogenization temperatures of fluid inclusions. All are liquid + vapor → liquid except as noted.

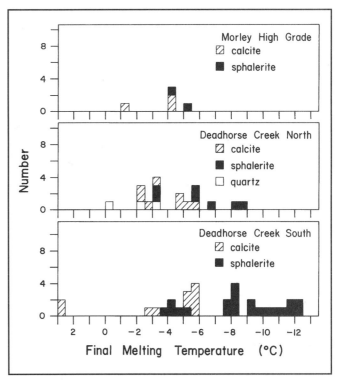

Figure 5. Final melting temperatures of fluid inclusions.

Studies of frozen inclusions reveal that Te of most inclusions is near the −20.8°C eutectic of the NaCl-H₂O system or its metastable eutectic at −28°C. A fluid inclusion in calcite from DHN displays a Te near −56.6°C, the eutectic of the CO₂-H₂O system. Two fluid inclusions in calcite from DHS, which homogenized at 120.0 and 96.6°C, exhibited positive Tm (Fig. 5). These are most likely clathrate melting temperatures, suggesting CO₂ enrichment in late-stage calcite. Tm for all inclusions (Fig. 5) are consistent with salinities in the undersaturated NaCl-H₂O system. Based on the equation of Potter et al. (1978) these salinities range from 16.2 to 0 equiv. wt% NaCl.

SULFUR ISOTOPIC STUDIES

Sulfur isotopic ratios were determined on paired galena and sphalerite specimens from the sulfide-rich Deadhorse Creek and Morley High Grade occurrences. The specimens were selected from grains in mutual contact, and fluid inclusion homogenization temperatures had previously been obtained

from the sphalerite grains. The results are given in Table 1 together with isotopic temperatures calculated from an equation of Rye and Ohmoto (1979) using Δ, the difference in $\delta^{34}S$ between sphalerite and galena. Although all $d^{34}S$ values are within the range +1 to +6 per mil, there are significant differences between the pairs of values. The related Deadhorse Creek occurrences have very similar isotopic equilibrium temperatures (Teq) and sphalerite fluid inclusion temperatures (Th), although the $\delta^{34}S$ values differ significantly between the two veins, and there is a discrepancy of 100°C between Teq and Th. This discrepancy is even larger in the case of the Morley High Grade occurrence. All of the foregoing suggests that the sulfide pairs are not in isotopic equilibrium, probably owing to sequential deposition of minerals under changing temperature conditions. Also, below 250°C fractionation of ^{34}S between sulfides and solution is strongly influenced by kinetic factors associated with mineral precipitation (Kyser, 1987). Calculated isotopic temperatures are thus often higher than fluid inclusion temperatures under these conditions.

The low positive $\delta^{34}S$ values for all sulfides may be suggestive of the source of the sulfur. On the one hand, the values are within the range of those of volcanogenic sulfide deposits (Rye and Ohmoto, 1979), suggesting possible derivation of sulfur without fractionation by leaching of volcanogenic sulfides, perhaps from the metavolcanic wall rock. However, by means of other mechanisms involving nonequilibrium processes, sulfate reduction may have produced the observed isotopic compositions of sulfides (e.g., Rye and Ohmoto, 1979).

**TABLE 1. SULPHUR ISOTOPIC DATA FROM THE DEADHORSE CREEK NORTH AND SOUTH
AND MORLEY HIGH GRADE VEINS**

Sample*		$\delta^{34}S$	$\Delta_{sp\text{-}gn}$†	$T_{eq.}(°C)$§	$T_{hom.}(°C)$**
DHS-X	Galena	+1.9	2.5	265	160
	Sphalerite	+4.4			
DHN-8	Galena	+3.4	2.6	275	175
	Sphalerite	+6.0			
MHG-b-1	Galena	+2.0	1.1	537	355
	Sphalerite	+3.1			
MHG-b-2	Galena	+1.0	2.4	275	355
	Sphalerite	+3.4			
MHG-c	Galena	+1.8	1.2	503	355
	Sphalerite	+3.0			

*DHN = Deadhorse Creek North; DHS = Deadhorse Creek South; MHG = Morley High Grade.
†Difference in $\delta^{34}S$ between sphalerite and galena. sp = sphalerite; gn = galena.
§Equilbration temperature from the equation of Rye and Ohmoto, 1979.
**Fluid inclusion homogenization temperature.

Interestingly, although $\Delta_{sp\text{-}gn}$ in DHS and DHN is nearly identical, $\delta^{34}S$ in sphalerite and galena is +1.5 higher at DHN than at DHS, even though the two occurrences are less than 1 km apart. This observation could be explained by the operation of the same kinetic factors at the two occurrences, thus giving rise to the same fractionation in sphalerite and galena, but with different source sulfur compositions. As DHN is hosted in metavolcanics whereas DHS is in metasediments, a variable contribution of local volcanogenic sulfur to a common solution could explain the difference.

DISCUSSION

Genesis of the Port Coldwell veins

As indicated earlier, cross-cutting relationships provide a maximum age of 1,108 Ma for the Port Coldwell veins. Lead isotopic evidence strongly suggests that the veins are of Keweenawan age and related to silver vein deposits of the Thunder Bay area, 250 km to the west (Franklin et al., 1986). The lead data from the Port Coldwell veins group closely with those of the Mainland Belt silver deposits at Thunder Bay; however, they are less radiogenic and lie close to a normal growth curve with a model age of 1,172 Ma. Their secondary isochron indicates an Archean source for lead, consistent with its derivation from the country rocks of the veins. The Pb-isotope evidence thus indicates that the veins are closely related temporally to the emplacement of the Coldwell Complex, as well as the Logan diabase and North Shore Volcanic Group to the west (Heaman and Machado, 1992); however, field relationships indicate that the veins postdate the Coldwell Complex.

The sulfur isotope data from this study are not definitive as to the source of sulfur. As discussed above, the data from the Deadhorse Creek occurrences yield a similar $\Delta_{sp\text{-}gn}$, indicating the operation of common thermal and kinetic controls on sulfur isotope fractionation at both localities, yet the sulfur isotope composition can be expalined by the introduction of sulfur and metals by a common SO_4^{2-}- or H_2S-bearing fluid, with local additions of reduced sulfur from sedimentary or volcanogenic sulfides in the country rocks.

The fluid inclusion data (Figs. 4 and 5) indicate a trend of declining temperature and salinity with progression through paragenetic sequence from early calcite to sphalerite to late calcite and quartz. Evidence for boiling is sparse and was thus not likely a major factor in precipitation of vein minerals. The shallow depth of deposition calculated from the limited boiling data together with the temperature-salinity relationship are suggestive of mixing of a hypogene solution with meteoric water. Late CO_2 enrichment of the solution may have been a result of interaction with wall rocks at the site of deposition.

The relatively high temperature solutions that formed these deposits have no obvious source within the present level of exposure. In particular, they postdate all igneous intrusions in the area and show no systematic spatial relationship to them. The veins occur throughout an area more than 60 km long, implying that they have arisen by means of a regional-scale mechanism.

Kissin (1988) proposed that five-element type veins of the Thunder Bay area were formed by processes associated with Keweenawan rifting. These include elevated heat flow, providing a means of heating and mobilizing connate fluids in sedimentary and volcanic rocks, as well as dilational faults, which supplied avenues for the solutions to rise to sites of deposition. The metals would have been scavenged by chloride-rich brines from deeper crustal levels. Evidence for this source of metals is available for the Deadhorse Creek occurrences in four lead isotopic analyses reported by Franklin et al. (1986). The analyses lie in a tight array on the lower crust curve in $^{207}Pb/^{204}Pb$ versus $^{206}Pb/^{204}Pb$ and $^{208}Pb/^{204}Pb$ versus $^{206}Pb/^{204}Pb$ diagrams of the plumbotectonics II model of Zartman and Doe (1981). Their lower crust curve is based on a model in which the lower crust is assumed to be granulite depleted in uranium and of low $^{233}U/^{204}Pb$ ratio and is characteristic of stable Precambrian terrane (Doe and Zartman, 1979), such as that forming the basement of this area. Kissin (1988) indicated that at depths of approximately 10 km, elevated heat flow associated with modern rifts would be sufficient to create temperatures as high as 400°C. Thus, origin of hydrothermal fluids at deep levels within the crust is consistent with both lead isotopic evidence and the availability of a heat source.

Relationship to other Pb-Zn-Ag vein deposits

Beaudoin and Sangster (1992) have recently reviewed various classifications of Pb-Zn-Ag veins and presented a descriptive model for what they believe to be a distinctive type of deposit: Ag-Pb-Zn veins in clastic metasedimentary terranes. Their classification does not include all types of silver-lead-zinc veins, others of which they categorize as Ag-Pb-Zn veins peripheral to porphyry Cu, Pb-Zn skarn, volcanic-hosted epithermal veins, and carbonate replacement and mantos.

The characteristics of deposits in the Beaudoin and Sangster (1992) model include galena and sphalerite and a suite of silver minerals and sulfosalts in siderite, calcite, and quartz gangue. The veins are enclosed in narrow zones of phyllic alteration that occur in a number of tectonic settings. Most veins are hosted by thick metasedimentary clastic sequences metamorphosed to at least greenschist facies, are located on or near crustal-scale faults, and are late features in the tectonic evolution of an orogen.

Fluid inclusion homogenization temperatures in these deposits range from 450 to 170°C in solutions ranging from 26 to 0 equiv. wt% NaCl. The occurrence of boiling and presence of CO_2 in inclusions is variable. Depths of mineralization inferred from boiling data range from 3 to 10 km and average 6 km. Mixing of a hydrothermal solution derived at depth with one or more meteoric solutions accounts for the variable salinities of the inclusions.

Sulfur isotopic compositions of galena, sphalerite, and pyrite range from 16 to 26 per mil and equal or are exceeded by the range of $\delta^{34}S$ seen in pyrite and pyrrhotite in country rocks. The sulfur in the vein sulfides is therefore believed to be locally derived. Lead isotopic data lie on or slightly above the orogene curve of the plumbotectonics model II of Zartman and Doe (1981), suggesting contribution of lead from both lower crustal and mantle sources.

The deposits are believed to have formed, according to the Beaudoin and Sangster (1992) model, by mobilization of deeply sourced fluids along deep crustal fault zones. Metamorphism is believed to have generated the fluids by dehydration reactions or wall rock/water exchange. Mixing between these deeply generated metamorphic fluids and shallow, meteoric fluids is believed to have played an important role in deposition. Extensional tectonics at the close of an orogenic cycle is believed to have been important in some districts.

The model outlined above bears a number of similarities to the Kissin (1988) hypothesis, especially in the deep origin of the hydrothermal fluid and its mixing with a shallow, crustal fluid. The Kissin hypothesis is related specifically to an anorogenic rifting environment in which heat is supplied by the anomalous heat flow associated with rifting.

The Beaudoin and Sangster model itself has some associated problems in that not all examples cited by them are in clastic metasedimentary terrane, but seem rather to lie in what might be better termed crystalline terrane, especially those of Freiburg, Germany, and the Kokanee Range, British Columbia. At the same time, the Kissin (1988) hypothesis is not compatible with some of the Beaudoin and Sangster examples because a rifting environment has not been identified in these cases. Possible examples associated with rifting include those of the Kokanee Range, British Columbia, and the Coeur d'Alene district, Idaho. In the Kokanee Range, late Eocene extensional faulting is associated with crustal thinning and elevated heat flow, accompanied by lamprophyre and diabase intrusion. Beaudoin et al. (1992) proposed that deep-seated fluids were responsible for mineralization in this district, in a process essentially the same as that proposed by Kissin (1988). According to Beaudoin and Sangster (1992), the ores of the Coeur d'Alene district formed at approximately 870 Ma and are roughly contemporaneous with the formation of a rifted margin in western North America (Stewart, 1972), although later deformation, metamorphism, and intrusion tend to obscure these relationships.

In many respects, the Port Coldwell veins resemble Ag-Pb-Zn veins in clastic metasedimentary terranes including ore, gangue, and alteration mineralogy and in salinity, homogenization temperatures, occurrence of boiling, CO_2 content, and mixing with meteoric water, as determined from fluid inclusions. However, the Deadhorse Creek veins were deposited at much shallower depths than those in the Beaudoin and Sangster model, although the MHG is likely a deeper as well as a higher temperature deposit. The narrow range of sulfur isotopic compositions seen in the Port Coldwell veins is in contrast to wider ranges of the Beaudoin and Sangster model; however, the small sample size for the Port Coldwell veins may contribute to this apparent discrepancy. Lead isotopic compositions of the Port Coldwell veins are clearly different from the Beaudoin and Sangster examples because they have lower crustal rather than orogene compositions.

The Port Coldwell veins may or may not belong to the class of Pb-Zn-Ag deposits defined by Beaudoin and Sangster. Only further study can resolve this point and, indeed, confirm or refute Beaudoin and Sangster's model. However, the lower formation temperatures and pressures of the Port Coldwell veins, the siting of these veins in crystalline Archean basement, and the clear spatial and temporal association of the veins with the Keweenawan Rift suggests that the Kissin (1988) hypothesis is a more viable model for the genesis of these Pb-Zn-Ag veins.

194

T. C. McCuaig and S. A. Kissin

ACKNOWLEDGMENTS

The following from the Department of Geology, Lakehead University, provided assistance in this work: A. L. Hammond and R. L. Viitala, thin- and polished-section preparation; S. T. Spivak, drafting; and W. K. Bourke, typing. A. D. Mackenzie of the Instrumentation Laboratory provided assistance in the operation of the analytical scanning electron microscope. B. R. Schnieders and A. A. Speed of the Ontario Ministry of Northern Development and Mines, Thunder Bay Office, graciously introduced us to the field area. The paper benefitted from the constructive reviews of K. D. Card and R. P. Sage and commentary by R. Kerrich and D. F. Sangster. This project was supported by Ontario Geological Research Fund Grant 300 to S. A. K. T. C. M. acknowledges receipt of a Lakehead University Alumni Award.

REFERENCES CITED

Barton, P. B., Jr., 1973, Solid solutions in the system Cu-Fe-S. Part I: The Cu-S and CuFe-S joins: Economic Geology, v. 68, p. 455–465.

Beaudoin, G., and Sangster, D. F., 1992, A descriptive model silver-lead-zinc veins in clastic metasedimentary terranes: Economic Geology, v. 87, p. 1005–1021.

Beaudoin, G., Taylor, B. E., and Sangster, D. F., 1992, Silver-lead-zinc veins and crustal hydrology during Eocene extension, southeastern British Columbia, Canada: Geochimica et Cosmochimica Acta, v. 56, p. 3513–3529.

Brown, P. E., 1989, FLINCOR: A microcomputer program for the reduction and investigation of fluid-inclusion data: American Mineralogist, v. 74, p. 1390–1393.

Doe, B. R., and Zartman, R. E., 1979. Plumbotectonics, the Phanerozoic, in Barnes, H. L., ed., Geochemistry of hydrothermal ore deposits (second edition): New York, John Wiley, p. 22–70.

Franklin, J. M., Kissin, S. A., Smyk, M. C., and Scott, S. D., 1986, Silver deposits associated with the Proterozoic rocks of the Thunder Bay District, Ontario: Canadian Journal of Earth Sciences, v. 23, p. 1576–1591.

Harcourt, G. A., and Bartley, M. W., 1938, Schreiber area, Thunder Bay District: Ontario Department of Mines, Map 47j, scale 1:31,680.

Heamon, L. M., and Machado, N., 1992, Timing and origin of Midcontinent Rift alkaline magmatism, North America: evidence from the Coldwell Complex: Contributions to Mineralogy and Petrology, v. 110, p. 289–303.

Kissin, S. A., 1988, Nickel-cobalt-native silver (five-element) veins: A rift-related ore type, in Kisvarsanyi, G., and Grant, S. K., eds., Proceedings Volume, North American Conference on the Tectonic Control of Ore Deposits and the Vertical and Horizontal Extent of Ore Systems, Rolla,

October, 1987: University of Missouri-Rolla, p. 268–279.

Kissin, S. A., 1992, Five-element (Ni-Co-As-Ag-Bi) veins: Geoscience Canada, v. 19, p. 113–124.

Kissin, S. A., and McCuaig, T. C., 1988, The genesis of silver vein deposits in the Thunder Bay area, northwestern Ontario, in Milne, V. G., ed., Geoscience Research Grant Program, Summary of research, 1987–1988: Ontario Geological Survey Miscellaneous Paper 140, p. 146–156.

Kyser, T. K., 1987, Equilibrium fractionation factors for stable isotopes, in Kyser, T. K., ed., Stable isotope geochemistry of low temperature fluids: Mineralogical Association of Canada Short Course, v. 13, p. 1–84.

Marmont, S., 1984, The Terrace Bay Batholith and associated mineralization: Ontario Geological Survey Open-File Report 5514, 95 p.

Patterson, G. C., Mason, J. K., and Schnieders, B. R., 1984, Report of activities 1983, Thunder Bay Resident Geologist Area, north central region, in Kustra, C. R., ed., Report of activities 1983, regional and resident geologists: Ontario Geological Survey Miscellaneous Paper 117, p. 47–106.

Patterson, G. C., Mason, J. K., and Schnieders, B. R., 1985, Thunder Bay Resident Geologist Area, north central region, in Kustra, C. R., ed., Report of activities 1984, regional and resident geologists: Ontario Geological Survey Miscellaneous Paper 122, p. 56–133.

Patterson, G. C., and 8 others, 1987, Thunder Bay Resident Geologist Area, north central region, in Kustra, C. R., ed., Report of activities 1986, regional and resident geologists: Ontario Geological Survey Miscellaneous Paper 134, p. 27–127.

Potter, R. W., II, Clynne, M. A., and Brown, D. L., 1978, Freezing point depression of aqueous sodium chloride solutions: Economic Geology, v. 73, p. 284–285.

Roedder, E., 1984, Fluid inclusions: Mineralogical Society of America, Reviews in Mineralogy, v. 12, 644 p.

Rye, R. G., and Ohmoto, H., 1979, Isotopes of sulfur and carbon, in Barnes, H. L., ed., Geochemistry of hydrothermal ore deposits (second edition): New York, John Wiley, p. 509–567.

Schnieders, B. R., Smyk, M. C., and McKay, D. B., 1992. Schreiber-Hemlo Resident Geologist's District—1991, in Fenwick, K. G., Newsome, J. W., and Pitt, A. E., eds., Report of activities 1991, resident geologists: Ontario Geological Survey, Miscellaneous Paper 158, p. 121–163.

Stewart, J. N., 1972, Initial deposits in the Cordilleran Geosyncline: Evidence of a late Precambrian (850 m.y.) continental separation: Geological Society of America Bulletin, v. 83. p. 1345–1360.

Walker, J. W. R., 1967, Geology of the Jackfish-Middleton area, District of Thunder Bay: Ontario Department of Mines, Geology Report 50, 41 p.

Williams, H. R., Stott, G. R., Heather, K. B., Muir, T. L., and Sage, R. P., 1991, Wawa Subprovince, Chapter 12, in Thurston, P. C., Williams, H. R., Sutcliffe, R. H., and Stott, G. M., eds., Geology of Ontario: Ontario Geological Survey Special Volume 4, Part 1, p. 485–539.

Zartman, R. E., and Doe, B. R., 1981. Plumbotectonics—The model: Tectonophysics, v. 74, p. 135–162.

MANUSCRIPT ACCEPTED BY THE SOCIETY JANUARY 16, 1996

Geological Society of America
Special Paper 312
1997

Petrography and sedimentation of the Middle Proterozoic (Keweenawan) Nonesuch Formation, western Lake Superior region, Midcontinent Rift System

Thomas Suszek*
Department of Geology, University of Minnesota, Duluth, Minnesota 55812

ABSTRACT

Detailed sedimentological descriptions and petrographic analysis of the upper Keweenawan Nonesuch Formation were accomplished for selected Bear Creek drill cores from Ashland, Bayfield, and Douglas Counties, Wisconsin. These data, coupled with the information from outcrops in northwestern Wisconsin and Upper Michigan, provide evidence on source rocks, environment of deposition, and the tectonic framework of the Nonesuch Formation in the Midcontinent Rift System.

Lower Keweenawan felsic, intermediate, and mafic volcanic units were the major contributors of detritus to the formation. Younger Keweenawan volcanic and granitic intrusive rocks were minor sources. Detritus from Early Proterozoic and Archean crystalline rocks increases in abundance upsection as older source rocks outside the rift were unroofed.

Sedimentary structures and stratigraphic facies relationships suggest that deltaic processes, sheetfloods, density and turbidity currents, and suspension settling were the primary mechanisms of deposition in a thermally stratified perennial lake. Rapid fluctuations in water levels were brought on by changes in tectonism or climate. The gradational contacts of the Nonesuch Formation with the underlying Copper Harbor Formation and the overlying Freda Formation, along with outcrop and drill core facies data, suggest that the site of Nonesuch deposition was adjacent to, and sometimes upon, a prograding alluvial fan complex.

The Nonesuch Formation in the Bear Creek drill cores was divided into six sedimentational intervals (1, 2, 3, 3a, 4, and 5, lowest to highest) based on textures, sedimentary structures, and color. Light gray to black rocks predominate in all of the intervals and indicate that deposition of Nonesuch sediments was in a predominantly euxinic environment. The facies assemblages represented in these sedimentational intervals record clastic deposition that occurred during the initial transgressive and final regressive stages of the Nonesuch Lake over the contemporaneous alluvial fan complex of the upper Copper Harbor Formation and the subaerial fluvial plain environment of the lower Freda Formation.

Examination of the genetic relationship of the Nonesuch sedimentational intervals, the textures within each facies type, petrographic data, paleocurrent analyses, and the regional interpretation of the western Lake Superior rift structure, suggest that most sediment was transported northward into the rift zone from the southern

*Present address: 4602 Jay Street, Duluth, Minnesota 55804.

Suszek, T., 1997, Petrography and sedimentation of the Middle Proterozoic (Keweenawan) Nonesuch Formation, western Lake Superior region, Midcontinent Rift System, *in* Ojakangas, R. W., Dickas, A. B., and Green, J. C., eds., Middle Proterozoic to Cambrian Rifting, Central North America: Boulder, Colorado, Geological Society of America Special Paper 312.

flank of the basin. Less important sources were present within the rift zone but on the northern side of the Nonesuch basin. These data also suggest that the Nonesuch Formation in the Bear Creek cores was deposited in a basin that was partially restricted, or perhaps completely isolated, from areas containing Nonesuch Formation farther east in Wisconsin and Upper Michigan.

INTRODUCTION

The 1.1-Ga Midcontinent Rift in the Lake Superior region contains a thick sequence of sedimentary rocks that overlies a thick sequence of flood basalts and associated volcanic rocks. The sedimentary rocks are as thick as 9 km and the volcanic flows as thick as 18,000 m (Allen et al., 1993). The sedimentary rocks are comprised of the Oronto Group and the overlying Bayfield Group.

The Oronto Group consists of, in ascending order, the Copper Harbor Conglomerate, the Nonesuch Formation, and the Freda Formation. The Copper Harbor and Freda are red bed units, whereas the Nonesuch is a gray to black unit. This paper is concerned with the Nonesuch Formation, and is a condensation of a Masters of Science thesis (Suszek, 1991).

From 1958 through 1961, Bear Creek Mining and Exploration Company conducted an exploration program in northwestern Wisconsin in search of sedimentary-hosted copper deposits similar to those found in the Middle Proterozoic (Keweenawan) Nonesuch and Copper Harbor Formations at the White Pine Mine approximately 145 km (90 mi) to the east of the study area in Upper Michigan. The project failed to discover any economic mineral potential, but did partially define the extent of the Oronto Group.

The purpose of this study was to examine the sedimentary rocks in selected Bear Creek drill cores (Fig. 1), with emphasis on the petrography and sedimentology of the Nonesuch Formation. Ten holes were measured, described, and sampled from the bottom of each hole in the Copper Harbor Formation to a short distance above the Nonesuch Formation–Freda Formation contact. To the best of my knowledge, this is the first comprehensive detailed sedimentologic and petrographic analysis of the Nonesuch Formation in the Bear Creek drill cores from northwestern Wisconsin. For the purpose of comparison with drill cores, six Nonesuch outcrops between Silver City, Michigan, and Mellen, Wisconsin (Fig. 1), were measured, described, and sampled.

SEDIMENTOLOGY

Outcrops

Sedimentary rocks of the Nonesuch Formation are dominantly light gray to black sandstones, siltstones, mudstones, and shales, with minor conglomeratic sandstones in the lower part of the section. The Nonesuch has a greater compositional and textural maturity than the underlying Copper Harbor red bed sequence. Grain sizes generally range from clay-sized material to medium-grained sand, with an average grain size of coarse silt (0.04 mm). Reddish brown rocks decrease in abundance upsection from the contact with the underlying Copper Harbor Formation. Conversely, there is a slight increase in reddish brown rocks as the contact with the overlying Freda Formation is approached. The Copper Harbor–Nonesuch contact was defined as the horizon where dark gray to black rocks dominate the section. The Nonesuch-Freda contact was defined as the horizon where red-brown beds again become dominant. The metamorphic grade in all three formations is zeolite to lower greenschist facies.

The Nonesuch Formation thins and coarsens to the west from 244 m (800 ft) at the Big Iron River to approximately 75 m (250 ft) at the Bad River (Copper Falls State Park). These two exposures are approximately 96 km (60 mi) apart. A typical section of Nonesuch Formation outcrop is illustrated in Figure 2.

The pattern of sedimentation, based on the outcrops examined, is dominated by numerous graded beds (Fig. 3) and fining-upward sequences (Fig. 4). Note that thin graded beds occur within the fining-upward sequences.

Paleocurrents

A total of 102 paleocurrent indicators were measured in the Nonesuch Formation at the Big Iron River, Presque Isle Park, and Parker Creek (Fig. 5). Each measurement was rotated to horizontal by rotation on an axis parallel to the strike of the bed. The data at the Big Iron River appear to support a general flow regime to the north-northeast, towards the rift zone axis, but to the west at the Presque Isle River the flow appears to be to the southwest parallel to the rift zone axis.

Drill Cores

The Nonesuch Formation in cores is similar to that in outcrops and is composed of unoxidized grayish black to brown sandstone, siltstone, and shale. Macroscopically, the coarse to fine-grained grayish black to brown sandstones are generally composed of quartz, feldspar, and lithic fragments in a micaceous or calcareous matrix/cement. The siltstones and shales are darker grayish green to black and are chloritic and micaceous. Calcium carbonate cement is generally confined to the sandstone and sandy siltstone beds. Minor reddish brown oxidized siltstone, mudstone, and shale are randomly distributed throughout the typically gray-black Nonesuch Formation, but increase markedly in abundance at the Copper Harbor–Nonesuch and Nonesuch-Freda contacts. Siltstone comprises the majority of the formation in these cores. Overall, the Nonesuch exhibits a general fining-upward trend.

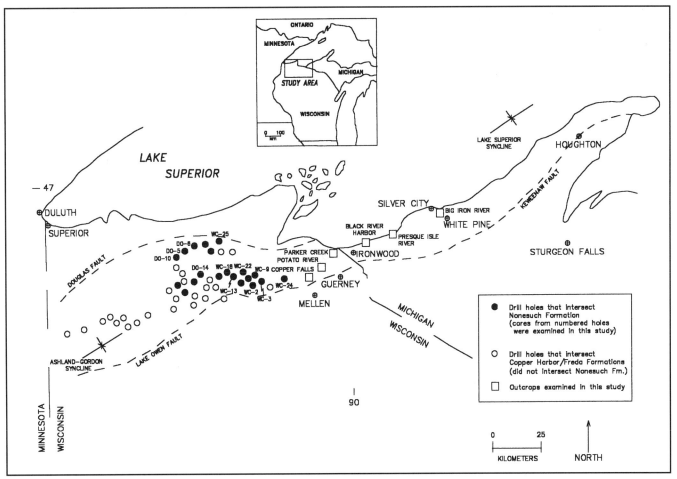

Figure 1. Location of Bear Creek drill holes and outcrops examined during this study.

The Nonesuch Formation cores were divided into six sedimentary intervals based on the occurrence of similar sedimentary structures and textural elements. The criteria used to develop these intervals were (1) significant changes in the percentages of coarse-grained and/or fine-grained sedimentary rocks; and (2) the occurrence of similar sedimentary structures, with particular emphasis on parallel-laminated beds, massive and graded beds, micro-trough cross-bedding, and carbonate laminite. All intervals are gradational and generally indicate gradual fluctuations in sedimentary processes. Figure 6 illustrates the sedimentational intervals in drill core WC-2.

PETROGRAPHY

Petrographic analysis of 27 thin sections from Nonesuch outcrop and drill core samples indicates that the average composition of the major framework clasts in the sandy and sandy-silt facies is: 36% volcanic rock fragments; 11.7% metamorphic, plutonic, and sedimentary rock fragments; 24% unit quartz grains; and 26.8% feldspar grains (predominantly plagioclase). Mafic volcanic rock fragments dominate in all of the outcrop samples. Mafic volcanic rock fragments dominate

in the cores examined from the southern and southeastern regions of the study area, whereas felsic volcanic fragments dominate in cores from the western and northwestern areas.

The Nonesuch is well indurated, but contains considerably less cement than the underlying Copper Harbor Formation. Cementing agents are, in order of abundance: calcite, quartz, and laumontite. Matrix materials are, in order of abundance: chlorite, epidote, sericitic clay, and silt-sized feldspar and quartz.

The sandstone samples from drill cores and outcrops indicate a general but subtle increase in mineralogical maturity from the upper Copper Harbor Formation (6 m thickness) through the Nonesuch to the lower Freda (25 m thickness). The majority of Nonesuch sandstone samples in drill cores and outcrops are lithic arenites, after Dott (1964; Fig. 7). Six samples, not included on the plot in Figure 7, contain greater than 15% micaceous matrix and are classified as lithic graywackes.

Mafic and intermediate volcanic rock fragments are most abundant in the cores along the southern part of the study area (cores WC-3, WC-2, WC-9, WC-13, WC-18, and WC-22) and decrease westward and northward. Felsic volcanic rock fragments increase to the west-northwest, as indicated by cores WC-25, DO-5, DO-6, DO-10 and DO-14 (see Fig. 1).

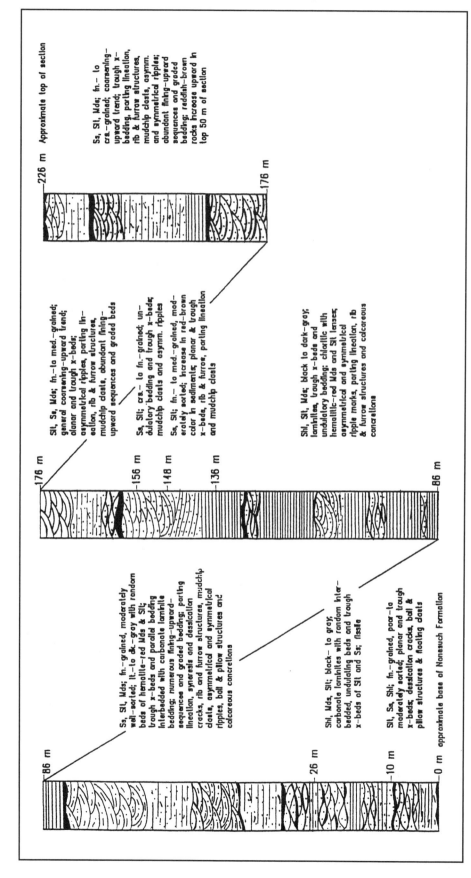

Figure 2. Diagrammatic illustration of the Nonesuch Formation as measured on the Big Iron River, Ontonagon County, Michigan. Numbers on the right of the columns are locations in meters from the base of the section. Abbreviations as follows: Ss, sandstone; Slt, siltstone; Mds, mudstone; Shl, shale; x, cross; fn., fine; crs., coarse; med., medium; dk., dark; lt., light.

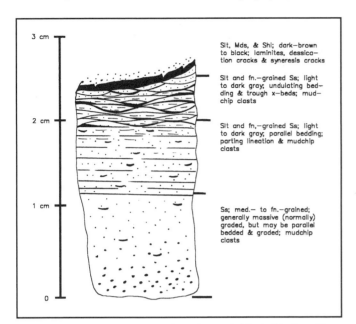

Figure 3. Diagrammatic illustration of a graded bed, Big Iron River measured section. This bed is 79 m above the base of the Nonesuch Formation. Abbreviations used are the same as in Figure 2.

Metasedimentary and plutonic rock fragments decrease to the north-northwest as indicated by cores WC-25, DO-6, and DO-10. Unstrained volcanic quartz shows a sharp increase to the northwest from WC-18 to WC-25 and appears to correlate with the felsic volcanic fragments that predominate in drill cores from this part of the study area. In the southern part of the core area, the highest percentages of strained quartz are in the cores that contain slate, schist, quartzite, iron-formation, and

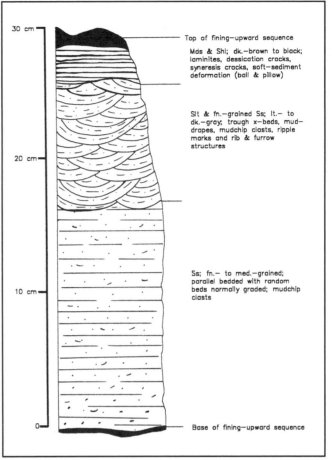

Figure 4. Diagrammatic illustration of a fining-upward-sequence, Big Iron River measured section. This sequence is 132 m above the base of the Nonesuch Formation. Abbreviations used are the same as in Figure 2.

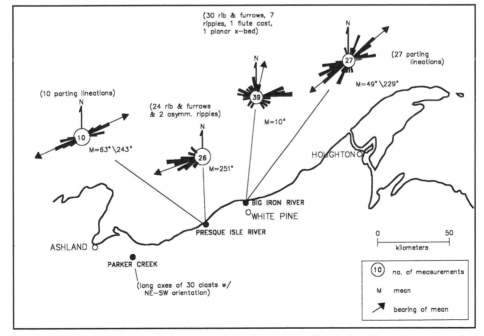

Figure 5. Paleocurrent indicators for the Nonesuch Formation at the Big Iron River measured section and the Presque Isle River exposure.

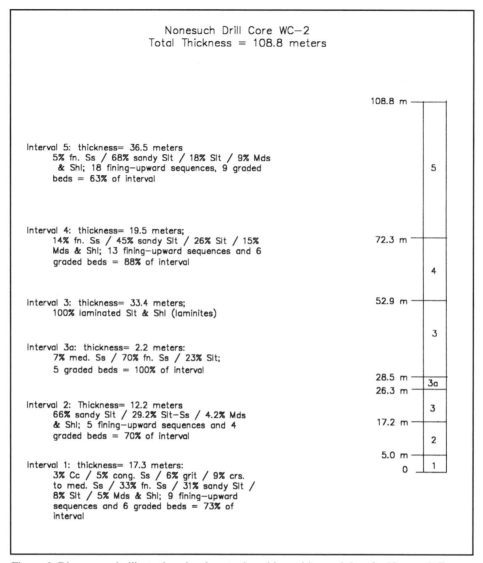

Figure 6. Diagrammatic illustration showing stratigraphic position and data for Nonesuch Formation sedimentational intervals in drill core WC-2. WC-2 is representative of cores exhibiting intervals 1 through 5. Abbreviations used are the same as in Figure 2.

plutonic rock fragments (WC-3, WC-2, WC-22, and WC-13). Multicycle calcite grains (reworked Copper Harbor Formation) and multicycle quartz grains occur in trace amounts in all cores.

Plagioclase shows a general decrease from east to west and from south to north, except in cores WC-3 and WC-13 where the highest percentages of plagioclase coincide with the highest percentages of intermediate volcanic rock fragments, suggesting derivation of the plagioclase was also from the intermediate volcanics. Potassium feldspar shows a slight decrease from east to west through the study area. The lateral variations in compositions of Nonesuch sandstones are illustrated in Figure 8.

An overall comparison of the lateral and vertical variations in composition in the Oronto Group in Bear Creek cores shows a general northward trend (WC-3 to WC-25) toward increased compositional maturity with an increase in felsic volcanic frag-

ments, unit strained quartz, and unit unstrained quartz, and a decrease in all other constituents. Compositional maturity increases upsection from a lithic (volcanic) arenite to a quartzo-feldspathic arenite. The same trend is observed in outcrops. Quartz, feldspars, and pre-Keweenawan metasedimentary and plutonic rock fragments generally increase upsection while volcanic rock fragments generally decrease.

ENVIRONMENT OF DEPOSITION

Recent investigations (Daniels, 1982; Elmore et al., 1988; Imbus et al., 1990; this study) suggest that the Nonesuch Formation represents deposition in a perennial lake located at the toe of a transgressing-regressing alluvial fan complex, within

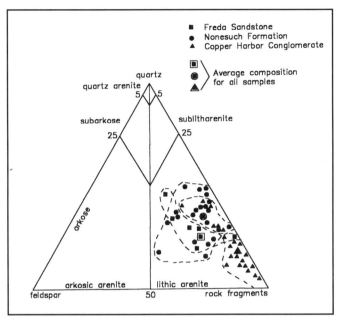

Figure 7. Compositions of Oronto Group rocks in Bear Creek drill cores of this study. A total of 47 thin sections provided the data; each symbol on the plot indicates the average of all samples from an individual core. Arenite classification after Dott (1964).

the Midcontinent Rift System. The sedimentational intervals observed in the Nonesuch drill cores contain vertical facies assemblages that record clastic deposition in a transgressing-regressing lacustrine environment. The principal evidence suggesting that the Nonesuch Formation was deposited in a lake as opposed to a marine environment is the close association of Nonesuch rocks with continental facies, especially the predom-

inantly red oxidized fluvial rocks of the Copper Harbor and Freda Formations, which have long been regarded as continental facies. The Copper Harbor and Freda Formations are gradational with and completely enclose the Nonesuch Formation (Elmore and Daniels, 1980; Daniels, 1982; Elmore and others, 1988; this study). It can also be demonstrated that the Nonesuch Formation is coeval in part with the upper Copper Harbor and the lower Freda Formations (Daniels, 1982; this study).

Figure 9 depicts the drill core study area as a local basin that is bounded on the north by the Douglas fault and on the south by the Lake Owen fault. The five sedimentational intervals observed in drill cores were stratigraphically superimposed in a model illustrating the progressive development of the upper Copper Harbor, Nonesuch, and lower Freda Formations as perceived from the integration of facies relationships observed in outcrops and drill cores.

Figure 9 illustrates the Copper Harbor Formation as a generally northward-prograding succession of middle to distal alluvial fan facies deposited in a subsiding basin. From south to north (basinward) the alluvial facies of the Copper Harbor consist of: (1) mid-fan facies, consisting of poorly sorted and crudely bedded clast-supported conglomerate and minor matrix-supported conglomerate interbedded with trough and planar cross-bedded pebbly sandstone; (2) distal-fan facies (sandflat facies), consisting of trough and planar cross-bedded pebbly sandstone to siltstone; and (3) distal-fan facies, central basin (mud-flat facies), consisting of small-scale trough cross-bedded and parallel laminated sandy siltstone, siltstone, and mudstone. The outcrops and drill cores record a gradual down-fan transition from deposition of poorly sorted, coarse alluvium, to deposition of finer-grained clastics by braided and meandering stream processes. Daniels and Elmore (1980) and Elmore

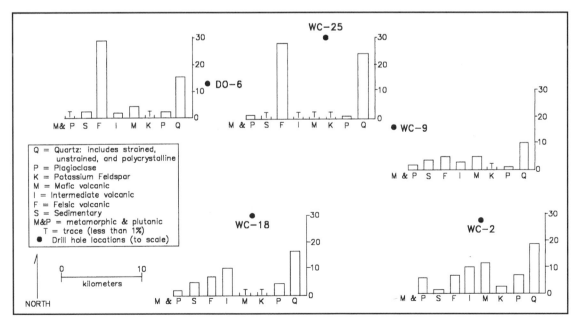

Figure 8. Lateral variations in the compositions of Nonesuch Formation sandstones from representative Bear Creek drill cores. (Units on the vertical axes indicate percent composition).

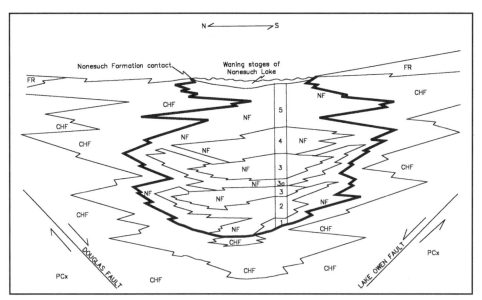

Figure 9. Diagrammatic illustration showing stratigraphic position and facies relationships for None-such Formation sedimentational intervals as inferred from drill hole WC-13. Similar inferences were made from drill cores WC-2, WC-18, WC-22, DO-6, and DO-14, which contain all six sedimentational intervals. FR, Freda Formation; NF, Nonesuch Formation; CHF, Copper Harbor Formation; PCx, Lower Proterozoic and Archean crystalline rocks. Not to scale.

(1984) interpreted these rocks, where exposed in outcrops in northwestern Wisconsin and Upper Michigan, to indicate a facies assemblage that represents deposition on a prograding alluvial fan.

The distal regions of the fan (sand flat to mud flat), contain abundant asymmetrical and symmetrical ripple marks and mud cracks, indicating active stream and beach processes, and periodic exposure and inundation of distal fan regions. Water levels fluctuated and were shallow enough to maintain oxidizing conditions, as indicated by the persistent reddish brown color of the Copper Harbor sediments.

The gradational nature of the Copper Harbor alluvial facies and the Nonesuch lacustrine facies (except in drill cores WC-2, WC-3, WC-22, Parker Creek, and Potato River Falls), suggests that the distal alluvial fan facies prograded directly into the None-such Lake, creating a fan-delta (Wescott and Ethridge, 1980). The fairly consistent gray-black color of Nonesuch detritus indicates that a subaqueous reducing environment was sustained in a perennial lake throughout deposition of the Nonesuch. Rapid, short-term fluctuations in water level probably due to brief climatic changes, suggest that the lake was probably confined to a closed or partially restricted basin. Periodic storm-generated sheetfloods and monsoon-type rainfalls within a closed basin, especially one devoid of vegetation on the alluvial slopes, could cause rapid fluctuations in water levels. This in turn would cause the rapid advance of the fine-grained lacustrine facies (i.e., dominantly reducing environment) onto the coarse-grained alluvial fan facies (i.e., subaerial oxidizing environment) as observed in cores WC-2, WC-3, WC-22, Parker Creek and Potato River Falls.

Drill core sedimentational interval 1

Sedimentational interval 1 consists of light gray massive, trough cross-bedded, and parallel laminated conglomeratic sandstones and coarse-grained sandstones (Fig. 6). The interval gradually fines upward to dark gray medium- to fine-grained sandstones, siltstones, and mudstones.

Figure 9 illustrates interval 1 as a fining-upward, shallow water marginal lacustrine facies deposited during the initial stages of development and growth of the Nonesuch Lake. This facies is gradational with the Copper Harbor distal alluvial fan, sand-flat to mud-flat facies. The sedimentary textures and structures are commonly similar across the Copper Harbor–Nonesuch contact zone in both drill cores and outcrops, and suggest that deposition was contemporaneous.

The most obvious physical change across the Copper Harbor–Nonesuch contact is that of a color change from reddish brown to gray and black. This generally coincides with a vertical facies change from coarse-grained shallow water sediments to fine-grained deeper water sediments. The color variation has been interpreted to represent a change in oxidation state due to a change in water chemistry within the environment of deposition (Daniels, 1982).

Sedimentational interval 1 consists predominantly of fining-upward sequences approximately 30 cm thick. Each sequence, from bottom to top consists of: (1) a basal, parallel laminated and commonly normally graded pebbly-sandstone to sandy-siltstone bed with mudchip rip-up clasts; (2) micro-trough cross-bedded fine-grained sandstones and siltstones; and

(3) a thin mudstone drape. Note that these sequences are similar to those in outcrops as illustrated in Figure 4.

According to Reading (1986, p. 11) "Where fining-upward sequences dominate an alluvial succession, interpretation is extremely difficult because both environmental switching and catastrophic flows can occur, each producing similar sedimentary sequences." The fining-upward sequences in the drill cores and in outcrops suggest that clastic deposition was predominantly by sporadic sheetfloods or by subaqueous fan-delta streams.

Sheetflood deposits are typically found on alluvial fans in semiarid climates (Bull, 1964; Denny, 1965; Hooke, 1967). The fining-upward sequences in interval 1 are interpreted to represent deposition by waning sheetfloods that originated high on the alluvial fan as short-lived storm events. Deposits of this nature occur where sediment carried both as bedload and in suspension in a high velocity channel flow, spreads out below the channel (Bull, 1964). The shallow sheet flows seldom persist far onto the alluvial sand flat, largely due to infiltration of the water (Rahn, 1967). This suggests that flooding events that deposited detritus as fining-upward sequences well beyond the toe of the alluvial fan into the Nonesuch depositional basin were probably of considerable magnitude and lateral extent. On an alluvial fan surface devoid of vegetation, even a moderate rainfall would probably have resulted in a sediment-laden sheetflood of considerable depth and velocity. This appears to be the case as indicated by the frequency and lateral extent of fining-upward intervals throughout all the Nonesuch Formation in drill cores and outcrops.

A sediment-laden sheet of water, upon entering the lake at the toe of the alluvial fan (i.e., fan-delta), would become a density current. The initial deposit from this current would be the basal parallel laminated sand and silt (±mica) with reddish brown mudchip clasts, from detritus that was transported as bedload (Collinson, 1978). Where the basal parallel laminations of each sequence are composed of fine-grained sand and reasonably free of mica, high-velocity currents and shallow water depth are indicated; where very micaceous sands occur, lower flow conditions may have applied (Collinson and Thompson, 1982, p. 97). With further velocity decrease, fine sand and silt would be deposited as small-scale cross-beds (rib-and-furrow structures) as current ripples migrated across the surface (Collinson and Thompson, 1982, p. 98). Asymmetrical ripple marks are present on numerous bedding surfaces. The current-generated ripples that were measured in the lower Nonesuch Formation at the Big Iron River and Presque Isle River sections have an average ripple index of 9, indicating that the micro-trough cross-beds were deposited by medium- to low-velocity currents (Reinick and Singh, 1980, p. 29). The mud drapes that top fining-upward sequences were deposited by suspension settling of mud dispersed in the water column.

Sharp contacts are common between the mudstone top of one fining-upward sequence and the overlying sandstone base of another. This indicates that deposition was sporadic, yet quite common, as might be expected with sheetflooding. The lack of gradation at sequence contacts suggests that extrinsic factors such as climate may have partially controlled the rate and style of deposition.

Randomly interbedded with the fining-upward sequences are ≤1-m-thick sandstone beds that are massive, ungraded, or normally graded, and very coarse to fine grained. The occurrence of these sandstone beds suggest deposition by turbidity currents that carried an abundance of coarse-grained detritus; they may well be coarse-grained channel sands deposited by turbidity currents in subaqueous channels that represent a lakeward extension of fan-delta fluvial processes.

Reddish brown oxidized sediments are interbedded with gray-black sediments throughout interval 1 and decrease in abundance upsection. The occurrence of red, oxidized sediments interbedded with gray to black sediments that were deposited in a reducing environment (Elmore et al., 1988) is interpreted as an interfingering relationship of the Copper Harbor and Nonesuch Formations. Oxidized sediments, when introduced into a reducing environment, can be preserved as such if they are rapidly deposited and buried (Ehrlich and Vogel, 1971).

Asymmetrical and symmetrical ripple marks, mud cracks, and synersis cracks indicate fluctuating water levels, variations in flow velocity and direction, and periodic subaerial exposure.

Drill core sedimentational interval 2

Sedimentational interval 2 consists of gray, massive, micro-trough cross-bedded and parallel-laminated, medium- to fine-grained sandstone, sandy siltstone, siltstone, and mudstone that record the continued expansion and deepening of the Nonesuch Lake. The vertical facies assemblage shows a gradual transition from shallow-water marginal lacustrine to deep-water marginal lacustrine, and represents a continuation of the fining-upward, basinward-fining trend observed in interval 1. An upsection increase in the percentage of fine-grained sandstone, siltstone, and mudstone, and the absence of pebbly sandstone and coarse-grained sandstone in drill cores, were the main criteria for separating out interval 2.

Fining-upward sequences continue to dominate the sedimentary pattern, but are not as frequent (Fig. 6). The number of massive and normally graded, medium-grained sandstone to sandy-siltstone beds increase upsection and basinward from interval 1. This suggests that as the lake became deeper, turbidity currents played a greater role in the depositional processes.

Interbedded reddish brown sandstones, siltstones, and mudstones decrease upsection. Near the top of the interval, carbonate laminite beds are sparsely interbedded with massive, normally graded, and micro-trough cross-bedded sandstones and siltstones. This is interpreted to represent a gradual basinward shift from current-generated deposits (sheetflood and turbidity currents), to deposition by suspension settling of mud and $CaCo_3$ from the water column in deeper and quieter areas of the basin.

The carbonate laminite consists of thin (≤1 mm), light gray, carbonate-rich fine-grained sandstones and siltstones and thinner (<1 mm), black to green, organic-rich mudstone/shale beds. Thicker sandstone-siltstone laminae commonly display normal grading or microtrough cross-bedding. Individual laminae commonly show evidence of loading and mixing. Sand and silt detritus was often rapidly deposited on top of soft mud laminae, as indicated by the occurrence of flame structures, load casts (ball and pillow variety), and the swirled appearance of mixed sandstone, siltstone, and mudstone. The black, organic-rich shale laminae were deposited by suspension settling of mud-sized detritus that was dispersed in the water column by the turbidity currents, and by the normal suspension settling of very fine grained detritus and organic material from higher up in the water column where density overflows and interflows may have occurred (Fig. 10).

Drill core sedimentational intervals 3 and 3a

Sedimentational interval 3 consists of gray micro-trough cross-bedded fine-grained sandstone, siltstone, mudstone, and carbonate laminite (Fig. 6). The Nonesuch facies in this interval records a transition from deep marginal lacustrine facies to central lake basin facies, and is interpreted to represent clastic deposition when the lake reached its highest water level. The facies transition from interval 2 to interval 3 is gradational. Percentages of siltstone and mudstone increase and become progressively darker and more organic-rich upsection (Elmore et al., 1988).

The dominant lithology is carbonate laminite. Interbedded with the laminite beds are what appear to be turbidity current deposits consisting of massive, graded, parallel-laminated, convolute bedded, and microtrough cross-bedded, medium- to fine-grained sandstone and siltstone beds. Turbidity currents generated by storms or initiated by gravity slumping along the steeper slopes flanking the central lake basin, continued to deposit medium-to fine-grained sandstone beds well into the lake basin. The turbidites in interval 3 are thinner and less frequent than the turbidites observed in interval 2.

The laminite is correlative in all drill cores. The question is posed as to what conditions were responsible for the widespread and fairly even distribution of carbonate silt and clayey mud detritus to form extensive carbonate laminite. A thermally stratified lake (Fig. 10) would have provided an effective means of sorting and distributing fine-grained detritus in the water column throughout the basin. Elmore (1983, 1984) and Elmore et al. (1988) suggested that the production and preservation of organics could have been best accomplished by the presence of a fluctuating thermocline in a stratified Nonesuch Lake. The thermocline is the layer of water characterized by a rapid decrease in temperature and an increase in density with depth that separates the overlying epilimnion from the underlying hypolimnion (Sturm and Matter, 1978). A thermocline, when encountered by turbidity currents, could have diverted a large amount of the finer-grained detritus to surface currents. A portion of the heavier detritus probably continued to move along the lake bottom whereas more of the lighter detritus was removed and distributed to interflows in the epilimnion. The epilimnion is the uppermost layer of water in a stratified lake, characterized by an essentially uniform temperature that is generally warmer than anywhere else in the lake (Sturm and Matter, 1978). Once in the epilimnion, pelagic sediments could have been distributed over a large area of the basin, and deposited over the deeper and quieter areas of the central lake bottom by suspension settling.

Mud-size detritus may have settled out of the epilimnion continually during the course of the year as dictated by climate

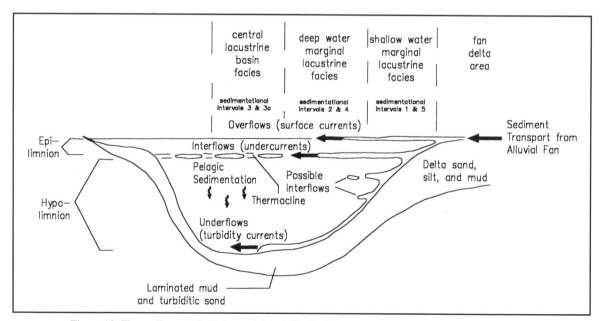

Figure 10. Thermally stratified lake model and distribution mechanisms proposed for the Nonesuch Formation in the Bear Creek drill cores (modified after Sturm and Matter, 1978).

and seasonal lake turnover. A nonseasonal lake turnover could have occurred if the epilimnion periodically became saturated with detritus due to an increase in the rate of sediment influx. The result would have been a decrease in the density variation between epilimnion and hypolimnion (the layer of cold, dense water at the bottom of the lake below the thermocline), which could have caused the lake to turn over and dump more mud and silt than normal into lower depths of the lake. Evidence for this is weak, but is suggested by the variation in thickness of mudstone laminae (0.1 to 0.75 mm) and siltstone laminae (0.5 to 1.0 mm).

The heavier, coarse-grained sediment may have been carried below the thermocline and distributed to interflow currents in the hypolimnion or deposited directly on the basin floor as turbidity flow deposits (Fig. 10). Microscopic trough cross-bedding randomly occurs in the coarser-grained carbonate-rich laminae, as seen in drill core WC-9 at 72 m (235 ft) above the base of the Nonesuch Formation. This suggests that low-velocity bottom currents were depositing and perhaps reworking detritus in the deepest areas of the lake basin.

The settling out of mud put into suspension by density currents and turbidity current flows along the lake bottom, and organic detritus derived from the algae that were produced during the warmer seasons (Elmore et al., 1988), accounts for most of the constituents in the black mud-rich laminae.

The origin of the carbonate in the light gray siltstone laminae is partially due to biogenic processes as suggested by the presence of algal-like filaments and an amorphous organic groundmass within the clastic laminae (Kelts and Hsu, 1978; Elmore et al., 1988; Imbus et al., 1990). In addition to biogenic carbonate, numerous well-rounded silt-sized limestone grains, consisting of fine-grained sparry calcite, were observed within the carbonate-rich laminae. The presence of these grains suggests that they may have been derived from subaerial evaporites or caliche, eroded by sheetfloods from nearby (Copper Harbor) alluvial fan sand-flat and mud-flat areas, reworked, and redeposited with silt-sized grains of quartz and feldspar by turbidity flows and by suspension settling. However, the presence of stromatolites in the Copper Harbor on the Keweenaw Peninsula (Daniels and Elmore, 1980; Elmore, 1983), and the presence of oncolites in the Copper Harbor (Suszek, 1991) shows that these are another possible source of carbonate.

Sedimentational interval 3a (Fig. 6) consists of massive, normally graded, and small-scale trough cross-bedded, coarse- to fine-grained sandstone and siltstone with only minor interbedded carbonate laminite near the bottom and top of the interval. Interval 3a occurs as a distinctly different facies stratigraphically enclosed by interval 3 (Fig. 9).

The facies assemblage in interval 3a records a temporary lake regression from deep-water, central lake facies to deep-water, marginal lake facies. Coarse-grained sandstones and siltstones increase near the top of interval 3, and prevail throughout most of interval 3a. The abundance of coarse-grained detritus rapidly decreases near the top of interval 3a as contact with upper portion of interval 3 is approached.

The sedimentary structures in interval 3a suggest that deposition was predominantly by turbidity and density flows. Graded beds range from 2 cm to 0.5 m in thickness. Fining-upward sequences deposited by sheetfloods are rare. This suggests that interval 3a represents a facies transition between deep-water marginal lacustrine and deep-water central lake, even though the texture of the rocks suggests deposition in areas of the lake basin close to the alluvial fan complex. The change in facies from 3 to 3a appears to have been short lived but widespread, and can be correlated in drill cores throughout the entire basin. Such a rapid facies change was probably brought on by a sudden, but brief, change in tectonics or climate. If the change was regional, one would expect to see a more gradual and stratigraphically thicker facies transition as was observed in intervals 1, 2 and 3.

Interval 3a fines upward where it is in gradational contact with the upper part of interval 3. The segment of interval 3 above interval 3a is again dominated by deep-water, central lake carbonate laminite that gradually coarsens upward to deep-water, marginal lacustrine facies.

Drill core sedimentational intervals 4 and 5

Sedimentational intervals 4 and 5 consist of gray, small-scale trough cross-bedded, massive and graded, coarse- to fine-grained pebbly sandstone, siltstone, and mudstone (Fig 6). These intervals record a gradational, coarsening-upward, and shallowing-upward facies trend as the Nonesuch drill core basin began to fill with detritus and the lake regressed (Fig. 9).

Interval 4 shows a decrease in carbonate laminite upsection. Medium- to fine-grained sandstones, siltstones, and mudstones occur as fining-upward sequences and turbidity flow deposits. Sheetflood deposits (fining-upward sequences) become more frequent upsection. Interbedded red oxidized sediments begin to occur again, and increase in abundance upsection.

Interval 5 records the final stages of lake regression. Gray, pebbly, coarse-grained, trough cross-bedded sandstones and fining-upward sequences increase upsection. The upper facies of interval 5 indicates a gradational transition from shallow-water marginal lacustrine to fluvial flood-plain deposits of the lower Freda Formation. The upper contact of interval 5 establishes the Nonesuch Formation–Freda Formation contact in drill cores WC-2, WC-3, WC-9, WC-13, WC-18, WC-22, and DO-14. Intervals 4 and 5 are missing in drill core WC-25, and interval 5 is missing in cores DO-5 and DO-6.

PROVENANCE AND BASIN ANALYSIS

Based on petrographic analysis, the most likely major source for Oronto Group sediments would have been Keweenawan igneous rocks within the Midcontinent Rift System. Heavy mineral suites show a relative increase in contributions from older Proterozoic and Archean rocks (i.e., extrabasinal rocks) at stratigraphically higher intervals in the

Oronto Group. These older metamorphic and plutonic rocks became important sources as streams reached farther afield as the rift was becoming filled, and as the Keweenawan volcanic rocks were reduced in quantity on the sides of the rift.

Hubbard (1972) suggested that lower Keweenawan and older rocks provided much, if not all, of the detritus for the Keweenawan sedimentary rocks in the Upper Peninsula of Michigan and in Wisconsin. He demonstrated structurally and petrographically that lower Keweenawan rocks were metamorphosed, uplifted, and eroded during later Keweenawan time. During that time, middle Keweenawan volcanic and sedimentary rocks were largely protected from erosion. Hubbard's (1972) conclusions are based on (1) the general lack of volcanic rock fragments with ophitic textures that are indicative of mafic rock types in the middle Keweenawan Portage Lake Volcanics, and (2) the predominance of volcanic rock fragments that have been metamorphosed to upper greenschist facies and are indicative of sedimentary and igneous rocks that occur in the lower Keweenawan Powder Mill Group, the oldest of the Keweenawan volcanic units.

The conglomerate and sandstone facies in drill cores contain minor fragments of granophyre and granite, which suggests that the Keweenawan Mellen Intrusive Complex and related intrusives to the southeast of the drill core locations were being unroofed during that time. Minor iron-formation, jasper, slate, and quartz-mica schist rock fragments indicate that Early Proterozoic Animikie Group and Archean rocks to the south were also being eroded.

The mafic, intermediate, and felsic volcanic rock fragments bear a strong resemblance to basalt, andesite, rhyolite, and quartz latite flows described by Hubbard (1975). The groundmass in the majority of the mafic and intermediate fragments has been extensively altered to epidote and chlorite, and feldspar phenocrysts are seriticized; these fragments are probably from the lower Keweenawan Siemens Creek Formation.

The felsic volcanic fragments exhibit distinctive snowflake textures and platy quartz textures that are commonly observed in the quartz latites and rhyolites of the lower Keweenawan Kallander Creek Formation (Hubbard, 1975), and the North Shore Volcanic Group (John C. Green, personal communication, 1992). Ophitic textures in basaltic fragments that would suggest the middle Keweenawan Portage Lake Volcanics as a possible source terrane were not observed.

Higher percentages of mafic to intermediate volcanic rock fragments in cores from the eastern and southern areas of the drill core study area suggest the presence of a mafic to intermediate volcanic source to the south and southeast (Fig. 11). Recent field investigations by Cannon et al. (1993), suggest that a lower Keweenawan broad central volcano centered near Mellen, Wisconsin (see Kallander Creek Volcanic Center, Fig. 11), affected the distribution and lithology of younger rocks.

The general increase in unit strained and unstrained volcanic quartz and a sharp increase in felsic volcanic rock fragments to the north and northwest (cores WC-25, DO-14, DO-10,

and DO-6), indicate that a felsic volcanic source probably existed in the area with the same composition as the felsic volcanic detritus observed directly below the lower Nonesuch Formation contact in core WC-25. The upsection increase in detritus from Early Proterozoic and Archean quartz-bearing crystalline rocks on the flanks of the rift was probably the major cause of the upsection increase in mineralogical maturity.

Analysis of paleocurrent structures in the Oronto Group from the Big Iron and Presque Isle Rivers obtained during this study (see Fig. 5), and from the work of previous investigators (White and Wright, 1960; Hamblin and Horner, 1961; Hite, 1968; Daniels, 1982), confirms that the transportation of detritus was generally towards the north, northeast, and northwest. However, paleocurrent structures show a change in current direction, from northeastward near the base of the Nonesuch Formation to southwestward near the top of the formation, at the Big Iron River and the Presque Isle River exposures.

Hite (1968) also noted that in the lower Freda Formation at the Bad River exposure (Copper Falls on Fig. 1), paleocurrent directions were to the southwest. Northeast of the Bad River, at the Potato River, the paleocurrents in the lower Freda Formation are toward the south and northeast (Hite, 1968). This suggests that an elevated highland existed between the Potato River and the Bad River sections, and diverted currents to the west and east. Local reversals in paleocurrent direction, according to Daniels (1982), may also be due to stream meandering, diversion by lava flows or alluvial fan lobes, or reversal of the slope of the basin floor.

Hite (1968) recorded an upsection increase in basaltic fragments in the upper Copper Harbor Formation exposed along the Bad River. Basaltic fragments also account for the largest detrital constituent in upper Copper Harbor Formation in cores WC-24, WC-2, and WC-3, which are the Bear Creek drill holes closest to the Keweenawan (Kallander Creek) central volcano of Cannon et al. (1993), and what Hite (1968) suggested was an elevated mafic volcanic highland in the vicinity of the Bad River. Note that the presence of a highland in this area is also indicated by paleocurrent data.

Drill hole WC-24, located at the far eastern edge of the drill core study area, intersected approximately 30 m (100 ft) of predominantly clast-supported basalt-rich conglomerate of the Copper Harbor Formation, directly below the Pleistocene glacial overburden. The absence of Nonesuch and Freda sedimentary rocks in this core could be the result of faulting and erosion. However, the proximity of WC-24 to the proposed volcanic center/highland of Hite (1968) and Cannon et al. (1993), suggests that these conglomerates were located on the upper slopes of a highland that was topographically high enough to be isolated from the Nonesuch and Freda sedimentary environments.

The Nonesuch Formation thins westward from the Big Iron River to the Bad River at Copper Falls State Park, and becomes thicker again westward from drill hole WC-24 along the southern part of the Nonesuch drill core basin in cores WC-3, WC-2, WC-9, WC-22, WC-13, and WC-18 (Fig. 11). These data also suggest

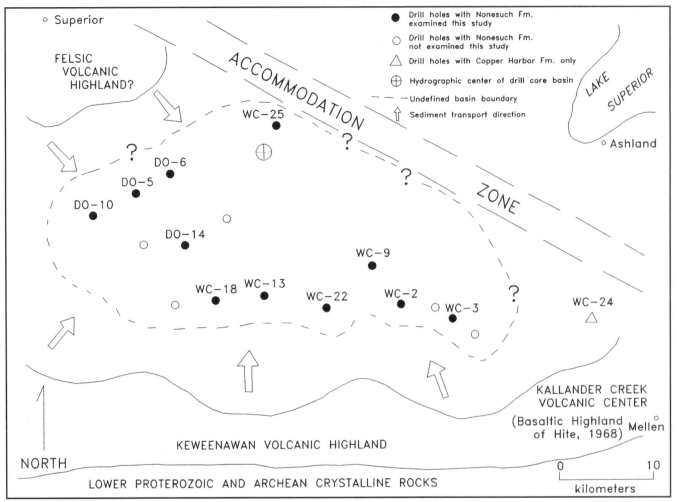

Figure 11. Generalized map of the drill core study area showing the Nonesuch depositional basin and source areas. (Proposed accommodation zone from Mudrey and Dickas, 1988).

that an elevated highland in the eastern part of the drill core study area may have separated or partially separated the study basin from a basin or basins containing Oronto Group sediments farther east in northwestern Wisconsin and Upper Michigan.

Drill holes WC-2, WC-9, and WC-22, along the southern perimeter of the Nonesuch drill core basin, indicate that the red-brown conglomeratic facies of the Copper Harbor Formation directly underlies gray-brown sandstones, siltstones, and mudstones of the overlying Nonesuch Formation. The Copper Harbor–Nonesuch contacts in these drill cores are similar to those observed in exposures to the east of the core study area at Parker Creek and Potato River Falls (see Fig. 1). Elmore et al. (1988) suggested that the abrupt transition from red-brown clast-supported conglomerates to fine-grained sandstones and siltstones is due to the rapid encroachment of the Nonesuch sediments onto a steeply dipping alluvial fan surface, which prohibited the development of a gradual proximal to distal subaerial fan facies–subaqueous fan-delta transition.

The facies changes at the Copper Harbor–Nonesuch con-

tacts become texturally less abrupt northward and westward from drill holes WC-2, WC-3, and WC-22. This indicates a gradual northward and westward (basinward) facies transition from conglomerate to sandstone to siltstone and mudstone.

Drill core sedimentological intervals 4 and 5 (dark gray, coarse- to fine-grained sandstones, siltstones, and mudstones as fining-upward sequences, and massive and graded beds), which overlie interval 3 (dark gray to black carbonate laminite), are stratigraphically thicker in the cores to the south (WC-3, WC-2, WC-9, WC-22, WC-13, WC-18, and DO-14), and totally absent or total less than 1 m thick in the northernmost cores WC-25 and DO-6 (see Fig. 1). This suggests that the sediments in intervals 4 and 5 were deposited in a deep- to shallow-water euxinic marginal lacustrine environment basinward (northward) of a prograding fluvial-deltaic environment.

The upper Copper Harbor Formation in drill cores WC-25, DO-6, and DO-10 contains grit and coarse-grained sandstones consisting predominantly of felsic volcanic rock fragments and volcanic quartz. Sedimentational interval 2 (gray fine-grained sandstone, siltstone, and mudstone) in these cores is in direct contact with this facies of the Copper Harbor Formation, which appears to be detritus derived mainly from a regolith that formed on a felsic volcanic feature, such as a lava flow or dome. This indicates the presence of an elevated felsic volcanic feature well within the margins of the rift zone and on the north-northwest side of the proposed Nonesuch drill core basin (see Fig. 11). Pettijohn (1957, p. 630) and Daniels (1982) indicated that the Nonesuch environment was probably initiated by tectonic disruption, damming by lava flows, or prograding alluvial fan lobes of the existing drainages. The petrography of drill cores WC-25, DO-6, and DO-10 support the suggestion that the Nonesuch lacustrine environment was in part formed by a felsic volcanic feature in the northwestern part of the drill core study area that restricted or totally disrupted regional drainage within the rift.

The only occurrence of an evaporite bed, 2.75 m of pinkish gray, massive crystalline calcite with anhydrite nodules ≤2 cm in diameter, was observed at the top of the Nonesuch Formation in drill core WC-25 near the contact with the reddish brown siltstone and mudstone of the Freda Formation (Elmore et al., 1988; this study). Carbonate laminite beds occur just below the calcite and anhydrite. The close relationship of carbonate laminite to evaporite to mudstone and siltstone facies suggests that this was topographically the lowest area in the basin where saline brines were concentrated in shallow waters as the Nonesuch Lake was drying up.

TECTONIC MODEL

Sedimentation of the Nonesuch Formation in the Bear Creek drill hole study area appears to have occurred in an east-west-trending, fault-bounded basin (see Fig. 9). The Douglas fault on the north and the Lake Owen fault on the south bounded the depositional basin in the study area within the Ashland syncline. Reverse movement along these faults sometime after Oronto Group deposition created the St. Croix Horst.

Recent geologic mapping southeast of the drill core study area near Mellen, Wisconsin (Cannon et al., 1993), data compiled on the Oronto Group exposures in northwestern Wisconsin (Hite, 1968), and the data examined during this study place an elevated volcanic highland south and east of drill hole WC-24 (see Fig. 11). The recent interpretation of seismic reflection and three-dimensional gravity data of the Lake Superior syncline (Allen et al., 1993), and the interpretation of gravity, magnetic, and geologic data by White (1966), indicate the presence of a prominent pre-Keweenawan ridge (White's Ridge) that crosses northern Wisconsin and southwest tip of Lake Superior north-northeast of the drill core study area. These data, along with information compiled from the Bear Creek drill cores, indicate that the Nonesuch drill core study area in northwestern Wisconsin (see Figs. 1 and 11) was bounded by uplands on the south, east, and north, and was either completely separate or partially isolated from the basin containing Oronto Group sedimentary rocks to the east in Wisconsin and Upper Michigan.

The presence of a marine environment of deposition, as opposed to an intracratonic rift-zone lake, during sedimentation of the Nonesuch Formation remains a possibility. Hieshima and Pratt (1991) indicate that sulfur/carbon ratios in the Nonesuch Formation provide evidence that suggest the Nonesuch depositional environment was a marine embayment. However, it appears that the nearest documented marine environment to the drill core study area and the areas of Nonesuch Formation exposure in northwestern Wisconsin and Upper Michigan during the general time of Nonesuch sedimentation, was located approximately 1,000 km (625 mi) from the west margin of the Grenville Front (Bickford et al., 1986), which presumably was accreted to North America about or shortly after the rift formed. The occurrence of a 1,000-km-long arm of the sea from the Grenville Front to the western Lake Superior region along the Midcontinent Rift System is certainly not impossible, but seems highly unlikely given the distance and the possibility that the Nonesuch may have been deposited in two separate basins.

The Brule basin, one of the proposed series of half-grabens along the Midcontinent Rift System (Mudrey and Dickas, 1988), coincides with the drill core study area (Fig. 12). The basin sloped to the south with a major fault south of the study area and a flexure to the north. An isopach map of the Nonesuch Formation in the Bear Creek drill cores shows that the Nonesuch is thicker in the southern part of the drill core study area. This places the Nonesuch depocenter approximately where the proposed deepest part of the Brule basin existed and is in general agreement with Mudrey and Dickas (1988). However, this appears to disagree with the facies relationships in the Nonesuch sedimentational intervals which show that sediments were generally deposited in a basin that sloped to the north toward a basin center in the vicinity of drill hole WC-25 where minor evaporites were deposited (Fig. 11). This contradiction can be explained if the rate of sediment influx exceeded the rate of basin subsidence along the southern boundary of the drill core study area during the time that Oronto Group sediments were being deposited. This places the deepest areas of the basin (not the thickest sequence) to the north and west of the drill holes that define the southern part of the basin (WC-3, WC-2, WC-9, WC-13, WC-18, WC-22). Thick accumulations of poorly sorted and crudely bedded Copper Harbor conglomerates derived from a source terrane south-southeast of the core study area indicate that deposition was rapid along what were probably steep, fault-bounded, southern and eastern boundaries of the Nonesuch drill core basin.

SUMMARY AND CONCLUSIONS

In general, the Nonesuch Formation in drill cores from northwestern Wisconsin is lithologically and petrographically similar to the Nonesuch in outcrops to the east in Wisconsin and Upper Michigan. Deposition occurred in a subaqueous environment. Although the marine versus lacustrine enigma may not be resolved, the close spatial and temporal relationship of the Nonesuch to the Copper Harbor Formation alluvial fan complex and the Freda Formation fluvial flood plain suggests that the Nonesuch was deposited in a lacustrine environment within the rift

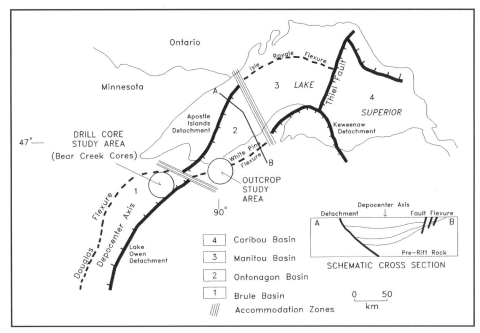

Figure 12. Interpretation of the western Lake Superior structure along the Midcontinent Rift System (modified from Mudrey and Dickas, 1988).

zone. Nonesuch deposition was gradational and diachronous with the upper Copper Harbor and lower Freda Formations.

The Bear Creek drill cores from northwestern Wisconsin contain a variety of lithologies and textures. Detailed sedimentologic and petrographic analysis of the cores determined that:

1. Provenance for the majority of Nonesuch detritus was the lower Keweenawan Powder Mill Group to the southeast of the drill core study area in northwestern Wisconsin. Pre-Keweenawan crystalline rocks and reworked upper Copper Harbor Formation (i.e., clasts of brown mudstone, multicycle calcite grains, and oncolites) were minor sources of detritus. Volcanic rock fragments constitute about 42% of framework grains in the coarse sandstones. The remaining framework components consist of grains of plagioclase, potassium feldspar, unit strained quartz, unstrained volcanic quartz, multicycle calcite grains, and multicycle quartz grains, granophyre and granite from lower Keweenawan sedimentary, volcanic and intrusive rocks, and slate, schist, quartzite, and iron-formation from Early Proterozoic and Archean crystalline rocks.

2. The lateral changes in the composition of Nonesuch detrital fragments in the core study area varies from predominantly mafic volcanic in the east to felsic volcanic in the west. Plagioclase and potassium-feldspars show no significant lateral variations. Metamorphic and intrusive fragments and unit quartz grains increase slightly upsection. The majority of the core samples are mineralogically immature and are classified as lithic arenites (<10% matrix).

3. Sediment transport was predominantly from south to north from an elevated highland south of the drill hole locations along the southern margin of the Midcontinent Rift System, and along a northeast-southwest trend approximately parallel to the rift zone axis. Some detritus was also derived from an elevated Keweenawan felsic volcanic terrane within the rift zone, north and west of drill hole WC-25. Sediment from this source area was probably transported to the south and east within the core study area.

4. Six sedimentational intervals in the Nonesuch Formation in the Bear Creek drill cores were established as follows: interval 1, dominantly shallow-water marginal lacustrine facies; interval 2, dominantly deep-water marginal lacustrine facies; interval 3, dominantly deep-water central lake basin facies; interval 3a, dominantly deep-water central lake basin or deep-water marginal lacustrine facies; interval 4, dominantly deep-water marginal lacustrine facies; and interval 5, dominantly shallow-water marginal lacustrine facies.

5. Deposits of the Nonesuch sedimentational intervals probably originated as ephemeral sheetfloods and subaerial fluvial-deltaic streams on Copper Harbor fans that developed into underflow turbidity currents, interflow density currents, and bottom currents in the Nonesuch basin. Finer-grained sediment was a product of suspension settling.

6. The environment of deposition for the Nonesuch Formation in cores was a fluctuating, thermally stratified, perennial lake located on the prograding Copper Harbor alluvial fan complex in a fault-bounded, subsiding basin within the Midcontinent Rift zone. Sedimentological data suggest that the Nonesuch in the core study area may have been deposited in a basin that was separate or partially restricted from Nonesuch depositional basins farther east in northwestern Wisconsin and Upper Michigan.

ACKNOWLEDGMENTS

The author would like to thank Dr. Richard W. Ojakangas, Dr. John C. Green, and Dr. Charles L. Matsch (University of Minnesota-Duluth) for their valuable advice, field and laboratory assistance, and encouragement during the course of this study. I am very thankful to Mr. Peter Jongewaard (Duluth, Minnesota) for providing his invaluable computer drafting skills and technical assistance. I would like to thank the reviewers, Mr. Paul Daniels and Dr. Doug Elmore, and the editors for their comments and criticisms. A special note of thanks to Mr. Russell Babcock (Kennicott Copper Corp.) for providing me with copies of the Bear Creek diamond drill core logs, and Mr. Tom Evans of the Wisconsin Geologic and Natural History Survey for permission to use the WGNHS drill core repository.

My sincere gratitude is extended to Mr. Gregg Anderson and Ms. Diana Friedhoff Miller (Amoco Production Company, Houston, Texas). Their valuable field and laboratory assistance, and generous financial support made this study possible.

REFERENCES CITED

Allen, D. J., Hinze, W. J., Dickas, A. B., and Mudrey, M. G., 1993, An integrated seismic reflection/three dimensional gravity investigation of the Midcontinent Rift System, USA: Western Lake Superior [abs.]: American Association of Petroleum Geologists Hedberg Research Conference on Basement and Basins of Eastern North America, Ann Arbor, Michigan, p.7–8.

Bickford, M. E., Van Schmus, W. R., and Zietz, I., 1986, Proterozoic history of the Midcontinent region of North America: Geology, v. 14, p.492–496.

Bull, W. B., 1964, Geomorphology of segmented alluvial fans in western Fresno County, California: U.S. Geologic Survey Professional Paper 352-E, p.89–129.

Cannon, W. F., Davis, D. W., Nicholson, S. W., Peterman, Z. E., and Zartman, R. E., 1993, The Kallander Creek Volcanics—A remnant of a Keweenawan central volcano centered near Mellen, Wisconsin [abs.]: Annual Institute on Lake Superior Geology, 39th, Eveleth, Minnesota, v. 39, p. 20–21.

Collinson, J. D., 1978, Vertical sequence and sand body shape in alluvial sequences, in Miall, A. D., ed., Fluvial sedimentology: Memoirs of the Canadian Society of Petroleum Ecologists, v. 5, p. 577–586.

Collinson, J. D., and Thompson, D. B., 1982, Sedimentary structures: London, Allen and Unwin, p. 97–98.

Daniels, P. A., Jr., 1982, Upper Precambrian sedimentary rocks: Oronto Group, Michigan-Wisconsin, in Wold, R. J., and Hinze, W. J., eds., Geology and tectonics of the Lake Superior basin: Geological Society of America Memoir 156, p. 107–133.

Daniels, P. A., and Elmore, R. D., 1980, Depositional setting of stromatolite-oolite facies on a Keweenaw alluvial fan [abs.]: Annual Institute on Lake Superior Geology, 26th, Eau Claire, Wisconsin, p. 27.

Denny, C. S., 1965, Alluvial fans in the Death Valley region, California and Nevada: U.S. Geological Survey Professional Paper 466, 62 p.

Dott, R. H., 1964, Wacke, graywacke and matrix—What approach to immature sandstone classification?: Journal of Sedimentary Petrology, v. 34, no. 3, p. 625–632.

Ehrlich, R., and Vogel, T. A., 1971, Depositional and diagenetic models for the lower Nonesuch Shale based on lithologic variation: Unpublished technical report, White Pine, Michigan, White Pine Copper Company, 121 p.

Elmore, R. D., 1983, Precambrian non-marine stromatolites in alluvial fan deposits, Copper Harbor Conglomerate, upper Michigan: Sedimentology, v. 30, p. 829–842.

Elmore, R. D., 1984, The Copper Harbor Conglomerate: a late Precambrian fining-upward alluvial fan sequence in northern Michigan: Geological

Society of America Bulletin, v. 95, p. 610–617.

Elmore, R. D., and Daniels, P. A., 1980, Depositional system model for upper Keweenawan Oronto Group sediments, northern Peninsula, Michigan [abs.]: Geological Society of America Abstracts with Programs, v. 12, no. 5, p. 225.

Elmore, R. D., Milavec, G., Imbus, S., Engel, M. H., and Daniels, P., 1988, The Precambrian Nonesuch Formation of the North American Midcontinent Rift: Sedimentology and organic geochemical aspects of lacustrine deposition: Precambrian Research, v. 43(3), no. 3, p. 191–213.

Hamblin, W. K., and Horner, W. J., 1961, Sources of Keweenawan conglomerates of northern Michigan: Journal of Geology, v. 69, p. 204–211.

Hieshima, G. B., and Pratt, L. M., 1991, Sulfur/carbon ratios and extractable organic matter of the Middle Proterozoic Nonesuch Formation, North American Midcontinent Rift: Precambrian Research, v. 54(1), no. 1, p. 65–79.

Hite, D. M., 1968, Sedimentology of the upper Keweenawan sequence of northern Wisconsin and adjacent Michigan [Ph.D. thesis]: Madison, University of Wisconsin, 217 p.

Hooke, R. B., 1967, Processes on arid-region alluvial fans: Journal of Geology, v. 75, p. 438–460.

Hubbard, H. A., 1972, Source of pebbles of volcanic rocks in the middle and upper Keweenawan conglomerates of northern Michigan: Journal of Geology, v. 80, p. 627–629.

Hubbard, H. A., 1975, Lower Keweenawan volcanic rocks of Michigan and Wisconsin: U.S. Geological Survey Journal of Research, v. 3, p. 529–541.

Imbus, S. W., Engel, M. H., and Elmore, R. D., 1990, Organic chemistry and sedimentology of Middle Proterozoic Nonesuch Formation—Hydrocarbon source rock assessment of a lacustrine rift deposit, in Katz, B. J., ed., Lacustrine basin exploration—Case studies and modern analogs: American Association of Petroleum Geologists, Memoir 50, p. 197–208.

Kelts, K., and Hsu, K. J., 1978, Freshwater carbonate sedimentation, in Lerman, A., ed., Lakes: Chemistry, geology, physics: New York, Springer-Verlag, p. 295–323.

Mudrey, M. G., Jr., and Dickas, A. B., 1988, Midcontinent Rift model based upon Gregory Rift tectonic and sedimentation geometries [abs.]: Annual Institute on Lake Superior Geology, 34th, Marquette, Michigan, v. 34, p. 73–75.

Pettijohn, F. J., 1957, Sedimentary rocks (second edition): New York, Harper and Row, p. 630.

Rahn, P. H., 1967, Sheetfloods, streamfloods and the formation of sediments: Annual Association of American Geographic, v. 57, p. 593–604.

Reading, H. G., 1986, Facies, in Reading, H. G., ed., Sedimentary environments and facies (2nd edition): Oxford, United Kingdom, Blackwell Scientific Publications, p. 11.

Reinick, H. E., and Singh, I. B., 1980, Depositional sedimentary environments: New York, Springer-Verlag, p. 29.

Sturm, M., and Matter, A., 1978, Turbidites and varves in Lake Brienz (Switzerland): deposition of clastic detritus by density currents, in Matter, A., and Tucker, M. E., eds., Modern and ancient lake sediments: Oxford, United Kingdom, Blackwell Scientific Publications, p. 145–166.

Suszek, T. J., 1991, Petrography and sedimentation of the Middle Proterozoic (Keweenawan) Nonesuch Formation, western Lake Superior region, Midcontinent Rift System [M.S. thesis]: Duluth, University of Minnesota, 198 p.

Wescott, W. A., and Ethridge, F. G., 1980, Fan-delta sedimentology and tectonic setting—Yallahs fan-delta, southeast, Jamaica: The American Association of Petroleum Geologists Bulletin, v. 64, p. 374–399.

White, W. S., 1966, Tectonics of the Keweenawan basin, western Lake Superior region: United States Geological Survey Professional Paper 524-E, p. E1-E23.

White, W. S., and Wright, J. C., 1960, Lithofacies of the Copper Harbor Conglomerate, northern Michigan: U.S. Geological Survey Professional Paper 400-B, p. B5–B8.

MANUSCRIPT ACCEPTED BY THE SOCIETY JANUARY 16, 1996

Geological Society of America
Special Paper 312
1997

Keweenawan Supergroup clastic rocks in the Midcontinent Rift of Iowa

Raymond R. Anderson
Iowa Department of Natural Resources Geological Survey Bureau, 109 Trowbridge Hall, Iowa City, Iowa 52242-1319

ABSTRACT

The Midcontinent Rift System (MRS) of North America is a failed rift that apparently formed in response to regionwide stresses associated with the Grenville Orogeny about 1,100 million years ago. In Iowa, the MRS is buried by about 660 to 1,650 m of Paleozoic and Mesozoic sedimentary rocks and Quaternary glaciogenic deposits. The structural configuration, nature of the major lithologic packages, and geologic history of the MRS in Iowa were investigated by examining and interpreting the limited data available. These data consisted of drill samples (including the M. G. Eischeid deep petroleum test), gravity and magnetic anomaly maps, and petroleum industry seismic reflection profiles. Investigations included petrographic examination of samples, mapping, and the production of a series of two-dimensional gravity profiles constrained by the seismic data. These studies reveal the MRS in Iowa to be characterized by a central horst (the Iowa Horst), dominated by mafic volcanic rocks, and thrust over thick sequences of younger Keweenawan Supergroup clastic rocks that fill flanking basins. At upper crustal depths, the MRS displays a relatively symmetrical structure, with sedimentary basins on both flanks of the central horst similar in depth and configuration. The clastic rocks that fill these basins appear to be dominated by two major sequences, the first apparently deposited shortly after the cessation of rift volcanism, and the second probably deposited during and shortly after the uplift of the Iowa Horst.

INTRODUCTION

The Midcontinent Rift System (MRS) is composed of several discrete tectonic segments (Dickas, 1986; Dickas and Mudrey, 1989). The Iowa zone of the western arm of the MRS extends from south-central Minnesota, across Iowa, and into southeast Nebraska (Dickas and Mudrey, 1989). The Iowa zone is dominated by an axial horst (named the Iowa Horst by Anderson, 1988), flanked by deep basins filled with Keweenawan Supergroup clastic rocks. Related clastic rocks were also deposited in the axial regions of the MRS, but were subsequently eroded during the uplift of the horst. Erosional remnants of these axial clastic rocks are preserved in several structural basins on the Iowa Horst. In this paper, the results of

a recent investigation of the MRS (Anderson, 1992) will be discussed, with specific emphasis on the Keweenawan Supergroup clastic rocks that are preserved on the Iowa Horst in Iowa and in the horst-flanking basins.

THE MIDCONTINENT RIFT SYSTEM IN IOWA

The Iowa zone of the MRS extends about 480 km southwestward from the southern end of the Belle Plaine Fault Zone, a major transfer fault zone that offsets the trend of the MRS in Minnesota, just north of the Iowa border (Fig. 1). The zone is dominated by the Iowa Horst (Anderson, 1988), an axial horst bounded by the Thurman-Redfield Structural Zone on the

Anderson, R. R., 1997, Keweenawan Supergroup clastic rocks in the Midcontinent Rift of Iowa, *in* Ojakangas, R. W., Dickas, A. B., and Green, J. C., eds., Middle Proterozoic to Cambrian Rifting, Central North America: Boulder, Colorado, Geological Society of America Special Paper 312.

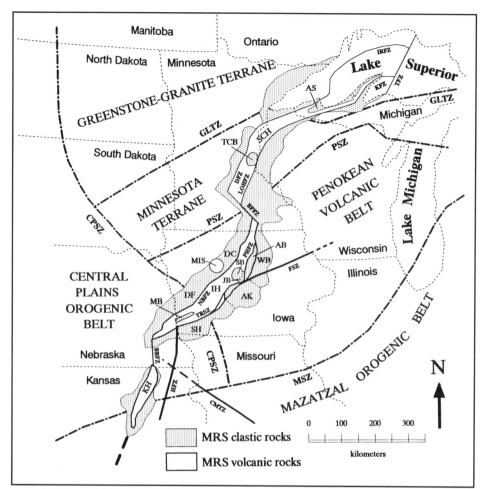

Figure 1. Structural components of the Midcontinent Rift System and related features. TFZ, Thiel Fault Zone; IRFZ, Isle Royale Fault Zone; KFZ, Keweenaw Fault Zone; DFZ, Douglas Fault Zone; LOHFZ, Lake Owen–Hastings Fault Zone; BPFZ, Belle Plaine Fault Zone; NBFZ, Northern Boundary Fault Zone; TRSZ, Thurman-Redfield Structural Zone; PHFZ, Perry-Hampton Fault Zone; BBFZ, Big Blue Fault Zone; HFZ, Humboldt Fault Zone; SCH, St. Croix Horst; IH, Iowa Horst; KH, Kansas Horst; GLTZ, Great Lakes Tectonic Zone; PSZ, Penokean Suture Zone; FSZ, Fayette Structural Zone; CPSZ, Central Plains Suture Zone; MSZ, Mazatzal Suture Zone; CMTZ, Central Missouri Tectonic Zone; AS, Ashland Basin; TCB, Twin City Basin; DC, Duncan Basin; DF, Defiance Basin; WB, Wellsburg Basin; AK, Ankeny Basin; SH, Shenandoah Basin; AB, Ames Block; SB, Stratford Basin; JB, Jewell Basin; MB, Mineola Basin; MIS, Manson Impact Structure. (Modified from Anderson, 1992.)

southeast and the Northern Boundary Fault Zone on the northwest, and flanked by basins that contain clastic sedimentary rocks with interpreted thicknesses in excess of 10 km.

The Iowa zone of the Midcontinent Rift developed entirely in Early Proterozoic structural terranes south of the Archean Superior Craton (Fig. 1). Most of the MRS formed within the Penokean Volcanic Belt, with the southern end of the zone crossing into the Central Plains Orogenic Belt. The contact between the Penokean and Central Plains belts is not well defined, but lies east of the Humboldt Fault System (Fig. 1). The trend of the MRS in the Iowa zone is subparallel to the southwesterly trend of the Late Proterozoic Penokean Suture Zone across south-central Minnesota and northwest Iowa (Fig. 1). The MRS may have developed along zones of weakness associated

the trend of island arcs accreted to the Superior Craton during the Early Proterozoic Penokean Orogeny.

The Iowa segment of the MRS is truncated in southeastern Nebraska by a second major transfer fault zone, identified by Anderson (1992) as the north-northwest-trending Big Blue Fault Zone, a structure with a trend generally coincident with the Big Blue River. The Central Missouri Tectonic Zone (Kisvarsanyi, 1984) also passes through the gap between the Iowa and Kansas segments of the MRS and may have been a factor in the formation of the offset.

Anderson (1992) concluded that the MRS in the Iowa zone was formed by a sequence of events generally similar to sequence described by Craddock (1972). Initial extensional deformation led to crustal thinning, fracturing, and axial graben formation, and extrusion of voluminous mafic-dominated flood basalts, generally confined to the axial graben. As extension diminished, volcanism waned; however, subsidence of the graben and flanking areas continued, driven by the isostatic imbalance created by the mass of dense basalts. Fluvioclastics were the dominant basin fill as subsidence continued. The regional deformation switched to contraction, reversing the direction of movement along the graben-bounding fault zones, and initiating the uplift of the axial graben to form a horst (the Iowa Horst). Clastic rocks eroded off the uplifted axial horst were mixed with fluvioclastic derived from outside the rift zone and deposited in horst-flanking basins. The volcanic, plutonic, and clastic rocks associated with the formation of the MRS in Iowa are considered generally coeval with similar rocks associated with the MRS in the Lake Superior region and are also assigned to the Keewenawan Supergroup (Anderson, 1992).

Clastic-filled basing flanking the Iowa Horst were identified and delineated primarily by their geophysical signatures. They are identified by their pronounced gravity minima and are interpreted as depressions in the crystalline rock basement, filled with relatively low density clastic rocks. Two basins are preserved northwest of the horst (Fig. 1); from north to south these are the Duncan and Defiance Basins (Anderson, 1988). Three basins are located southeast of the Iowa Horst (Fig. 1); from north to south these are the Wellsburg, Ankeny, and Shenandoah Basins (Anderson, 1988). The two northwest basins apparently are interconnected, forming a continuous sequence of clastic rocks. The three southeasterly basins are also connected in a similar fashion. The areal extent of the southeastern basins of the Iowa segment appears to be greater than that of the northwestern basins, but the system is basically symmetrical in interpreted cross section.

Most of the Keewenawan Supergroup clastic rocks that were deposited in axial regions of the MRS in Iowa were subsequently eroded as the Iowa Horst was activated and uplifted. Clastic rocks are, however, preserved on a step-faulted block on the eastern edge of the horst in central Iowa (the Ames Block), in the Stratford and Jewell Basins in the center of the horst, in the Mineola Basin on the axis of the horst at the Nebraska border (Anderson, 1992), and at several areas in Nebraska (Sims, 1990).

WELLS PENETRATING KEEWENAWAN ROCKS IN IOWA

The drilling of the Amoco M. G. Eischeid #1 deep petroleum test well in 1987 sampled a much greater thickness of MRS rocks than all previous Iowa wells combined. Prior to the Eischeid well, 71 drill holes had penetrated to the MRS in Iowa (see Anderson, 1992). Most of these wells are clustered (Fig. 2) in three areas, the Redfield and Dallas City gas storage structures in central Dallas County (34 wells), and the Vincent gas storage structure in northeast Webster County (15 wells). These areas were drilled by natural gas pipeline companies during the development of gas storage facilities in domal structures in

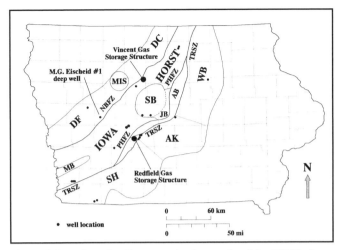

Figure 2. Location of wells penetrating into rocks of the MRS in Iowa (modified from Anderson, 1992). See Figure 1 for explanation of abbreviations.

basal Paleozoic sandstones above the major MRS fault zones. Of these 49 gas storage structure wells, 44 penetrate into mafic igneous rocks (total of 116 m), and only 6 of the 49 include partial cores (totaling only 29 m). One core, the #1 Sharp in Webster County, penetrates 17 m of mafic igneous volcanics, including portions of three individual lava flows, separated by weathered, vesicular flow-tops. The remaining 5 gas storage structure wells (total of 33 m) penetrated sandstone-dominated MRS clastic rocks. These include 4 cores, totaling 24 m, and were described by Anderson and McKay (1996). Because of the concentration of these wells, regional information about MRS geology in Iowa is very limited. One deep well is an especially important source of information about the MRS in Iowa, the M. G. Eischeid #1 well.

M. G. Eischeid #1 well

The deepest penetration and most informative samples of MRS clastic rocks in Iowa were obtained during the drilling of the M. G. Eischeid #1 well, located about 8 km west of the Iowa Horst, in the northern end of the Defiance Basin (Anderson, 1990a). The well, drilled by Amoco Production Company in 1987 near the town of Halbur in Carroll County, completely penetrated the MRS clastic rocks, reaching a total depth of 5,492 m. Below 862 m of Phanerozoic strata (Fig. 3), the Eischeid well penetrated 4,584 m of Midcontinent Rift clastic rocks, with 2,372 m of Upper Red Clastic Group rocks, including 250 m of unit H, not originally assigned to the sequence by Witzke (1990), and 2,212 m of the Lower Red Clastic Group. This is the deepest penetration of the clastics anywhere along the entire trend of the MRS. An excellent suite of cutting samples was recovered (generally at 3.1-m intervals), and cores of the clastic rocks were recovered at 4 intervals.

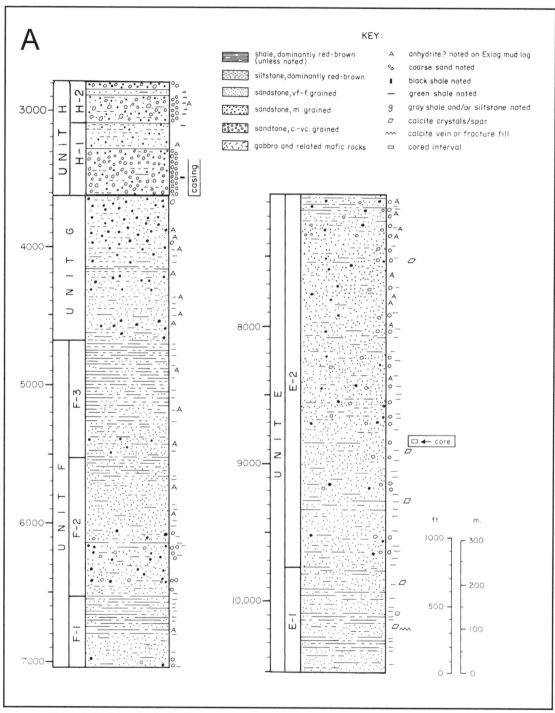

Figure 3. Well log displaying lithology and stratigraphy of the red clastic rocks encountered during the drilling of the M. G. Eischeid #1 deep petroleum test, Carroll County, Iowa (modified from Witzke, 1990). Unit H has been tentatively included with the Upper Red Clastic Group. 3A, shows Upper Red Clastic Group rocks penetrated in the Eischeid well. 3B (facing page), describes Lower Red Clastic Group rocks penetrated. Depth values in feet.

Stratigraphy of the Eischeid #1 samples

Beneath the Phanerozoic sequence, the M. G. Eischeid well encountered 4,584 m of clastic rocks associated with the Midcontinent Rift System and assigned to the Keweenawan Supergroup (Anderson, 1990b). These clastic rocks were divided into two groups, informally named the Upper Red Clastic sequence and Lower Red Clastic sequence (Witzke, 1990). Since both Red Clastic sequences were subdivided into formation-level stratigraphic subdivisions by Witzke (1990), the Upper and Lower Red Clastic sequence are afforded group status.

Upper Red Clastic Group. The clastic rock sequence, encountered at depths from 862 to 3,243 m in the M. G. Eischeid well, was assigned to the Upper Red Clastic Group by

Anderson (1992). The lithologies of this sequence were described by Witzke (1990) from down-hole logs and petrographic and other studies. He informally called them, from the top, Unit H, Unit G, Unit F, and Unit E. Witzke (1990) assigned units G, F, and E to the Upper Red Clastic Group, but was unsure of the affinity of Unit H. Unit H was tentatively assigned to the Upper Red Clastic Group by Anderson (1992).

Unit H. Unit H (depth 862 to 1,112 m) is dominated by two primary lithologies (Fig. 3a), a clear to light reddish brown, fine- to very coarse-grained arkosic sandstone (using the classification of Folk, 1968), and a reddish brown, micaceous, shaly to sandy siltstone (McKay, 1990). The common occurrence of loose quartz and feldspar grains among the well samples implies that some of the sandstones are poorly indurated. A pet-

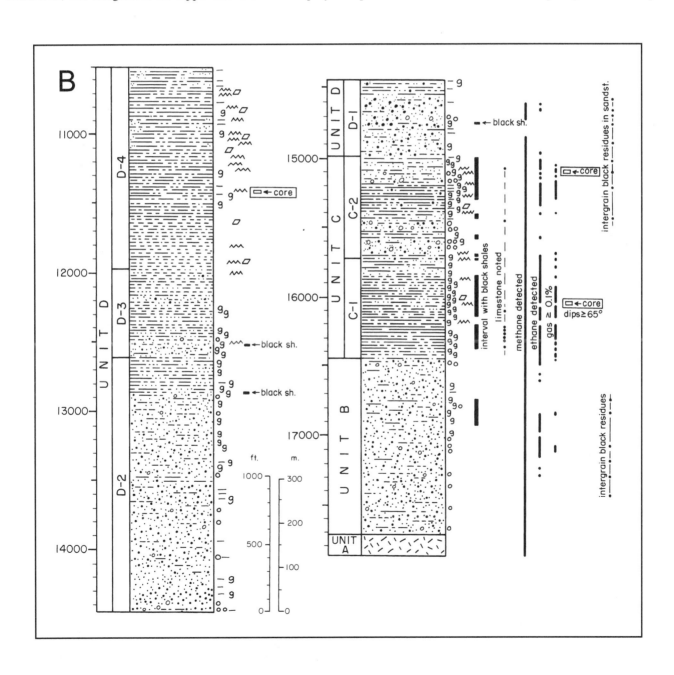

**TABLE 1. MEAN FRAMEWORK GRAIN COMPOSITION
OF #1 EISCHEID RED CLASTICS**

| Unit | Depth | Modal Analysis | | |
| | | Quartz | Felsdpar | Lithic Fragments |
	(m)	(%)	(%)	(%)
Upper Red Clastic Group				
H	862–1112	79	17	4
G	1112–1443	76	17	7
F	1443–2160	69	22	9
E	2160–3234	67	20	13
Lower Red Clastic Group				
D	3234–4609	76	21	3
C	4609–5062	72	24	3
B	5062–5446	87	12	1

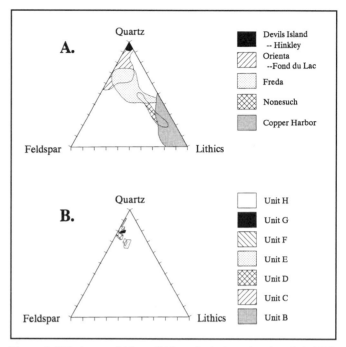

Figure 4. QFL (quartz-feldspar-lithic grains) triangular plot of: A, Mid-continent Rift System clastic rocks from the Lake Superior region (modified from Ojakangas, 1986); and B, Eischeid well data (modified from Ludvigson et al., 1990).

rographic study of thin sections produced from cuttings collected during the drilling of the Eischeid well was conducted as a part of a preliminary study of MRS clastic rocks (Ludvigson et al., 1990; McKay, 1990). The mean composition of samples analyzed (Table 1) identified it as a subarkose (Fig. 4). All of the feldspar grains encountered were potassium feldspar (Ludvigson et al., 1990).

McKay (1990) discussed the stratigraphic position of Unit H, presenting compelling petrographic evidence for a pre–Mount Simon emplacement. He reported the presence of trace amounts of microbrecciated rock fragments in Unit H, identical to clasts found down-hole throughout a large part of the red clastic section. The presence of microbreccia chips probably implies that Unit H rocks were affected by the same structural movements that produced the tectonized fabrics observed throughout much of the Eischeid drill-hole sequence (Ludvigson et al., 1990). The presence of the microbreccia and other data led McKay (1990) to suggest a "Late Keweenawan" (Middle Proterozoic) age for Unit H and Anderson (1992) to tentatively assign Unit H to the Upper Red Clastic Group. Although Unit H is poorly consolidated (markedly less well consolidated than the other Upper Red Clastic units), and may have been deposited in the ~400 m.y. period between deposition of the Keweenawan Supergroup clastics and the basal Paleozoic sandstones, in this paper Unit H is tentatively assigned to the Upper Red Clastic Group.

Unit G. Rocks penetrated at depths from 1,112 to 1,443 m (Fig. 3A) were informally named Unit G and described as dominated by pale gray (clear) to red-brown, fine- to medium-grained sandstone interbedded with red-brown (and minor green) siltstone and shale (Witzke, 1990). The sandstones are dominated by quartzose grains, with a mean framework grain composition (Table 1) indicative of subarkose (Fig. 4). The volcanic rock component of the lithic grain population in this interval (35%) is the highest observed in any clastic unit in the Eischeid well (Ludvigson et al., 1990).

Unit F. The third clastic unit encountered in the Eischeid well (Fig. 3B), Unit F (depth 1,443 to 2,160 m), was subdivided into three generally fining-upwards sequences labeled (top down) F-3, F-2, and F-1 by Witzke (1990). These sequences grade from basal red-brown to pale gray (clear), very fine- to fine-grained sandstones upward to red-brown (in part mottled light green to gray-green) to red (in part mottled light green to gray-green) shales and siltstones (Witzke, 1990). These fluvial sandstones have a mean framework grain composition (Table 1) identifying it as a subarkose (Fig. 4B).

Unit E. The basal unit of the Upper Red Clastic Group, Unit E (depth 2,160 to 3,234 m), is a thick, sandstone-dominated unit that is characterized by the highest percentage of lithic fragments (Table 1), with the mean of framework grain composition of a lithic arkose (Fig. 4B; Ludvigson et al., 1990). The unit was subdivided into two descriptive intervals (Fig. 3A), E-2 (upper) and E-1. Interval E-2 is dominated by red to brown, generally very fine to fine-grained sandstones with red to brown (some mottled grayish green) interbedded siltstones and shales (Witzke, 1990). Core 1 sampled 1 m of this interval (2,718.2 to 2,719.3 m), identified as a feldspathic litharenite by Barns (1990), and displayed horizontal bedding and two facies, one well stratified with low-angle cross-stratification and red mudstone rip-up clasts, the second a massive, nonstratified sequence with smaller mud clasts (Ludvigson et al., 1990). Interval E-1 contains more shale than E-2 and is composed of four relatively thin fining-upward sequences, each grading from a basal very fine to fine-grained sandstone upward to a red to dark brown shale and siltstone (Witzke, 1990).

Lower Red Clastic Group. The rocks encountered at depth between 3,234 and 5,446 m (Fig. 3B) in the Eischeid well were assigned to the Lower Red Clastic Group (Anderson, 1990a). The series was divided into three component units by Witzke (1990), from the top down, these are Unit D, Unit C, and Unit B (Fig. 3B).

Unit D. The uppermost unit in the Lower Red Clastic Group, Unit D (depth 3,234 to 4,609 m), was described by Witzke (1990) as a thick, generally fining-upward sequence that is dominated by sandstones in the lower half and siltstone to shale in the upper half. He subdivided the unit into four intervals (Fig. 3B); from the top down, these are D-4, D-3, D-2, and D-1. The sandstones in Unit D are quartz dominated, with a mean framework grain composition of a subarkose (Table 1). The lithic component includes the highest percentage of sedimentary and metamorphic rock fragments (average 74% of lithic grains) observed in any unit (Fig. 4B) in the Eischeid well (Ludvigson et al., 1990).

Interval D-4 (depth 3,234 to 3,680 m) is dominated by red to brown (with minor light green to gray) shales in its upper half (Fig. 3b) and red-brown siltstones in the lower half (Witzke, 1990). Core #2 (depth 3,501.8 to 3,506.2 m) was recovered from this interval and was described by Ludvigson et al. (1990) as displaying two facies, one composed of red, very fine to fine-grained, cross-stratified sandstones, and the second a very fine grained, horizontally stratified sandstone with minor mudstone. These sediments are interpreted as representative of a fluvial setting with shallow channel fills, subaerial exposure, inundation by overbank deposits, and then a return to shallow channel deposition (Ludvigson et al., 1990). The cored interval also displays folds, slickensides, and other fault-related deformational features.

The underlying interval, D-3 (depth 3,680 to 3,877 m), is siltstone dominated with minor interbedded sandstone and shale. This interval also contains gray to dark gray siltstones and shales and the first occurrences of black siltstones and black carbonaceous specks. The lower portion of this interval is dominated by light gray, fine- to medium-grained sandstones with minor shales and siltstones (Witzke, 1990).

The upper portion of interval D-2 (depth 3,877 to 4,446 m), is dominated by varicolored siltstone with dark gray to black shaley laminations (some pyritic). The majority of the interval is sandstone-dominated with minor red-brown to gray-green siltstone and shale interbeds. Coarse sand grains are abundant in the basal portions of interval D-2.

The basal interval in Unit D, D-1 (depth 4,446 to 4,609 m), was described by Witzke (1990) as a sandstone-dominated sequence with an upper varicolored shale and siltstone package. The sandstone is primarily light gray to red-brown and fine- to medium-grained with minor coarse grains. The shales are red-brown to gray (some black) and contain siltstone interbeds. The mud log indicates traces of methane and ethane were detected in this interval. Witzke (1990) reported intergranular black residues, possibly hydrocarbon residues indicative of past petroleum movement through this unit.

Unit C. Unit C (depth 4,609 to 5,062 m) is the most distinctive Keweenawan Supergroup clastic unit encountered in the M. G. Eischeid well (Fig. 3b). It is unique in its abundance of gray to black siltstones and shales, calcite cements, calcite vein-fills, and structural deformation. The unit is also the most thoroughly cored, with two cores totaling 11.7 m taken during the drilling. The unit was subdivided into two intervals, an upper interval (C-2) and a lower interval (C-1).

Interval C-2 (depth 4,349 to 4,710 m) is an interbedded sequence of sandstones, siltstones, and shales, with gray to black siltstones and shales more common in the upper portion of the interval (Fig. 3b) and red-brown to green-gray colors more common in the basal portions (Witzke, 1990). Methane and ethane were detected throughout the interval and black intergranular residues, possibly relict hydrocarbons, were reported. Core #3 (depth 4,644.9 to 4,652.3 m), taken in the upper part of the interval, consists of horizontally laminated, millimeter to decimeter thick interlayered medium to dark gray shales and lighter gray siltstones to fine-grained sandstones (Ludvigson et al., 1990). Barnes (1990) reported the unit to be moderately to poorly sorted, fine- to very coarse grained, and mostly subangular feldsarenites to lithic feldsarenites. He also reported carbonate lithoclasts in this interval. These strata were interpreted to have been deposited in a lake or other body of standing water with a fluctuating water depth and intermittent influxes of coarse detritus. Petrologic study of two samples of coarse detritus from interval C-2 yielded a mean subarkose composition (Table 1) as reported by Ludvigson et al. (1990). The 24% feldspar observed in this interval is the greatest concentration observed in any interval in the Eischeid well (Fig. 4B).

Interval C-1 (depth 4,831 to 5,062 m) is dominated by light to dark gray siltstones and shales with dark brown to black shaley interbeds (Witzke, 1990). Calcite cements, spar, and vein fills are common in this interval and a carbonate-rich region near the bottom of the interval was identified as limestone on the mud log. Core #4 (depth 4,936.3 to 4,940.9 m) shows a steep tectonically derived dip ranging from 65° to vertical and slightly overturned. The cored interval is composed of well-laminated, interbedded light gray, micaceous siltstone and medium gray to black shale (Ludvigson et al., 1990). Calcite veinlets and slickensided fault surfaces with calcitic rough facets are common in the core and indicate reverse faulting, requiring lateral shortening (Ludvigson and Spry, 1990).

Unit B. The basal clastic unit in the Eischeid well (depth 5,062 to 5,446 m) was informally called Unit B (Fig. 3B) by Witzke (1990), who described it as dominated by white to light gray and red, very fine to fine-grained sandstone with possible

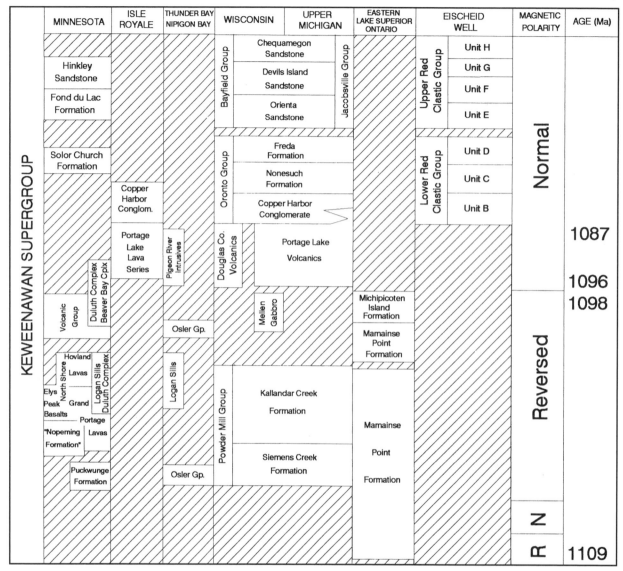

Figure 5. Comparative Keweenawan stratigraphy of the Midcontinent Rift System in the Lake Superior region (modified from Green, 1982; and Anderson, 1992).

siltstone and shale partings. Petrographic studies of Unit B sandstones by Ludvigson et al. (1990) indicated that it was a subarkose (Table 1) and contained the highest percentage of quartzose grains (87%) observed in the Keweenawan Supergroup of Iowa (Fig. 4B), with a small lithic component dominated by sedimentary and metamorphic clasts (74%).

Discussion of Upper and Lower Red Clastic Groups

The Upper Red Clastic Group occupies the same relative stratigraphic position as the Bayfield Group in the Lake Superior region (Fig. 5). Both units are apparently dominated by fluvial deposition, with the possible exception of the Devils Island Formation of the Bayfield Group, which may be a lacustrine deposit (Thwaites, 1912). There are, however, compositional differences

between the clastic rocks of the two groups. The Bayfield Group is more mature, both in texture and mineralogy, than the underlying Oronto Group (Ojakangas, 1986), whereas the rocks of the Upper Red Clastic Group are mineralogically less mature than the underlying Lower Red Clastic Group rocks. Volcanic rock fragments contribute less to Bayfield Group rocks than to the Oronto Group (Ojakangas and Morey, 1982). However, in the Eischeid well volcanic rock fragments are more common in the Upper Red Clastic Group than the lower sequence. Finally, Eischeid drilling did not penetrate any Proterozoic quartzarenites, such as those seen in the Devils Island and Hinckley Sandstones in the Lake Superior area.

The Bayfield Group and Upper Red Clastic Group do, however, share many characteristics. Both groups apparently overlie the initial MRS clastic sequence (Oronto Group and

Lower Red Clastic Group). The Bayfield Group, as described by Ojakangas and Morey (1982), is dominated by fluvial deposition (excluding the possible lacustrine Devils Island Sandstone). The lithologies present in the Upper Red Clastic Group as described by Witzke (1990) and the petrology and sedimentary structures as established from the cored intervals by Ludvigson et al. (1990) also suggest a fluvial origin. The differences in the composition of the units of the Upper Red Clastics Group and the Bayfield Group can be explained by differences in the lithologies of source terranes.

Although no quartzarenites were encountered in the Eischeid well, Unit F of the Upper Red Clastic Group displays the highest abundance of siltstone and shale. These lower energy deposits may be analogous to the possible lacustrine deposits reported by Thwaites (1912) in the middle unit (Devils Island) of the Bayfield Group, perhaps deposited by a low gradient river that flowed into a nearby lake. It is more probable that Unit F represents overbank and crevasse splay deposits in a fluvial meander belt environment.

Several trends are evident in the composition of the sandstone component of the Upper Red Clastic Group. Quartzose grain content increases upward in the sequence, from 67% in Unit E to 79% in Unit H. The QFL composition of the framework grains is comparable to compositions of units in the Lake Superior area (Fig. 4). In addition, the average lithic fragment percentage decreased upsection from 13% in Unit E to 4% in Unit H, with the volcanic rock component of the lithic rocks increasing from 22% in Unit E to 62% in Unit G and sedimentary and metamorphic rock component decreasing from 65% in Unit E to 37% in Unit G. The compositions of the lithic fragments in the two intervals examined in Unit H were quite disparate and additional analyses are needed.

The relative increase in the concentration of volcanic rock fragments upsection in the Upper Red Clastic Group in the Eischeid well may record the local proximal erosional unroofing of the basalts on the Iowa Horst. Well data from the trend of the MRS in Iowa and seismic interpretations indicate that Keweenawan sedimentary rocks were erosionally removed from most areas of the Iowa Horst, exposing underlying volcanic rocks (Anderson, 1988, 1990c).

The Lower Red Clastic Group occupies the same stratigraphic position as the Oronto Group in the Lake Superior area (Fig. 5). The biggest differences between the two groups are the coarse clastic facies preserved in the Copper Harbor Conglomerate, the basal unit in the Oronto Group, compared to the dominance of quartz sand in Unit B, the basal unit of the Lower Red Clastic Group. The Copper Harbor is dominated by a conglomeratic facies (Daniels, 1982) composed primarily of mafic and felsic volcanic rock clasts. Associated sandstone facies also are dominated by volcanic rock fragments. Unit B is dominated by a sandstone facies that displays the highest quartzose grain content (87%) of any Proterozoic clastic unit in the Eischeid well (Table 1; Fig. 4). This disparity can be explained by the positions of the two units relative to the central horst of the Mid-

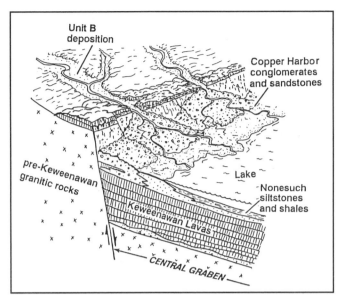

Figure 6. Depositional environment of Unit B and strata correlative with the lower Oronto Group (Anderson, 1992).

continent Rift. The Copper Harbor Conglomerate exposures are located on the central horst, which was an axial graben at the time of Copper Harbor deposition. The volcanic-dominated rocks of the Copper Harbor Conglomerate were deposited by alluvial fans that were derived in large part from the erosion of Keweenawan volcanic rocks capping the footwalls of the graben-bounding normal faults. In contrast, the Eischeid well, and Unit B, are located in one of the clastic basins that flank the central horst, and Unit B was therefore deposited outside of the central graben. Unit B is dominated by quartzose sediment derived from granitic rocks, the most common basement rock lithology in the MRS region, and was probably deposited by rivers that generally flowed towards the axis of the rift (Fig. 6), similar to and probably feeding the fluvial and lacustrine environments within the subsiding axial grabben similar to the depositional systems that deposited the Copper Harbor Conglomerate as described by Daniels (1982). As these rivers approached the axial graben they may have eroded footwall-capping volcanic rocks, depositing Copper Harbor Conglomerate–like lithologies in the graben.

Unit C is lithologically and sedimentologically very similar to the Nonesuch Formation of the Oronto Group. The fine-grained components of Unit C display unidirectional, bidirectional, symmetrical, and trough cross-stratification, and small-scale amalgamated hummocky cross-stratification. Shrinkage cracks are also present in this interval (Ludvigson et al., 1990). Milavec (1986) identified similar features in the Nonesuch Formation.

The Nonesuch Formation has been interpreted as a lake deposit by most workers (e.g., Elmore, 1981; Daniels, 1982; Milavec, 1986). However, some workers have suggested that it may have been deposited in a marine environment (Burnie et al., 1972; Hieshima et al., 1989; and Hieshima and Pratt, 1991). In either case, the Nonesuch appears to have been deposited in a standing body of water. Structures observed in cores of Unit C also suggest deposition in a standing body of water (Ludvigson et al., 1990). The presence of these slack-water deposits in the Eischeid well indicate that lacustrine environments were not confined to the central horst, the position where correlative Nonesuch deposits are observed. Interpretation of petroleum industry seismic data suggests that Unit C may extend at least 11 km west of the Iowa Horst in the Defiance Basin, and perhaps as much as 24 km or more. Recent drilling on the central peak of the Manson Impact Structure in Pocahontas County, Iowa, has recovered abundant dark gray and black pyritic shales and mudstones associated with crystalline basement rocks and impact melt rocks (Anderson, 1993). These dark shales, tentatively identified as Unit C rocks, contain pyrite laminae similar to those reported by Barnes (1990) in Unit C dark shales in the Eischeid well. The lithologic association of Unit C clastic rocks and crystalline basement rocks at Manson, and the lack of Unit B red clastic sandstones, suggest that the body of water in which Unit C was deposited may have been very extensive, extending beyond the distal depositional margin of Unit B clastics. The Manson cores are located 30 km northwest of the Northern Boundary Fault Zone, the northwestern margin of the Iowa Horst. If the body of water in which Unit C was deposited was continuous across the MRS axial graben, and if the body continued for an equal distance east of the graben, it would have been at least 140 km wide, probably extending for the entire 480 km length of the Iowa zone of the MRS, conceivably the entire length of the MRS.

The black coloration in Unit C, like the Nonesuch Formation, is primarily the product of disseminated organic carbon and pyrite. Palacas et al. (1990) reported total organic carbon (TOC) values as high as 1.4% (averaging 0.6%) in the dark shales and siltstones from cores of Unit C. The Nonesuch Formation has yielded maximum TOC values of almost 4% (Hieshima et al., 1989). The petroleum source-rock potential of Unit C was discussed by Palacas et al. (1990) and Anderson (1992) and will be summarized later in this paper.

Unit D occupies the same stratigraphic position in the Eischeid well as the Freda Formation in the Oronto Group of the Lake Superior area. Interpretation of Core #1, recovered from Unit D, suggested to Ludvigson et al. (1990) that the unit was deposited in a fluvial environment. A similar interpretation has been proposed for the Freda Formation by many workers (e.g., Daniels, 1982; Morey and Ojakangas, 1982). Petrographic studies of Unit D reveal a mean detrital composition of $Q_{76}F_{21}L_3$ compared to a mean composition of $Q_{52}F_{18}L_{30}$ for seven Freda exposures reported by Daniels (1982). The lithic component of the Freda is dominated (72%) by volcanic rock fragments. The higher volcanic lithic component indicates that volcanic rocks were exposed in the Freda source area (along the hanging wall of the graben-bounding faults), and similar lithologies were not exposed in the source area of Unit D (the

plutonic and metamorphic rocks in which the MRS developed). Dark shale interbeds in the basal portions of Unit D suggest a gradational transition from Unit C depositional conditions.

Although there are petrographic differences between the Lower Red Clastic Group and the Oronto Group (and equivalent units in the Lake Superior region), these two rock sequences probably represent sediments deposited in different regions of a continuous depositional system. The Lower Red Clastic Group was deposited in a more distal region of the depositional system, outside of the MRS axial graben, sourced primarily from eroded pre-MRS granitic terranes. The Oronto Group was deposited in the more proximal region of the depositional system, in the MRS axial graben, with clastic sediments (especially the basal Copper Harbor Conglomerate) derived largely from MRS volcanic rocks on the footwall of the graben-bounding faults (Fig. 6). The lacustrine or estuarine deposits of the second formation of each group (Unit C of the Lower Red Clastic Group and Nonesuch Shale of the Oronto Group) are probably directly correlative, part of a continuous system (possibly marine) that may have run the length of the MRS. The upper units of the Lower Red Clastic Group (Unit D) and the Oronto Group (Freda Formation) are both fluvial sandstones apparently conformably overlying the lacustrine or estuarine units. The higher concentrations of volcanic rock fragments in the Freta Formation as compared to Unit D may represent erosion of local exposures of MRS volcanic rocks.

Unit A

Beneath the sequence of Keweenawan Supergroup clastic rocks, the Eischeid drill hole passed into a virtually undeformed, medium-grained metagabbro, consisting of relatively fresh augite, altered plagioclase, and lesser amounts of hypersthene, all described by Van Schmus et al. (1990). The unit, probably intruded as a dike or sill into Penokean gneissic basement, was encountered at the basement surface, a depth of 5,446 m and penetrated to a total depth of 5,493 m. The penetration includes one cored interval, from 5,456 to 5,459 m. Zircons were recovered from the cored interval and yielded a U-Pb age of 1,281 ± 50 Ma, an age significantly older than age of Keweenawan Supergroup igneous rocks (1,086 to 1,108 Ma) and probably associated with the coeval Mackenzie dike swarm of Canada (Van Schmus et al., 1990). Although this dike was emplaced prior to the formation of the Midcontinent Rift and deposition of Keweenawan Supergroup rocks, it and other Mackenzie dikes are manifestations of continentwide extension that ultimately led the opening of the Grenville Ocean. The closing of this ocean was responsible for the contraction that led to the formation of the MRS.

Additional Red Clastic unit

One additional lithologic unit has been encountered in several wells in southwest Iowa. This unit, a possible Keweenawan Supergroup red clastic formation, is a sandstone first described by Anderson and McKay (1995) and here informally named Unit I. It was encountered just beneath the basal Paleozoic sandstones in the State Center #5 (Marshall County) and the Gulf Energy 16-13 Poetker and Wisnom #1 (Fremont County) wells. This unit displays a higher average quartz grain content (85%) and lower content of lithic grains (1%) than other red clastic units. It is distinguished from the basal Paleozoic sandstones by its larger (sand-sized) feldspar grains, lack of feldspar overgrowths, thin interbeds of red mudstone, and the presence of rare cataclastic grains. In the Poetker well this unit overlies a sandstone with a typical red clastic QFL composition. This juxtaposition indicates that Unit I lies above the previously described red clastic units and is probably apparently the uppermost known red clastic unit in Iowa.

HYDROCARBON POTENTIAL OF RED CLASTIC ROCKS IN IOWA

It appears that some intervals of the Lower Red Clastic Group, especially Unit C, once contained high levels of organic carbon and may have produced large volumes of petroleum. However, hydrocarbon generation probably reached its peak relatively early in the rock's history, about 800 Ma (Palacas et al., 1990). Black intergranular residue reported in the Eischeid #1 borehole mud log may identify paths of petroleum migration through surrounding rocks, but no areas of past petroleum accumulations were identified. The organic-rich rocks in the Eischeid well have realized most of their petroleum potential and are presently over mature and retain no potential for producing commercial petroleum. Porosity is also very low in the Keweenawan Supergroup clastic rocks in the Eischeid well. Much of the original porosity was destroyed by compaction and the precipitation of intergranular cements.

The absence of commercial petroleum or even petroleum potential in the Keweenawan Supergroup clastic rocks in the M. G. Eischeid #1 well does not preclude the possible presence of commercial quantities of petroleum in other areas of the Midcontinent Rift System. The advanced state of maturity observed in Unit C by Palacas et al. (1990) is the result of high temperatures experienced by the unit, probably early in its history. The methane-filled fluid inclusions in tectonic veinlets described by Ludvigson and Spry (1990) indicate that late Keweenawan deformation was coeval with hydrocarbon evolution and migration in the Defiance Basin. Palastro and Finn (1990) estimated minimum paleotemperatures between 175 and 197 °C for this thermal event, considerably higher than the present estimated bottom-hole temperature of 105 °C (Barker, 1990). It seems unlikely that this thermal regime was the product of deeper burial alone. It probably was the product of latent heat from igneous activity along the trend of the Midcontinent Rift, and possibly hydrothermal activity associated with later stages of this activity. This elevated heat flow and geothermal activity would have been most pronounced near the axis of the central graben.

Interpretation of seismic data acquired over the Defiance Basin suggests that reflectors interpreted as Unit C lithologies continue for at least 11 km and possibly as far as 24 km northwest of the Iowa Horst (originally the central graben). If Unit C also contained significant amounts of organic carbon at distance from the central graben, these organic-rich strata may not have been exposed to the high paleotemperatures experienced by rocks closer to the axis of the rift. Consequently, these rocks might not have lost their petroleum as early as rocks closer to the center of the MRS, and petroleum may still remain in stratigraphic or structural traps. Citing fluid inclusion and isotopic evidence, Ludvigson and Spry (1990) suggested that during uplift of the axial horst, fluid transport in the Keweenawan sedimentary basins was away from the rift axis. The petroleum would have migrated updip, generally away from the axis of the rift. The best potential exploration targets may exist on the outer flanks of the Midcontinent Rift System (Ludvigson and Spry, 1990).

STRUCTURAL CONFIGURATION OF THE MRS IN IOWA

The structural configuration of the Iowa zone of the MRS was recently investigated (Anderson, 1992) using geophysical data and sparse well control. This study utilized industry reflection seismic data to constrain a sequence of two-dimensional gravity profiles, tied together by interpretation of magnetic and gravity anomaly maps of Iowa. These data allowed production of the most detailed model to date of the MRS in Iowa.

MRS gravity anomaly in Iowa

The "Bouguer Gravity Anomaly Map of Iowa" (Anderson, 1981; Fig. 7) clearly displays anomalies produced by the major structures that characterize the MRS in Iowa (Fig. 8). The strongly positive axial anomaly is the result of the thick sequence of dense mafic-dominated volcanic and plutonic rocks that comprise the Iowa Horst, and its shallow position within the Precambrian surface. The flanking gravity minima are the product of the thick sequences of less dense, Keweenawan Supergroup clastic sedimentary rocks that are preserved in the basins marginal to the horst. Five clastic-filled basins are defined by closed contours of gravity minima along the flanks of the Iowa Horst (Fig. 8). The very steep gravity gradient between the axial high and the flanking lows attests to the abrupt contact between the two features, their density contrast, and the very large volumes of the disparate lithologies.

MRS magnetic anomaly in Iowa

Over most of its trend in Iowa, the magnetic signature of the MRS, as seen on the "Aeromagnetic Map of Iowa" (Zietz et al., 1976), is characterized by a series of linear anomalies following the trend of the gravity anomaly. Magnetic intensities are gener-

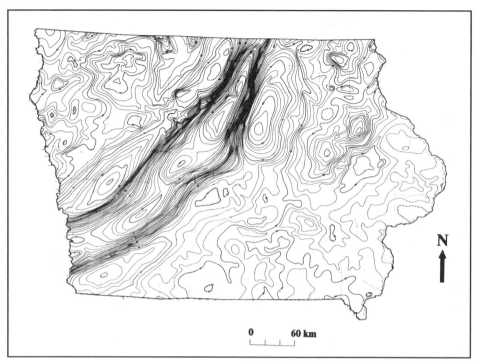

Figure 7. Bouguer gravity anomaly map of Iowa (modified from Anderson, 1981). contour interval = 5 milligals.

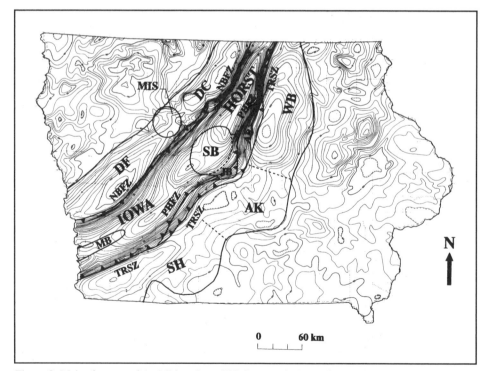

Figure 8. Major features of the Midcontinent Rift Sysytem in Iowa displayed on the Bouguer gravity anomaly map of Iowa (modified from Anderson, 1992). See Figure 1 for explanation of abbreviations.

ally within a few hundred gammas (nanoteslas) of regional values, as compared to anomalies for typical intrusions that can exceed 1,000 gammas (Carmichael and Black, 1986). The lack of a strong magnetic signature associated with the generally highly magnetic mafic igneous rocks of the Iowa Horst can probably be attributed to a combination of the strong remanent components in these rocks and the vertical sequence of normal and overlying reverse remnant polarities. Keweenawan igneous rocks from the Lake Superior region display Köenigsberger ratios (ratios of the remanent to induced magnetism) as high as 10 (Hinze et al., 1982) and display normal paleolongitude around 180° E and paleolatitudes between 0 and 30° N (Halls and Pesonen, 1982). Since most major MRS volcanic rock sequences display only a single normal or reverse remanent polarity, the basal units reverse and the upper normal, the net magnetic anomaly observed at the land surface is low. In much of Iowa, it appears that the lower, reversely polarized rocks are more abundant on the central horst than are the upper, normally polarized rocks, producing a net magnetic signature at the surface that is very near regional intensities. In northern Iowa and Minnesota, the signature of the central horst becomes strongly positive. This suggests a northward increase in the thickness of the upper, normally polarized rocks compared to the underlying, reversely polarized rocks.

The clastic rocks that fill the flanking basins are magnetically transparent. The increase in the depth of the underlying magnetic basement results in a lowering of the magnetic surface and observed aeromagnetic field intensity. This creates, in effect, a filtering out of the shorter-wavelength components of the field. On the "Aeromagnetic Map of Iowa" (Zietz et al., 1976), this produces a series of smooth, low-intensity magnetic minima with a general trend parallel to the axis of the rift and flanking the Iowa Horst. A close examination of detailed aeromagnetic data in this region (e.g., Henderson and Vargo, 1965) clearly shows the limits of the magnetic (igneous) rocks of the Iowa Horst and their lateral contact with the nonmagnetic (clastic) rocks of the flanking basins. The shallow depth of the magnetic basalts on the Iowa Horst produces a high-relief anomaly surface characterized by closely spaced, tightly curving contours. Off the horst, the greater depth of the magnetic basement beneath the clastics produces much more subdued relief with more widely spaced, smoothly curving contours.

Similarly, the zero edges of the clastic rocks in the flanking basins of the MRS in Iowa can be approximated by the increase in the complexity of the contours as the distance between the magnetic basement and the magnetometer decreases and the shorter wavelength components of the magnetic field are detected. This zero edge can only be approximated, because the variability of crystalline basement lithologies, their magnetic characteristics, and their elevations makes it impossible to identify the precise edge of the clastic rocks. The zero edge of Keweenawan Supergroup clastic rocks in Iowa were approximated using a series of preliminary aeromagnetic anomaly maps (with a 20-gamma contour interval) produced by the U.S. Geological Survey and Iowa Geological Survey.

Reflection seismic data over the MRS in Iowa

Reflection seismic data originally collected by Petty-Ray Geophysical Geosource, Inc., were released by Halliburton Geophysical Services, Inc., for investigation by Anderson (1992). Data released included final stack and migrated profiles and profile location maps along seven lines (Fig. 9). Three east-west lines (profiles 7, 12, and 9) completely cross the Iowa Horst and extend over parts of both flanking basins (although profile 7 displayed no useful information below Phanerozoic reflectors). Four north-south or east-west lines (profiles 8, 11, 10, and 13) cross most of the Iowa Horst and portions of one flanking basin. In his study, Anderson (1992) utilized interpretive stick diagrams that were manually constructed from the migrated profiles. Interpretations were constrained by general rift geometry, lithologies, and seismic velocities described by earlier workers in the Lake Superior outcrop belt, Iowa drill data (especially information from the M. G. Eischeid well), magnetic data displayed on unpublished preliminary survey maps and maps by Henderson and Vargo (1965) and Zietz et al. (1976), the "Bouguer Gravity Anomaly Map of Iowa" (Anderson, 1981), and by modeling in conjunction with two-dimensional gravity models along the seismic profiles extended.

Two-dimensional gravity modeling along seismic profiles

To better understand the structure of the MRS in Iowa, Anderson (1992) modeled a series of cross-sectional profiles across the feature. The locations of these profiles were based on the locations of five of the reflection seismic lines. Initial visual interpretation of the migrated seismic data was used as the basis for the development of a series of coincident two-dimensional gravity profiles.

Gravity modeling was conducted using GRAVMAG, a two-dimensional modeling program created in part from GRAV2D (a two-dimensional gravity modeling program obtained from Purdue University where it had been adapted by Shanabrook, Ciolek, and Locher from equations developed by Talwani and Worzel, 1959).

KEEWEENAWAN SUPERGROUP CLASTIC BASINS IN IOWA

Clastics-filled basins flanking the Iowa Horst

Western flanking basins. The thick sequence of Keweenawan Supergroup red clastics preserved on the western flank the Iowa Horst forms a belt from 35 to 51 km wide that runs continuously across the state of Iowa. Although the belt is apparently a continuous blanket of sedimentary rocks, it is divided into two depocenters, best delineated on the Bouguer gravity anomaly map of Iowa (Fig. 7). The central areas of these basins, the Duncan Basin on the north and the Defiance Basin on the south, are located by closed gravity minima. The Manson Impact Structure is located on the

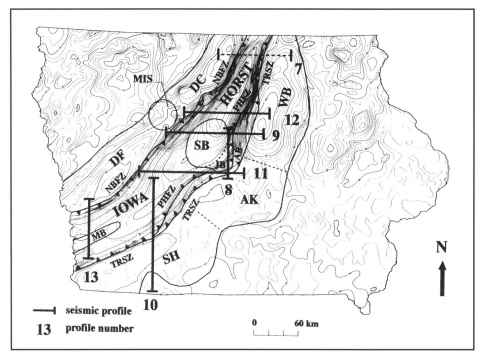

Figure 9. Locations of Petty-Ray Geosource, Inc., reflection seismic profiles and associated gravity model profiles on Bouguer gravity anomaly map of Iowa (modified from Anderson, 1992). Profile 7 provided no useful information about Proterozoic geology. See Figure 1 for explanation of abbreviations.

saddle that separates the basins. Model calculations (Fig. 10) indicate that a total of about 49,500 km^3 of Keweenawan Supergroup clastic rocks are present in the western basins (Table 2).

Duncan Basin. The northernmost of the MRS clastic basins flanking the west side of the Iowa Horst in Iowa, the Duncan Basin (Fig. 8), extends in a southwesterly direction for about 110 km from the Minnesota border. It ranges in width from about 35 to 41 km, although the zero edge of the clastics is difficult to define and may extend as much as 16 km beyond the mapped limits. The basin reaches a maximum modeled depth (Fig. 10) of 8,500 m (Table 2) against the Iowa Horst. Modeled overthrusting of the clastic sequence by the volcanic rocks of the Iowa Horst varies from 5 to 13 km, generally greater on the southern end of the basin. No drill holes penetrate into Keweenawan Supergroup rocks in the Duncan Basin.

The Lower Red Clastic Group in the Duncan Basin displays a relatively uniform model width of about 32 km (Fig. 11). It covers an area of about 3,800 km^2, and has a maximum thickness of about 5,200 m adjacent to the Iowa Horst (Table 2).

The Upper Red Clastic Group in the Duncan Basin (Fig. 12), reaches a maximum thickness of about 2,800 m adjacent to the Iowa Horst.

Defiance Basin. The southernmost of the western flanking basins, the Defiance Basin, includes an area of about 9,000 km^2 in Iowa, extending northwest for about 190 km from the Nebraska border (Fig. 8). The basin ranges in width from about 35 km on the north to about 54 km in more southerly areas. The zero edge is difficult to define and may extend as much as

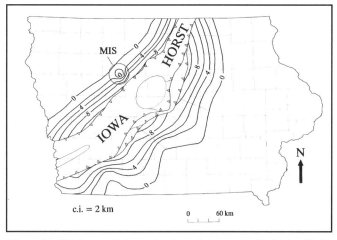

Figure 10. Thickness of the total red clastic groups in the basins flanking the Iowa Horst (Anderson, 1992).

16 km west of its model location. The Keweenawan Supergroup clastic rocks that fill the basin reach a modeled maximum thickness in excess of 10,000 m near the Iowa Horst in the southern portions of the basin (Fig. 10). The volcanic rocks of the Iowa Horst are thrust horizontally from 8 to 16 km over Defiance Basin clastics.

Only two wells penetrate into these Keweenawan Supergroup clastic rocks in the Defiance Basin. One well, the M. G. Eischeid #1, has the thickest continuous section of MRS clastic rocks (4,584

**TABLE 2. SURFACE AREA AND MODEL VOLUMES OF
RED CLASTICS IN KEWEENAWAN SUPERGROUP ROCKS IN IOWA**

Basin	Surface Area (km²)	Maximum Depth (m)	Volume of Clastics		
			Upper Group (km³)	Lower Group (km³)	Total (km³)
Horst-Flanking Basins	**38,100**	**10,500**	**60,000**	**89,200**	**149,200**
Western Basins	**12,800**	**10,500**	**18,100**	**31,400**	**49,500**
Duncan	3,800	8,500	6,200	11,600	15,500
Defiance	9,000	10,500	11,900	19,800	31,700
Eastern Basins	**25,300**	**10,500**	**41,900**	**57,800**	**99,700**
Wellsburg	9,200	10,500	13,200	20,300	30,800
Ankeny	5,900	10,500	9,700	16,400	27,400
Shenandoah	9,200	10,000	19,000	21,100	40,100
Super-Horst Basins	**5,530**	**7,000**	**8,300**	**9,900**	**18,200**
Ames Block	2,250	7,000	7,800	6,200	14,000
Stratford	1,650	3,500	0	2,050	2,050
Jewell	1,090	3,500	0	1,650	1,650
Mineola	540	1,200	500	0	500
Total Red Clastics	**43,630**	**10,500**	**68,300**	**99,100**	**167,400**

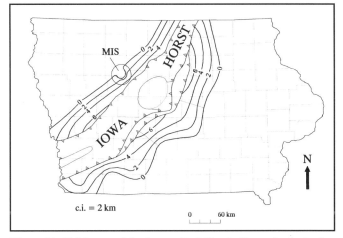

Figure 11. Thickness of the Lower Red Clastic Group in the basins flanking the Iowa Horst (Anderson, 1992).

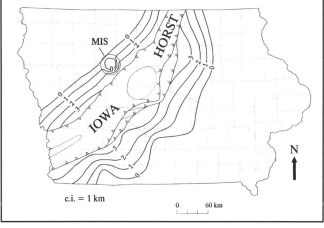

Figure 12. Thickness of the Upper Red Clastic Group in the basins flanking the Iowa Horst (Anderson, 1992).

m) sampled anywhere along the trend of the rift. The second well, the Martin Augustine #1, penetrated 417 m of Keweenawan Supergroup red clastics but produced only poor samples.

The Lower Red Clastic Group in the Defiance Basin (Fig. 11) forms a belt ranging in width from 27 to 45 km, covers an area of about 6,900 km², and displays a maximum model thickness in excess of 6,000 m in the southern area of the basin near the gravity minimum. The Upper Red Clastic Group in the Defiance Basin (Fig. 12) ranges in width from 35 to 56 km and reaches a maximum model depth of about 3,400 m. Model calculation suggests that about 11,900 km³ of the Upper Red Clastic Group are preserved in the Defiance Basin in Iowa.

Eastern flanking basins. On the eastern edge of the Iowa

Horst, an apparently continuous belt of Keweenawan Supergroup red clastic rocks stretches across Iowa from Minnesota to Nebraska. It is larger than the western basin, with clastics covering an area of 25,300 km². The eastern basin clastics have been divided into three basins (Fig. 8), which are, from the north to the south, the Wellsburg, Ankeny, and Shenandoah Basins, their centers defined by closed isogal minima on the Bouguer gravity anomaly map of Iowa (Fig. 7).

Wellsburg Basin. The northernmost of the three eastern clastic basins, the Wellsburg Basin (Fig. 8), extends south in Iowa from the Minnesota border for about 160 km to central Marshall County where it abuts the Ankeny Basin. The basin ranges in width from 21 to 75 km, although its eastern zero edge is poorly constrained and may extend up to 16 km farther east than is mapped. Modeling (Fig. 10) indicates that Keweenawan Supergroup red clastic rocks reach a maximum depth of about 10,500 m in the basin, and these rocks are penetrated by only one well, #1 Huntley in Butler County, which sampled 392 m of the clastics. The red clastics in the Wellsburg Basin are overthrust by the Iowa Horst (as observed on seismic profiles 9 and 12), from about 5 km on the north to as much as 13 km in the area of the Ames Block.

The Lower Red Clastic Group in the Wellsburg Basin (Fig. 11) forms a belt about 18 to 56 km wide. The Lower Red Clastic in the basin reach a maximum model thickness of almost 7,500 m.

The Upper Red Clastic Group in the Wellsburg Basin displays a maximum thickness in excess of 4,800 m where the central portions of the basin abut the Iowa Horst (Fig. 12).

Ankeny Basin. The central of the three eastern flanking MRS basins in Iowa is the Ankeny Basin (Fig. 8). The basin extends for about 110 km between the Wellsburg and Shenandoah Basins. It ranges in width from about 37 to 80 km, although the eastern zero edge of the clastics is poorly defined. Modeling (Fig. 10) indicates a maximum thickness of Keweenawan Supergroup red clastics rocks in excess of 7,500 m. The red clastics in the Ankeny Basin were sampled by five wells, the Nevada City Town Well #3, which penetrated 50 m of the strata at the north end of the basin, and four wells of the Dallas City gas storage structure on the south end. The Dallas City gas storage structure wells include three cores, the McCallum A-1 (6 m of red clastics), Berdie Lehman #1 (12 m), and the Rhinehart A-1 (5 m), and one well with cuttings samples only, Broderick #6 (4 m). Iowa Horst rocks are thrust from 8 to 13 km over the clastic rocks in the Ankeny Basin.

Lower Red Clastic Group rocks in the Ankeny Basin (Fig. 11) range in width from about 40 to 56 km and cover an area of about 4,100 km². Modeling indicates that the Lower Red Clastic Group rocks in the Ankeny Basin reach a maximum thickness of about 7,000 m.

The Upper Red Clastic Group in the Ankeny Basin reaches a maximum thickness (Fig. 12) of about 3,900 km near the Iowa Horst in the center of the basin.

Shenandoah Basin. The Shenandoah Basin, the southernmost of the eastern flanking basins in Iowa (Fig. 8). The basin is divided into two subbasins by an area of positive relief on the crystalline basement, roughly coincident with the Central Plains Suture Zone as mapped by Anderson (1988) and shown on Figure 1. The basin extends for about 160 km from the Nebraska border to the Ankeny Basin. It displays a maximum width, in the northern lobe, of about 75 km and a minimum width of 42 km between the two lobes, although the eastern zero edge is not well constrained. Keweenawan Supergroup red clastics rocks in the Shenandoah Basin range up to about 10,000 m in thickness (Fig. 10). The rocks of the Iowa Horst are thrust over the red clastic rocks of the Shenandoah Basin from 8 to 16 km.

Two, and possibly three, wells penetrate into the Keweenawan Supergroup red clastics of the Shenandoah Basin. The #1 Wilson in Page County sampled 182 m of red clastic rocks, and the Gulf Energy 16-13 Poetker in Mills County penetrated 24 m of red clastics. Additionally, a third well, the Ohio Oil Company Wisnom #1 penetrated 60 m of quartzarenite below 945 m. The stratigraphy of this quartzarenite was initially uncertain, since the only quartzarenite known in this portion of the geologic column is in the basal Cambrian sandstone, which rarely exceeds 30 m in thickness in this region. The anomalous quartzarenite in the Wisnom well may represent Keweenawan Supergroup rocks deposited in a depositional environment similar to the Hinckley or Devils Island Sandstone (Bayfield Group) in the Lake Superior region.

Lower Red Clastic Group rocks in the Shenandoah Basin (Fig. 11) form a belt ranging from 24 to 59 km in width and covering an area of about 7,200 km². The unit reaches a maximum depth of about 6,000 m in the basin.

Upper Red Clastic Group rocks in the Shenandoah Basin (Fig. 12) reach a maximum thickness of almost 3,500 m near the Iowa Horst in the southern part of the basin.

MRS clastic rocks on the Iowa Horst

Keweenawan Supergroup red clastic rocks are known to be preserved in four places on the Iowa Horst in Iowa. These rocks were apparently structurally preserved from the erosion that removed related rocks from other areas of the horst. They are preserved on the Ames Block and in the Stratford, Jewell, and Mineola Basins (Fig. 8).

Ames Block. Although the uniformity of the uplift of the Iowa Horst, as documented by the long, continuous, horizontal reflectors seen on the seismic profiles of the volcanic rocks, is remarkable (Anderson, 1992), several major structural features have been identified on the horst. Key among these is the Ames Block, an elongate feature that trends along the eastern edge of the Iowa Horst (Fig. 8). The Ames Block is bounded on the west by the Perry-Hampton Fault Zone (Anderson, 1992) and on the east by the Thurman-Redfield Structural Zone.

The Ames Block is floored by Keweenawan volcanic rocks an is considered a part of the Iowa Horst. Both Lower Red Clastic Group and Upper Red Clastic Group rocks have been identified on the block (Anderson, 1992), and both units have thicknesses over much of the block that are nearly identical to thicknesses modeled in the adjoining Wellsburg and Ankeny Basins. The Ames Block, as a part of the Iowa Horst, has been uplifted with respect to the flanking basins. This uplift displays a maximum displacement of about 3 km at its northern and south-central regions, but is reduced to only about 1.6 km between.

The Ames Block is down-dropped with respect to the main body of the Iowa Horst, which lies west of the Perry-Hampton Fault Zone. Modeling of the apparent displacement, controlled by seismic interpretations, ranges from a minimum value of 3.4 km along the southern profile 11 (Fig. 13C), to a maximum of over 7 km along profile 9, and 4.3 km near the northern end of the block along seismic profile 12. The Keweenawan Supergroup red clastics cover 2,250 km² (Table 2) of the Ames Block.

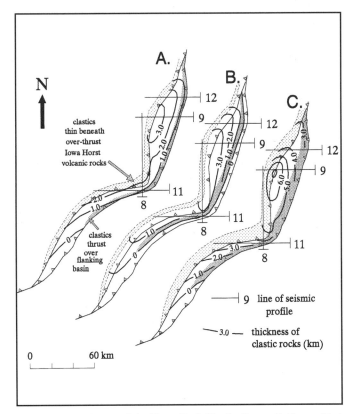

Figure 13. Thickness of: A, Upper Red Clastic Group; B, Lower Red Clastic Group; and C, the total red clastic groups on the Ames Block (Anderson, 1992).

The Lower Red Clastic Group rocks on the Ames Block range in thickness from zero on the southern end of the block, increasing rapidly to about 1.3 km along seismic profile 11 and to maximum thickness in excess of 3 km along profile 9 (Fig. 13B). Like the Lower Red Clastic Group rocks of the adjoining basins, the unit increases in thickness to the west.

The Upper Red Clastic Group rocks on the Ames Block are also similar in thickness to related rocks in the adjoining flanking basin (Fig. 13A). The unit ranges in thickness from zero on the southern end of the block to a maximum of about 3.5 km along profile 9 and about 2 km along profile 12. Both Upper and Lower Red Clastic Group rocks are apparently erosionally beveled along the southern end of the Ames Block, where drill data confirm the presence of MRS volcanic rocks at the Precambrian surface.

Available information suggests that the uplift of the Iowa Horst included reactivation and reversal of normal faults that bounded the central graben of the MRS in Iowa. However, along the central area of the eastern margin of the graben, a new reactivation structure formed, the Perry Hampton Fault Zone. The thick sequence of Upper Red Clastic Group rocks on the Ames Block suggests that deposition of the unit on the block occurred at the same time as in the adjoining flanking basins, so the Ames Block was not uplifted with the Iowa Horst. Then, at

some late stage, the Ames Block was uplifted, probably along with the last stage of Iowa Horst movement. Additional complexity is added by the absence of clastic rocks at the southern end of the block, as confirmed by well data. This southern end may have remained attached to the main portion of the Iowa Horst, and moved steadily upward with it.

Stratford Basin. The Stratford Basin (Fig. 8), and the associated Stratford Geophysical Anomaly, was named for the centrally located town of Stratford, on the Hamilton-Webster county line by Osweiler (1982), who interpreted the feature as an elliptical, clastic-filled basin with an area of 1,650 km² that formed along the axis of the MRS. Osweilder demonstrated that an earlier interpretation of the feature as a clastic-roofed granitic lopolith (Cohen, 1966) was not consistent with the gravity and magnetic anomalies and known MRS geology. She suggested that a clastic-filled basin, floored by MRS volcanic rocks and similar to the Twin City Basin in Minnesota (King and Zietz, 1971), was the best interpretation.

Using a gravity contrast of 0.6 g/cc between the clastics and the underlying mafic volcanic rocks, Osweiler (1982) modeled the basin as elongate, about 56 km long, trending in a northeasterly direction (along the axis of the MRS), and about 37 km wide. Her model basin reached a maximum depth of about 950 m below the base of overlying Phanerozoic rocks. The basin was also investigated by Anderson (1992) who modeled it using seismic profile 9, on which it displays a maximum depth of 0.7 seconds (about 1,660 m assuming a velocity of 4,740 m/sec for the basin-filling clastics). This depth required a density of 2.65 g/cc (density contrast of only 0.2 g/cc with the underlying basalts) in order to produce a gravity model consistent with the seismic interpretations. Seismic profile 9 crosses

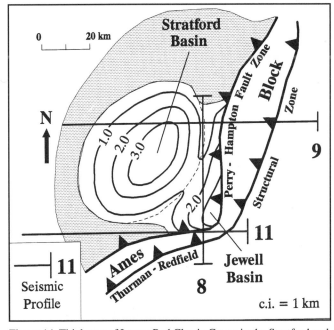

Figure 14. Thickness of Lower Red Clastic Group in the Stratford and Jewell Basins (Anderson, 1992).

the Iowa Horst north of the deepest portions of the Stratford Basin (as modeled with gravity data). The Stratford Basin as delineated by Anderson (1992) has a width and length similar to those determined by Osweiler (1982), but using a density contrast of 0.2 g/cc, his model (Fig. 14) reaches a greater maximum depth, about 3,600 m.

The density (2.65 g/cc) used in modeling the Stratford Basin was greater than the average density (2.55 g/cc) used in modeling the Lower Red Clastic Group rocks in the adjoining Wellsburg Basin along profile 9. The density of the Lower Red Clastic Group rocks in the Stratford Basin may be increased by a large component of mafic volcanic rock fragments (similar to clasts in the Copper Harbor Conglomerate). This theory is supported by a description of the rocks encountered during the drilling of the Ogden City #2 well in 1929. The Ogden City #2 well penetrated 31 m of Lower Red Clastic Group rocks. Although no samples exist for the basal 2 m of the well, the lowest samples collected, from 854 m, were described by Norton (1928) as litharenite with abundant feldspar, mafic igneous rock fragments, and mafic minerals. This lithology is similar to the Copper Harbor Conglomerate, and may be the graben-fill equivalent of Unit B of the Lower Red Clastic Group in the M. G. Eischeid well. Another possible explanation for the higher densities of rocks in the Stratford Basin is the deep burial experienced by Lower Red Clastics Group rocks along the axis of the MRS. This burial may have produced a more thorough cementation, similar to cementation observed in the dense clastic rocks encountered in the Amerada Schroeder well that lies in a similar position on the Iowa Horst in Nebraska.

In addition to the Ogden City well, the Boone City #1 well, drilled in 1890, also apparently penetrated into the clastics of the Stratford Basin. It encountered a sequence of red shales and soft red sandstones, here interpreted as Lower Red Clastic Group rocks. A total of 90 m of these rocks was penetrated in the Boone City #1 well.

Jewell Basin. The Jewell Basin (Fig. 8) was defined by Anderson (1992) as the area of the Iowa Horst that is capped by Lower Red Clastic Group rocks between the elliptical Stratford Basin and the Perry-Hampton Fault Zone. In the area of the Jewell Basin, some of the structural complexities associated with the uplifting of the Iowa Horst appear to be related to its intersection with the Fayette Structural Zone (Fig. 1). As the horst was forced upward, the area of the Jewell Basin displayed a lesser amount of uplift, structurally preserving Lower Red Clastic Group rocks. Seismic profile 8 (Fig. 14) crosses the Jewell Basin and displays a continuously thickening clastic sequence ranging from about 450 m in thickness on the north to about 3,800 m on the south. Interpretation of available data suggests that the Jewell Basin was never a separate depositional basin. It apparently contains Lower Red Clastic Group rocks that were originally deposited uniformly within the central graben of the MRS in Iowa, then were structurally preserved by differential uplift of the Iowa Horst. No drill holes penetrate Keweenawan Supergroup clastic rocks in the Jewell Basin.

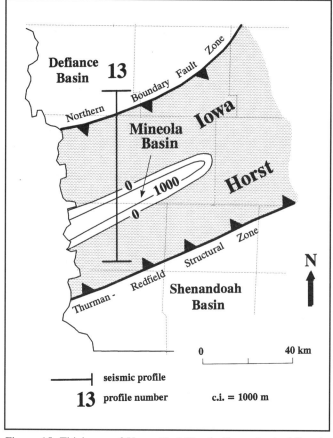

Figure 15. Thickness of Upper Red Clastic Group in the Mineola Basin (Anderson, 1992). Contour interval = 2 km.

Mineola Basin. The southernmost of the clastic basins identified on the Iowa Horst in Iowa is the Mineola Basin (Fig. 8). It was originally referred to as the Mineola Graben, after early gravity modeling suggested a clastic-filled graben with its center near the northern Mills County town of Mineola (Anderson, 1988). However, interpretation of seismic profile 13 clearly displays a basin structure with no prominent bounding faults. On seismic profile 13 the basin appears to be about 14 km wide, with a maximum depth of 0.8 seconds. Modeling to match observed gravity values with geometeries interpreted from seismic data along profile 13 using Lower Red Clastic Group densities proved difficult. However, using Upper Red Clastic Group densities in a basin about 1,500 m deep (assuming Upper Red Clastic Group velocity of 3,960 m/sec), the model fit the observed data very well. Therefore, it appears that the Mineola Basin is filled with sedimentary rocks displaying densities typical of Upper Red Clastic Group rocks.

The limits of the Mineola Basin are best delineated on the "Aeromagnetic Map of Iowa" (Zietz et al., 1976). Interpretation of these data suggests that the basin is about 65 km long (Fig. 15), extending up the axis of the rift from the Nebraska border to eastern Pottawattamie County, Iowa. The basin displays a relatively continuous width of about 14 km over most of its length, with a maximum depth of about 1,500 m.

The interpreted presence of Upper Red Clastic Group sedimentary rocks in the Mineola Basin suggests that it must have formed in the late stages of the uplifting of the Iowa Horst, after the erosion of Lower Red Clastic Group rocks and the unroofing of horst volcanic rocks. This suggests that the Mineola Basin is the youngest Keweenawan age structural feature on the Iowa Horst. It is also possible the Mineola Basin contains Lower Red Clastic Group sediments that display an anomalously low density, perhaps due to lithologies, cementation, or alteration characteristics different from other Lower Red Clastic Group rocks. The much younger Paleozoic (Pennsylvanian) Glenwood Syncline (Hershey et al., 1960) is nearly coincident with the Mineola Basin and may represent structural reactivation of the feature.

CONCLUSIONS

Although a total of about 167,400 km^3 of Keweenawan Supergroup Upper and Lower Red Clastic Groups sedimentary rocks are preserved within an area of 43,630 km^2 of the Precambrian surface of Iowa, these rocks remain largely unknown. The M. G. Eischeid #1 deep petroleum test well penetrated 4,584 m of these clastic rocks, but only 12 additional wells, totaling 966 m, also sample Red Clastics rocks in the state.

A large volume of red clastic groups rocks are preserved in the basins that flank the axial Iowa Horst within the Midcontinent Rift System in Iowa. Gravity modeling, controlled by seismic interpretations, indicates a cumulative volume of about 167,400 km^3 of these clastic rocks in the five basins that flank the Iowa Horst in Iowa. These clastic rocks have been subdivided into two groups, the Upper Red Clastic Group with four component formations and the basal Lower Red Clastic Group and its three component formations. These two groups are closely similar in many ways to the Keweenawan Supergroup Bayfield and Oronto Groups, in the Lake Superior region of Wisconsin, and the units are probably nearly coeval.

Keweenawan Supergroup clastic rocks are also preserved in three structural basins on the Iowa Horst and on a horst-marginal block. A total of about 18,200 m^3 of Keweenawan Supergroup clastic rocks are preserved on the Iowa Horst. These rocks are interpreted as the Lower Red Clastic Group in the Stratford and Jewell Basins, Upper Red Clastic Group in the Mineola Basin, and both Upper and Lower Red Clastic Groups on the Ames Block.

Dark gray and black shales and mudstones were observed over a 900-m interval in the Lower Red Clastic Group of the M. G. Eischeid #1 well. The identification of similar (and probably related) units on the central peak of the Manson Impact Structure indicates that the body of standing water in which these organic-rich clastic rocks were deposited was probably at least 140 km wide and possibly extended along the entire 480 km length of the Iowa segment of the MRS. The body of water may have been marine, and may have continued even farther to connect with the ocean to the east or south and may have extended the entire length of the MRS.

ACKNOWLEDGMENTS

The author wishes to thank and acknowledge Richard Ojakangas and Bert Dickas for their reviews of this manuscript and helpful suggestions.

REFERENCES CITED

Anderson, R. R., 1981, Bouguer gravity anomaly map of Iowa: Iowa Geological Survey Miscellaneous Map Series 7, 1:100,000 scale.

Anderson, R. R., 1988, Precambrian geology of Iowa and adjacent areas: Iowa Department of Natural Resources Geological Survey Bureau, Open-File Map 88-3, 1:100,000 scale.

Anderson, R. R., editor, 1990a, The Amoco M. G. Eischeid #1 deep petroleum test, Carroll County, Iowa; Report of preliminary investigations: Iowa Department of Natural Resources Geological Survey Bureau, Special Report Series no. 2, 185 p.

Anderson, R. R., 1990b, Review of the Precambrian geological history of the central United States and the Midcontinent Rift System, *in* Anderson, R. R., ed., The Amoco M. G. Eischeid #1 deep petroleum test, Carroll County, Iowa; Report of preliminary investigations: Iowa Department of Natural Resources Geological Survey Bureau, Special Report Series no. 2, p. 1–26.

Anderson, R. R., 1990c, Interpretation of geophysical data over the Midcontinent Rift System in the area of the M. G. Eischeid #1 oil test, Carroll County, Iowa, *in* Anderson, R. R., ed., The Amoco M. G. Eischeid #1 deep petroleum test, Carroll County; Iowa: Report of preliminary investigations: Iowa Department of Natural Resources Geological Survey Bureau, Special Report Series no. 2, p. 27–38.

Anderson, R. R., 1992, The Midcontinent Rift of Iowa [Ph.D thesis]: Iowa City, University of Iowa, 324 p.

Anderson, R. R., 1993, Overview of petroleum potential of Proterozoic clastic rocks of the Midcontinent Rift System in Iowa: Geological Society of America Abstracts with Program, v. 25, no. 3, p. 2–3.

Anderson, R. R., and McKay, R. M., 1996, Clastic rocks associated with the Midcontinent Rift System in Iowa: U.S Geological Survey Bulletin 1989i, in press.

Barker, C. E., 1990, Reconnaissance fluid inclusion study, Amoco Eischeid #1 well, Defiance Basin, Midcontinent Rift System, Iowa, *in* Anderson, R. R., ed., The Amoco M. G. Eischeid #1 deep petroleum test, Carroll County, Iowa: Report of preliminary investigations: Iowa Department of Natural Resources Geological Survey Bureau, Special Report Series no. 2, p. 169–174.

Barnes, D. A., 1990, Sandstone petrology from conventional core in the M. G. Eischeid #1 well, Carroll County, Iowa. *in* Anderson, R.R., ed., The Amoco M. G. Eischeid #1 deep petroleum test, Carroll County, Iowa; Report of preliminary investigations: Iowa Department of Natural Resources Geological Survey Bureau, Special Report Series no. 2, p. 113–118.

Burnie, S. W., Schwartz, H. P., and Crocket, J. H., 1972, A sulfur isotope study of the White Pine mine, Michigan: Geological Society of America, Abstracts with Program, v. 4, no. 7, p. 462.

Carmichael, R. S., and Black, R. A., 1986, Analysis and use of MAGSAT satellite magnetic data for interpretation of crustal structure and character in the U.S. Midcontinent: Physics of Earth and Planetary Interiors, v. 44, no. 4, p. 333–347.

Cohen, T. J., 1966, Explosion seismic studies of the Mid-Continent Gravity High [Ph.D. thesis]: Madison, University of Wisconsin, 329 p.

Craddock, C., 1972, Regional geologic setting, *in* Sims, P. K., and Morey, G. B., eds., Geology of Minnesota: A centennial volume: Minnesota Geological Survey, p. 281–291.

Daniels, P. A., 1982. Upper Precambrian sedimentary rocks: Oronto Group, Michigan-Wisconsin, *in* Wold, R. J., and Hinze, W. J., eds., Geology and tectonics in the Lake Superior Basin: Geological Society of America Memoir 156, p. 107–134.

Dickas, A. B., 1986, Comparative Precambrian stratigraphy and structure along the Mid-Continent Rift: American Association of Petroleum Geologists Bulletin, v. 70, no. 3, p. 225–238.

Dickas, A.B., and Mudrey, M. G., Jr., 1989, Central North American case for segmented rift development: Abstracts, 28th International Geological Congress, v. 1, p. I-396–I-397.

Elmore, R. D., 1981, The Copper Harbor Conglomerate and Nonesuch Shale: Sedimentation in a Precambrian intercratonic rift, Upper Michigan [Ph.D. thesis]: Ann Arbor, University of Michigan, 200 p.

Folk, R. L., 1968, Petrology of Sedimentary Rocks: Austin, Texas, Hemphill's University of Texas Geology, 170 p.

Green, J. C., 1982, Geology of Keweenawan extrusive rocks, in Wold, R. J., and Hinze, W. J., eds., Geology and tectonics of the Lake Superior Basin: Geological Society of America Memoir 156, p. 47–56.

Halls, H. C., and Pesonen, L. J., 1982, Paleomagnetism of Keweenawan rocks, in Wold, R. J., and Hinze, W. J., eds., Geology and tectonics of the Lake Superior Basin: Geological Society of America Memoir 156, p. 173–202.

Henderson, J. R., and Vargo, J. L., 1965, Aeromagnetic map of central Iowa: U.S. Geological Survey, Geophysical Investigations Map GP-476, 1:250,000 scale.

Hershey H. G., Brown, C. N., van Eck, O. J., and Northup, R. C., 1960, Highway construction materials from the consolidated rocks of southwestern Iowa: Iowa Highway Research Board, Bulletin 15, 151 p.

Hieshima, G. B., and Pratt, L. M., 1991, Sulfur/carbon ratios and extractable organic matter of the Middle Proterozoic Nonesuch Formation, North American Midcontinent Rift System: Precambrian Geology, v. 54, no. 1, p. 65–79.

Hieshima, G. B., Zaback, D. A., and Pratt, L. M., 1989, Petroleum potential of Precambrian Nonesuch Formation, Mid-Continent Rift System: American Association of Petroleum Geologists Bulletin, v. 73, no. 3, p. 363.

Hinze, W. J., Braile, L. W., and O'Hara, N. W., 1982, Gravity and magnetic anomaly studies of Lake Superior, in Wold, R. J., and Hinze, W. J., eds., Geology and tectonics of the Lake Superior Basin: Geological Society of America Memoir 156, p. 203–222.

King, P. B., and Zietz, I., 1971, Aeromagnetic study of the Midcontinent Gravity High of central United States: Geological Society of America Bulletin, v. 82, p. 2187–2208.

Kisvarsanyi, E. B., 1984, The Precambrian tectonic framework of Missouri as interpreted from magnetic anomaly map: Missouri Division of Geology and Land Survey, Contributions to Precambrian Geology, v. 14, part B, 19 p.

Ludvigson, G. A., and Spry, P. E., 1990, Tectonic and hydrologic significance of carbonate veinlets in the Keweenawan sedimentary rocks of the Amoco M. G. Eischeid #1 drillhole, in Anderson, R. R., ed., The Amoco M. G. Eischeid #1 deep petroleum test, Carroll County, Iowa; Report of preliminary investigations: Iowa Department of Natural Resources Geological Survey Bureau, Special Report Series no. 2, p. 153–168.

Ludvigson, G. A., McKay, R. M., and Anderson, R. R., 1990, Petrology of Keweenawan sedimentary rocks in the M. G. Eischeid #1 drillhole, in Anderson, R. R., ed., The Amoco M. G. Eischeid #1 deep petroleum test, Carroll County, Iowa; Report of preliminary investigations: Iowa Department of Natural Resources Geological Survey Bureau, Special Report Series no. 2, p. 77–112.

McKay, R. M., 1990, The problem of the Cambrian–"Red Clastics" contact in the M. G. Eischeid #1 well, in Anderson, R. R., ed., The Amoco M. G. Eischeid #1 deep petroleum test, Carroll County, Iowa; Report of preliminary investigations: Iowa Department of Natural Resources Geo-

logical Survey Bureau, Special Report Series no. 2, p. 59–66.

Milavec, G. J., 1986, The Nonesuch Formation: Pre-Cambrian sedimentation in an intercratonic rift [M.S. thesis]: Norman, University of Oklahoma, 142 p.

Morey, G. B., and Ojakangas, R. W., 1982, Keweenawan sedimentary rocks of eastern Minnesota and northwestern Wisconsin, in Wold, R. J. and Hinze, W. J., ed., Geology and tectonics of the Lake Superior Basin: Geological Society of America, Memoir 156, p. 135–146.

Norton, W. H., 1928, Deep wells of Iowa (A supplementary report): Iowa Geological Survey Annual Report 33, p. 9–447.

Ojakangas, R. W., 1986, Reservoir characteristics of the Keweenawan Supergroup, Lake Superior region, in Mudrey, M. G., Jr., ed., Precambrian petroleum potential, Wisconsin and Michigan: Geoscience Wisconsin, v. 11, p. 25–31.

Ojakangas, R. W. and Morey, G. B., 1982, Keweenawan sedimentary rocks of the Lake Superior region, in Wold, R. J., and Hinze, W. J., ed., Geology and tectonics of the Lake Superior Basin: Geological Society of America Memoir 156, p. 157–164.

Osweiler, D. J., 1982, A geological interpretation of the Stratford Geophysical Anomaly on the Midcontinent paleorift zone, central Iowa [M.S. thesis]: Iowa City, University of Iowa, 166 p.

Palacas, J. G., Schmoker, J. W., Pawlewicz, M. J., Davis, T. A., and Anderson, R. R., 1990, Petroleum source-rock assessment of Middle Proterozoic (Keweenawan) sedimentary rocks, Eischeid #1 well, Carroll County, Iowa, in Anderson, R. R., ed., The Amoco M. G. Eischeid #1 deep petroleum test, Carroll County, Iowa; Report of preliminary investigations: Iowa Department of Natural Resources Geological Survey Bureau, Special Report Series no. 2, p. 119–134.

Polastro, R. M., and Finn, T. M., 1990, Clay mineralogy and bulk rock composition of selected core and cutting samples from the Amoco Eischeid #1 well, Carroll County, Iowa—A reconnaissance study from X-ray powder diffraction analysis, in Anderson, R. R., ed., The Amoco M. G. Eischeid #1 deep petroleum test, Carroll County, Iowa; Report of preliminary investigations: Iowa Department of Natural Resources Geological Survey Bureau, Special Report Series no. 2, p. 119–134.

Sims, P. K., 1990, Precambrian basement map of the northern midcontinent, U.S.A.: U.S. Geological Survey, Miscellaneous Map Series Map I-1853-A, 1:1,000,000 scale, and accompanying text, 10 p.

Talwani, M., and Worzel, J. L., 1959, Rapid computation of the gravitational attraction of three-dimensional bodies of arbitrary shape: Journal of Geophysical Research, v. 64, p. 49–59.

Thwaites, F. T., 1912, Sandstones of the Wisconsin coast of Lake Superior: Wisconsin Geological and Natural History Survey Bulletin, v. 25, 117 p.

Van Schmus, W. R., Wallin, E. T., and Shearer, C. K., 1990, Age relationships for gabbro from the Amoco M. G. Eischeid #1 well, in Anderson, R. R., ed., The Amoco M. G. Eischeid #1 deep petroleum test, Carroll County, Iowa; Report of preliminary investigations: Iowa Department of Natural Resources Geological Survey Bureau, Special Report Series no. 2, p. 67–76.

Witzke, B. J., 1990, General stratigraphy of the Phanerozoic and Keweenawan sequences, M. G. Eischeid #1 drillhole, Carroll County, Iowa, in Anderson, R. R., ed., The Amoco M. G. Eischeid #1 deep petroleum test, Carroll County, Iowa; Report of preliminary investigations: Iowa Department of Natural Resources Geological Survey Bureau, Special Report Series no. 2, p. 32–58.

Zietz, I., Gilbert, F. P., and Keys, J. R., 1976, Aeromagnetic map of Iowa: color coded intensities: U.S. Geological Survey, Geophysical Investigations Map GP-911, 1:1,000,000 scale.

MANUSCRIPT ACCEPTED BY THE SOCIETY JANUARY 16, 1996

Geological Society of America
Special Paper 312
1997

Tectonic implications and influence of the Midcontinent Rift System in Nebraska and adjoining areas

Marvin P. Carlson
*Nebraska Geological Survey, University of Nebraska, Conservation and Survey Division, 901 North 17th Street,
113 Nebraska Hall, Lincoln, Nebraska 68588-0517*

ABSTRACT

Southeastern Nebraska is underlain by Early Proterozoic rocks of the Central Plains (1,600 to 1,800 Ma) and Penokean (1,830 to 1,890 Ma) Provinces. The Midcontinent Rift System was imposed upon and reflects the preexisting tectonic grain of the older basement. During Paleozoic time several periods of structural change occurred related to the Midcontinent Rift System. In the same geographic area, there was successive development of the Southeast Nebraska arch, the North Kansas basin, the Nemaha uplift, and the Table Rock arch. The locations of these regional features were controlled by the patterns established by rift tectonics. Features such as the Eastern Nebraska high, the Abilene Anticline, the Union fault, and the Burchard fault are reactivations of specific rift features.

INTRODUCTION

Nebraska is underlain by Early Proterozoic rocks of the Central Plains (1,600 to 1,800 Ma) and Penokean (1,830 to 1,890 Ma) Provinces (Sims, 1990). These rocks were intruded by pre-rift Middle Proterozoic epizonal plutons (1,340 to 1,480 Ma). The Midcontinent Rift System (MRS; 1,100 Ma) was imposed upon and reflects the preexisting tectonic grain of this older basement. This paper summarizes the relationships between the MRS and the structures that developed during the Phanerozoic.

RIFT FRAMEWORK

The correspondence of the trends of the rift segments with older tectonic zones has been noted by several authors including Klasner et al. (1982), Carlson (1988a), and Sims (1990; Fig. 1). On the western limb of the MRS, the segment partially exposed near Lake Superior is parallel to the Great Lakes tectonic zone. The Minnesota and Iowa segments subparallel the Penokean suture, termed the Niagara fault zone in Wisconsin and the Storm Lake tectonic zone in Iowa. An apparent offset between these segments occurs as the rift crosses this suture. The Nebraska segment is an extension of the southwest-trending structures within the

Penokean orogen noted in Wisconsin, which in turn are reflected in the Paleozoic rocks across Iowa. The Kansas segment reflects preexisting tectonic grain within the Central Plains Province. The separation and difference in trends between the Nebraska and Kansas segments are probably indicative of the boundary between the Central Plains and Penokean Provinces.

The boundaries and internal framework of the Nebraska segment are defined by both geophysical information and data from drill holes. Figure 2 illustrates the location of drill holes that penetrate into Precambrian basement in southeastern Nebraska. Most samples are from rotary cuttings and penetration is typically only several meters. The rock types were determined from petrographic studies of these samples by Lidiak (1972) and by Carlson and Treves (1988). Additional data are on open file at the Nebraska Geological Survey. The gneiss and granite represent samples of the older Proterozoic basement of the Central Plains and Penokean Provinces. The distribution of the basalt illustrates the location of both the Nebraska and Kansas rift segments. The rift-related sedimentary rocks are present both in grabens preserved within the central horst and in flanking basins. The Elk Creek Carbonatite (Fig. 2) is dated at 545 Ma (Z. E. Peterman, personal communication, 1985), much younger than formation of the MRS. However, the loca-

Carlson, M. P., 1997, Tectonic implications and influence of the Midcontinent Rift System in Nebraska and adjoining areas, *in* Ojakangas, R. W., Dickas, A. B., and Green, J. C., eds., Middle Proterozoic to Cambrian Rifting, Central North America: Boulder, Colorado, Geological Society of America Special Paper 312.

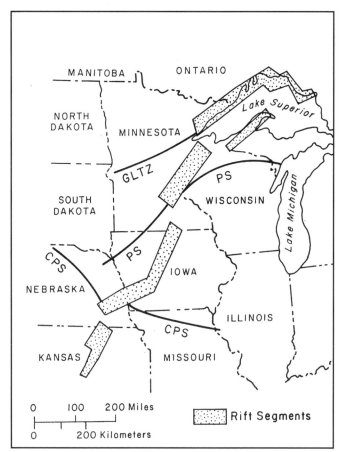

Figure 1. General trends and major rift segments of the western limb of the Midcontinent Rift System and the locations of the Great Lakes tectonic zone (GLTZ), the Penokean suture (PS), and the Central Plains suture (CPS).

tion of this intrusive body was probably controlled by the tectonic grain established during rifting.

RIFT DEVELOPMENT

The general pattern of evolution of the MRS is consistent along its western limb. Initial filling of the rift graben by volcanic and overlying sedimentary rocks was followed by structural inversion that created an axial horst of mafic rocks. Basins adjacent to this horst were filled with sediment derived regionally and from erosion of the exposed former rift rocks. However, on the volcanic horst, some of the early post-volcanic sediments were preserved in small faulted basins. The data for Nebraska support this pattern of evolution of the rift system. A core taken in Saunders County, Nebraska, contains basalt from the horst (Fig. 2) and is characterized by multiple flow units. These flows were dated at 990 Ma (Lidiak, 1972), a K-Ar age that is in general agreement with ages of volcanic rocks of the Keweenawan Supergroup as exposed in the Lake Superior region. A core taken in Cass County, Nebraska (Fig. 2), contains sedimentary rocks that were preserved in one of the small basins within the

horst. The sedimentary package consists of fluvial silty sands, mudstones, granitic gravels and conglomerates, and eolian sheet sands. Clasts and thin intervals of dark mudstone also are present that might be depositional equivalents to the Nonesuch Shale. Although not dated, the sedimentary rocks preserved in this and other similar basins within the horst are genetically equivalent to units such as the Oronto Group deposited within the rift after volcanism ceased and prior to the structural inversion. The tectonic reversal from a major rift graben to a central horst created flanking basins that filled with sediments probably equivalent to the Bayfield Group. These sediments are estimated by Anderson (1992) to be as thick as 4 km in Nebraska although maximum penetration by drilling in Nebraska has been only about 500 m. Lithologies as described in the rotary drill cuttings range from silt to medium quartz sand. The rift-related sedimentary rocks, both within and adjacent to the horst, are less mature than the initial clastic deposits of the Paleozoic.

PALEOZOIC HISTORY

The relationships of the geometry of the Midcontinent Rift segments to the tectonic features of both the early and middle Paleozoic rocks are illustrated in Figure 3. There is no rock record for the Early to Middle Cambrian but some tectonic activity is suggested by the 545-Ma intrusion of the Elk Creek Carbonatite. Depositional patterns of the initial Paleozoic transgression (Late Cambrian to Early Ordovician) illustrate that little relief existed upon the Precambrian surface across the southeast Nebraska area (Carlson et al., 1990). For the most part, the major unconformities and facies created during Early and Middle Paleozoic time are regional and do not represent local tectonic events (Carlson, 1970). There were, however, several periods of localized structural activity during the Paleozoic that were directly related to the Midcontinent Rift System (Carlson, 1988b). In the same geographic area, there was successive development of the Southeast Nebraska arch, the North Kansas basin, and the Nemaha uplift. Later rift-related tectonic features include the Table Rock arch, the Eastern Nebraska high, the Abilene Anticline, the Union fault, and the Burchard fault.

EARLY PALEOZOIC FEATURES

The local absence of Late Cambrian to Early Ordovician sedimentary rocks in parts of southeast Nebraska and northeast Kansas defines the location of the Southeast Nebraska arch (Fig. 3). Uplift and erosion along this arch occurred prior to the deposition of the St. Peter Sandstone in Middle Ordovician time. The lack of bounding faults suggests that rather than reactivation of rift fractures by horizontal maximum compressive stress, the Southeast Nebraska arch is the result of vertical maximum compressive stress, perhaps a thermal event that most strongly affected the complex juncture of the Kansas and Nebraska rift segments.

Although complicated by later uplift and erosion in the Pennsylvanian, thickness patterns of the Middle and Upper Ordovician and Silurian strata define the history of formation of the North Kansas basin. For example, Silurian rocks thicken to over 150 m toward the original axis of the North Kansas basin (Carlson, 1970). The similarity in location, trend, and non-faulted tectonics between the North Kansas basin and the earlier uplift of the Southeast Nebraska arch indicates similar vertical movements in the development of both features.

Subsidence of the North Kansas basin lessened during Devonian and Mississippian time as evidenced by the thickness patterns of these rocks (Carlson, 1988b). The lithology of the uppermost Mississippian and lowermost Pennsylvanian strata demonstrates the general shallowing of the seas that presaged the Ouachita Orogeny along the southern margin of the continent.

LATE AND POST-PALEOZOIC FEATURES

The major effect of the Ouachita orogeny occurred during Middle Pennsylvanian time as maximum horizontal compressive stress was transmitted across the Midcontinent. A variety of tectonic features were created by this contraction superimposed on preexisting structures (Carlson, 1988a). The geometry of the Nemaha trend (Fig. 3) reflects the interaction between both the Kansas and Nebraska rift segments and the contraction generated from the southeast by the Ouachita orogeny. Left-lateral reactivation of the internal faults of the Kansas rift segment, and possible underthrusting, created a series of northwest-trending horsts and grabens that comprise the Nemaha uplift (Berendsen and Blair, 1986). Vertical displacement on the east-bounding fault of the Nemaha uplift (Humboldt fault) exceeds 850 m on the Precambrian surface near the Nebraska-Kansas border (Carlson, 1970).

Geometry of the Phanerozoic structures that resulted from reactivation of the Nebraska rift segment is somewhat different from the geometry related to the Kansas segment due to the more principal compression of the north-northwest-trending Ouachita maximum horizontal forces against the east-northeast-trending rift segment in Nebraska. Within the Nebraska rift segment, reactivation of the original fault-bounded rift partitions propagated outward as wrench faults in the adjacent basement rocks. The northwest-trending wrench faults in the basement, in turn, affected the overlying Paleozoic strata by creating a series

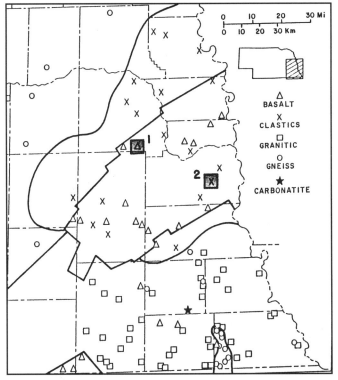

Figure 2. Well location and the rock type of the Precambrian within each well in southeastern Nebraska. Highlighted wells 1 (Saunders County) and 2 (Cass County) are discussed in text.

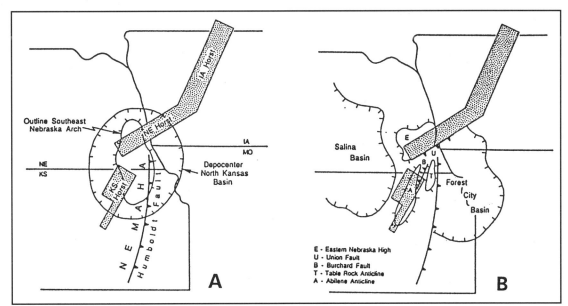

Figure 3. Early Paleozoic structures (A) and middle Paleozoic structures (B) in southeastern Nebraska and adjoining areas.

of southeast-trending, fault-bounded anticlines within the Forest City basin (Carlson, 1989). Some of these fault-bounded anticlines host oil reservoirs. This kind of structural geometry is common in sedimentary rocks that overlie deep-seated wrench faults (Harding, 1974).

In addition to the subsidiary structures composing the Nemaha-Humboldt complex, secondary structures were developed by reactivation of the bounding faults on the Kansas rift segment (Fig. 3). The Abilene Anticline in northern Kansas and the related Burchard fault in Nebraska express reactivation of the eastern boundary fault of the Kansas rift segment. The Union fault is a reactivation of the southern boundary fault of the Nebraska Rift segment. Basin filling and continued downwarping are demonstrated by the thickness patterns of Pennsylvanian strata in the Forest City basin and to a lesser extent in the Salina basin. In post-Permian time, uplift again occurred with smaller displacements (<100 m) than the movements that occurred during the Pennsylvanian. For the most part, the same rift-related trends were activated notably superimposing the Table Rock arch over the Nemaha and emphasizing the Eastern Nebraska high (Fig. 3). Seismicity studies indicate that activity continues along the tectonic framework imposed by the Midcontinent Rift System in southeastern Nebraska and adjoining areas (Steeples, 1989).

CONCLUSIONS

The Midcontinent Rift System has been a dominant feature in creating the structural framework of central North America (Carlson, 1990). Reactivation of both the bounding and internal structures of the rift controlled tectonics during the Phanerozoic in southeastern Nebraska and adjacent areas. For the older Paleozoic features the distribution of sediments and their thickness patterns delineate the rift-related Southeast Nebraska arch and the North Kansas basin. Structure contour maps on Paleozoic horizons delineate the reactivation that created the Nemaha uplift and a variety of other Late Paleozoic features. A knowledge of the preexisting features in the basement rocks assists in defining the structures created by reactivation in the overlying Phanerozoic sediments. The geometry of these younger structures can, in turn, be used to decipher in more detail the deeply buried features of the basement.

ACKNOWLEDGMENTS

The paper was reviewed by G. B. Morey, R. F. Diffendal, D. C. Gosselin, and R. R. Burchett. S. B. Treves provided petrographic descriptions and classifications that assisted in subdivision of the Precambrian rocks.

REFERENCES CITED

Anderson, R. R., 1992, The Midcontinent Rift of Iowa [Ph.D. thesis]: Iowa City, Iowa, University of Iowa, 324 p.

Berendsen, P., and Blair, K. P., 1986, Subsurface structural maps over the central North American rift system (CNARS), central Kansas, with discussion: Kansas Geological Survey Subsurface Geology Series 8, 16 p.

Carlson, M. P., 1970, Distribution and subdivision of Precambrian and lower and middle Paleozoic rocks in the subsurface of Nebraska: Nebraska Geological Survey Report of Investigations Number 3, 25 p.

Carlson, M. P., 1988a, Interpretation of the geometry of the Midcontinent Rift System: Geological Society of America Abstracts with Programs, v. 20, no. 7, p. A205.

Carlson, M. P., 1988b, Tectonic history of southeastern Nebraska and adjoining areas: Geological Society of American Abstracts with Programs, v. 20, no. 2, p. 93.

Carlson, M. P., 1989, The influence of the Midcontinent Rift System on the occurrence of oil in the Forest City basin [abs.]: American Association of Petroleum Geologists Bulletin, v. 73, no. 3, p. 340.

Carlson, M. P., 1990, Genetic relationships between the Precambrian basement and Phanerozoic tectonics, Midcontinent region, North America: 9th International Conference on Basement Tectonics, Geological Society of Australia Abstracts, no. 26, p. 23

Carlson, M. P., and Treves, S. B., 1988, Precambrian framework of southeastern Nebraska: Geological Society of America Abstracts with Programs, v. 20, no. 2, p. 93.

Carlson, M. P., Burchett, R. R., and Maroney, D. G., 1979, Transform faulting: a reinterpretation of the tectonics of southeastern Nebraska: Geological Society of America Abstracts with Programs, v. 11, no. 5, p. 227.

Carlson, M. P., McKay, R. M., and Palmer, A. R., 1990, Faunal zones and transgression within the Sauk Sequence (Upper Cambrian–Lower Ordovician), Nebraska-Iowa region: Geological Society of American Abstracts with Programs, v. 22, no. 7, p. A320.

Harding, T. P., 1974, Petroleum traps associated with wrench faults [abs.]: American Association of Petroleum Geologists Bulletin, v. 58, no. 7, p. 1290–1304.

Klasner, J. S., Cannon, W. F., and Van Schmus, W. R., 1982, The pre-Keweenawan tectonic history of southern Canadian Shield and its influence on formation of the Midcontinent Rift, in Geology and tectonics of the Lake Superior basin: Geological Society of America Memoir 156, p. 27–46.

Lidiak, E. G., 1972, Precambrian rocks in the subsurface of Nebraska: Nebraska Geological Survey Bulletin 26, 41 p.

Sims, P. K., Kisvarsanyi, E. B., and Morey, G. B., 1987, Geology and metallogeny of the Archean and Proterozoic basement terranes in the northern Midcontinent, U.S.A.—An overview: U.S. Geological Survey Bulletin 1815, 51 p.

Sims, P. K., compiler, 1990, Precambrian basement map of the northern Midcontinent, U.S.A., with contributions from individual states: U.S. Geological Survey Miscellaneous Investigation Series Map I-1853-A, scale 1:1,000,000.

Steeples, D. W., 1989, Structure of the Salina–Forest City interbasin boundary from seismic studies, in Steeples, D. W., ed., Geophysics in Kansas: Kansas Geological Survey Bulletin 226, p. 31–52.

MANUSCRIPT ACCEPTED BY THE SOCIETY JANUARY 16, 1996

Geological Society of America
Special Paper 312
1997

Tectonic evolution of the Midcontinent Rift System in Kansas

Pieter Berendsen
Kansas Geological Survey, University of Kansas, 1930 Constant Avenue, Campus West, Lawrence,
Kansas 66047

ABSTRACT

Structure and lithology of basement rocks composing the Kansas segment of the Midcontinent Rift System, buried beneath a Paleozoic sedimentary cover, are discussed and used to interpret its tectonic history. The Keweenawan development of the Kansas segment of the rift is believed to have progressed in a similar manner as the rest of the Midcontinent Rift System. Structures are often difficult to interpret because they were repeatedly activated throughout the Proterozoic and Paleozoic. The north-northeast-tending rift cuts through an older Proterozoic terrane, in which at present three regionally important northwest-trending structures are recognized. They are the Bolivar-Mansfield, the Chesapeake, and the Fall River Tectonic Zones. The Fall River Tectonic Zone in central Kansas constitutes a major tectonic boundary. Keweenawan age mafic volcanic and igneous rocks are only encountered north of this boundary beneath the Paleozoic cover along the trend of the Midcontinent Rift System. Rift structures were reactivated during the Late Proterozoic. This resulted in the preservation of Keweenawan rocks in fault-bounded basins as for example along the trace of the Permian Abilene Anticline, and erosion of Keweenawan rocks from uplifted areas, including the Nemaha Uplift or Tectonic Zone. The eroded material is believed to have been transported in an easterly direction and deposited in fault-bounded basins in eastern Kansas and western Missouri.

INTRODUCTION

The approximately 1,500 km long (930 mi) western arm of the Midcontinent Rift System, extending from the Lake Superior region southwestward into Kansas, may be divided into three distinct segments or zones (Dickas and Mudrey, 1989), each separated from the others by accommodation structures. This paper discusses the tectonic evolution of the southernmost or Kansas segment of the rift. The Kansas segment shows an apparent left-lateral offset with respect to the Iowa segment (Dickas and Mudrey, 1989), situated north of an accommodation structure, separating the two segments. This regionally significant accommodation structure, striking northwest at right angles to the general trend of the rift, occurs in southeastern Nebraska and is characterized by the absence of rift-related rocks.

Emplacement of mafic intrusives and outpouring of voluminous amounts of basalt characterize the initial extensional phase of rifting of the Kansas segment at about 1,100 Ma (Van

Schmus et al., 1990). In the later, postigneous stages of rifting, initially coarse and immature clastic rocks, followed by increasingly finer and more mature strata, collected in the axial and flanking basins that developed along the trend of the rift (McSwiggen et al., 1987; Cannon et al., 1989). This general scenario of rift development appears to be true for the rift as a whole (Woelk and Hinze, 1991), despite later tectonic processes that combine to produce present-day complex structural patterns.

Even though subtle gravity anomalies on strike with the trend of the Midcontinent Rift System can be traced in to Oklahoma (Lyons, 1959), magnetic anomalies related to Keweenawan mafic intrusive and extrusive rocks, exposed at or near the Precambrian bedrock surface, terminate in central Kansas (King and Zietz, 1971; Yarger et al., 1981). In this region a structural interpretation and lithology of the basement along the trend of the rift and its immediate surroundings are obtained from drill-hole information. This paper discusses the

Berendsen, P., 1997, Tectonic evolution of the Midcontinent Rift System in Kansas, *in* Ojakangas, R. W., Dickas, A. B., and Green, J. C., eds., Middle Proterozoic to Cambrian Rifting, Central North America: Boulder, Colorado, Geological Society of America Special Paper 312.

structure and distribution of rift-related rocks in the Kansas segment of the Midcontinent Rift System.

RIFT-RELATED STRUCTURES

In the search for hydrocarbons in Pennsylvanian and older rocks, thousands of holes have been drilled within the general area of the rift. Drill-hole density is especially high along the trend of the Nemaha Tectonic Zone and the Central Kansas Uplift (Fig. 1), where it is not uncommon for tens of wells to be drilled in one section. Using this information, interpretive structure contour maps on top of major Paleozoic formations and the top of the Precambrian basement can be constructed. Analyses of such maps show that major structures, composed of faults and fault trends, exist in the subsurface. These structures have been reactivated sporadically throughout geologic time. Renewed movement often resulted in anastomosing faults or complicated fault slices, in which case the term "Tectonic Zone" is used. Without exception, all major faults interpreted from drill hole information are high-angle normal or reverse faults.

North-northeast- and northwest-trending faults dominate this scenario and break the area into orthogonal blocks. Three northwest-trending tectonic zones of regional significance are recognized in Kansas and Missouri. They are the Bolivar-Mansfield, Chesapeake, and Fall River Tectonic Zones (Fig. 1). The Missouri Gravity Low (Fig. 1) geophysically defines a fourth northwest-trending tectonic zone (Arvidson et al., 1982; Kerr, 1982), one of great significance because it coincides with the accomodation zone separating the Kansas and Nebraska segments of the rift. Related northwest-trending structures in Missouri can be traced beneath 1,480-Ma rhyolites of the St. Francois Mountains and thus they predate the extrusives (Sims and Peterman, 1986). According to Clendenin et al. (1989), the lateral component of movement on these tectonic zones in Missouri is always greater than the vertical displacement. Berendsen and Blair (1986) recognize a similar pattern in Kansas. Even though the influence of the Fall River and Bolivar-Mansfield Tectonic Zones on the development of the rift is not as profound as that of the Missouri Gravity Low, they nevertheless differentially affect the development of the rift as shown by the distribution of

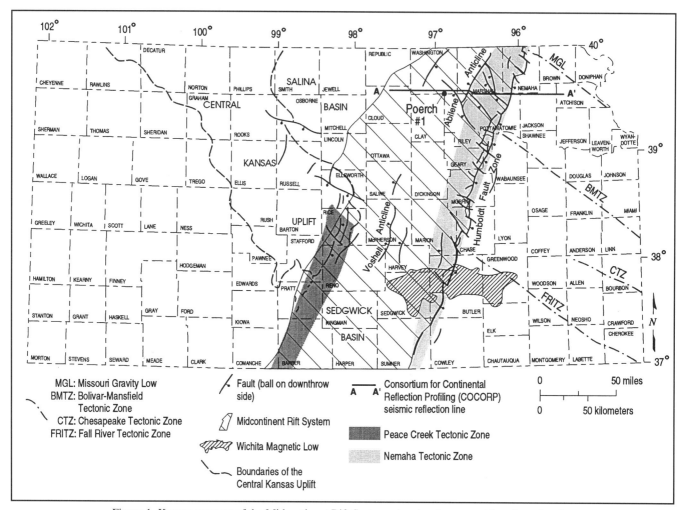

Figure 1. Kansas segment of the Midcontinent Rift System, showing the general location of major tectonic zones, Paleozoic and Proterozoic uplifts and basins, and other features discussed in the text.

Keweenawan igneous and sedimentary rocks and the relief of the structural block north and south of each tectonic zone (Figs. 2 and 3). The Chesapeake Tectonic Zone (Fig. 3) is a well-recognized structure in Missouri (McCracken, 1971), but lack of data prevents definition of this zone in eastern Kansas.

North-northeast-trending, rift-related structures are prevalent across Kansas in an area stretching from Nemaha and Washington Counties at the Nebraska border to Barber and Sumner Counties at the Oklahoma border (Fig. 1). Numerous faults are recognized in the central or axial portion of the Kansas rift segment. Structures associated with the Abilene Anticline (Jewett, 1951), stretching from southern Nebraska through Marshall and into Dickinson Counties (Fig. 1) are recognizable in Permian surface rocks. In the subsurface this anticline is a zone of complex faulting similar to the Nemaha Uplift

or Tectonic Zone to the east (Fig. 1). The Voshell Anticline (Fig. 1) located in McPherson and Reno Counties, may be regarded as a southern extension of the Abilene Anticline. In the Salina basin (Fig. 1), centered to the west of the axial rift zone, the paucity of drill holes does not permit interpretation of structures. East of the axial portion of the rift, the Nemaha Uplift or Tectonic Zone (Berendsen et al., 1989) in combination with the zone of complex faulting on the east side of the uplift known as the Humboldt fault zone, are well-known structural elements (Fig. 1). The Nemaha Tectonic Zone is broken up into orthogonal blocks by a series of regularly spaced northwest-trending faults, some of which are the earlier mentioned regional Bolivar-Mansfield and Fall River structures. Normal and high-angle reverse faults are common along the trend of the Nemaha Tectonic Zone. The Humboldt fault zone consists in

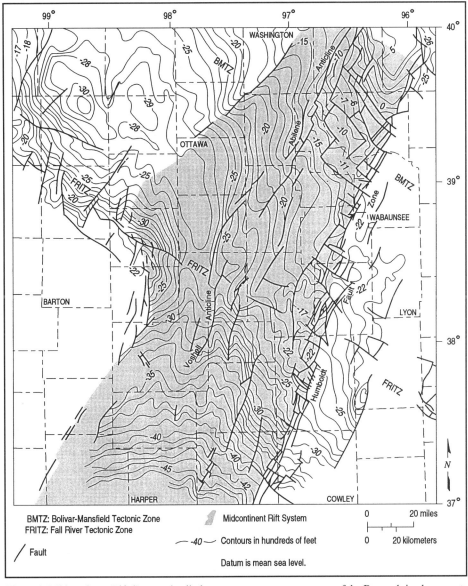

Figure 2. Midcontinent Rift System detailed structure contour map on top of the Precambrian basement.

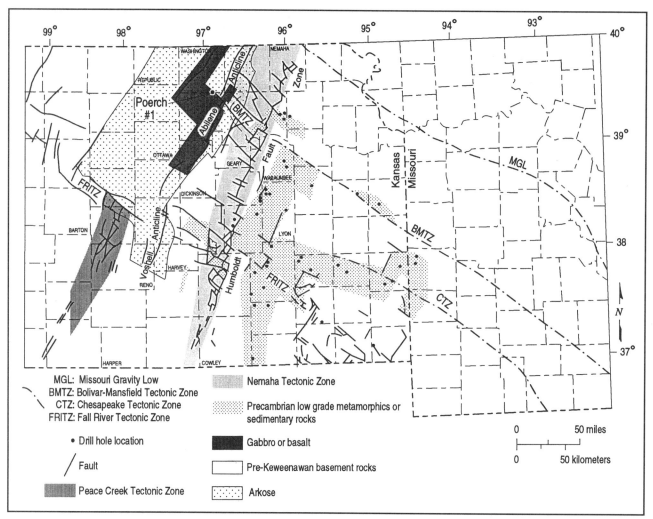

Figure 3. Interpreted distribution of Keweenawan rocks in the Midcontinent Rift System, and the generalized distribution and location of drill holes in which pre-Paleozoic clastic sedimentary rocks are encountered in eastern Kansas and western Missouri.

most places of a series of anastomosing stepped-down-to-the-east faults showing compound vertical offset of as much as 915 m (3,000 ft). Abutting the Central Kansas Uplift and extending southward into the Sedgwick Basin, the Peace Creek Tectonic Zone is structurally similar to the Nemaha Tectonic Zone. The area bounded by the Peace Creek Tectonic Zone to the west, the Nemaha Tectonic Zone to the east, and the area of intense faulting in between is considered to represent the Midcontinent Rift System in Kansas.

PRECAMBRIAN BASEMENT AND ROCK TYPE DISTRIBUTION

Details of the structure of the Precambrian basement in the Kansas rift segment of the Midcontinent Rift System and adjacent areas is shown in Figure 2. Depth to the basement varies from about 230 m (750 ft) in Nemaha County in northern Kansas, to about 1,920 m (6,300 ft) in southern Harper County,

Kansas, near the Oklahoma state line. Generally speaking, the basement slopes gradually to the south and southwest, except in the northwestern part of the area where the basement slopes west into the deep part of the Salina basin.

The distribution of Keweenawan and younger pre-Paleozoic rock types is shown in Figure 3. They consist of middle Keweenawan mafic intrusive and volcanic rocks, and younger clastic sedimentary rocks similar in nature to those found in the Lake Superior region (Berendsen and Barczuk, 1993). Mafic rocks are encountered in 19 drill holes and are restricted to the northern half of the rift zone in Kansas, from Dickinson County northward to the Nebraska border (Fig. 3). Surrounding the area underlain by mafic rocks and extending farther to the south into Harvey and Reno Counties, Kansas (Fig. 3), clastic sedimentary rocks, presumed to be upper Keweenawan in age, are encountered in 51 drill holes at the basement surface. Based on heavy mineral content and petrographic characteristics of available sample cuttings, the clastic sedimentary rocks have been tentatively correlated with rocks of the Oronto (Daniels, 1982) and Bayfield Groups, and equivalent formations of the same age of the Lake Superior region (Berendsen and Barczuk, 1993). The clastic sedimentary rocks of presumed Keweenawan age in Kansas are assigned to the Rice Series (Berendsen, 1994) and generally consist of conglomerate, arkose or feldspathic sandstone, and red and green siltstone and shale. The rocks are mostly oxidized and in places subjected to low-grade metamorphism. Rocks similar in nature to the Rice Series, but of uncertain stratigraphic position, occur widespread in eastern Kansas and western Missouri beyond the eastern limit of the Midcontinent Rift System. These pre-Paleozoic clastic sedimentary rocks, some of which have been drilled to thicknesses in excess of 460 m (1,500 ft), appear to be preserved in fault-bounded blocks or basins within a predominant, 1,350 to 1,800 Ma, granite-rhyolite terrane (Bickford et al., 1981).

The Precambrian basement was beveled (Cole, 1976) before Upper Cambrian sedimentary rocks were deposited (Zeller, 1968). While the amount of erosion is difficult to estimate, younger granitic plutons intruded into the older granitic terrane, and where exposed on the basement surface, show no topographic expression. North-northeast-trending faults, including faults associated with the Abilene Anticline (Fig. 1), which at the basement is expressed as a horst and graben structure, are instrumental in delineating the distribution pattern of the volcanics. Volcanic rocks are located in the center of the rift and are not exposed at the Precambrian basement surface in central Kansas south of the Fall River Tectonic Zone (Fig. 3), a regionally significant northwest-trending zone of discontinuous faults. The area immediately to the south and partially coinciding with the Fall River Tectonic Zone is also marked by a pronounced linear trend of magnetic lows, including the large east-west-trending Wichita Magnetic Low (Fig. 1; Yarger, 1983).

During the later stages of rifting, when thermal collapse started to play a role, clastic sedimentary rocks of the Rice Series were deposited in rift-flanking basins. They are widely distributed west of the central part of the rift and north of the Fall River Tectonic Zone, and show a lower magnetic susceptibility than the volcanic rocks (Yarger et al., 1981). South of the Fall River Tectonic Zone clastic sedimentary rocks occur only in the central part of the rift. No Rice Series rocks are found in the Wichita Magnetic Low (Yarger, 1983) or farther to the south. Farther north in Marshall County, northeastern Kansas, Keweenawan age clastic sedimentary rocks are preserved in fault-bounded basins (grabens) along the trace of the Paleozoic Abilene Anticline (Fig. 3). These rocks have been drilled to thicknesses in excess of 517 m (1,700 ft). Based on the heavy mineral content and petrographic characteristics, they can tentatively be correlated with rocks of the Oronto and Bayfield Groups and other strata of the presumed same age in the Lake Superior region (Berendsen and Barczuk, 1993). East of the subsurface expression of the Abilene Anticline clastic sedimentary rocks are only found in a narrow strip, generally less than 16 km (10 mi) wide. No clastic sedimentary rocks occur along the length of the Nemaha Tectonic Zone (Fig. 1).

DISCUSSION

It is difficult to decipher the tectonic history of the Kansas rift segment of the Midcontinent Rift System due to folding and faulting by younger tectonic events that mask the effects of older deformations. From surface and subsurface mapping of Paleozoic formations it is apparent that the Late Mississippian to Early Pennsylvanian Ouachita orogeny had a pronounced effect on the region. This is especially apparent along the trend of the Nemaha Tectonic Zone where structures common to a wrench fault environment are observed (Berendsen and Blair, 1986).

Inasmuch that Woelk and Hinze (1991) argued for homogeneity in structural style and evolution along the length of the rift, one would expect the Kansas rift segment to resemble other segments of the rift to the north. The similarity in petrographic composition and heavy mineral content of clastic sedimentary rocks from well cuttings in Kansas to Keweenawan clastic sedimentary rocks in the Lake Superior region (Berendsen and Barczuk, 1993), as well as a similar tectonic style as shown by mapped basement structures (Berendsen and Blair, 1992) and interpretation of seismic reflection profiles along the Kansas segment of the rift, support the conclusion reached by Woelk and Hinze (1991). Thus the general scenario of clastic-filled basins formed laterally of a central graben at a time when extensional processes waned and thermal collapse took place (Cannon et al., 1989) is assumed to be a valid model for the Kansas segment of the rift.

During the last ten years samples and geophysical records from a number of oil exploration wells drilled in the rift zone in northern Kansas have become available for study. In addition, an east-west seismic reflection profile has been obtained by the Consortium for Continental Reflection Profiling (COCORP; Fig. 1). Traversing the rift through the northern tier of counties, it provides much information about the deep structure of the rift (Serpa et al., 1984). These authors interpret the central part of the rift to consist of a 40 km wide (25 mi), asymmetric basin, reaching a depth of 8 km (5 mi) on the west and 3 km (1.9 mi) on the east. Texaco U.S.A. drilled a 3,444 m deep (11,300 ft) hole (Poersch #1) in Washington County, a few miles south of the COCORP line (Figs. 1 and 3), on what appeared on the seismic profile to be a large anticlinal feature. Instead of intersecting Keweenawan clastic sedimentary rocks at the basement surface as anticipated, the drill hole encountered in descending order 91 m (300 ft) of gabbro, 1,307 m (4,300 ft) of mostly basalt flows, and 1,186 m (3,900 ft) of clastic sedimentary rock. This reverse sequence was interpreted by Woelk and Hinze (1991) as being the result of a high-angle reverse fault that juxtaposes older volcanics on top of the younger clastics. A regionally normal lower Paleozoic section overlies the basement, showing no indication of having been displaced. Thus, prior to beveling of the basement by pre-Paleozoic erosion and subsequent deposition of Paleozoic sedimentary rocks, a fault having a minimum vertical displacement of 1,216 m (4,000 ft) is interpreted to have cut the rocks drilled in the Poersch #1 bore hole. Woelk and Hinze (1991) interpret a northeast-striking, high-angle reverse fault on the COCORP seismic profile and believe that this fault is responsible for the displacement in the Poersch #1 drill hole. However, this author favors the interpretation that the fault in the Poersch #1 drill hole trends northwest, and is related to the set of faults associated with the Bolivar-Mansfield Tectonic Zone (Fig. 1).

The trend of the Humboldt fault zone suggests that this structure was associated with the formation of the rift. Based upon interpreted faulting of the pre-Paleozoic sequence (Serpa et al., 1984; Woelk and Hinze, 1991) and the surface distribution of Keweenawan rocks in the rift zone (Berendsen and Blair, 1992), it appears that several thousands of feet of clastic sedimentary and volcanic rocks were removed from the rift. Merriam (1963) showed that Late Mississippian–Early Pennsylvanian erosion caused the granitic rocks on the Nemaha Tectonic Zone to be exposed. If prior to that erosional event, pre-Paleozoic clastic sedimentary rocks were present on the uplift, one might expect to find erosional wedges or other evidence in the Paleozoic section east of the Humboldt fault zone. Thus it is likely that Keweenawan-age clastic sedimentary rocks that were originally deposited in the area now occupied by the Nemaha Tectonic Zone were eroded before deposition of Cambrian carbonate rocks.

East of the Nemaha Tectonic Zone, throughout eastern Kansas and western Missouri, several dozen drill holes record penetrations of over a thousand meters of pre-Paleozoic clastic sedimentary rocks (Fig. 3). None of these drill holes bottomed out in Precambrian igneous rocks indicating that the full sedimentary section was not penetrated. Interpretation of potential field data indicates that substantial amounts of these rocks are preserved in fault-bounded basins within a largely granitic terrane. The rocks have been variously described as conglomerate, sandstone, shale, arkose, schist, slate, metamorphics, and phyllite. Descriptions (Haworth and Bennett, 1908; Skilman, 1948) and examination of well cuttings by the author indicate that all of these clastic sedimentary rocks are unmetamorphosed or at the most subjected to low grade metamorphism, and resemble rocks of the Rice Series (Berendsen, 1994). These rocks have to be younger than the 1,350 to 1,485 m.y. old, silicic volcanic and plutonic rocks (Denison et al., 1984; Bickford et al., 1981) that surround them.

Based upon available information this author would suggest that these clastic sedimentary rocks were eroded from the area of the rift and deposited in fault-bounded basins to the east. Much of the material was probably derived from the Nemaha Tectonic Zone, after it became a positive feature in post-Keweenawan, pre-Paleozoic time. The history of the Nemaha Uplift was one of repeated activation lasting throughout the Paleozoic era, most notably during the Middle Ordovician (Berendsen and Speczik, 1991; Berendsen et al., 1992), Late Devonian–Early Mississippian (Lambert, 1992), and Late Mississippian–Early Pennsylvanian (Merriam, 1963).

Except for the initial extensional and later contractional tectonics involved with the formation of the rift (Serpa et al., 1984; McSwiggen et al., 1987; Behrendt et al., 1988; Cannon et al.,

1989; Chandler et al., 1989; Hinze et al., 1990), much of the later tectonics involves wrenching imposed upon older rift structures (Berendsen and Blair, 1986; Berendsen et al., 1989). En echelon structures, pull-apart basins, or rhomb grabens and horsts are recognized, and are commonly associated with horizontal slip along faults (Aydin and Nur, 1982). Antithetic fault systems and propellor and scissor faults, along which throw is reversed, are other common structures in a wrench environment.

CONCLUSION

The Kansas segment of the Midcontinent Rift System is separated from the Nebraska and Iowa segments by an approximately 48 km (30 mi) wide accommodation zone in southeastern Nebraska, in which northwest-trending structures associated with the Missouri Gravity Low are dominant. In the Kansas segment, Keweenawan rocks exposed at the basement surface are almost entirely restricted to the area north of the northwest-trending, regionally significant, Fall River Tectonic Zone. To the south of this area potential field data suggest that Keweenawan igneous rocks occur at much greater depth. Other regionally significant northwest-trending tectonic zones cross the rift. Apparent left-lateral offset is suggested along some of these structures. The Humboldt fault zone on the east and the Peace Creek Tectonic Zone on the west are major north-northeast-trending rift-related structures that define the lateral extent of the Kansas segment of the Midcontinent Rift System. The Abilene Anticline, recognized in Permian rocks in Nemaha and Riley Counties, is expressed as a horst and graben structure in the basement. The distribution of Keweenawan rocks along the Abilene Anticline is influenced by the northwest-trending Bolivar-Mansfield Tectonic Zone.

During the development of the Kansas segment of the Midcontinent Rift System, the area now occupied by the Nemaha Tectonic Zone probably was part of a clastic sedimentary basin east of the axial zone of the rift. This area became a positive feature during the later Proterozoic resulting in the removal of the clastic sediments from the Nemaha Tectonic Zone. Clastic sedimentary rocks preserved in fault-bounded basins in eastern Kansas and western Missouri are believed to have been derived from this source.

ACKNOWLEDGMENTS

Much of the documentation and many of the ideas contained in this paper were developed over the years while conducting rift-related studies. Therefore, I would like to thank the following persons for their many contributions and stimulating discussions: K. P. Blair, K. D. Newell, J. Doveton, and F. W. Wilson of the Kansas Geological Survey, and A. Barczuk and S. Speczik of the University of Warsaw (Poland). The help of graphic arts designer P. Acker in preparing the figures is greatly appreciated. The thoughtful reviews by A. B. Dickas and P. A. Daniels as well as the help of the editor for this volume, R. W. Ojakangas, improved the paper significantly.

REFERENCES CITED

Arvidson, R. E., Guinness, E. A., Strebeck, J. W., Davies, G. F., and Schultz, K. J., 1982, Image processing applied to gravity and topography data covering the continental U.S.: Eos (Transactions, American Geophysical Union), v. 63, no. 18, p. 261–265.

Aydin, A., and Nur, A., 1982, Evolution of pull-apart basins and their scale independence: Tectonics, v. 1, no. 1, p. 91–105.

Behrendt, J. C., Green, A. G., Cannon, W. F., Hutchinson, D. R., Lee, M., Milkereit, B., Agena, W. F., and Spencer, C., 1988, Crustal structure of the Midcontinent Rift System–Results from GLIMPCE deep seismic reflection profiles: Geology, v. 16, p. 81–85.

Berendsen, P., 1994, Review of Precambrian rift stratigraphy: Revision of stratigraphic nomenclature in Kansas: Kansas Geological Survey Bulletin 230, p. 1–4.

Berendsen, P., and Barczuk, A., 1993, Petrography and correlation of Precambrian clastic sedimentary rocks associated with the Midcontinent Rift System: U.S. Geological Survey Bulletin 1989-E, 20 p.

Berendsen, P., and Blair, K. P., 1986, Subsurface structural maps over the Central North American rift system (CNARS), central Kansas, with discussion: Kansas Geological Survey, Subsurface Geology Series 8, 7 maps, 16 p.

Berendsen, P., and Blair, K. P., 1992, Midcontinent Rift System, Precambrian structure map: Kansas Geologiacal Survey Open-File Report 92-41 A, scale 1:500,000.

Berendsen, P., Blair, K. P., and Newell, K. D., 1989, Structural aspects of the Midcontinent Rift System in Kansas: Transaction volume of the 1989 American Association of Petroleum Geologists Midcontinent Section Meeting, Oklahoma City: Oklahoma City, Oklahoma Geological Society, 26 p.

Berendsen, P., Doveton, J. H., and Speczik, S., 1992, Distribution and characteristics of a Middle Ordovician oolitic ironstone in northeastern Kansas based on petrographic and petrophysical properties: A Laurasian ironstone case study: Sedimentary Geology, v. 76, p. 207–219.

Berendsen, P., and Speczik, S., 1991, Sedimentary environment of Middle Ordovician iron oolites in northeastern Kansas, U.S.A.: Acta Geologica Polonica, v. 41, no. 3–4, p. 215–226.

Bickford, M. E., Harrower, K. L., Hoppe, W. J., Nelson, B. K., Nusbaum, R. L., and Thomas, J. J., 1981, Rb-Sr and U-Pb geochronology and distribution of rock types in the Precambrian basement of Missouri and Kansas: Geological Society of America Bulletin, part I, v. 92, p. 323–341, 26 figs., 1 table.

Cannon, W. F., 1989, The North American Midcontinent Rift beneath Lake Superior from GLIMPCE seismic reflection profiling: Tectonics, v. 8, no. 2, p. 305–332.

Chandler, V. W., McSwiggen, P. L., Morey, G. B., Hinze, W. J., and Anderson, R. R., 1989, Interpretation of seismic reflection, gravity, and magnetic data across Middle Proterozoic Mid-Continent Rift System, northwestern Wisconsin, eastern Minnesota, and central Iowa: American Association of Petroleum Geologists Bulletin, v. 73, p. 261–275.

Clendenin, C. W., Niewendorp, C. A., and Lowell, G. R., 1989, Reinterpretation of faulting in southeast Missouri: Geology, v. 17, p. 217–220.

Cole, V. B., 1976, Configuration of the top of the Precambrian rocks in Kansas: Kansas Geological Survey Map M-7, scale 1:500 000, 1 sheet.

Daniels, P. A., 1982, Upper Precambrian sedimentary rocks, Oronto Group, Michigan-Wisconsin, *in* Wold, R. J., and Hinze, W. J., eds., Geology and tectonics of the Lake Superior basin: Geological Society of America Memoir 156, p. 107–133.

Denison, R. E., Lidiak, E. G., Bickford, M. E., and Kisvarsanyi, E. B., 1984, Geology and geochronology of Precambrian rocks in the central interior region of the United States: U.S. Geological Survey Professional Paper 1241-C, 20 p.

Dickas, A. B., and Mudrey, M. G., Jr., 1989, Central North American case for segmented rift development: International Geological Congress, 28th, Washington, D.C., Abstracts, v. 1 of 3.

Haworth, E., and Bennett, J., 1908, General stratigraphy; *in* Special report on oil and gas, Chapter III: Geological Survey of Kansas, The University of Kansas, p. 57–160.

Hinze, W. J., Braile, L. W., and Chandler, V. W., 1990, A geophysical profile of the southern margin of the Midcontinent Rift System in western Lake Superior: Tectonics, v. 9, p. 303–310.

Jewett, J. M., 1951, Geologic structures in Kansas: Kansas Geological Survey, Bulletin 90 (1951 Reports of Studies), pt. 6, p. 105–172.

Kerr, R. A., 1982, New gravity anomalies mapped from old data: Science, v. 215, p. 1220–1222.

King, E. R., and Zietz, I., 1971, Aeromagnetic study of the Midcontinent Gravity High of central United States: Geological Society of America Bulletin, v. 82, p. 2187–2208.

Lambert, M. W., 1992, Lithology and geochemistry of shale members within the Devonian-Mississippian Chattanooga (Woodford) Shale, Midcontinent, USA: Department of Geology, University of Kansas, Lawrence, Kansas, 160 p.

Lyons, P. L., 1959, The Greenleaf anomaly, a significant gravity feature, *in* Hambleton, W. W., ed., Symposium on Geophysics in Kansas: Kansas Geological Survey Bulletin, no. 137, p. 105–120.

McCracken, M. H., 1971, Structural features of Missouri: Missouri Geological Survey and Water Resources, Report of Investigations, no. 49, 99 p.

McSwiggen, P. L., Morey, G. B., and Chandler, V. W., 1987, New model of the Midcontinent Rift in eastern Minnesota and western Wisconsin: Tectonics, v. 6, p. 677–685.

Merriam, D. F., 1963, The geologic history of Kansas: Kansas Geological Survey Bulletin 162, 317 p.

Serpa, L., Setzer, H., Farmer, L., Brown, J., Oliver, S., Kaufman, S., Sharp, J., and Steeples, D. W., 1984, Structure of the southern Keweenaw rift from COCORP surveys across the Midcontinent Geophysical Anomaly in northeastern Kansas: Tectonics, v. 3, p. 367–384.

Sims, P. K., and Peterman, Z. E., 1986, Early Proterozoic Central Plains orogen: A major buried structure in the north-central United States: Geology, v. 14, p. 488–491.

Skillman, M. W., 1948, Pre-Upper Cambrian sediments of Vernon County, Missouri: Missouri Department of Business and Administration, Division of Geological Survey and Water Resources, Report of Investigations, no. 7, 17 p.

Van Schmus, W. R., Martin, M. W., Sprowl, D. R., Geissman, J., and Berendsen, P., 1990, Age, Nd and Pb isotopic composition, and magnetic polarity for subsurface samples of the 1100 Ma Midcontinent rift: Geological Society of America Abstracts with Programs, v. 22, no. 7, p. A174.

Woelk, T. S., and Hinze, W. J., 1991, Model of the Midcontinent Rift System in northeastern Kansas: Geology, v. 19, p. 277–280.

Yarger, H., Robertson, R., Martin, J., Ng, K., Sooby, R., and Wentland, R., 1981, Aeromagnetic map of Kansas: Kansas Geological Survey Map M-16, scale 1:500,000.

Yarger, H. L., 1983, Regional interpretation of Kansas aeromagnetic data: Kansas Geological Survey Geophysics Series 1, 35 p.

Zeller, D. E., editor, 1968, The stratigraphic succession in Kansas: Kansas Geological Survey, Bulletin 189, p. 81 p.

MANUSCRIPT ACCEPTED BY THE SOCIETY JANUARY 16, 1996

Geological Society of America
Special Paper 312
1997

Pre–Mount Simon basins of western Ohio

Benjamin H. Richard
Department of Geological Sciences, Wright State University, Dayton, Ohio 45435
Paul J. Wolfe
Department of Physics, Wright State University, Dayton, Ohio 45435
Paul E. Potter
Department of Geology, University of Cincinnati, 500 Geology/Physics Building, Cincinnati, Ohio 45221-0013

ABSTRACT

Pre–Mount Simon sedimentary rocks, now known as the Middle Run Formation, have been identified in western Ohio, eastern Indiana, and northern Kentucky in ten wells and in Ohio on 213 km of seismic data. These sedimentary rocks are remnants of an extensive alluvial plain and consist for the most part of lithic arenite. One well, the Friend No. 1 Mattinson Well, penetrated 366 m of carbonate rocks that appear to have been deposited in a marine basin.

Pre–Mount Simon basins have been identified on seismic data. We believe these are structural basins that preserved the thickest sedimentary rocks in structural lows. Evidence suggests that these sediments were deposited after the Grenville orogeny. Most of the basin formation occurred after sedimentation and before Mount Simon deposition. We feel that these sedimentary rocks and basins are either Eocambrian or Early Cambrian in age and not part of the Midcontinent Rift System. The lack of consistent magnetic and gravity signatures lends support to this conclusion. However, we also recognize that in the 400 million years between the Grenville orogeny and deposition of the Mount Simon, there may have been several periods of deposition and erosion and not just a single sedimentary event.

We believe the Middle Run Formation represents deposition on a semiarid alluvial plain by braided south-flowing streams, focused along a weak north-south axis, that drained southward toward the margin of the Laurentian continent.

INTRODUCTION

Pre–Mount Simon sedimentary rocks have been identified on seismic sections and in deep wells in western Ohio, eastern Indiana, and northern Kentucky (Fig. 1, Table 1). We interpret these rocks using 213 km of seismic data in Ohio and 10 wells that penetrate pre–Mount Simon sedimentary rocks. These data come from a wide range of industry, government, and academic sources. From the data, basins have been identified at five locations in western Ohio. In this paper they are designated by the seismic lines on which they appear: Consortium–Wright State University (WSU) in northern Warren and Clinton Counties, Selma in southern Clark County, and three on the COCORP line (west, middle, east). We describe the relationship of these basins to the gravity and magnetic fields.

There is disagreement about the lithology of the sediments filling pre–Mt. Simon basins, the origin of the basins, and their age. We believe these basins formed after the Grenville orogeny, though some researchers disagree. Important background information valuable to our study are papers by Lucius and von Frese (1988), Shirley (1990), Shrake (1991), Shrake et al. (1991), Pratt et al. (1989, 1992), Dickas et al. (1992), Green (1983), Hinze et al. (1975), Hinze et al. (1992), and the broad regional reviews by Harris (1992) and Drahovzal et al. (1992) that cover four states.

SEISMIC CHARACTERISTICS

The seismic characteristics shown on the Consortium–WSU seismic line are the basis for our seismic identification of the Middle Run Formation elsewhere in Ohio. The seismic section

Richard, B. H., Wolfe, P. J., and Potter, P. E., 1997, Pre–Mount Simon basins of western Ohio, *in* Ojakangas, R. W., Dickas, A. B., and Green, J. C., eds., Middle Proterozoic to Cambrian Rifting, Central North America: Boulder, Colorado, Geological Society of America Special Paper 312.

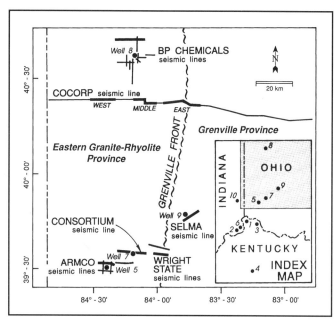

Figure 1. Index map of western Ohio showing location of boreholes and seismic profiles used in this study (after Wolfe et al., 1993).

TABLE 1. WELLS PENETRATING MIDDLE RUN FORMATION AND POSSIBLE EQUIVALENTS

No.	Well	County and State	Year	Thickness (m)
1	Ford No. 1 Connor	Boone, Kentucky	1960	37
2	Ashland No. 1 Eichler	Switzerland, Indiana	1965	2
3	Ashland No. 1 Wilson	Campbell, Kentucky	1966	22
4	Texaco No. 1 Sherrer	Jessamine, Kentucky	1966	241*
5	Armco No. 1 Armco	Butler, Ohio	1967	1
6	Ashland No. 1 Sullivan	Switzerland, Indiana	1985	9
7	Ohio Geological Survey	Warren, Ohio	1988	582†
8	BP Chemicals No. 4§	Allen, Ohio	1991	24†
9	Friend No. 1 Mattinson	Clark, Ohio	1926	366**
10	Gulf Oil No. 1 Scott	Fayette, Indiana	1965	2

*Underlies 150 m of basalt.
§Three earlier wells of the same property reached the Middle Run.
†Continuous core.
**Contains 35 m of lithology similar to the Middle Run and 331 m of carbonate rock.

for this line is shown in Figure 2 below our line-drawing interpretation. (Due to the difficulty of reproducing seismic sections on such a small scale, only the line-drawing interpretation is presented for the other seismic lines). There is more control on this seismic line than all others because of the Ohio Geological Survey Drill Hole 2627 (Shrake et al., 1991; Shrake, 1991) where 582 m of the basin sedimentary rocks (lithic arenite) were cored. The seismic character of these sedimentary rocks indicates they are relatively homogeneous with only a few weak reflectors that are traceable for several kilometers. Total thickness, as based on seismic interpretation, is probably over 2,000 m. The bottom of

the basin appears to show only a slight discordance with the units below. Rocks beneath the basin on the western part produce strong continuous reflections whereas on the east the reflections are weak and discontinuous. Although faulting is prevalent in the lower units, it is not clear to what extent the faults continue into the overlying basin, but if they do, there is no significant offset. This basin is an asymmetric syncline. The western part overlies the Eastern Granite-Rhyolite Province mapped by Pratt et al. (1992). The nature of the strong, continuous and widespread reflectors that are part of the Eastern Granite-Rhyolite Province is puzzling. While it is tempting to visualize these as basalts, which are widespread and continuous, the lack of magnetic signatures argues against Keweenawan basalts. The Grenville Province does not have the strong, continuous seismic reflections that characterize the Eastern Granite-Rhyolite Province. The units underlying the eastern part of this basin are seismically similar to those Grenville Province characteristics. We used this similarity, in addition to the change in magnetic characteristics, to draw the location of the Grenville Front, the boundary of the two provinces (Fig. 1).

We believe the Selma line is not a true dip line and, therefore, all the dips are subdued (Fig. 3). This basin has the second greatest penetration of pre–Mount Simon sedimentary rocks of which only a small part may be Middle Run. The Friend No. 1 Mattinson well was drilled into 366 m of pre–Mount Simon carbonate rock in 1926 (Wasson, 1933; Colony, 1931; Norris et al., 1952). The seismic character of these rocks suggests that they are relatively uniform except for several strong, relatively continuous reflectors. Seismic signature shows that sedimentary rocks in this basin are about 2,500 m thick. Unlike the other basins, the base of the Mount Simon Sandstone appears quite irregular—possibly representing deposition on a karst surface. The units beneath the basin exhibit weak, discontinuous faulted reflections characteristic of the Grenville Province. The structure of the basin is more difficult to define, but may represent graben formation with the seismic section showing only subdued apparent dip.

Three basins are identified on the Ohio COCORP line (Fig. 4). The east basin clearly has a Middle Run seismic signature—weak, discontinuous reflectors—similar to those of the Consortium–WSU line. Seismically, the base of this basin appears conformable to the strong reflectors beneath it. This basin seems to be a true graben and contains about 3,800 m of sedimentary rock, the thickest sequence of layered rocks in any of the identified basins. Although not as clearly shown as in the Consortium–WSU line, the pre–Middle Run rocks on the western part of this line have characteristics of the Eastern Granite-Rhyolite Province (strong, continuous) reflections, whereas to the east they are similar to the Grenville Province (discontinuous) reflections. The seismic character of the middle basin is similar to that of the east basin except for several relatively continuous reflectors. The middle basin is several kilometers south of the BP Chemical well that cored about 100 m into Middle Run sedimentary rocks. The basal sedimentary rocks of the

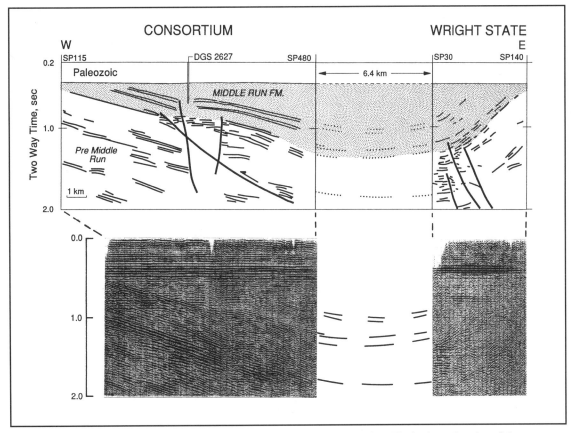

Figure 2. Consortium–Wright State University line and interpretation. The horizontal scales of the seismic are different; therefore, the Consortium line was reduced to make each time scale the same. Note interpreted synclinal structure on the line and typical discontinuous Grenvillian structure below the Middle Run Formation in the WSU portion of the line and continuous Eastern Granite-Rhyolite below the Middle Run Formation in the Consortium portion of the line.

middle basin appear conformable to the strong continuous reflectors of the underlying units that we interpret to be Eastern Granite-Rhyolite Province rocks. Seismic analysis suggests the middle basin is a half graben with listric faults on its west limb. The west basin contains up to 1,000 m of pre–Mount Simon sedimentary rocks at its deepest part and is symmetrical with normal faulting on both sides. There is some evidence of onlap. These sedimentary rocks overlie units that have an Eastern Granite-Rhyolite seismic signature. The west basin, with its strong continuous reflectors, has the most diverse seismic character. Several possibilities exist, for example, the Middle Run may contain some volcanic flows or these reflectors could be carbonate rocks. In either case, collectively the basins of the COCORP line seem to show facies change from east to west.

Even though the BP Chemical and Armco Steel seismic data do not easily define any basins, they surround wells that have cored Middle Run Formation (Table 1). We have drawn the base of the Middle Run as the first strong continuous reflector (Figs. 5 and 6). It is possible that the strong reflector on the BP line is an igneous flow within the Middle Run sandstones. If so, the Middle Run sedimentary rocks are quite thick. If the areas represented by

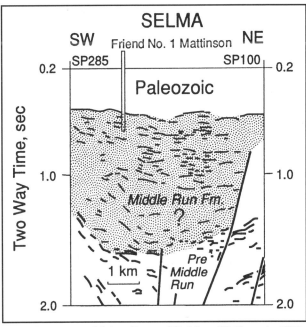

Figure 3. Interpreted Selma line modified from Wolfe et al., 1993.

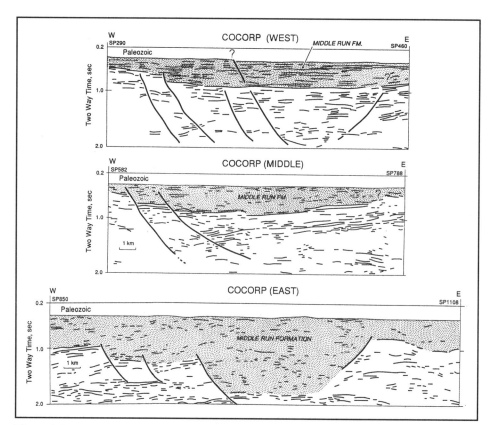

Figure 4. Interpreted COCORP Lines—East, Middle, and West basins. Note the increasing amplitude and continuity of the Middle Run Formation reflections from east and west (modified from Wolfe et al., 1993).

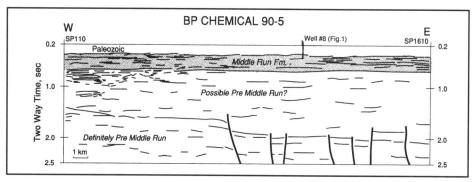

Figure 5. Interpreted BP Chemical line. The base of the Middle Run is difficult to pick, and may be much deeper (modified from Wolfe et al., 1993).

BP Chemical and Armco Steel data are in fact parts of basins, the basins are much larger and deeper than the basins previously discussed. Again, the rocks beneath the Middle Run are represented by strong, continuous reflectors interpreted to be rocks of the Eastern Granite-Rhyolite Province.

GRAVITY AND MAGNETIC EXPRESSIONS OF BASINS

It has been tempting to correlate the pre–Mount Simon sandstones of western Ohio with the sedimentary and volcanic rock sequences that occur in the Midcontinent Rift (Shrake et al., 1990, 1991; Harris, 1992; Drahovzal et al., 1992; Dickas et al., 1992) especially since Hinze et al. (1975) and Wollensak (1988) extended the Midcontinent Rift into central Michigan. In the Midcontinent Rift in Iowa, Wisconsin, and Minnesota there is a correlation between sedimentary fill and the Bouguer gravity and magnetic maps (Fig. 7). Gravity and magnetic lows generally coincide with the sedimentary basins flanking the central volcanic mass. Green (1983), Wollensack (1988), Hinze et al. (1992) and Milkereit et al. (1992) suggested the Midcontinent

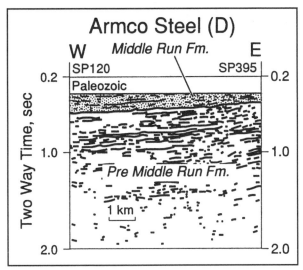

Figure 6. Interpreted Armco Steel line. The base of the Middle Run is difficult to pick (modified from Wolfe et al., 1993).

Rift System is overridden by the Grenville Front in southeastern Michigan and does not continue into Ohio (Fig. 7). If these Ohio basins are associated with the Midcontinent Rift, they should possess its magnetic signature. Figure 8 is a total-field magnetic map of Ohio. On this map we have located the Grenville Front, which is the boundary between the Grenville Province and the Eastern Granite-Rhyolite Province. Because there is a marked difference from low-grade metasedimentary rocks to the west and high-grade metamorphic and intrusive rocks to the east, we have drawn this boundary along the change from low to high magnetic gradients. Seismic data of this study support this conclusion. To the west of the Grenville Front, as drawn, are strong continuous deep reflectors, whereas to the east, there are weak discontinuous deep reflectors.

The Consortium–WSU basin generally has low magnetic relief except for its eastern end, which we consider to overlie the Grenville Front. The Selma basin is on the southwest side of a high-gradient magnetic high within the Grenville Province. Based on seismic signature the east COCORP basin overlies the Grenville Front and is on the north side of a magnetic high. The magnetic field is nearly constant in the region of the middle

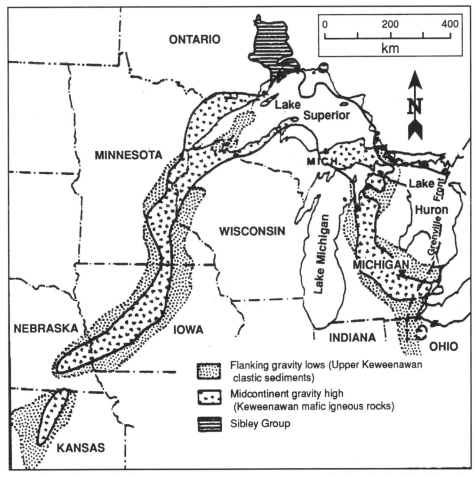

Figure 7. Generalized map of the Midcontinent Rift System showing the relationship of gravity anomalies to stratigraphy (after Green, 1983).

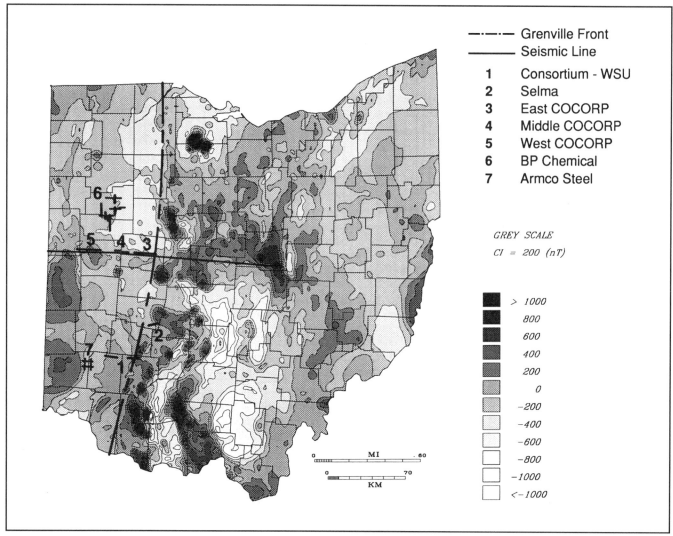

Figure 8. Reduced-to-pole magnetic anomalies of Ohio showing the location of seismic lines used in this study and the Grenville Front.

COCORP basin. The west COCORP basin is on a low-gradient magnetic high. Consequently, we see no consistent correlation between magnetic field data and pre–Mount Simon basins.

Is there a relation between these basins and the gravity field of Ohio (Fig. 9)? The Consortium–WSU basin crosses from the west side of a gravity high to the center of the high. On the other hand, the Selma basin is on the east side of a gravity low, the east COCORP basin is on the flank of a gravity high, the middle COCORP basin is on the east side of a gravity high, and the west COCORP basin is centered over a gravity high. Consequently, there appears to be no consistent spatial correlation between gravity anomalies and known pre–Mount Simon basins.

We conclude from the lack of consistent correlation of these basins with gravity and magnetic anomalies and the few strong connection of basins along the Midcontinent Rift with gravity and magnetic anomalies that the basins are not part of the Midcontinent Rift System.

PETROLOGY, SOURCE, AND ENVIRONMENT

Two different types of sedimentary rocks have been identified in basins delineated by seismic data. These are lithic arenites and organic-rich, phosphatic limestones. Most of the wells penetrating these sediments have sampled the lithic arenite that has been identified as the Middle Run Formation, whereas the Friend No. 1 Mattinson well has sampled 366 m of sediments that were reported as mostly organic-rich, phosphatic limestones (Colony, 1931; Wasson, 1933).

The core of the Middle Run associated with the Consortium–WSU line was described in detail by Shrake et al. (1991), Dickas et al. (1992) and Wolfe et al. (1993). It consists of more than 95% densely compacted, grayish red sandstone with a few silty, brownish red shales or argillaceous siltstones. It has many thin fining-upward fluvial cycles, typically with a sharp base and a few scattered lags of pebbles. Much of the sandstone is

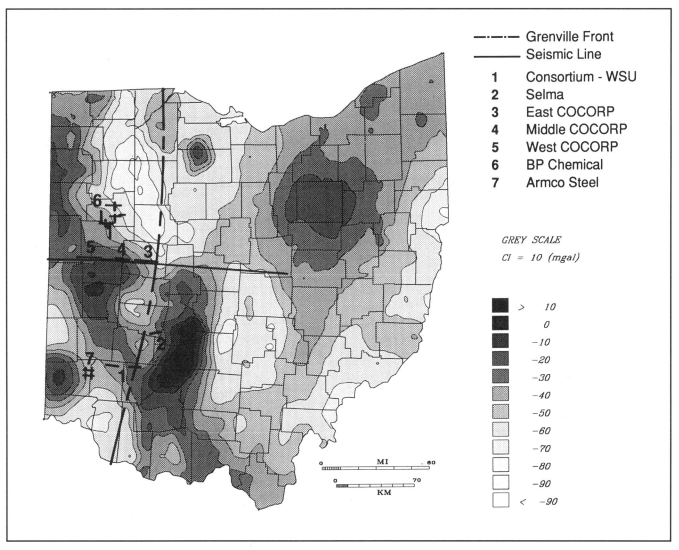

Figure 9. Bouguer gravity anomalies of Ohio showing the location of seismic lines used on this study and the Grenville Front.

massive; the scattered clasts in the Middle Run are never abundant and rarely form a true conglomerate. The pebbles are mostly volcanic rocks, but include some red shale, siltstone, and sandstone. Poorly defined cross-bedding, rarely abundant, typically passes upward to parallel lamination or small-scale ripple lamination in siltstone. Minor small calcite-cemented masses, with an open framework of grains, suggest incipient caliche and thus, a dry climate.

Petrographic study of 19 thin sections from the Middle Run in the Warren County core (Table 2) shows that the sandstone consists mostly of tightly compacted, moderately well sorted framework of subangular grains with only a few percent detrital matrix. Early calcite varies from nearly zero to more than 30% and averages 16%. Quartz, feldspar, and rock fragments constitute most of the framework grains. Monocrystalline or unit quartz is about six times more abundant than polycrystalline or com-

TABLE 2. PETROGRAPHIC SUMMARY OF MIDDLE RUN FORMATION, WARREN COUNTY, OHIO, DRILL HOLE DGS2627*

Texture		Normalized Framework Composition	
Framework	82 ± 13	Quartz	39 ± 13
Matrix	2 ± 4	Feldspar	17 ± 4
Cement	16 ± 15	Rock Fragments	44 ± 14
Feldspar		Rock Fragments	
Potash	14 ± 3	Plutonic	7 ± 7
Plagioclase	3 ± 2	Metamorphic	46 ± 20
		Volcanic	35 ± 20
		Sedimentary	12 ± 7

*Mean and standard deviation, %.

posite quartz. Feldspars are dominantly potash. Rock fragments generally dominate, with a few samples containing nearly 50%. The dominant rock fragments are metamorphic rocks, which average 46%, followed by volcanic rocks, which average 35%, but there is much variability between thin sections—in some sections metamorphic rocks predominate whereas in other volcanic rocks clearly predominate. Nonetheless, use of the t-test for the difference between paired means clearly showed that metamorphic rock fragments are more abundant than volcanic rock fragments. The volcanic rock fragments include many dark, dense, fine-grained volcanic rocks with a few laths of plagioclase. Also present are acid (potash-rich) aphanitic chert-like volcanic rocks of altered glass and similarly textured, but low potash, siliceous grains, as well as typical volcanic (andesitic?) grains with a well-defined fabric of plagioclase laths. The most abundant of these are fine-grained siliceous volcanic rocks that are potash-rich. A few plutonic and argillaceous grains, as well as grains of sandstone and siltstone, are also present. Cherts, opaques, and heavy minerals rarely exceed 1%.

Judging by the feldspar and rock fragments, both the Granite-Rhyolite and Grenville Provinces contributed sand-sized detritus to the Middle Run, although most of the scattered small pebbles are volcanic rocks. Contributions from both provinces to the Middle Run harmonizes well with the seismic evidence that it overlies both provinces. We believe deposition was by small streams flowing across a semiarid alluvial plain.

The pre–Mount Simon rocks in the Friend No. 1 Mattinson well drilled in 1926 are controversial. R. J. Colony (1931; of Columbia University, a well-established and recognized chemically trained petrologist) and I. B. Wasson (1933; the well-site geologist and wife of Theron Wasson, Chief geologist for Pure Oil Company) described these rocks as mostly black, organic-rich phosphatic limestone, whereas Harris (1992) described them as altered basalt. This difference must be resolved. Having looked at the samples and several thin sections and the descriptions of Wasson and Colony, we follow Wasson and Colony. The samples were collected from a cable tool rig and apparently only small grains were recovered. The absence of large grains made it impossible to identify fossils or allochems. The lack of fossils or allochems was a main argument used by Harris to suggest the fragments are altered basalt. The first unit described by Wasson (1933) is about 27 m (from 1,015 to 1,042 m) of fine red sandstone and shale fragments, which based on lithology, may be correlative with the Middle Run Formation. Wasson (1933) did consider that this unit could underlie the Mount Simon and could be either Cambrian or Proterozoic in age. It is a fine red sandstone with a coarse sandstone containing pink feldspar at its base. From 1,058 to 1,170 m she described a black dolomite that Colony (1931) called a high magnesium limestone on the basis of his laboratory analysis. According to Colony, insoluble residues from this unit yielded considerable gray feldspar and at least one thin sandstone was described as either an arkose or graywacke. From 1,170 to 1,381 m Wasson (1933) described a sequence of black, carbonaceous limestone. Colony (1931)

described this unit as containing numerous round phosphate nodules within which there is considerable pyrite. He burned the black limestone and found it contained considerable organic material giving off a pyroligneous odor. A zone in this well that extends from 1,328 to 1,336 m is important when comparing the description of this well by Wasson (1933) and Colony (1931) to that of Rast contained in the Harris (1992) paper where it is described as "fragments of considerably altered basalt." Colony (1931) classified this unit as "not a good arkose" and later put arkose in quotes. This would probably be called a lithic arenite in modern terminology.

Rast (in Harris, 1992), described the black dolomite of Wasson as a felsic, volcanic rock that is partially altered and replaced by carbonate and clays. He interpreted the lower black limestone of Wasson (1933) as pervasively replaced mafic volcanics. He and his collaborators identified the material as calcite microspar, but dismiss a sedimentary origin because they were not able to identify primary sedimentary features in the chips. For some reason the "arkose" zone was not converted to calcite. Norris et al. (1952) described this well as having an oil show in the black limestone at 1,329 m, which is the location of the "arkose." It is really surprising that this very unusual well has been lost in the literature for so many years.

The presence of the pyrite and phosphate nodules suggests to us that this carbonate is probably of marine origin and its organic carbon holds promise as a possible source rock for petroleum. These features, along with the lack of metamorphism, lead us, as they led Colony and Wasson, to believe that this unusual carbonate is post-Grenvillian in age.

DISCUSSION

We have used the change in seismic reflection and magnetic field character to establish the boundary between Proterozoic provinces of Ohio. The Eastern Granite-Rhyolite Province typically has weak magnetic anomalies and strong seismic reflections. In contrast, the Grenville Province has strong magnetic anomalies and weak seismic reflections. Both the Consortium–WSU basin and the eastern COCORP basin of our study appear to be on top of the Grenville Front.

The basins containing layered rocks below the Mount Simon appear to be structural basins that formed for the most part after the Granite-Rhyolite and Grenville Provinces were in place. Structure displayed in the basins mimics the structure beneath, though possibly somewhat subdued. There is no apparent change in the seismic character of the sedimentary rocks at the edges of the basin and hence all basin fill appears to be preserved as structural remnants. The sandstones and the siltstones of the Middle Run are widespread and we believe represent alluvial plain deposits with both the Eastern Granite-Rhyolite and Grenville Provinces contributing detritus. Most of the clasts have either a granitic or rhyolitic origin. The Selma basin, which is possibly a part of the same basin identified on the Consortium–WSU line, contains some arkose that could be a Middle Run equivalent. The major rock is an organic-rich, phosphatic limestone, probably deposited in a marine basin whose relation to the Middle Run has yet to be determined.

We believe the ages of all of the Middle Run rocks to be younger than Grenville. The western part of the Consortium–WSU basin rests on distinctive high amplitude seismic reflections characteristic of what has been considered a part of a large Proterozoic basin (Pratt et al., 1992) whereas the eastern end of this seismic profile has weaker, discontinuous reflections similar to the Selma line. The Selma line is well within the Grenville Province. The structures of units within these basins are nearly parallel to the underlying rocks. These data suggest the Middle Run sediments were deposited in a basin that formed on top of the Grenville Front. Though we have no absolute dates of deposition of sediments or formation of basins, the above information suggests the basins are younger than Grenville and older than Middle Cambrian (the age of the Mount Simon; Babcock, 1994). We see no relation of the basin location to potential field anomalies as in the Midcontinent Rift. Dalziel (1991) and Easton (1992) proposed that post-Grenville separation of North and South America produced rifting in eastern North America. Armstrong (1992) proposed that this continental margin continued into Early Cambrian time.

Drahovzal et al. (1992) suggested red clastic rocks and theoletic basalt in the Texaco No. 1 Sherrer well southwest of Lexington, Kentucky, are similar to Keweenawan units in Michigan and Wisconsin and were probably the same age. They believe the Ohio basins relate to the Kentucky basin that the Texaco No. 1 Sherrer well penetrated. Regionally we find many red clastic rocks preserved in basins that have diverse ages. Rankin (1967, 1975, 1976) and Schwab (1986) discussed a Late Proterozoic opening and closing of the proto-Atlantic Iapetus Ocean with post-Grenvillian, immature, red sandstone filling rift or riftlike basins. There are also basalt flows within these sedimentary rocks. Milkereit et al. (1992) have recognized sedimentary rocks deposited in post-Grenvillian riftlike basins in peninsular Ontario. Williams and Telford (1986) described a series of intracontinental rift zones in the Ottawa, Canada, area that have had intermittent activity since the Precambrian. This is ample evidence that rifting has occurred in eastern North America from Midcontinent Rift time through early Paleozoic. Sedimentary rocks and basalt flows of similar composition are found in basins of differing ages. For these reasons we feel that for determining relative age, structural relationships suggested by seismic reflection and potential field data are safer to use than lithologic similarity. Using these criteria, the sedimentary rocks below the Mount Simon are younger than Grenvillian rocks. We, therefore, favor an Eocambrian or Early Cambrian age for the Middle Run, although we recognize that in the 400 million years from the Grenvillien Orogeny to the Middle Cambrian Mount Simon, there surely could have been several periods of major erosion and deposition that only more drilling, age dating, and seismic study can fully discern. Thus several sandstones of different ages may be present in western Ohio,

eastern Indiana, and north-central Kentucky. There may also be other undiscovered related lithologies in addition to those found in the Mattinson well.

The relationship of the Selma basin to the Consortium–WSU basin leaves four unresolved questions. What is the relation between the "arkose" and the carbonate units in the Friend No. 1 Mattinson well? Is it depositional or in fault contact? Is the "arkose" unit in this well the same as the Middle Run Formation? Could the Selma basin be the same as the Consortium–WSU basin based on similar seismic signatures? Additional drilling and seismic data are needed to answer these important questions.

ACKNOWLEDGMENTS

This paper would not have been possible without data supplied by a number of organizations. The well data from Warren County, Ohio, came from Ohio Department of Natural Resources, Division of Geological Survey. The Warren County seismic line was acquired by a consortium that included Ohio Department of Natural Resources, Stocker and Sitler, Inc., University of Cincinnati, and Wright State University. This Consortium was monetarily supported by Cincinnati Gas and Electric, Dayton Power and Light, Columbia Natural Resources, Ohio Oil and Gas Association, Paragon Geophysical, and Wright State University. The data were processed by Hosking Geophysical, Strata Search, Stroder and Unruh Geophysical, and Woods Geophysical. The Wright State University seismic lines were supported by grants from Amoco, Conoco, British Petroleum, Unocal, and Arco oil companies and some of the shot hole drilling was done by Duncan Exploration. These lines were processed by Lauren Geophysical and Woods Geophysical. Grants from the above oil companies paid for the COCORP tapes that were reprocessed by Lauren Geophysical. Dr. R. Carleton of the Ohio Geological Survey contributed to the petrology.

Armco Steel and B.P. Chemical provided us with the seismic lines in Middletown and Lima. These lines are quality control lines in support of their disposal wells in the Mount Simon Sandstone. Ralph von Frese of Ohio State University provided the magnetic and gravity maps used in Figures 8 and 9.

Finally, we would like to thank Eileen Porter and Cindy Harrison of Wright State University who patiently worked with us preparing this manuscript.

REFERENCES CITED

Armstrong, D. K., 1992, Paleozoic and Mesozoic sedimentation: Tectonic influences on a "stable" craton, *in* Thurston, P. C., Williams, H. R., Sutcliffe, R. H., and Stott, M. G., eds., Geology of Ontario: Ontario Geologic Survey, Special Volume 4, Part 2, p. 1314–1323.

Babcock, L. E., 1994, Biostratigraphic significance and paleogeographic implications of Cambrian fossils from a deep core, Warren County, Ohio: Journal of Paleontology, v. 68, no. 1, p. 24–30.

Colony, R. J., 1931, Report on Friend well, Clark County, Ohio: Report to Pure Oil Company (obtained from the Ohio Division of Geological Survey,

Department of Natural Resources), 6 p.

Dalziel, W. D., 1991, Pacific margins of Laurentia and East Antarctica—Australia as a conjugate rift pair: Evidence and implications for an Eocambrian supercontinent: Geology, v. 19, p. 598–601.

Dickas, A. B., Mudrey, M. G., Jr., Ojakangas, R. W., and Shrake, D. L., 1992, A possible southeastern extension of the Midcontinent Rift System located in Ohio: Tectonics, v. 11, p. 1406–1414.

Drahovzal, J. A., Harris, D. C., Wickstrom, L. H., Walker, D., Baranoski, M. T., Keith, B. D., and Furer, L. C., 1992, The east continent rift basin: A new discovery: Kentucky Geological Survey, Special Publication 18 Ser. XI, (Cincinnati Arch Consortium), 25 p.

Easton, R. M., 1992, Mesoproterozoic evolution of the southeast margin of Laurentia, *in* Thurston, P. C., Williams, H. R., Sutcliffe, R. H., and Stott, M. G., eds., Geology of Ontario: Ontario Geologic Survey, Special Volume 4, Part 2, p. 1302–1314.

Green, J. C., 1983, Geologic and geochemical evidence for the nature and development of the Middle Proterozoic (Keweenawan) Mid-Continental Rift of North America: Tectonophysics, v. 94, p. 413–437.

Harris, D. C., 1992, Regional stratigraphy of the Middle Run, *in* Drahovzal, J. A., Wickstrom, L. H., and Keith, B. D., eds., Cincinnati Arch Consortium, the geophysics and geology of the East Continent Rift Basin: Indiana Geological Survey Open-File Report 92-4, 337 p., 21 pl.

Hinze, W. J., Kellogg, R. L., and O'Hara, N. W., 1975, Geophysical studies of basement geology of southern peninsula of Michigan: American Association of Petroleum Geologists Bulletin, v. 59, p. 1562–1584.

Hinze, W. J., Allen, D. J., Fox, A. J., Sunwood, D., Woelk, T., and Green, A. G., 1992, Geophysical investigations and crustal structure of the North American Midcontinent Rift System: Tectonophysics, v. 213, p. 17–32.

Lucius, J. E., and von Frese, R. R. B., 1988, Aeromagnetic and gravity anomaly constraints on the crustal geology of Ohio: Geological Society of America Bulletin, v. 100, p. 104–116.

Milkereit, B., Forsyth, D. A., Green, A. G., Davidson, A., Hanmer, S. Hutchinson, D. R., Hinze, W. J., and Mereu, R. F., 1992, Seismic images of a Grenvillian terrane boundary: Geology, v. 20, p. 1027–1030.

Norris, S. E., Cron, W. P., Goldthwait, R. P., and Sanderson, E. E., 1952, The water resources of Clark County, Ohio: Ohio Division of Water, Bulletin 22, 82 p.

Pratt, T., Culotta, R., Hauser, E., Nelson, Brown, L., Kaufman, S., Oliver, J., and Hinze, W. 1989, Major Proterozoic basement features of the eastern Midcontinent of North America revealed by recent COCORP profiling: Geology, v. 17, p. 505–509.

Pratt, T. L., Hauser, E. C., and Nelson, K. D., 1992, Widespread buried Precambrian layered sequences in the U.S. Midcontinent: Evidence for

large Proterozoic depositional basins: American Association of Petroleum Geologists, v. 76, p. 1384–1401.

Rankin, D. W., 1967, Guide to the geology of the Mt. Rogers area, Virginia, North Carolina and Tennessee: Carolina Geological Society, 48 p.

Rankin, D. W., 1975, The continental margin of eastern North America in the southern Appalachians—the opening and closing of the proto-Atlantic oceans: American Journal of Science, v. 275A, p. 298–336.

Rankin, D. W., 1976, Appalachian salients and recesses: Late Precambrian continental breakup and the opening of the Iapetus Ocean: Journal of Geophysical Research, v. 81, p. 5605–5019.

Schwab, F. L., 1986, Upper Precambrian–Lower Paleozoic clastic sequences, Blue Ridge and adjacent areas, Virginia and North Carolina: Initial rifting and continental margin development, Appalachian orogen, *in* Textoris, D. A., ed., Society of Economic Paleontologists and Mineralogists, Field Guidebooks, southeastern United States, 3rd Annual Midyear Meeting: Society of Economic Paleontologists and Mineralologists, p. 1–42.

Shirley, C., 1990, Ohio buried basin challenges theory: American Association of Petroleum Geologists Explorer (March), p. 8–11.

Shrake, D. L., 1991, The Middle Run Formation: A subsurface stratigraphic unit in southwestern Ohio: Ohio Journal of Science, v. 91, p. 49–55.

Shrake, D. L., Wolfe, P. J., Richard, B. H., Swinford, E., Wickstrom, L. H., Potter, P. E., and Sitler, G. W., 1990, Lithologic and geophysical description of a continuously cored hole in Warren County, Ohio, including description of the Middle Run Formation (Precambrian?) and a seismic profile across the core site: Ohio Department of Natural Resources Division of Geological Survey Information Circular, no. 56, 11 p.

Shrake, D. L., Carlton, R. W., Wickstrom, L. H., Potter, P. E., Richard, B. H., Wolfe, P. J., and Sitler, G. W., 1991, A pre–Mount Simon basin under the Cincinnati Arch: Geology, v. 19, p. 139–142.

Wasson, I. B., 1933, Sub-Trenton formations in Ohio: Journal of Geology, v. 40, p. 673–687.

Williams, D. A., and Telford, P. E., 1986, Paleozoic geology of the Ottawa area, Field Trip 8: Geological Association of Canada, Joint Annual Meeting, 1986, Ottawa, Ontario, Guidebook, 25 p.

Wolfe, P. J., Richard, B. H., and Potter, P. E., 1993, Potential seen in Middle Run basins of western Ohio: Oil and Gas Journal, April 5, p. 68–73.

Wollensak, M. S., editor, 1988, Upper Keeweenawan rift-fill sequence, Mid-Continent Rift System, Michigan: Michigan Basin Geological Society, Fall Field Trip Guidebook, Michigan State University, East Lansing, 150 p.

MANUSCRIPT ACCEPTED BY THE SOCIETY JANUARY 16, 1996

Geological Society of America
Special Paper 312
1997

The East Continent Rift Complex:
Evidence and conclusions

T. Joshua Stark
Equitable Resources Energy Company, 1601 Lewis Avenue, Billings, Montana 59102

ABSTRACT

The interpretation of a wide variety of data has led to the postulated existence of a major multiphase rift complex in the eastern Midcontinent, here referred to as the East Continent Rift Complex (ECRC). Initial extension and associated transtensive deformation is interpreted to be Keweenawan in age and related to the Midcontinent Rift System. Major thrusting and associated transpressive deformation during Grenvillian collision appears to have significantly modified the ECRC, resulting in the development of an orthogonal mosaic of basement faulting. Epeirogenic remobilization appears to have occurred during Eocambrian extension, resulting in the development of additional rift segments. Subsequent remobilization tectonism of rift features greatly influenced the ensuing structural and stratigraphic development of the Paleozoic strata in the eastern Midcontinent region.

INTRODUCTION

Prior to the final decade of the twentieth century, conventional geoscientific theory maintained that certain structural and stratigraphic aspects of the eastern Midcontinent area were axiomatic. It was widely assumed that the Cincinnati Arch was underlain by a basement ridge of the Granite-Rhyolite Province. Within the Illinois Basin, primary extensional deformation was believed to have begun during the Eocambrian (650 to 560 Ma), forming a protobasin that subsequently "sagged" due to cooling and contraction. Overlying the crystalline basement, the Mount Simon Sandstone was recognized as the basal unit of the sedimentary sequence.

Recent events, including the discovery of pre-Paleozoic strata in numerous wells and reevaluation of industry and COCORP regional geophysical data, demonstrate that geoscientists need to reevaluate existent working models of the eastern Midcontinent. Geological theories pertaining to the eastern Midcontinent may have to account for an additional 0.5 billion years of structural and stratigraphic development. Previously unrecognized Proterozoic sedimentary strata are now interpreted to reach thicknesses in excess of 7,000 m in areas of the eastern Midcontinent. Proterozoic structural mechanisms are likely to be the formative processes responsible for many of the Paleozoic structures of the area.

The intent of this paper is to present data and interpretations that suggest the existence of a major polyphase rift, herein referred to as the East Continent Rift Complex (ECRC). Rift complex formation is postulated to have initiated during the Keweenawan (1.05 to 1.3 Ga), with subsequent modification during Grenvillian contraction (880 to 990 Ma), and additional extensional deformation during the Eocambrian (650 to 560 Ma). In addition to exploring geopotential, structural, and stratigraphic data relating to the ECRC, preliminary formative models and mechanisms of this area will be proposed.

REGIONAL SETTING

The proposed ECRC consists of a series of polyphase rift segments extending from the Ouachita orogen in Arkansas, through portions of Missouri, Tennessee, Illinois, Kentucky, Indiana, and Ohio, to ultimately disappear beneath the Grenville Front Tectonic Zone (Grenville Province, Fig. 1). The eastern margin of the ECRC lies beneath overthrusted Grenville meta-

Stark, T. J., 1997, The East Continent Rift Complex: Evidence and conclusions, *in* Ojakangas, R. W., Dickas, A. B., and Green, J. C., Middle Proterozoic to Cambrian Rifting, Central North America: Boulder, Colorado, Geological Society of America Special Paper 312.

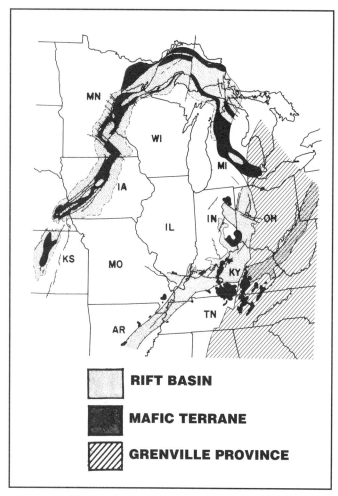

Figure 1. Rift complexes of the North American Midcontinent.

signature as a continuation of mafic suites beneath the Grenville Front implies that rift extension predated Grenvillian collision, and suggests that rift clastics and volcanics may extend a considerable distance eastward beneath the allochthon.

The northeasternmost portion of the ECRC disappears beneath the Grenville Front Tectonic Zone immediately adjacent to the southeastern terminus of the Mid-Michigan Gravity Anomaly (Hinze et al., 1975), the southernmost extension of the Midcontinent Rift System (Dickas, 1986). Based upon lithologic similarities of volcanic rocks (McCormick, 1961; Greenberg and Vitaliano, 1962; Lucius and Von Frese, 1988; Walker and Misra, 1992) and siliciclastic rocks (Wasson, 1932; Fettke, 1948; Shrake et al., 1990; Potter and Carlton, 1991; Dickas et al., 1992a; Drahovzal et al., 1992; Harris, 1992; Wickstrom, 1992), it has been proposed that the rift basin of western Ohio shares a similar developmental history with the Midcontinent Rift System. While the contact between the ECRC and the Midcontinent Rift System (MRS) remains problematic, a sufficient body of evidence exists to suggest that, during the late Precambrian, a great belt of extensional half grabens structurally differentiated the 1.5-Ga Granite-Rhyolite Province.

In order to establish a hierarchy of rift morphology in the Midcontinent craton, the chart in Figure 2 is proposed. Discrete rift components (principally half grabens) are referred to as rift segments. Individual segments are both Keweenawan and Eocambrian in age, and are linked spatially by remobilized margins. Segments may be continuous or separated by accommodation structures, rift passes, border fault systems, or transcurrent faults. A series of rift segments defines a rift complex; the proposed rift complex is the East Continent Rift Complex (ECRC), postulated to be associated with the Midcontinent Rift System.

A preliminary map illustrating the structure of crystalline basement rocks is presented in Figure 3. The map is a synthesis of the efforts of numerous researchers, assembled in a patchwork manner and subsequently modified to reflect the ECRC model. Whereas borehole data, seismic time-to-depth conversion (Hester, 1983, Cincinnati Arch Consortium, 1992; J. A. Drahovzal, personal communication, 1992), and geopotential modeling (Gibson, 1991; Stark, this volume, Chapter 17) have been utilized in map construction, many of the structures and depth to basement estimates are theoretical and speculative. Numerous

morphic rocks in central Kentucky and western Ohio, complicating the location of the eastern margin of the complex. The East Continent Gravity High (Bryan, 1975) partially underlies the Grenville thrust sheets, and has been interpreted to be caused by a basaltic rift feature by numerous authors (Rudman et al., 1965; Bryan, 1975; Keller et al., 1982, 1983; McPhee, 1983; Black, 1986; Lucias and Von Frese, 1988; Dickas et al., 1992a; Drahovzal et al., 1992). The interpretation of this geopotential

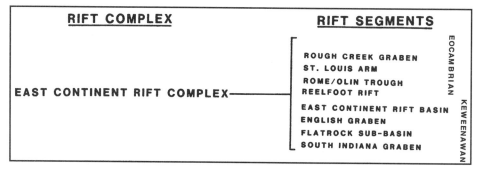

Figure 2. Differentiation of the East Continent Rift Complex into named discrete rift segments.

published subsurface interpretive basement maps were utilized in the construction of Figure 3. Considerable interpretive license has been utilized by the author et al. in the generation of this illustration, and the reader must bear in mind that the figure is more appropriate to indicate tectonic process than to clearly define structural morphology.

Through the utilization of geopotential, structural, and stratigraphic data, an orthorhombic mosaic of basement structure is suggested. Extensive transcurrent offset of the ECRC in a northwest-southeast orientation may have occurred as a consequence of Keweenawan transtension or Grenvillian transpression. Whereas orthogonal basement architecture has previously been recognized (Black, 1986, 1989), a regional model of deformation integrating the concepts of Appalachian-Ouachita transcurrent offset (Thomas, 1989), wrench deformation of the Midcontinent Rift System (Dickas, 1986), and cross faulting of the English Graben (Stark, 1990, this volume, Chapter 17; Dickas et al., 1992b; Hendricks et al., 1994) and the East Continent Rift Basin is required to integrate the dynamics of structural evolution and remobilization tectonism in the eastern North American craton.

In several areas, it appears that post-Keweenawan uplift and erosion may have removed Proterozoic rift sediments prior to the deposition of the Paleozoic sequence, adding to the complexity and segmentation of the ECRC. These eroded areas may be spatially linked with rift accommodation structures and border fault systems.

The southwestern segments of the ECRC are currently expressed by areas of more recent subsidence, as evidenced by the downwarped Mississippian Embayment and the Moorman Syncline, associated respectively with the Reelfoot Rift and the Rough Creek Graben (Fig. 4). Conversely, the northeastern segments of the ECRC are currently characterized by areas of positive structural relief. The formation of these arch areas may be related to the subsidence of surrounding features or to positive isostatic adjustment, where rift sediments may have been exhumed from the adjacent crystalline terrane. In this manner, the positive structural elements of the Cincinnati Arch and the Lexington Dome (Fig. 4) may have formed. The areas of relative subsidence and uplift appear to be separated by a high-relief, fault-bounded structural feature, here referred to as the Louisville Accommodation Structure (LAS: Fig. 4) This major mafic igneous feature was first delineated by a statewide aeromagnetic analysis of Indiana (Henderson and Zietz, 1958, p. 32) and referred to as "a major tectonic element which extends a considerable distance south and east ... into Kentucky." Subsequently, portions of the LAS were referred to as the Louisville High (Black, 1989).

The extreme structural diversity exhibited in the ECRC may be a function of rift complex longevity. Sedimentologic comparisons, igneous petrography, and radiometric dating (Zartman et al., 1967; Cincinnati Arch Consortium, 1992) suggest that rifting in portions of the ECRC began prior to or during Midcontinent Rift System extension. The 1990 earthquake

epicenters of Meade County, Kentucky (Street et al., 1991) are interpreted to represent modern remobilization events that occurred within the English Graben. Throughout the proposed 1.3-billion-year developmental history of the ECRC, the mechanisms of extension, collision, thermal subsidence, isostasy, and epeirogenic remobilization have exerted broad control on the structural and stratigraphic differentiation of the eastern Midcontinent.

A preliminary assessment of available data has resulted in the development of the following evolutionary sequence for the ECRC: (1) Keweenawan extension, (2) Grenvillian contraction, (3) Pre-Eocambrian uplift/erosion, (4) Eocambrian extension.

These major deformation events established the principal zones of structural instability in the eastern Midcontinent. Subsequent deformation occurring throughout the Paleozoic appears to be related to epeirogenic remobilization of these zones of weakness. A similar sequence has been proposed for the evolution of the East Continent Rift Basin (Cincinnati Arch Consortium, 1992; Drahovzal et al., 1992). Additional segments of the ECRC, proposed to be Precambrian in age, may have evolved through a similar cycle of events.

KEWEENAWAN EXTENSION (1.05 to 1.3 Ga)

Keweenawan extension appears to have rifted the Granite-Rhyolite Province along multiple segments of the ECRC. Industry seismic data is interpreted to suggest that primary extension formed internally complex half grabens along rift margins, possibly providing local sources for syntectonic deposition of sediments in alluvial, fluvial, and lacustrine systems. Continued extension appears to have emplaced extensive flows of flood basalts within the eastern Midcontinent region, penetrated in several deep boreholes located in Indiana and Kentucky (Greenburg and Vitaliano, 1962; Cincinnati Arch Consortium, 1992; M. Bass, personal communication, 1992). These basalts, similar to certain basalts of the North Shore Volcanic Group (Walker and Misra, 1992) may have eventually formed a series of overlapping basalt plateaus, similar to flood basalt accretion noted in the Keweenawan Lake Superior Basin of the MRS (Green, 1977, 1982). The basalts are primarily alkalic in composition (Walker and Misra, 1992; Drahovzal et al., 1992), and associated felsic extrusives (rhyolites and trachytes) have been noted in numerous boreholes. Felsic rocks are found in many of the Keweenawan lava plateaus (White; 1972; Annells, 1974; Green, 1982). Geochemical analysis of the basalts indicate similarity with mafic magma generated in a continental rift setting (Wilson, 1989; Drahovzal et al., 1992; Walker and Misra, 1992).

Some researchers suggest that continuing extension (Black, 1989; Drahovzal et al., 1992) led to major mafic additions to the crust (Keller et al., 1975, 1982), which gives geophysical expression as the East Continent Gravity High. Opposing viewpoints suggest that the ECGH may be a shallow feature occurring within the Grenville allochthon (D. J. Allen, personal communication, 1993). Transtensive wrench systems, possibly similar in

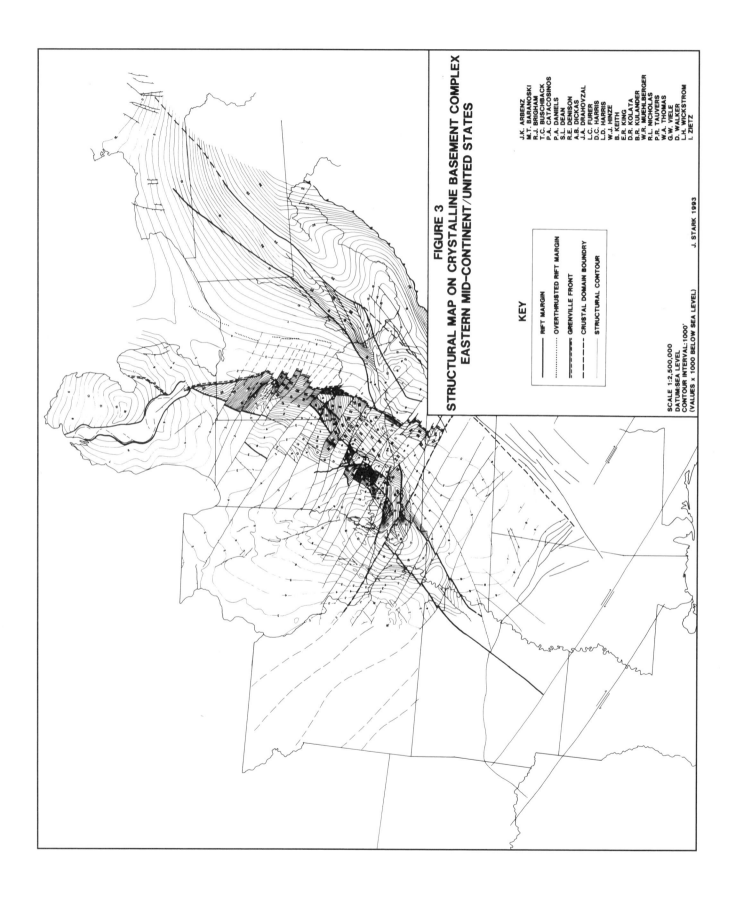

FIGURE 3

STRUCTURAL MAP ON CRYSTALLINE BASEMENT COMPLEX
EASTERN MID-CONTINENT/UNITED STATES

KEY

——————— RIFT MARGIN

·················· OVERTHRUSTED RIFT MARGIN

–··–··–··– GRENVILLE FRONT

– – – – – CRUSTAL DOMAIN BOUNDRY

——————— STRUCTURAL CONTOUR

SCALE 1:2,500,000
DATUM:SEA LEVEL
CONTOUR INTERVAL:1000'
(VALUES x 1000 BELOW SEA LEVEL)

J. STARK 1993

J.K. ARBENZ
M.T. BARANOSKI
R.J. BRIGHAM
T.C. BUSCHBACK
P.A. CATACOSIMOS
P.A. DANIELS
S.L. DEAN
R.E. DENISON
A.B. DICKAS
J.A. DRAHOVZAL
L.C. FURER
D.C. HARRIS
L.D. HARRIS
W.J. HINZE
B. KEITH
E.R. KING
D.R. KOLATA
B.R. KULANDER
W.R. MUEHLBERGER
R.L. NICHOLAS
P.R. TAUVERS
W.A. THOMAS
G.W. VIELE
D. WALKER
L.H. WICKSTROM
I. ZIETZ

Figure 3. Crystalline basement structure of Eastern North American craton. Referenced (basement structural map) authors cited in legend.

◄────────────────────

morphology to modern mid-oceanic spreading rifts, may have evolved during this period of rift development. Sedimentary basins may have developed and filled with Middle Proterozoic sediments, apparently deposited by alluvial and fluvial systems (Harris, 1992). Borehole penetrations of the Proterozoic sediments in the area have revealed the existence of a lithic arenite sequence referred to as the Middle Run Formation (Shrake et al.,1990). The potentially equivalent upper Keweenawan clastics (Dickas, 1986) are the Oronto Group of the Midcontinent Rift System in the Lake Superior region. Late-stage thermal subsidence of the Proterozoic rift complex may have downwarped

areas surrounding the ECRC, resulting in the accumulation of thick Proterozoic sequences over a large portion of the eastern Midcontinent region (Pratt et al., 1992; Drahovzal et al., 1992).

A brief summary of the postulated Proterozoic segments of the ECRC are discussed in the following section. A map locating the individual segments is presented in Figure 4. The following segments are interpreted as Proterozoic in age: East Continent Rift Basin/Fort Wayne Rift, English Graben/Flatrock subbasin, and South Indiana Graben

The Kentucky-Indiana Trough and Kentucky-Ohio Trough (Black, 1986, 1989) have been interpreted as major Precambrian features that appear to control many structural elements of the later-forming Rome Trough and are considered to be earlier names for portions of segments of the ECRC (J. A. Drahovzal, personal communication, 1993). Based upon geopotential, seis-

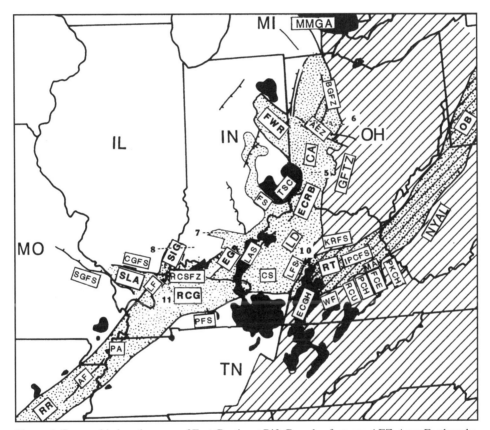

Figure 4. Geographic location map of East Continent Rift Complex features. AEZ, Anna Earthquake Zone; AF, Axial Fault; BGFZ, Bowling Green Fault Zone; CA, Cincinnati Arch; CGFS, Cottage Grove Fault System; CS, Cumberland Saddle; ECGH, East Continent Gravity High; ECRB, East Continent Rift Basin; EG, English Graben; FCE, Floyd County Embayment; FS, Flatrock subbasin; FWR, Fort Wayne Rift; GFTZ, Grenville Front Tectonic Zone; IPCFS, Irvine–Paint Creek Fault System; KRFS, Kentucky River Fault System; LAS, Louisville Accommodation Structure; LD, Lexington Dome; LF, Lusk Fault; LFS, Lexington Fault System; MA, Magoffin Arch; MMGA, Mid-Michigan Gravity Anomaly; NYAL, New York–Alabama Lineament; OB, Olin Basin; PA, Pascola Arch; PCH, Perry County High; PFS, Penneyrile Fault System; PKCH, Pike County High; RCG, Rough Creek Graben; RCU, Rockcastle Uplift; RCSFZ, Rough Creek–Shawneetown Fault Zone; RR, Reelfoot Rift; RT, Rome Trough; TSC, Tri-State Caldera; SGFS, Saint Genevieve Fault System; SIG, South Indiana Graben; SLA, Saint Louis Arm; WF, Warfield Fault. Numbered dashed lines denote the figure numbers of exhibited seismic lines.

mic, and limited borehole data, the ECRC rift segments appear to be characterized by considerable thickness of rift fill sediments and volcanics, with basement depths reaching 7,600 to 9,150 m (Hester, 1983; Hester and Kohl, 1983; Howe, 1985; Shrake et al., 1990, Drahovzal et al., 1992).

East Continent Rift Basin/Fort Wayne Rift

The recently described East Continent Rift Basin (ECRB; Drahovzal et al., 1992) has long been suspected to exhibit rift-related characteristics (Rudman et al., 1965; Bryan, 1975; Halls, 1978; Keller et al., 1975, 1981; Cable and Beardsley, 1984; Shumaker, 1986; Lucius and Von Frese, 1988; Pratt et al., 1989), but not until recently has a comprehensive, multidisiplinary evaluation been performed (Cincinnati Arch Consortium, 1992). The impetus for this evaluation was the discovery of thick, pre–Mount Simon sediments in the Warren County, Ohio, deep hole DGS 2627. These strata were subsequently named the Middle Run Formation (Shrake et al., 1990). Approximately 580 m of sediments interpreted as Proterozoic strata were cored in this well. Technical problems resulted in the termination of drilling while still within the Proterozoic sequence. Ensuing efforts led to the acquisition of an 12.9-km seismic traverse centered upon the borehole (Fig. 5). The seismic data are interpreted to show the flank of an inclined and beveled half-graben structure, reaching depths to crystalline basement approaching 5,490 m. Due to the position of the well almost precisely upon the axis of the Cincinnati Arch, it became immediately evident that earlier ideas regarding basement architecture in western Ohio were inaccurate, and that

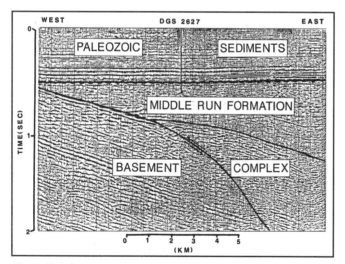

Figure 5. Seismic traverse across Ohio Geological Survey well DGS 2627 illustrating the development of the Proterozoic East Continent Rift Basin beneath the axis of the Cincinnati Arch in Warren County, Ohio. Note the thrusted basement complex overlain by Middle Run Formation (Proterozoic) sediments, the angular unconformity at the pre-Paleozoic erosion surface, and the flat-lying Paleozoic sequence (Shrake et al., 1990).

more rigorous evaluation should be performed. As a result, the oil industry funded the Cincinnati Arch Consortium (CAC); their efforts have led to definition and recognition of the East Continent Rift Basin, a major segment of the ECRC.

In the currently proposed model, the ECRB formed through extension of the Granite-Rhyolite Province. The Grenville Front occurs to the east of the East Continent Rift Basin, and industry seismic data are interpreted to show that the proposed Proterozoic depocenter extends a considerable distance beneath the thrusted Grenville basement rocks (Drahovzal et al., 1992). Geopotential data indicates that the East Continent Gravity High is partially common with allochthonous sheets of the Grenville Front Tectonic Zone. Similarly, broad gravity lows have been correlated with seismic data to indicate the position of thick sequences of Proterozoic clastic rocks and intercalated volcanic flows. The absence of abundant metamorphic rock fragments in these lithic arenite sequences demonstrate that these clastics were not derived from a Grenvillian source area and did not accumulate in a foreland basin setting. Rather, the volcanic/plutonic rock fragments contained within the clastics and continental-style basalt flows demonstrate that the Proterozoic fill accumulated prior to the approach of the Grenville allochthon (Harris, 1992).

In central western Ohio, tests holes in the ECRC along the fault system associated with the Anna Earthquake Zone penetrated Proterozoic basalts, gabbros, and rhyolites, whereas deep holes to the north and south encountered Middle Run lithic arenites. This would imply that the Anna Earthquake Zone structures were uplifted prior to Late Proterozoic erosion, ultimately resulting in a subcrop pattern of proposed MRS equivalent volcanic and intrusive rocks (or the underlying Granite-Rhyolite Province crystalline basement) flanked to the north and south by red-bed strata interpreted to be lithologically equivalent to rocks comprising the Oronto Group of the Lake Superior district (Dickas et al., 1992a). This pre-Paleozoic subcrop pattern is also evident from truncation features noted in the reprocessed COCORP OH-1 regional seismic line (J. A. Drahovzal, personal communication, 1992; Fig. 6). The once structurally positive basement blocks have remobilized throughout subsequent geologic time and today appear to be associated with the active Anna Earthquake Zone, which is second only to the New Madrid area in seismicity of the eastern Midcontinent region. The proposed erosional deflation of the Proterozoic sequence and the continuing mobility of this area may indicate that the Anna Earthquake Zone is associated with an accommodation structure within the East Continent Rift Basin.

The fault blocks of the Anna Earthquake Zone are common to the deep fault systems that border the Fort Wayne Rift (McPhee, 1983). The Fort Wayne Rift projects from western Ohio into northeastern Indiana, and appears to be coeval with the East Continent Rift Basin. It is here suggested that the Fort Wayne Rift contains a northern mafic basalt fill sequence associated with the East Continent Gravity High, and a southern peripheral basin, expressed by a gravity minimum and a subdued, low-amplitude magnetic mini-

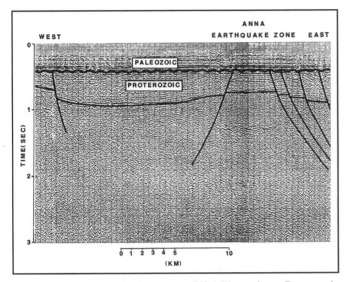

Figure 6. COCORP seismic traverse OH-1 illustrating a Proterozoic depocenter and structural horst beneath flat-lying Paleozoic strata in the Fort Wayne Rift/Anna Earthquake Zone area (Shelby County, Ohio). Upon the pre-Paleozoic angular unconformity, volcanic strata subcrops on the horst block, whereas Middle Run sediments subcrop in the adjacent depocenters. This seismically active area occurs within the East Continent Rift Basin (Cincinnati Arch Consortium, 1992).

mum that presumably represents a depocenter filled primarily with Proterozoic clastic rocks and minor volcanic flows. Both the Fort Wayne Rift/ECRB and the fault blocks interpreted to be associated with the Anna Earthquake Zone are evident in the reprocessed COCORP OH-1 seismic traverse (J. A. Drahovzal, personal communication, 1992; Fig. 6).

In northwesternmost Ohio, reported lithologies from basement tests within the projected trend of the ECRC lack sediments presumed to be Proterozoic in age. Instead, deep boreholes report penetrating crystalline basement consisting of granite and granitic gneiss at relatively shallow depths. These reported depths to basement disagree with recent geopotential interpretations that the East Continent Rift Basin deepens significantly in northwestern Ohio (Gibson, 1991). Wells in northernmost Ohio were not examined during the Cincinnati Arch Consortium study, and it has been suggested that careful study of the deep borehole lithologies may show that the reported "granite" may actually be misidentified Middle Run sediments (L. H. Wickstrom, personal communication, 1993). This uncertainty increases the difficulty in demonstrating an association between the Midcontinent Rift System and the East Continent Rift Complex, and shows that additional research will be required to resolve the complexities of the area.

Additional information pertaining to the ECRB is provided in great detail in the Cincinnati Arch Consortium Open-File Report (1992), and the interested reader will find there a more comprehensive analysis of this newly discovered rift segment.

English Graben

The proposed (possibly Keweenawan-aged) rift segment that may link the South Indiana Graben area with the recently described East Continent Rift Basin is the English Graben (Stark, 1990, this volume, Chapter 17; Dickas et al., 1992b; Hendricks et al., 1994). The English Graben has primarily been defined by geopotential data. Proprietary industry seismic data are interpreted to indicate that portions of the rift segment reach depths to basement exceeding 6,100 m (S. H. Rowley, personal communication, 1992), suggesting the existence of approximately 4,270 m of layered reflectors that are interpreted as Proterozoic strata unconformably underlying approximately 1,830 m of Paleozoic section. Similar to the Proterozoic depocenters of the East Continent Rift Basin (Cincinnati Arch Consortium, 1992), the thick sequences of proposed Precambrian strata within the English Graben are characterized by broad negative gravity anomalies. Additional confirming seismic data are provided by the COCORP seismic traverse of Indiana (IN-1), which crosses the western margin of the English Graben and images approximately 38 km of the rift interior. Half-graben asymmetry is evident (Fig. 7), and east of VP1200, it was noted that "dipping reflectors and a ramplike feature indicate deformation in the layering, perhaps indicative of a fundamental lateral change" (Pratt et al., 1989, p. 506). A broad negative gravity anomaly and associated low-amplitude magnetic anomaly in the area of the postulated English Graben are interpreted to suggest that rift fill may consist primarily of clastic rocks

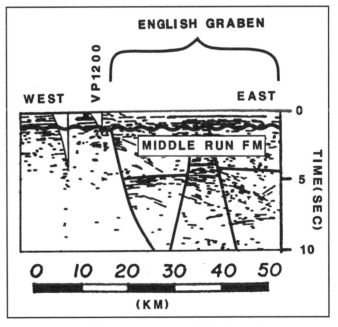

Figure 7. COCORP seismic traverse IN-1 illustrating horizontal reflectors and major defractions beneath flat-lying Paleozoic strata. Note the contrast between the western nonrifted area and the reflectors within the English Graben. The line is located along the western margin of the English Graben in Orange County, Indiana (modified from Pratt et al., 1989).

and associated volcanic flows, similar to the strata found within other segments of the ECRC.

A more detailed analysis of the newly postulated English Graben and associated Flatrock subbasin is presented in a companion paper within this volume.

South Indiana Graben

The South Indiana Graben was first referred to as a portion of the New Madrid Rift Complex (Braile et al., 1982), a four-armed rift junction including the St. Louis Arm, the South Indiana Graben, and the Rough Creek Graben (see Fig. 4). Occurring upon the margins of the South Indiana Graben is the Wabash Valley Fault System, an apparently remobilized belt of faults that offset Paleozoic strata. A 68-km seismic reflection reconnaissance of the Wabash Valley Fault System (Sexton et al., 1986) was interpreted to indicate that the shallow, minor displacements (120 m) in the Paleozoic rocks could be traced to basement faults of approximately 500 m offset. A series of northeast-southwest trending rift grabens, possibly filled with pre–Mount Simon clastic and volcanic rocks, was also proposed. The largest investigated structure, the Grayville Graben, was determined to be 15 km wide, and to contain approximately 3 km of pre–Mount Simon synrift fill (Sexton et al., 1986; Fig. 8).

COCORP midcontinent seismic traverses in Illinois (IL-1) and in Indiana (IN-1) also identified possible Proterozoic sediments in the area of the Wabash Valley Fault System (Pratt et al., 1992); their interpretation, however, suggested that the pre–Mount Simon sequence in this area is related to a regional Proterozoic depocenter and not confined to the rift segment. The COCORP interpretation of the extent of the Proterozoic depocenter in part agrees with the Cincinnati Arch Consortium study (Fig. 9). It is proposed that the eastern Midcontinent experienced

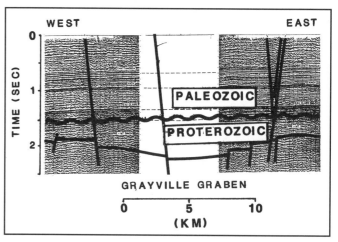

Figure 8. Seismic traverse showing the Grayville Graben as a portion of the Proterozoic-aged South Indiana Graben in White County, Illinois, and Gibson County, Indiana. The seismic data was interpreted to exhibit a graben 15 km wide, containing approximately 3 km of pre-Paleozoic rift fill. Additionally, the survey indicated that shallow Paleozoic faults of the Wabash Valley Fault System were continuous with basement faults of greater magnitude (Sexton et al., 1986).

regional downwarping derived from post-Keweenawan thermal subsidence, similar to the subsidence event interpreted to have effected the Illinois Basin area subsequent to Eocambrian extension (Treworgy et al., 1989).

GRENVILLIAN CONTRACTION (880 to 990 Ma)

Grenvillian contraction resulted in overthrusting the eastern margin of the ECRC and appears to have emplaced an allochthon of metamorphic strata above and adjacent to the rift

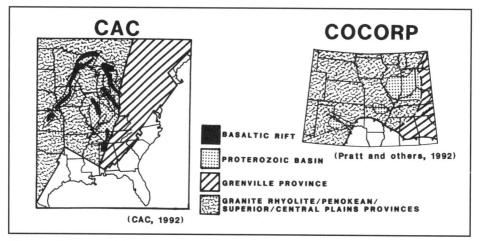

Figure 9. Two regional interpretations of Proterozoic basins of the eastern Midcontinent suggest broad depocenters covering vast lateral areas. Potential field and seismic data, however, indicate the existence of deep Proterozoic rift segments within these broad depocenters. An analogue would be the position of the Rough Creek Graben (rift) within the Illinois Basin (sag depocenter; Cincinnati Arch Consortium, 1992; Pratt et al., 1992).

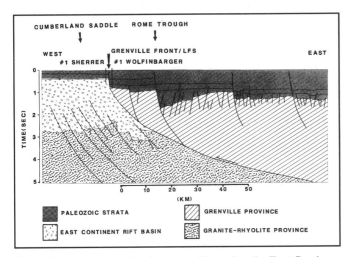

Figure 10. Interpreted seismic section illustrating the East Continent Rift Basin and the Grenville Front Tectonic Zone, overlain by the Cumberland Saddle and the Rome Trough in Kentucky. The border fault for the Eocambrian-aged Rome Trough (Lexington Fault System) is related to relaxation of the allochthonous Grenville Front. Interrift faults are related to relaxation of deep sole faults. The Texaco #1 Sherrer drilled a thinned Paleozoic section, passed into a Proterozoic basalt flow unit, and reached total depth in the Middle Run Formation. The Texaco #1 Wolfinbarger penetrated growth fault-thickened Cambrian strata within the Rome Trough, reaching total depth within the metasediments of the Grenville Front. The Cumberland Saddle (positive arch feature) is interpreted to have formed due to isostatic compensation of underlying low-density rift sediments. The Rome Trough formed during Eocambrian trailing edge extension, remobilizing a Proterozoic-aged suture between two microplates.

sediments. Proterozoic structural relationships are evident in a 111-km, east-west industry seismic interpretation (Fig. 10) in central Kentucky, trending from the Cumberland Saddle/East Continent Rift Basin area on the west across the Lexington Fault System/Grenville Front to terminate above thrusted Grenville metamorphic rocks to the east. A synthetic seismogram constructed from the nearby Texaco #1 Sherrer well (position shown on Fig. 10) ties the seismic line to well data and suggests that, immediately west of the Grenville Front, an approximately 7,930-m section of sediments consists of about 915 m of Paleozoic strata resting upon approximately 7,015 m of postulated Proterozoic rift fill. The lack of a Mount Simon section and thinned Rome section in the Sherrer well indicates that during earliest Paleozoic deposition this area was structurally positive. Immediately beneath the Paleozoic strata, a 155-m-thick series of basalt flows was penetrated; geochemical analysis suggests that the lavas were continental flood basalts (Walker and Misra, 1992; Drahovzal et al., 1992) and are compositionally similar to the Keweenawan flows of the Midcontinent Rift System. Beneath the basalts, the well penetrated 122 m of lithic arenites assigned to the Middle Run Formation (Cincinnati Arch Consortium, 1992).

The Texaco #1 Wolfinbarger also penetrated the deep section, 5.8 km to the east across the Lexington Fault System/

Grenville Front. The well penetrated the Grenvillian complex at a depth of 1,834 m, and did not encounter lithic arenites of a Middle Run lithology (Drahovzal et al., 1992).

The interpretation of the regional seismic line suggests that the Grenville Front represents the terminal edge of a thrusted allochthonous metamorphic terrane; the Lexington Fault System (and farther to the north, the Bowling Green Fault System) may be the shallow manifestation of the thrust complex.

Interpreted preexisting northwest-southeast trending transtensive wrench systems may have become remobilized as transpressive detachment zones during Grenvillian contraction, possibly resulting in the differential thrusting of the Grenville Front (see Fig. 3). Transcurrent features that may have evolved during the Keweenawan and Grenvillian events appear to extend over a substantial portion of the craton and form a major component of basement orthogonal mosaic architecture. Concomitant fold systems and the formation of a Grenvillian foreland basin may have evolved as a consequence of the contraction (Drahovzal et al., 1992). The foreland basin may have been extensively peneplanated during pre-Eocambrian erosion. Boreholes penetrating the Proterozoic lithic arenites of the region to date have not contained significant (if any) metamorphic rock fragments. This suggests that, as yet, sediments deposited within remnants of a Grenvillian foreland basin have not been encountered.

PRE-EOCAMBRIAN UPLIFT/EROSION (prior to 650 to 560 Ma)

Numerous published (Sexton et al., 1986; Shrake et al., 1990; Heigold, 1990) and proprietary seismic lines exhibit defined subparallel rhythmic reflectors previously attributed to "intrabasement layers" or seismic multiples. Comparison of GLIMPCE (Cannon et al., 1989) with COCORP data indicates that these pre-Phanerozoic reflectors are likely to be volcanic flows and assorted interflow sedimentary rocks equivalent to the middle Keweenawan of the Lake Superior district. Whereas the stacked basalt flows of the ECRC were probably deposited in planer horizontal sheets, the seismic sections indicate the subparallel flows to be inclined and successively beveled by pre-Paleozoic erosion. This suggests a substantial component of syn- and/or post-depositional deformation prior to the formation of the pre-Sauk unconformity. The different rock types that occur beneath the unconformity upon the basement fault complex in western central Ohio suggests the regional aspect of this erosive episode.

Further evidence of a subsurface angular unconformity is the mechanical deviation of a borehole at the point of penetration at the contact between strata of differing dip orientation. In the Warren County, Ohio, deep core test DGS 2627, an immediate drillstring deviation to the northwest was noted as the bit passed from Mount Simon sediments into the Middle Run Formation, implying a southeasterly dip of the Proterozoic strata (see Fig. 5).

The Middle Run lithic arenites were deposited in an arid,

continental setting, in distal alluvial fans and braid-plain environments (Harris, 1992). Lithic rock fragments are almost exclusively volcanic and plutonic, with characteristic compaction and associated hematite, calcite, feldspar, quartz, and illite cements. These diagenetic characteristics are distinctly different from those of the overlying Mount Simon sandstones, indicating that the Middle Run diagenetic history was essentially completed prior to the deposition of the overlying Paleozoic sequence.

Extensive deformation and subsequent pre-Paleozoic erosion may have removed most or all Grenvillian foreland basin sediments (Drahovzal et al., 1992), as well as removing a substantial portion of Upper Proterozoic strata in certain areas. Overt structural evidence of the primary rift phase was greatly subdued by this episode of erosional planation.

EOCAMBRIAN EXTENSION (650 to 560 Ma)

Eocambrian extension rifted the protocontinent Laurentia, forming a series of half grabens (Harris, 1978) associated with the opening of the Iapetus Ocean (Rankin, 1976; Thomas, 1989). As Laurentia separated, post-Grenvillian sediments accumulated upon the eastern passive margin, now recognized in the Blue Ridge area of the Appalachians. Sediments were primarily alluvial deposits with associated volcanics; the basalt-free Ocoee Supergroup also accumulated upon the trailing margin as marine continental shelf and submarine-fan deposits (Rankin et al., 1989).

Inboard within the craton, ancestral zones of weakness responded to renewed extension. Rift-margin boundary faults appear to have developed along remobilized transcurrent features (Rough Creek–Shawneetown Fault System, Kentucky River Fault System), relaxed Grenvillian décollement and sole faults (Lexington Fault System, Bowling Green Fault Zone), and probable continental sutures (Warfield Fault System along the New York–Alabama Lineament; see Fig. 4). A major difference from earlier rift-fill sedimentation is the general absence of volcanics from the clastic Eocambrian rift sequence. Radical thickening of Early to Middle Cambrian sediments occurred along the escarpment border fault boundaries of the newly formed half grabens, resulting in the accumulation of as much as 5,185 m of synrift Eocambrian strata (Hester, unpublished data). Areas of Cambrian-aged rifting may have been superposed upon thick sequences of Proterozoic strata, creating intrarift "deeps" reaching an interpreted maximum depth of 8,845 m in Webster and Union Counties, Kentucky (Hester, unpublished data). Examination of industry seismic data indicates that faulting that began in the Eocambrian remained active until Middle Cambrian time. Whereas some intrarift faults became stable and were buried by thick sequences of Paleozoic strata, many border fault systems were remobilized throughout the Paleozoic and have exerted considerable control upon subsequent structural and stratigraphic development.

As extension waned, a second episode of thermal subsidence began in the Middle Cambrian, downwarping the areas surrounding the Eocambrian rifts and allowing for the widespread deposition of Sauk Sequence carbonates (Treworgy et al., 1989).

A brief summary of the postulated Eocambrian segments of the ECRC is discussed in the following section. A map locating the individual segments is presented in Figure 4. The following segments are interpreted as Eocambrian in age: Reelfoot Rift, Rough Creek Graben, and Rome Trough.

Reelfoot Rift

The existence of the Reelfoot Rift was first postulated based upon the borehole evidence of extensive Cambrian and Ordovician basinal sedimentation (Schwalb, 1969) beneath the broad downwarp of the Mississippian Embayment. The emplacement of anomalous mantle material at the base of the crust resulted in approximately 1 to 2 km of uplift, crustal extension, and the development of an axial zone of rotational block fault rifting (Ervin and McGinnis, 1975). The end of mafic emplacement noted in the flanking Saint Francis Mountains may represent the cessation of rift-related igneous activity (Tikrity, 1968). Geopotential studies (Hildenbrand et al., 1977; Kane et al., 1981) indicated boundary-fault positions of the Reelfoot Rift and inferred the existence of synrift sediments. Based upon industry seismic data, rift sediments in excess of 9,150 m were inferred (Howe, 1985; Howe and Thompson, 1984). Howe defined all lower sediments within the Reelfoot Rift to be Lamotte (Cambrian) in age, utilizing seismic data to illustrate immense thicknesses of the formation.

Pre-Lamotte sediments were subsequently noted in area boreholes. These sediments were subdivided into two formations, the lower Reelfoot Arkose, deposited in alluvial fans and fluvial braided environments, and the Saint Francis Formation, described as a unit grading from the underlying arkoses into marine clastics and carbonates (Houseknecht, 1989). Geophysical log correlations indicate that the Lamotte Sandstone grades laterally into a marine shale that occurs above the Saint Francis Formation within the margins of the rift. Recent paleontologic evidence from southwest Ohio suggests that the Mount Simon (Lamotte equivalent) may be Middle Cambrian (Babcock, 1992). The Mount Simon/Lamotte rests unconformably upon crystalline rocks exterior to the rift, demonstrating that the rift is at least pre–Middle Cambrian in age. The thickened Cambro-Ordovician sediments found within the rift (Schwalb, 1969) similarly suggests synrift deposition during the Eocambrian.

Rough Creek Graben

The Rough Creek Graben (Soderberg and Keller, 1981) constitutes a major Eocambrian rift segment, oriented east-west in western Kentucky and southeasternmost Illinois (Fig. 4). Lying beneath the shallow structural sag of the Moorman Syncline, the Rough Creek Graben exists as an asymmetric half graben with intrarift "basement rocks" dipping to the north into the axial depocenter situated along the northern border fault system (Fig. 11).

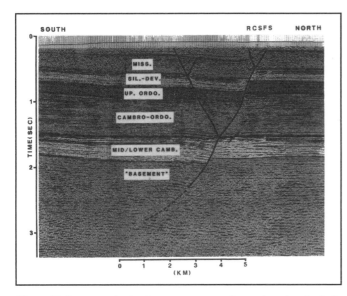

Figure 11. Interpreted seismic traverse across the northern border fault of the Rough Creek Graben (Rough Creek–Shawneetown Fault System) in western Kentucky. Cambrian growth faulting and lithofacies changes define this late-phase border fault system. Note the prominent antithetic fault and curved listric fault plane in this complex, braided fault system (Bertagne and Leising, 1990).

Geopotential analysis (Lidiak and Zietz, 1976; Schwalb, 1982; Hildenbrand et al., 1982) indicates the northern border fault, the Rough Creek–Shawneetown Fault System (RCSFS), is manifested at depth by the escarpment margin of the rift segment. Interpretation of seismic lines crossing the RCSFS indicate a complex braided fault zone as wide as 2.4 km (Nelson and Lumm, 1984), characterized by listric growth faults at depth, defining a depocenter with a maximum depth to basement of 8,845 m (time to depth conversion; Hester, unpublished data). Deep boreholes within the Rough Creek Graben penetrated thickened synrift Cambrian sediments (Schwalb, 1982; Houseknecht and Weavering, 1983), represented predominantly as marine shale sequences that are absent upon the adjoining craton. Eocambrian alluvial and fluvial continental sandstone sediments are speculated to occur beneath the marine sequence (Schwalb, 1982).

The easternmost extension of the RCSFS dies out upon the southwestern flank of the Cincinnati Arch; the western portion of the fault system merges with the Lusk Fault, the northwestern border fault of the Reelfoot Rift (Nelson and Lumm, 1984; see Fig. 4). Rift segment continuity indicates that, during the Eocambrian, the southern portion of the ECRC was a narrow embayment, open southward to the sea (Kolata and Nelson, 1991).

The southern border fault system of the Rough Creek Graben is considerably less defined, occurring as a series of normal, relatively small-scale extensional faults referred to as the Pennyrile Fault System (Schwalb, 1982). Both the Pennyrile and the RCSFS deep border fault systems may have initiated as portions of much larger wrench systems during Keweenawan/Grenvillian transcurrent deformation; these zones of weakness were subse-

quently modified into extensional features during the breakup of Laurentia.

The east-west epicratonic transcurrent system that formed the RCSFS, originally referred to as the 38th Parallel Lineament (Heyl, 1972), continues beyond the rift segments as the Cottage Grove Fault System, trending across southern Illinois and into Missouri. Similarly, the Penneyrile Fault System can be tentatively extended to the west, ultimately intercepting the Saint Genevieve Fault System. These two facing fault systems define the St. Louis Arm, a proposed rift segment defined by a series of linear positive geopotential anomalies, interpreted to be caused by mafic intrusives (Braile et al., 1984). The lack of known synrift sediments (Nelson and Zhang, 1991) and the abrupt termination of rhythmic Proterozoic reflectors imaged by COCORP data (Nelson et al., 1988) within the St. Louis Arm suggests that, west of the Reelfoot/Rough Creek segments, the area within the ancestral transcurrent faults remained structurally positive and never experienced synrift infill by volcanic and sedimentary rocks. Tectonism may have been expressed as simple wrench deformation; these ancestral zones may subsequently have provided the locus for mafic intrusive emplacement and late-stage remobilization wrenching during the Permian.

Rome Trough

Within the proposed ECRC, another major Eocambrian rift segment is the Rome Trough (McGuire and Howell, 1963). Whereas the Rome Trough and the Rough Creek Graben segments appear to form a coeval extensional trough inboard approximately 300 to 400 km from the passive margin shelf edge of the ancestral Cambrian craton, the rift segments are discontinuous across the Cumberland Saddle area of the Cincinnati Arch (Fig. 4). This synform structure appears to be underlain by the preexistent East Continent Rift Basin; it is suggested that, while the Rough Creek Graben and the Rome Trough responded to trailing edge extension, the East Continent Rift Basin accommodated the deformation without additional synrift sedimentation.

As in the Rough Creek Graben, major thickening of Cambrian sediments occurs upon the downthrown side of the northern border fault system of the Rome Trough (Thomas, 1960), and again growth faults were cited as the mechanism for synrift sedimentation (Webb, 1969, 1980). In reference to the regional seismic traverse (see Fig. 10), the interval from the base of the Cambrian Copper Ridge Dolomite to Precambrian strata increases from 120 m in the Texaco #1 Sherrer well (exterior and to the west of the Rome Trough) to 765 m in the Wolfinbarger well (within the Rome Trough), demonstrating major thickening of Middle and Lower Cambrian sediments associated with growth faulting in the Rome Trough. The interpretation suggests that Cambrian growth faults may have developed due to the relaxation of steeply ramped thrust faults generated during Grenvillian contraction. The intrarift areas of thick sediment accumulation are also defined by negative gravity anomalies (Ammerman and Keller, 1979). The northern border fault, the Kentucky River Fault

System, is also a recognized portion of the 38th Parallel Lineament. An intrarift feature, the Irvine–Paint Creek Fault System exhibits down-to-the-south displacement and trends essentially parallel to the northern border fault system. The facing southern border fault system, the Warfield Fault Zone, is poorly defined and is indicated by the margins of basement prominences and subtle arches, rather than by a coherent fault system. Apparently related to the preexistent cross-cutting Kentucky–Indiana Graben (Black, 1986), the southern boundary structures include the Rockcastle Uplift, the Pike County High, and the Perry County High (see Fig. 4). Differential Cambrian sedimentation onlapping these prominences indicates that these features were paleotopographic highs during Cambrian extension. Separating the Pike and Perry County Highs, the Floyd County Embayment reaches depths comparable to the adjacent Rome Trough; whereas the basement of the Floyd County Embayment dips to the south towards the deepening continental margin, basement inclination within the Rome Trough is to the north, towards the Irvine–Paint Creek Fault System. The subtle deep arch at the confluence of the Floyd County Embayment and the Rome Trough is herein named the Magoffin Arch (see Fig. 4). Immediately to the southeast of the Pike County High, industry seismic traverses are interpreted to indicate left-lateral offset of the New York–Alabama Lineament, a feature postulated to be a suture of an accretionary terrain that acted as a buttress during Grenvillian collision (King and Zietz, 1978). Current morphology of the lineament in the area is interpreted to be an extensional graben with a central horst block feature.

East of the Pike County High along the Kentucky/West Virginia border, the structural polarity of the Rome Trough reverses, resulting in a depocenter on the south side of rift segment. Polarity reversal appears to occur where the Warfield Fault intercepts the New York–Alabama Lineament; the lineament then appears to be essentially subparallel with the southeastern border fault of the Rome Trough across West Virginia and into the Olin Basin of Pennsylvania. The relationship suggests that, in response to extension generated during the rifting of Laurentia, the preexistent continental suture became remobilized and may have acted as the locus of inboard rifting for a significant portion of the Rome Trough.

SUMMARY

Based upon the data and interpretations suggested in this study, it is proposed that the ECRC represents a major extensional feature of the eastern Midcontinent. The scale of the ECRC is similar to that of the Midcontinent Rift Complex, and the processes and sediments contained within these systems suggests a related developmental history.

Initial Keweenawan-aged extension and associated transtensive deformation, followed by subsequent thermal subsidence, are proposed to have created the primary architecture of the ECRC. Grenvillian collision and extensive transpressive deformation appears to have modified the craton, creating a

mosaic of orthogonal basement zones of weakness. Extensive pre-Paleozoic erosion planated the craton and subdued rift features. Renewed Eocambrian extension interpreted to be related to the rifting of Laurentia remobilized a variety of preexistent faults and sutures, and appears to have resulted in growth-fault type synrift sedimentation within half-graben structures, followed by a second episode of thermal subsidence. Subsequent epeirogenic remobilization of rift features has greatly influenced the ensuing structural and stratigraphic development of the Paleozoic strata in the eastern Midcontinent region.

REFERENCES CITED

Ammerman, M. L., and Keller, G. R., 1979, Delineation of the Rome Trough in eastern Kentucky by gravity and deep drilling data: American Association of Petroleum Geologists Bulletin, v. 63, no. 3, p. 341–353.

Annells, R. N., 1974, Keweenawan volcanic rocks of the Michipicoten Island; Lake Superior, Ontario; an eruptive center of Proterozoic age: Geological Survey of Canada Bulletin 218, 141 p.

Babcock, L. E., 1992, Rift-related sedimentation in southwestern Ohio and northern Kentucky: influence on Proterozoic to Cambrian paleogeography: Geological Society of America Abstracts with Programs, v. 24, no. 7, p. A364.

Bertagne, A. J., and Leising, T. C., 1990, Interpretation of seismic data from the Rough Creek Graben of western Kentucky and southern Illinois, *in* Leighton, M. W., Kolata, D. R., Oltz, D. F., and Eidel, J. J., eds., Interior cratonic basins: American Association of Petroleum Geologists Memoir 51, 819 p.

Black, D. F. B., 1986, Basement faulting in Kentucky, *in* Aldrich, M. J., and Laughlin, A. W., eds., Proceedings of the Sixth International Conference on Basement Tectonics, 1985, Santa Fe, New Mexico: Salt Lake City, Utah, International Basement Tectonics Association, p. 125–139.

Black, D. F. B., 1989, Tectonic evolution in central and eastern Kentucky: a multidisciplinary study of surface and subsurface structure: U.S. Geological Survey Open-File Report OF89-0106, 151 p.

Braile, L. W., Keller, G. R., Hinze, W. J., and Lidiak, E. G., 1982, An ancient rift complex and its relation to contemporary seismicity in the New Madrid seismic zone: Tectonics, v. 1, no. 2, p. 225–237.

Braile, L. W., Hinze, W. J., Sexton, J. L., Keller, G. R., and Lidiak, E. G., 1984, Tectonic development of the New Madrid seismic zone: U.S. Geological Survey Open-File Report 84-77D, p. 204–233.

Bryan, B. K., 1975, The Greenwood Anomaly of the Cumberland Plateau: a test application of GRAVMAP, a system of computer programs for reducing and automatic contouring land gravity data [M.S. thesis]: Knoxville, University of Tennessee, 118 p.

Cable, M. S., and Beardsley, R. W., 1984, Structural controls on late Cambrian and early Ordovician carbonate sedimentation in eastern Kentucky: American Journal of Science, v. 284, p. 797–823.

Cannon, W. F., and 11 others, 1989, The North American Midcontinent Rift beneath Lake Superior from GLIMPCE seismic reflection profiling: Tectonics, v. 8, p. 305–332.

Cincinnati Arch Consortium, 1992, The geology and geophysics of the East Continent Rift Basin: Ohio Department of Natural Resources Division of Geological Survey, Kentucky Geological Survey, Indiana Geological Survey, Open-File Report OFR 92-4, 130 p.

Dickas, A. B., 1986, Comparative Precambrian stratigraphy and structure along the Mid-Continent Rift: American Association of Petroleum Geologists Bulletin, v. 70, no. 3, p. 225–238.

Dickas, A. B., Mudrey, M. G., Ojakangas, R. W., and Shrake, D. L., 1992a, A possible southeastern extension of the Midcontinent Rift System located in Ohio: Tectonics, v. 11, no. 6, p. 1406–1414.

Dickas, A. B., Kholodkov, Y. I., Zaichikov, G. I., Chernikov, B. A., and Stark,

T. J., 1992b, Intercontinental rift systems of the United States of America and associated mineral resources: North Caucasus Scientific Center of Higher Institutions Journal, v. 1 & 2, p. 48–60.

Drahovzal, J. A., Harris, D. C., Wickstrom, L. H., Walker, D., Baranoski, M. T., Keith, B., and Furer, L. C., 1992, The East Continent Rift Basin: A new discovery: Kentucky Geological Survey Special Publication 18, Series XI, 25 p.

Ervin, C. P., and McGinnis, L. D., 1975, Reelfoot Rift: Reactivated precursor to the Mississippian Embayment: Geological Society of America Bulletin, v. 86, p. 1827–1895.

Fettke, C. R., 1948, Subsurface Trenton and sub-Trenton rocks in Ohio, New York, Pennsylvania, and West Virginia: American Association of Petroleum Geologists Bulletin, v. 32, p. 1457–1492.

Gibson, R. I., 1991, Cincinnati Arch Consortium, gravity and magnetic interpretation: Report by Gibson Consulting for Cincinnati Arch Consortium, 8 p.

Green, J. C., 1977, Keweenawan plateau volcanism in the Lake Superior region, *in* Baragor, W. R. A., Coleman, L. C., and Hall, J. M., eds., Volcanic regimes in Canada: Geological Association of Canada Special Paper 16, p. 407–422.

Green, J. C., 1982, Geology of the Keweenawan extrusive rocks, *in* Wold, R. J., and Hinze, W. J., eds., Geology and tectonics of the Lake Superior region: Geological Society of America Memoir 156, p. 47–55.

Greenburg, S. S., and Vitaliano, C. J., 1962, Basalt from a deep well in Lawrence County, Indiana: Journal of Geology, v. 70, p. 482–488.

Halls, H. C., 1978, The late Precambrian North American rift system—A survey of recent geological and geophysical investigations, *in* Ramberg, I. B., and Neumann, E. R., eds., Tectonics and geophysics of continental rifts: The Netherlands, Reidel Publishers, p. 111–123.

Harris, D. C., 1992, Petrology of the Middle Run Formation (Precambrian), East Continent Rift Basin, western Ohio and north-central Kentucky: Geological Society of America Annual Meeting, Abstracts with Programs, v. 24, no. 7, p. A231.

Harris, L. D., 1978, The Eastern Interior Aulacogen and its relation to Devonian shale-gas production: Second Eastern Gas Symposium, U.S. Department of Energy, METC/SP-78/6, 163 p.

Heigold, P. C., 1990, Crustal character of the Illinois Basin, in Leighton, M. W., Kolata, D. R., Oltz, D. F., and Eidel, J. J., editors., Interior Cratonic Basins, American Association of Petroleum Geologists Memoir 51, p. 247–261.

Henderson, J. R., and Zietz, I., 1958, Interpretation of anaeromagnetic survey of Indiana: Indiana Geological Survey Professional Paper 316-B, 37 p.

Hendricks, R. T., Ettensohn, F. R., Stark, T. J., and Greb, S. F., 1994, Geology of the Devonian strata of the Falls of the Ohio area, Kentucky-Indiana: Stratigraphy, sedimentology, paleontology, structure, and diagenesis: Annual Field Conference of the Geological Society of Kentucky, September 10–11, 1993, 65 p.

Hester, N. C., and Kohl, P.R., 1983, Evolution and hydrocarbon potential of the pre-Knox rocks of the Moorman Trough of western Kentucky: Northeast Geology, v. 67, no. 3/4, p. 192–193.

Heyl, A. V., 1972, The 38th Parallel Lineament and its relationship to ore deposits: Economic Geology, v. 67, p. 879–894.

Hildenbrand, T. G., Kane, M. F., and Stauder, W., 1977, Magnetic and gravity anomalies in the northern Mississippian Embayment and their spatial relation to seismicity: U.S. Geological Survey Map MF-914, scale 1:1,000,000.

Hildenbrand, T. G., Kane, M. F., and Hendricks, J. D., 1982, Magnetic basement in the upper Mississippian Embayment region—A preliminary report, *in* McKeown, F. A., and Pakiser, L. C., eds., Investigations of the New Madrid, Missouri, earthquake region: U.S. Geological Survey Professional Paper 1236, p. 39–53.

Hinze, W. J., Kellogg, R. L., and O'Hara, N. W., 1975, Geophysical studies of basement geology of Southern Peninsula of Michigan: American Association of Petroleum Geologists Bulletin, v. 59, p. 1562–1584.

Houseknecht, D. W., 1989, Earliest Paleozoic stratigraphy and facies, Reelfoot Basin and adjacent craton: Oil and Gas Journal, p. 331–371.

Houseknecht, D. W., and Weaverling, P. H., 1983, Early Paleozoic sedimentation in the Reelfoot Rift [abs.]: American Association of Petroleum Geologists, Annual Meeting of the Eastern Section, 12th, Carbondale, Illinois: American Association of Petroleum Geologists Bulletin, v. 67, p. 1456.

Howe, J. R., 1985, Tectonics, sedimentation, and hydrocarbon potential of the Reelfoot Aulacogen [M.S. thesis]: University of Oklahoma, 109 p.

Howe, J. R., and Thompson, T. L., 1984, Tectonics, sedimentation, and hydrocarbon potential of the Reelfoot Rift: Oil and Gas Journal, Nov. 12, p. 174–190.

Kane, M. F., Hildenbrand, T. G., and Hendricks, J. D., 1981, Model for the tectonic evolution of the Mississippian Embayment and its seismicity: Geology, v. 9, p. 563–568.

Keller, G. R., Ammerman, M. L., and Bland, A. E., 1981, A geophysical and tectonic study of east-central Kentucky with emphasis on the Rome Trough, *in* Luther, M. K., ed., Proceedings, Kentucky Oil and Gas Association 49th and 40th Annual Meetings, Technical Sessions, 1975 and 1976: Kentucky Geological Survey, Series 11, Special Publication 4, p. 41–48.

Keller, G. R., Bland, A.E., and Greenberg, J. K., 1982, Evidence for a major Precambrian tectonic event (rifting?) in the eastern Midcontinent region, United States: Tectonics, v. 1, p. 213–223.

Keller, G. R., Lidiak, E. G., Hinze, W. J., and Braile, L. W., 1983, The role of rifting in the tectonic development of the Mid-continent, U.S.A.: Tectonophysics, v. 94, p. 391–412.

Keller, G. R., Bryan, B. K., Bland, A. E., and Greenberg, J. K., 1975, Possible Precambrian rifting in the southeast United States [abs.]: Eos (Transactions, American Geophysical Union), v. 56, p. 602.

King, E. R., and Zietz, I., 1978, The New York–Alabama lineament: Geophysical evidence for a major crustal break in the basement beneath the Appalachian basin: Geology, v. 6, p. 312–318.

Kolata, D. R., and Nelson, W. J., 1991, Tectonic history of the Illinois Basin, *in* Leighton, M. W., Kolata, D. R., Oltz, D. F., and Eidel, J. J., eds., Interior cratonic basins: American Association of Petroleum Geologists Memoir 51, p. 263–285.

Lidiak, E. G., and Zietz, I., 1976, Interpretation of aeromagnetic anomalies between latitudes 37 degrees N and 38 degrees N in the eastern and central United States: Geological Society of America Special Paper 167, 37 p.

Lucius, J. E., and Von Frese, R. R. B., 1988, Aeromagnetic and gravity anomaly constraints on the crustal geology of Ohio: Geological Society of America Bulletin, v. 100, no. 1, p. 104–116.

McCormick, G. R., 1961, Petrology of Precambrian rocks of Ohio: Ohio Division of the Geological Survey, Report of Investigations 41, 60 p.

McPhee, J. P., 1983, Regional gravity analysis of the Anna, Ohio, seismogenic region [M.S. thesis]: West Lafayette, Indiana, Purdue University, 100 p.

McGuire, W. H., and Howell, P., 1963, Oil and gas possibilities of the Cambrian and Lower Ordovician in Kentucky: Spindletop Research Center, Lexington, Kentucky, 216 p.

Nelson, D., Pratt, T., Culotta, R., Zhang, J., Hauser, E., Brown, L., Kaufman, S., and Oliver, J., 1988, The first deep seismic reflection images of eastern U.S. Midcontinent: new COCORP profiling results: Geological Society of America, Abstracts with Programs, v. 20, no. 7, p. A233.

Nelson, K. D., and Zhang, J., 1991, A COCORP deep reflection profile across the buried Reelfoot Rift, south central United States: Tectonophysics, v. 197, p. 271–293.

Nelson, W. J., and Lumm, D. K., 1984, Structural geology of southeastern Illinois and vicinity: Illinois Geological Survey Contract/Grant Report 1984-2, 127 p.

Potter, P. E., and Carlton, R., 1991, Late Precambrian Middle Run Formation of adjacent Indiana, Kentucky, and Ohio [abs.]: Proceedings, Annual Appalachian Petroleum Geology Symposium, 22nd, p. 71–73.

Pratt, T., Culotta, R., Hauser, E., Nelson, D., Brown, L., Kaufman, S., Oliver, J., and Hinze, W., 1989, Major Proterozoic basement features of the eastern Mid-continent of North America revealed by recent COCORP profiling: Geology, v. 17, p. 505–509.

Pratt, T., Hauser, E. C., and Nelson, K. D., 1992, Widespread buried Precambrian layered sequences in the U.S. Mid-continent: Evidence for large Proterozoic depositional basins: American Association of Petroleum Geologists Bulletin, v. 76, no. 9, p. 1384–1401.

Rankin, D. W., 1976, Appalachian salients and recesses; late Precambrian continental breakup and the opening of the Iapetus Ocean: Journal of Geophysical Research, v. 81, p. 5605–5619.

Rankin, D. W., and 9 others, 1989, Pre-orogenic terrains, *in* Hatcher, R. D., Thomas, W. A., and Viele, G. W., eds., The Appalachian-Ouachita orogen in the United States: Boulder, Colorado, Geological Society of America, The Geology of North America, v. F-2, p. 7–100.

Rudman, A. J., Summerson, C. H., and Hinze, W. J., 1965, Geology of basement in midwestern United States: American Association of Petroleum Geologists Bulletin, v. 49, p. 894–904.

Schwalb, H. R., 1969, Paleozoic geology of the Jackson Purchase region, Kentucky: Kentucky Geological Survey Report of Investigations 10, 40 p.

Schwalb, H. R., 1982, Paleozoic geology of the New Madrid area: U.S. Nuclear Regulatory Commission, NUREG CR-2909, 61 p.

Sexton, J. L., Braile, L. W., Hinze, W. J., and Campbell, M. J., 1986, Seismic relection profiling studies of a buried Precambrian rift beneath the Wabash Valley fault zone: Geophysics, v. 51, no. 3, p. 640–660.

Shrake, D. L., Wolfe, P. J., Richard, B. H., Swinford, E. M., Wickstrom, L. H., Potter, P. E., and Sitler, G. W., 1990, Lithologic and geophysical description of a continuously cored hole in Warren County, Ohio, including description of the Middle Run Formation (Precambrian?) and a seismic profile across the core site: Ohio Department of Natural Resources, Division of the Geological Survey, Informational Circular 56, 11 p.

Shumaker, R. C., 1986, Structural development of Paleozoic continental basins of eastern North America, *in* Aldrich, M. J., and Laughlin, A. W., eds., Proceedings of the Sixth International Conference on Basement Tectonics, 1985, Santa Fe, New Mexico: Salt Lake City, Utah, International Basement Tectonics Association, p. 82–95.

Soderberg, R. K., and Keller, G. R., 1981, Geophysical evidence for deep basin in western Kentucky: American Assocociation of Petroleum Geologists Bulletin, v. 65, p. 226–234.

Stark, T. J., 1990, Tectonic controls on Silurian reef growth in southwestern Indiana [abs.]: Abstracts of the Technical Sessions, Kentucky Oil and Gas Association, Fifty-Fourth Annual Meeting, Louisville, Kentucky, 1990, p. 15.

Street, R., Zehulin, A., and Harris, J., 1991, The Meade County, Kentucky, earthquakes of January and March 1990: Seismological Research Letters, v. 62, no. 2, p. 105–111.

Thomas, G. R., 1960, Geology of recent deep drilling in eastern Kentucky: Kentucky Geological Survey Special Publication 3, Series 10, p. 10–28.

Thomas, W. A., 1989, The Appalachian-Ouachita Orogen beneath the Gulf Coastal Plain between the outcrops in the Appalachian and Ouachita Mountains, *in* Hatcher, R. D., Thomas, W. A., and Viele, G. W., eds., The Appalachian-Ouachita orogen in the United States: Boulder, Colorado, Geological Society of America, The Geology of North America, v. F-2, p. 537–554.

Tikrity, S. S., 1968, Tectonic genesis of the Ozark uplift [Ph.D. thesis]: St. Louis, Missouri, Washington University, 196 p.

Treworgy, J. D., Sargent, M. L., and Kolata, D. R., 1989, Tectonic subsidence history of Illinois Basin [abs.]: American Association of Petroleum Geologists, Eastern Section Meeting, v. 73, no. 8, p. 1040–1041.

Walker, D., and Misra, K. C., 1992, Tectonic significance of basalts of the Middle Run Formation (Upper Proterozoic) of the East Continent Rift Basin, Indiana and Kentucky: Geological Society of America Annual Meeting, Abstracts with Programs, v. 24, no. 7, A330.

Wasson, I. B., 1932, Sub-Trenton formations in Ohio: Journal of Geology, v. 40, p. 673–687.

Webb, E. J., 1969, Geologic history of the Cambrian system in the Appalachian Basin: Proceedings, Kentucky Oil and Gas Association, Kentucky Geological Survey, Technical Sessions, Special Publication 18, Series 10, p. 7–15.

Webb, E. J., 1980, Cambrian sedimentation and structural evolution of the Rome Trough in Kentucky [Ph.D. thesis]: Cincinnati, Ohio, University of Cincinnati, 98 p.

White, W. S., 1972, The base of the upper Keweenawan, Michigan and Wisconsin: U.S. Geological Survey Bulletin 1354-F, 23 p.

Wickstrom, L. H., 1992, Stratigraphy, structure, and extent of the East Continent Rift Basin: Geological Society of America Annual Meeting, Abstracts with Programs, v. 24, no. 7, p. A330.

Wilson, M., 1989, Igneous petrogenesis: London, Unwin Hyman Publishing, 466 p.

Zartman, R. E., Brock, M. R., Heyl, A. V., and Thomas, H. H., 1967, K-Ar and Rb-Sr ages of some alkalic intrusive rocks from central and eastern United States: American Journal of Science, v. 265, no. 10, p. 848–870.

MANUSCRIPT ACCEPTED BY THE SOCIETY JANUARY 16, 1996

Geological Society of America
Special Paper 312
1997

The English Graben:
A possible Precambrian rift segment
of the eastern Midcontinent

T. Joshua Stark
Equitable Resources Energy Company, 1601 Lewis Avenue, Billings, Montana 59102

ABSTRACT

The English Graben is interpreted as a Proterozoic Keweenawan-aged rift segment, forming a portion of the East Continent Rift Complex in areas of Kentucky, Indiana, and Ohio. The western margin of the graben is bounded by basement-involved extensional fault blocks, presently expressed by shallow faults offsetting Paleozoic strata. The eastern margin is defined by the Louisville Accommodation Structure (LAS), a high-relief fault-bounded basement feature expressed in Paleozoic strata by a series of shallow fault systems. Potential field data is interpreted to indicate deformation of the graben, with several kilometers of transcurrent displacement apparent along the margins of the rift segment. Earthquake epicenters recorded in the area may be localized along these wrench systems, responding to current east-west horizontal maximum principle compression. Hydrocarbon fields occur along the margins of the rift and appear to be associated with structural remobilization of bounding fault systems. Both proprietary and published seismic sections demonstrate the existence of a complex structural depocenter with a depth to crystalline basement exceeding 6,100 m. Four borehole penetrations of the Proterozoic sequence in the area of the rift suggest that the depocenter is filled with strata similar in lithology to the Precambrian Middle Run Formation described in Ohio and Kentucky. Variations in stratigraphic thickness and lithology are noted in association with the graben margins and it is suggested that the epeirogenic remobilization of the rift margins has controlled the structural and stratigraphic development of the area.

INTRODUCTION

Since the discovery of new Proterozoic depocenters in 1988 (Shrake et al., 1990; Cincinnati Arch Consortium, 1992: Drahovzal et al., 1992), long-accepted concepts regarding the basement architecture of the eastern Midcontinent have been challenged, and newly emerging theories of Proterozoic cratonic differentiation have revolutionized the regional geology. The discovery of previously unknown sedimentary rocks beneath the Paleozoic strata of the eastern Midcontinent region has accelerated research of the area, culminating in the recognition of discrete Proterozoic

depocenters and the postulated existence of a major epicratonic extensional system, the East Continent Rift Complex (Stark, this volume, Chapter 16).

This paper focuses upon an individual rift segment of the East Continent Rift Complex, referred to as the English Graben (Stark, 1990; Dickas et al., 1992a; Hendricks et al., 1994; Hendricks, 1996). It is interpreted that the English Graben evolved prior to or during Keweenawan extension (1.32 to 1.05 Ga; Cincinnati Arch Consortium, 1992), and has undergone extensive remobilization during ensuing tectonic events. This pervasive remobilization tectonism has acted as a major control upon

Stark, T. J., 1997, The English Graben: A possible Precambrian rift segment of the eastern Midcontinent, *in* Ojakangas, R. W., Dickas, A. B., and Green, J. C., Middle Proterozoic to Cambrian Rifting, Central North America: Boulder, Colorado, Geological Society of America Special Paper 312.

the structural and stratigraphic evolution of the region, and has provided the geological anomalies that allow for the delineation of the English Graben.

A substantial body of interrelated evidence suggests the existence of the English Graben. Geological and geophysical data demonstrate that southeastern Indiana and surrounding regions exhibit a more complex morphology and developmental history than previously described. The following sections will summarize these data and suggest a model that conforms with both local and regional settings.

REGIONAL SETTING

Situated in central northern Kentucky, southeastern Indiana, and southwestern Ohio (Fig. 1), the English Graben appears to be a segment of a much larger rift complex. The graben extends from the eastern end of the Eocambrian-aged Rough Creek Graben (Soderburg and Keller, 1981) northward into Indiana. The western margin of the English Graben is interpreted to consist of a steplike array of major extensional basement fault blocks that are downthrown to the east and offset by extensive transcurrent fault deformation. The eastern margin of the English Graben is bounded by the Louisville Accommodation Structure (LAS), a major high-relief fault-bounded mafic complex defined by surficial geology, a basement corehole test, anomalous potential field data, and surface reflection information.

Near the surface termination (northwestern Washington County, Indiana) of the Mount Carmel Fault (Ashley, 1898) and the northern margin of the LAS (central Harrison County, Indiana), a bend in the English Graben changes the orientation of the depocenter to the northeast. Penetration of pre-Paleozoic strata in two boreholes located in Switzerland County of southeastern-most Indiana defines the confluence of the English Graben with the Proterozoic-aged East Continent Rift Basin (Drahovzal et al., 1992; Cincinnati Arch Consortium, 1992). Farther to the southwest, the East Continent Rift Basin and the English Graben occur as subparallel depocenters, separated by the high-relief LAS.

Branching to the north from the English Graben in southeastern Indiana, the Flatrock Sub-basin (Hendricks, 1996) is separated from the English Graben by a positive archlike structure referred to as the Ripley Island positive feature.

GEOPHYSICAL EVIDENCE

Potential field data

In the late 1940s, a statewide aeromagnetic survey of Indiana was conducted to provide insights into the composition and structure of the basement. That survey (Henderson and Meuschke, 1951) was compared with a statewide gravity map, with a variety of structural horizons, and with oil and gas production maps. The study indicated that basement lithology was heterogeneous, and characterized by distinct magnetic provinces. These provinces include (1) broad areas of low-frequency, low-amplitude anomalies accentuated by widely separated intense

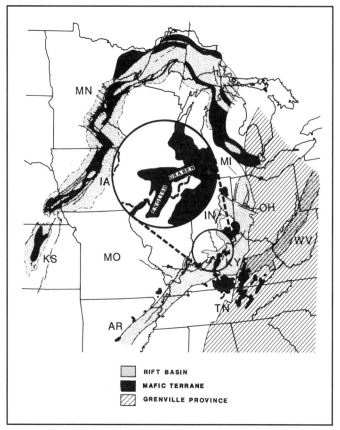

Figure 1. Regional setting of the English Graben.

positive anomalies; (2) broad negative anomalies bounded by linear, high gradient margins; and (3) several large areas of positive magnetic anomaly (Fig. 2).

The broad low-frequency, low-amplitude areas correspond with basement penetrations of the Middle Proterozoic Granite-Rhyolite Province. The small, isolated positive magnetic anomalies situated within these areas may be produced by mafic plutons (Henderson and Zietz, 1958). This interpretation is supported by subsequent studies (Braile et al., 1984; Howe, 1985) which suggest that the locations of such plutons are related to zones of instability within the basement.

Comparison of the magnetic data with Bouguer gravity anomaly data (Fig. 3) reveals that the broad negative aeromagnetic anomalies correspond with negative gravity anomalies. Based upon proprietary seismic surveys (Steven Rowley, 1992, personal communication), these negative gravity anomalies are interpreted as a thick Precambrian rift-fill sequence of continental sediments and intercalated volcanic basalt flows. It is here suggested that the section may be similar to the Precambrian Middle Run Formation (Shrake et al., 1990), found within the East Continent Rift Basin (Drahovzal et al., 1992; Cincinnati Arch Consortium, 1992).

The larger positive magnetic anomalies (i.e., the Louisville Accommodation Structure; LAS) may predate Keweenawan-

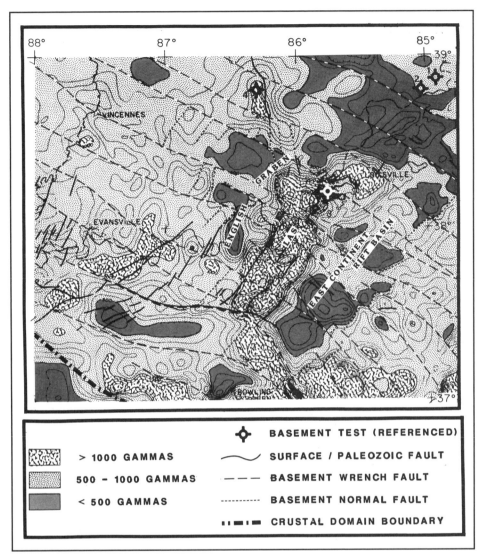

Figure 2. Interpreted total aeromagnetic intensity map of English Graben area. The major structural elements exhibited are the Louisville Accommodation Structure (LAS; mottled texture), the English Graben, and the East Continent Rift Basin. The Rough Creek Graben is an east-west trending minimum located to the south of these three structural elements. Conspicuous offset of the LAS (gabbroic intrusive) and left-lateral offset upon the western margin of the English Graben is evident along regional magnetic trends. Upon the southern reaches of the English Graben, both rift margins are defined by fault systems that offset Paleozoic strata. (Basemap after Johnson et al., 1980.) Well 1, Ashland Oil #1 Collins; Well 2, Ashland Oil #1 Eichler; Well 3, DuPont #1 Wadd/Fee; Well 4, Farm Bureau #1 Brown.

aged extension or may be related to major rift features. Surface faults coincide with several of the sharp magnetic linear gradients that define the margins of large positive anomalies. In addition, these magnetic gradients often delineate the margins of adjacent negative gravity anomalies. The LAS may be associated with the East Continent Gravity High (Bryan, 1975), a major potential field anomaly attributed to Keweenawan-aged rifting (Rudman et al., 1965; Bryan, 1975; Keller et al., 1982; McPhee, 1983; Black, 1986; Lucius and Von Frese, 1988; Dickas et al., 1992a; Drahovzal et al., 1992).

The LAS is the most prominent positive magnetic anomaly in southern Indiana and adjacent Kentucky. As shown in Figure 4, upon the western margin of the LAS, "gas fields are found to be aligned along a strong north-south positive magnetic trend. ... The most acceptable theory ... is that the reversals or interruptions in the normal dip [are] perhaps related to zones of weakness associated with the emplacement of a large [mafic] basement intrusive" (Henderson and Zietz, 1958, p. 32). The naturally fractured shale gas production in these fields coincides with the linear margin of the LAS, indicating structural defor-

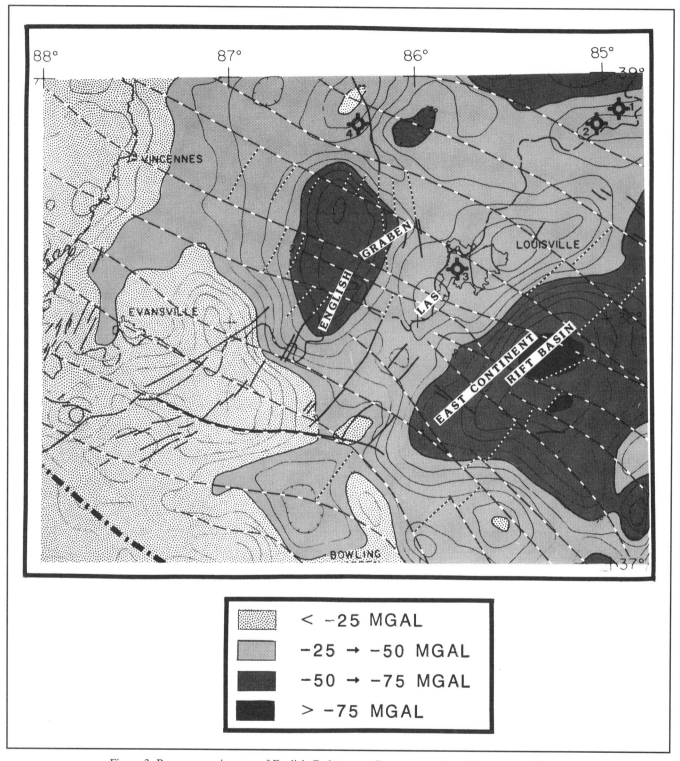

Figure 3. Bouguer gravity map of English Graben area. Bouguer gravity minimum anomalies are evident both in the English Graben and in the East Continent Rift Basin, indicating thick columns of sediments in these depocenters. These interpretations are confirmed by seismic data, which indicates a depth to crystalline basement in excess of 6,100 m within the depocenters. The Louisville Accommodation Structure (LAS) divides these rift segments with a relatively positive gravity anomaly. Borehole core data places the top of the LAS mafic complex at a depth of 1,815 m. Wells 1–4 are the same as on Figure 2. (Basemap after Keller et al., 1980.)

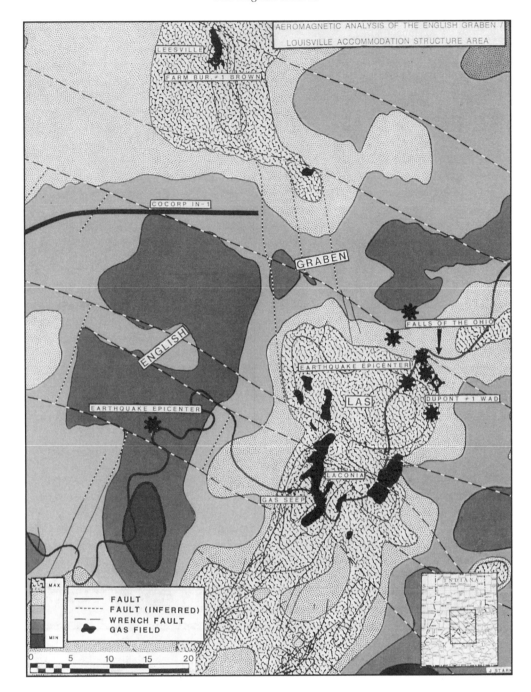

Figure 4. Seismicity, shallow faulting, and hydrocarbon fields versus aeromagnetic interpretation of English Graben area. The western margin of the Louisville Accommodation Structure (LAS) is bounded by the Locust Hill Fault System, which offsets Paleozoic strata and demonstrates that this mafic complex is structurally defined by faults. Fault displacement decreases to the northeast, creating a gas seep in the Ohio River and naturally fractured shale production in the New Albany Shale. Contemporary tectonism (earthquake epicenters) and control of modern drainage is evident in Crawford/Meade Counties (within the English Graben) and at Falls of the Ohio State Park (see Fig. 7 in a following section). The western margin of the English Graben is similarly fault-bounded by large extensional block faults that exhibit left-lateral offset. The location of the COCORP IN-1 seismic line (see Fig. 5 in the next section) shows position relative to the English Graben. Proterozoic well penetrations are denoted by dry hole symbols. Earthquake epicenters are denoted by asterisks.

mation of overlying strata along the boundary of this mafic intrusive.

The LAS extends southwest to the northern margin of the Rough Creek Graben. The western margin of the LAS is defined by the Locust Hill Fault System. Large offsets of the linear margins of the LAS are evident. In Harrison County, Indiana, the trend of the LAS changes abruptly to a east-west orientation. Offset and rotation of the LAS are interpreted to occur upon regional zones of transcurrent faulting, trending northwest to southeast for hundreds of kilometers. In a regional aspect, the wrench systems provide the northwest-southeast component of orthogonal basement morphology (Stark, this volume, Chapter 16), whereas the shallow Paleozoic faults and the interpreted extensional basement faults are oriented northeast-southwest. Superposition indicates that the mafic LAS complex was emplaced prior to initiation of transcurrent (northwest-southeast) fault deformation.

In summary, the gravity and magnetic data suggest that structural and compositional variations within the basement are associated with structural trends occurring within the overlying sedimentary strata along the margins of the English Graben. Sharp magnetic gradients delineate the margins of the proposed rift segment. Bouguer gravity minimums within the rift suggest the presence of a substantial sedimentary package. Lateral offsets of both compositional anomalies and structural margins suggest substantial post-rift transcurrent deformation.

Surface reflection data

Proprietary seismic data indicate that the sedimentary fill within the English Graben may reach thicknesses exceeding 6,100 m (Steven Rowley, 1992, personal communication), suggesting as much as 4,270 m of Proterozoic strata. Interpreted basement growth faults are associated with gravity and magnetic gradients. Within the rift, large fault blocks form major structural features, complicating the asymmetry of the half graben. Rhythmic, sharply defined seismic horizons may represent basalt flows and sills, similar to those within the Proterozoic strata penetrated in Ohio and Kentucky (Cincinnati Arch Consortium, 1992; Walker and Misra, 1992; Drahovzal et al., 1992). Broad depocenters are commonly characterized by negative gravity anomalies. The wrench faults and sedimentary depocenters interpreted from the potential field data are apparently substantiated by seismic data.

A COCORP seismic profile in southern Indiana (Line IN-1; Fig. 5) traverses the northern margin of the English Graben, revealing approximately 38 km of the rift interior. Beneath the thin veneer of Paleozoic sedimentary strata, an asymmetric half-graben is interpreted within the Proterozoic strata. East of VP1200, it was noted that, "dipping reflectors and a ramplike feature indicate deformation in the layering, perhaps indicative of a fundamental lateral change" (Pratt et al., 1989, p. 506). This lateral change is recognized as the thickening of a pre–Mount Simon section referred to as the Centralia sequence (Pratt et al., 1992). Future drilling will determine if this nomenclature is

appropriate or whether the Middle Run Formation (Shrake et al., 1990) should be utilized. The existence of a thick pre–Mount Simon section within the English Graben demonstrates that significant sedimentation occurred prior to peneplanation at the end of the Proterozoic. The reflectors within the English Graben are similar to seismic profiles crossing the East Continent Rift Basin, suggesting coeval development of the two basins.

An analogue to the English Graben occurs within the Central African Rift System (Genik, 1993). This study proposes that the African rift system exhibits a multiphase geologic history, and contains both extensional and transtensional tectonic elements. Specifically, the subparallel Doseo and Salamat Basins of Chad are separated by the fault-bounded Gongo High. The Doseo and Salamat Basins are transtensional basins and contain complex internal structures generated over an extended geologic history. Cross sections of the Chad basins and the English Graben/Louisville Accommodation Structure/East Continent Rift Basin profile (Fig. 6) exhibit remarkable similarities. In both examples, two subparallel half grabens are separated by a high-relief fault-bounded structural prominence. The English Graben may have formed in a manner similar to the Doseo and Salamat Basins.

Earthquake seismology data

Within the East Continent Rift Complex, recent seismic events may correlate with zones of basement instability. The two most active earthquake zones in the eastern Midcontinent, the New Madrid Zone and the Anna Earthquake Zone, may be related to accommodation structures within the ECRC (Stark, this volume, Chapter 16). Less frequent, spatially isolated seismic events have occurred within individual rift segments. These events may represent motion along strike-slip faults within the basement. As the present stress pattern is one of east-west horizontal maximum principle compression (Zoback and Zoback, 1980), northwest-southeast–trending faults may experience present-day reactivation.

A series of earthquakes, culminating in the largest earthquake on January 24, 1990, was recorded near Meade County, Kentucky, along the Indiana-Kentucky border. This area, which had no history of previous seismic activity, occurs along an extremely sinuous stretch of the Ohio River, about 48 km (30 mi) west of the Falls of the Ohio in Louisville, Kentucky. The epicenter was apparently located along a major basement strike-slip fault within the English Graben. Left-lateral offset of basement fault blocks on the western margin of the English Graben is interpreted along the same wrench system. From analysis of shock and aftershock data, it was noted that, "Our preferred mechanism … has a NW trending nodal plane with a strike of 318 degrees and a dip of 60 degrees to the NE. … The distribution of aftershock data would suggest that the NW trending nodal plane is … the rupture plane (i.e., a left-lateral strike-slip event)" (Street et al., 1991, p. 108).

Additional seismic activity has historically been noted

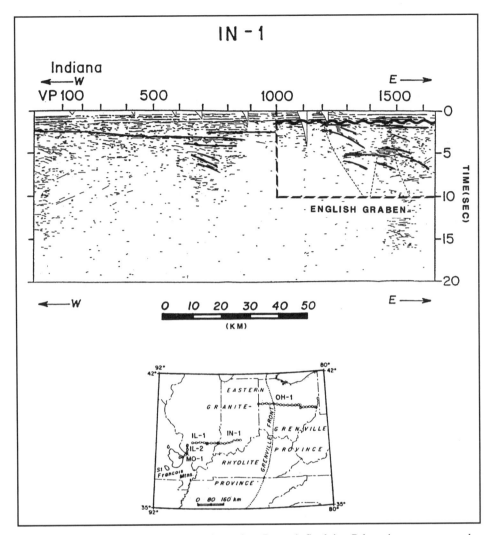

Figure 5. COCORP IN-1 regional seismic section. Beneath flat-lying Paleozoic strata, an angular unconformity is evident upon Proterozoic rocks. In the area of the English Graben (boxed), strong defraction patterns suggest major listric faults positioned upon the margin of the rift segment and considerable thickening of Proterozoic sediments (after Pratt et al., 1989). A COCORP seismic line location map is shown at bottom.

along the northeastern margin of the LAS. At least six earthquake epicenters have been localized in the area of the Falls of the Ohio (Fig. 7), a series of rapids and falls (currently bypassed by a system of dams, spillways, and canals) forming the single portage of the Ohio River. These epicenters may have similar focal mechanisms and represent additional strike-slip events along a basement wrench system (Hendricks et al., 1994).

GEOLOGICAL EVIDENCE

Structural data

Epeirogenic remobilization has apparently modified both the eastern and western border fault systems of the English Graben (Fig. 8). Along the western margin, the Curdsville–Little Hurricane Island, Africa, Hawesville, German Ridge, and Indian Creek Faults (Ault and Sullivan, 1982) offset Paleozoic strata. Vertical displacement of Mississippian Chesterian strata along these faults is generally less than 30 m. Potential field, structural, isopachous, and seismic analyses suggest that these western faults may be remobilized rift margin extensional faults. Outcrops and over 600 well logs suggest that a minor graben, the Owensboro Graben, occurs between two bounding faults along the western margin of the English Graben (Greb, 1989; Fig. 9). The areas overlying the inclined corners of these fault-bounded basement blocks display isopachous thins in both the Devonian New Albany Shale and in the Mississippian Valmeyerian strata.

Along the eastern margin of the English Graben, the Locust Hill Fault System (Fig. 8) is related to the western flank of the LAS. This fault system is thought to be associated with the gas

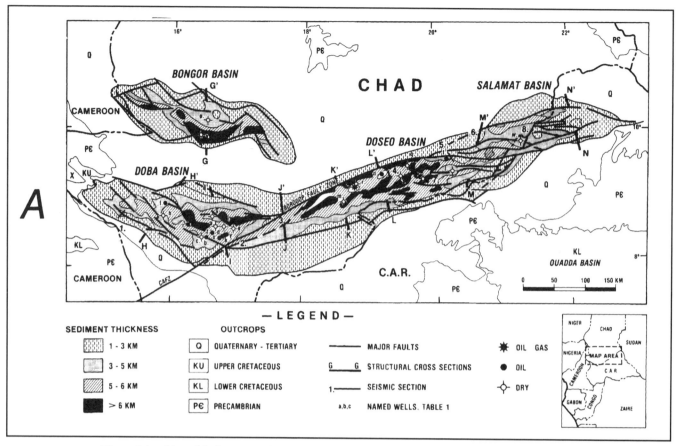

Figure 6 (on this and facing page). Comparative cross-sectional analysis East Continent Rift Complex versus Central African Rift System. A comparison of transtensive (pull-apart) rift basins in the Central African Rift System (A) with the rift complex in the eastern Midcontinent exhibit remarkable similarities. The subparallel Doseo and Salamat Basins of Chad (B) are separated by the fault-bounded Gongo High. Similarly, the English Graben and the East Continent Rift Basin (C) are separated by the Louisville Accommodation Structure (LAS). The Doseo and Salamat Basins are transtensional basins and contain complex internal structures generated over an extended geologic history. The English Graben may have formed in a manner similar to these African rift basins (Genik, 1993).

seep located at Tobacco Landing, Indiana, and with the naturally fractured New Albany Shale production found in Laconia/Doe Run Field (Fig. 10). Along the northern margin of the LAS, the Crandall and Georgetown faults correlate with distinct linear magnetic trends. It is interpreted that these shallow faults are related to the deep-seated block faults defining the margin of the remobilized structural prominence.

Several kilometers of transcurrent offset normal to border block faults are interpreted from potential field data analysis of the English Graben. Attributed to Keweenawan transtensive or Grenvillian transpressive tectonism, this predominantly left-lateral deformation appears to offset the margins of both the LAS and the western margin block fault systems. Some of these wrench fault features (e.g., the Bardstown Monocline) exhibit sedimentologic variation across the projected fault plane in Siluro-Devonian strata (Todd Hendricks, 1993, personal communication). Palinspastic reconstruction of the English Graben indicates nonuniform offset along individual wrench systems on

opposing rift margins. These structural relations suggest that the English Graben may have been formed as a transtensive "pull-apart" feature, rather than as a simple extensional graben.

Deformation of the overlying Paleozoic sequence demonstrates that post-Grenvillian fault block movement also significantly modified local structural architecture. Taconic contraction, scissor-block rotation, and minor shear tectonism exerted a significant control upon sedimentological processes (Mescher, 1980; Fisher, 1981, Wickstrom and Gray, 1989; Wickstrom et al., 1992). Portions of the English Graben are overlain by the Ordovician-aged Sebree Trough (Schwalb, 1980; Keith, 1985, 1989; Wickstrom et al., 1992), an organic-rich black shale basin in facies relationship with the northwestern Trenton carbonate bank and the southeastern Lexington carbonate shelf (Keith, 1989; Fig. 11; also see Stratigraphic Data). Syndepositional remobilization of basement faults may have created a downwarped depocenter where the organic-rich Kope Formation/Point Pleasant Formation accumulated in a silled, nonoxygenated environment.

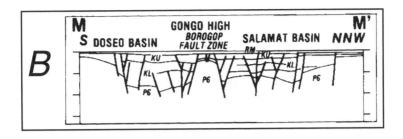

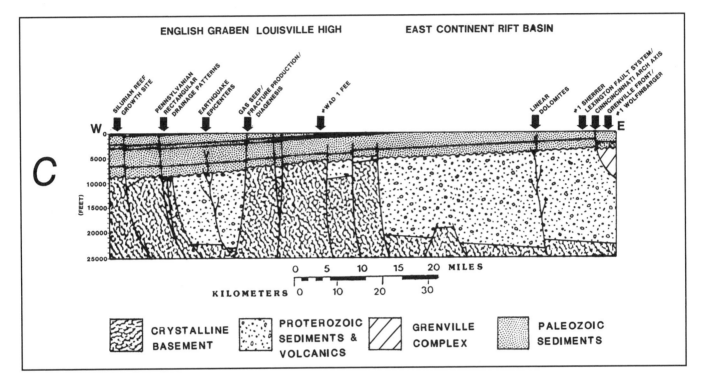

Paleotopography characterized by differentially raised fault block corners also provided suitable growth sites for pinnacle and patch reefs during the Silurian (Stark, 1990).

The current structural morphology of the area overlying the eastern rift margin appears to be directly linked to the development of forced folds above basement block corners, and the development of linear zones of intense natural fracturing. Situated upon the western margin of the LAS, the Doe Run/Laconia gas storage field occurs as a doubly plunging asymmetric anticlinal trap, characterized by counter-regional dip, an underlying Silurian pinnacle reef, and by the occurrence of vugular dolomite in the Siluro-Devonian carbonates.

Current evaluation of the Doe Run/Laconia field suggests that structural development in the area is attributable to episodic remobilization of the western margin of the LAS. Isopachous and dolomite distribution maps suggest that syndepositional mobility of underlying fault zones may be related to the evolution of the LAS. A subsequent remobilization event along the western bounding fault of the dome resulted in the formation of the fractured New Albany Shale trend upon the flank of the structure.

The western border fault of the LAS appears to exhibit a continuous linear trend linked northward to the Mount Carmel Fault System (Ashley, 1898). Delineated by outcrop data, the Mount Carmel Fault System is downdropped to the west and forms the eastern border of the Illinois Basin. Both industry seismic and potential field data suggest that the system is rooted in the basement, and that the 75 m of fault displacement noted in the Paleozoic strata is likely to reflect the remobilization of a much larger fault at depth. Recognized as a listric fault (Tanner, 1985), the Mount Carmel Fault System displays a complex morphology, including antithetic dolomitized graben structures and "bends" in the otherwise linear fault trend.

The subparallel Leesville Anticline is located to the west on the down side of the fault complex. Characterized by en-echelon doubly plunging anticlinal closures, the Leesville Anticline also yields gas production and gas storage structures from the Middle Devonian Geneva Dolomite. Silurian pinnacle reefs are commonly associated with the structures. The gas saturated domal closures appear to occur at the intersection of the Mount Carmel Fault and anomalous fault "bends." Surface expression of the Mount Carmel Fault in the area of the anomalous "bends" is characterized by a change in surface structural dip and the occurrence

Figure 7. Falls of the Ohio, Kentucky. The Falls of the Ohio River consists of a series of falls and rapids developed in the Middle Devonian–aged Jeffersonville Limestone riverbed, creating the single portage of the Ohio River. Potential field and seismic analysis indicates that the Falls of the Ohio may be located upon a major wrench fault, and upon the northern flank of the LAS. Remobilization of the proposed basement wrench system is suggested by the occurrence of at least six earthquake epicenters in the vicinity of Falls of the Ohio.

of shear zones within the fault plane (George Tanner, personal communication, 1988). Potential field data suggests that the "bends" are positioned above major basement wrench fault zones. During late-stage remobilization of the Mount Carmel Fault System, polygonally offset basement blocks may have been reattached to the master fault system by small shear faults. The individual gas saturated domal closures appear to represent forced folds above basement block corners. The en-echelon arrangement of folds upon the Leesville Anticline and the presence of dolomitization in the antithetic faulted grabens of the system suggests a complex deformation history. Cross-cutting relations suggest that the Mount Carmel Fault System either predates the Keweenawan extensional event or formed as a syntectonic feature genetically related with the English Graben.

In summary, it is evident that both margins of the interpreted Precambrian rift are overlain by fault complexes that offset Paleozoic strata. Potential field and seismic data indicate the area within

these opposing fault systems contains a thick, layered, pre–Mount Simon sedimentary section. Major structural as well as compositional changes in the basement are interpreted across these bounding fault trends. It is interpreted that primary extension occurred during the Precambrian Keweenawan episode of continental rifting. During or after rift extension, predominantly left-lateral transcurrent offset modified the rift segment. It is interpreted that primary wrench faulting may have been initiated as early as the Keweenawan, in a transtensive tectonic setting. Subsequent compression during the Grenville event may have continued wrench movement under transpressive deformation. Paleozoic sediments exhibit evidence of syndepositional variation above interpreted basement rift features. Remobilization along basement rift faults appears to have occurred during the Taconian, Acadian, and Alleghanian orogenies, as indicated by variations in formation thickness and character (see Stratigraphic Data), surficial fault traces, and naturally fractured shale production trends (see Hydrocarbon Production Data).

Hydrocarbon production data

The best example of the correlation between positive magnetic anomalies and hydrocarbon productive structural features occurs upon the LAS. Discovered in 1863 in Meade County, Kentucky (Jillson, 1922), the Doe Run/Laconia field extends across the Ohio River into Harrison County, Indiana. Many of the early wells were drilled adjacent to a gas seep recognized in the Ohio River. Originally drilled to distill salt from produced brine, these wells produced natural gas from the fractured New Albany Shale (Hasenmueller and Woodard, 1981). In Laconia field, the fractured production occurred along a linear trend, with a central "sweet spot" of higher volume gas wells localized along a minor dip slope flexure on the structural flank. Boreholes drilled updip to the east lacked the presence of fractured New Albany Shale gas, encountering instead gas saturated reservoirs in the underlying dolomitized Siluro-Devonian carbonates. It was recognized that the Siluro-Devonian reservoir was configured as a doubly plunging asymmetric anticlinal dome, exhib-

Figure 8. Shallow faults associated with the rift margins of the English Graben. Epeirogenic remobilization appears to have modified both the eastern and western border fault systems of the English Graben. Along the western margin, numerous normal faults offset Paleozoic strata with vertical displacements generally less than 30 m. Analyses suggest that these western faults may be remobilized rift margin extensional faults. Along the eastern margin of the English Graben, the Locust Hill Fault System is related to the western flank of the Louisville Accommondation Structure (LAS). Several kilometers of transcurrent offset normal to border block faults is interpreted from potential field data analysis of the English Graben. The western border fault of the LAS appears to exhibit a continuous linear trend linked northward to the Mount Carmel Fault System, which forms the eastern border of the Illinois Basin. Data suggest that the system is rooted in the basement, and that the 75 m of fault displacement noted in the Paleozoic strata is likely to reflect the remobilization of a much larger fault at depth. Well 1, Ashland Oil #1 Collins; Well 2, Ashland Oil #1 Eichler; Well 3, DuPont #1 Wadd/Fee; Well 4, Farm Bureau #1 Brown.

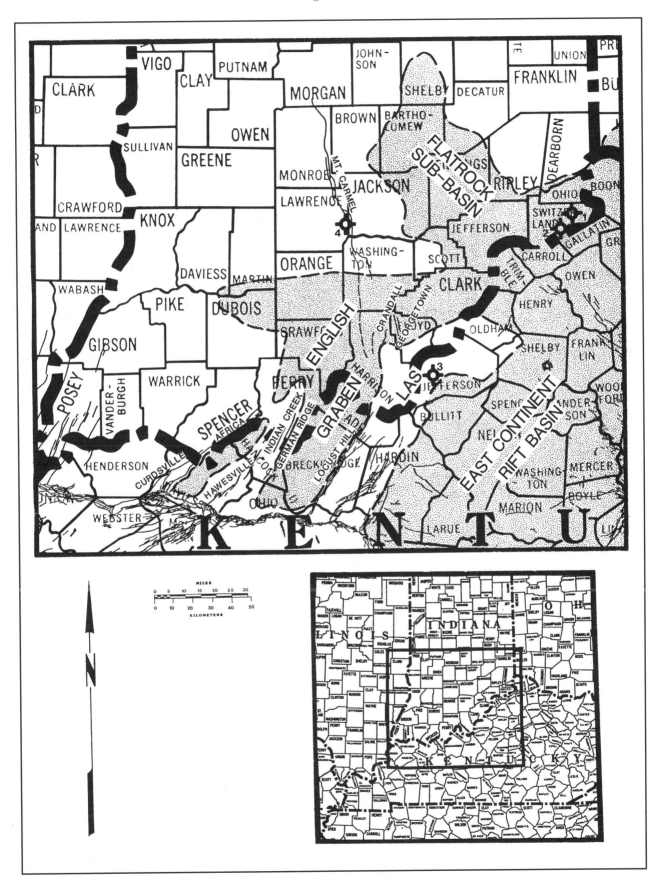

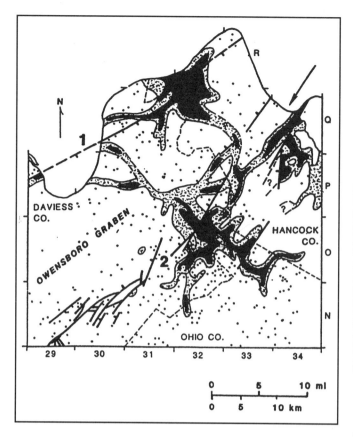

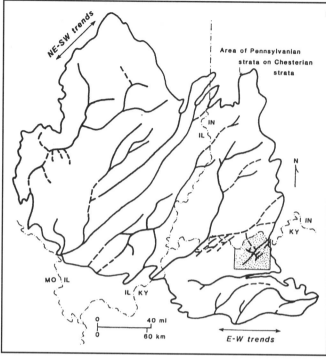

Figure 9. Drainage patterns upon the Absarokan unconformity related to remobilization of English Graben margins. The development of rectilinear drainage patterns upon the Absarokan unconformity surface is noted along the western border fault system of the English Graben. The oriented drainage exhibits a close correlation with interpreted basement extensional fault systems. Outcrops and more than 600 well logs suggest that a minor depocenter, the Owensboro Graben, occurs between two bounding faults (1, Africa/Waitman-Griffith fault zone, and 2, Knottsville Fault Zone) along the western margin of the English Graben (Greb, 1989). Fine stipled pattern indicates Mississippian Lower Kinkaid/Degonia Formations upon unconformity surface. Dense stipled pattern indicates erosional downcutting to Mississippian Clore/Palestine Formations. Dark pattern indicates erosional downcutting to Mississippian Menard Formation or lower.

iting extreme contra-regional dip upon the eastern updip margin (Fig. 10). The Laconia/Doe Run field is today considered a single continuous feature, and is operated as a gas storage facility by Louisville Gas & Electric Co.

In addition to the Laconia/Doe Run field, commercial development of adjacent areas (in the years 1885 to 1935) discovered six gas fields in Harrison County, Indiana, and Meade County, Kentucky. Production was established in both the fractured New Albany Shale and in the underlying dolomitized carbonates; these fields also exhibit structural/diagenetic trap mechanisms.

To the north along the Mount Carmel Fault, additional gas storage fields have been established in the Middle Devonian Geneva Dolomite strata. First recognized by a preserved "limestone strip" (Hopkins and Siebenthal, 1897), the complexly folded and faulted surficial strata adjacent to the Mount Carmel Fault have long provided a source of debate for Indiana geologists. The first description of folding that accompanied the faulting (Esarey, 1923) demonstrated the viability of hydrocarbon trapping structures along the fault zone. Occurring as doubly plunging anticlinal domes, these fields are locate on the downthrown side of the Mount Carmel Fault, upon the subparallel Leesville Anticline. Consisting of a series of en-echelon anticlinal closures, this fault/anticline system may be related to deformation within the adjacent English Graben.

The Unionville and Leesville gas storage fields, developed upon the Leesville Anticline, exhibit a similar structural mor-

phology to the Laconia/Doe Run field (Fig. 10). It is interpreted that these closures may have formed as forced folds above the corners of rotated blocks on the downthrown side of the Mount Carmel Fault.

The western border fault system of the English Graben is located well within the confines of the Illinois Basin. Numerous hydrocarbon-productive fields occur in stratistructural traps along faults offsetting Paleozoic strata. It is interpreted that these shallow faults reflect rift-related extensional faults at depth.

In addition to stratistructural traps, fault block rotation appears to have created several types of stratigraphic traps. Growth-style sedimentation of the Mississippian Jackson Sandstone (Stevensport Group) was noted (Droste and Keller, 1989b) along the English Graben margin in Spencer and Perry Counties, Indiana. The stacked sand body pattern was attributed to "locally increased rates of subsidence controlled by basement mechanics" (Droste and Keller, 1989b). Additional trap mechanisms attributed to English Graben remobilization are the growth of Silurian

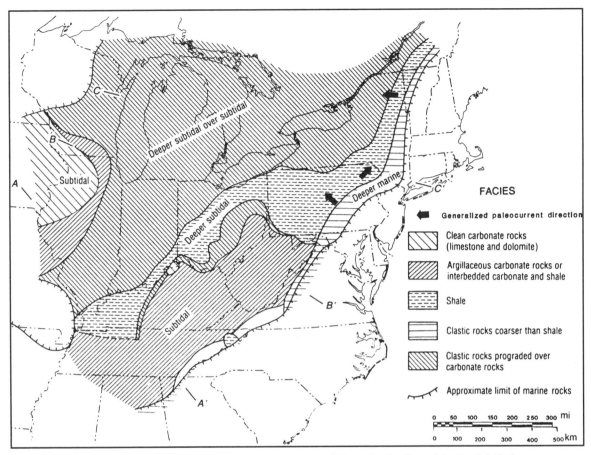

Figure 10. Ordovician Sebree Trough, Ohio, Indiana, and Kentucky. Portions of the English Graben are overlain by the Ordovician-aged Sebree Trough, an organic-rich black shale basin separating the northwestern Trenton carbonate bank from the southeastern Lexington carbonate shelf. Within the trough, black shales of the Kope Formation/Point Pleasant Formation rest directly upon Black Riverian carbonates, and interfinger with the adjacent carbonate platforms in facies relationship. In northwestern Ohio, the Sebree Trough margin is interpreted to be coincident with a series of northeast trending basement fault systems. Associated with these faults, the lithologic transition from carbonate to shale in the Trenton Formation suggests syndepositional remobilization of the basement systems (Keith, 1989).

pinnacle reefs upon upraised corners of rift margin rotational blocks (Stark, 1990). The Troy Field of Spencer County, Indiana, is underlain by a complex of Silurian pinnacle reefs, creating reef drape closure in seven hydrocarbon productive formations in the overlying Mississippian strata.

In summary, it is significant to consider that the majority of hydrocarbons found in rift settings are usually associated with accommodation structures (Rosendal et al., 1992). The mobility of accommodation features, coupled with the considerations of local source area, associated facies patterns, and potential for diagenetic modification, provide a wealth of opportunities for hydrocarbon entrapment. Within the English Graben, hydrocarbon traps associated with the LAS include the development of forced folds, dolomitization of the Siluro-Devonian carbonates, and natural fracture trends. Stratistructural traps are common upon the upthrown sides of Paleozoic faults, interpreted to be related to rift margin features.

Stratigraphic data

Lithologic changes in Upper Ordovician Champlainian and Cincinnatian Series strata (Rocklandian through mid-Edenian) support the interpretation that remobilization of basement features in the English Graben acted as a control for differential sedimentation.

During the Late Ordovician, a black shale basin referred to as the Sebree Trough evolved, separating the northwestern Trenton carbonate shelf from the southeastern Lexington carbonate bank in Tennessee, Kentucky, Indiana, and Ohio (Schwalb,1980; Keith, 1985, 1989; Wickstrom et al., 1992; Wickstrom and Gray, 1989; Bergstrom and Mitchell, 1990; Fig. 11). Within the trough, black shales of the Kope Formation/Point Pleasant Formation rest directly upon Black Riverian carbonates, and interfinger with the adjacent carbonate platforms in facies relationship.

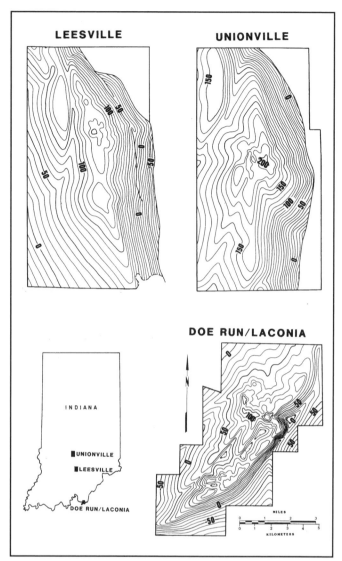

Figure 11. Structural analysis of hydrocarbon fields in the English Graben area. Associated with the English Graben, hydrocarbon traps include the development of stratistructural fields, forced folds, dolomitization, and natural fracture trends. Hydrocarbon fields occur along the margins of the rift, and appear to be associated with structural remobilization of bounding fault systems. Numerous hydrocarbon-productive fields occur in stratistructural traps along faults offsetting Paleozoic strata. The area overlying the eastern rift margin appears to be characterized by the development of forced folds above basement block corners, and the development of linear zones of intense natural fracturing that affect Paleozoic strata. Mapped upon top of New Albany Group. Contour interval is 10 ft.

In northwestern Ohio, the Sebree Trough margin is interpreted to be coincident with a series of northeast-trending basement fault systems. Based upon several hundred kilometers of seismic reconnaissance data, these basement fault systems are predominantly downthrown to the southeast into the proto-Appalachian Basin. Associated with these faults, a major lithologic transition from carbonate to shale is evident in the Trenton Formation, suggesting syndepositional remobilization of the

basement systems. Wickstrom et al. (1992) postulate that the facies transition and downwarping of the Sebree Trough are coeval events that occurred as a result of the Taconic orogeny. Conodont and graptolite biogeographic studies demonstrate that trough development may have been coeval in the Ouachita region, and closely coincides with regional Taconic orogeny events (Mitchell et al., 1991). The relationship between structural remobilization and the evolution of the Sebree Trough in northwestern Ohio appears to extend into Indiana. It is here suggested that the remobilization of rift margin faults of the English Graben may have significantly effected the position and development of Sebree Trough in Indiana.

In southwestern Ohio, the trough crosses the East Continent Rift Basin immediately north of the deep test DGS 2627 (Shrake et al., 1990). Trending into Indiana, the trough parallels the northwestern margin of the English Graben, to the north and exterior to the rift. West of the Mount Carmel Fault extension, the position of the Sebree Trough appears to be common with the English Graben. This localization of shale sedimentation may indicate an increasing magnitude of Taconic deformation along rift margins, adjacent to the deep depocenter of the English Graben.

In nearby Ripley, Jennings, and Jefferson Counties, Indiana, the Sebree Trough is discontinuous across a localized positive structural feature referred to as Ripley Island uplift (Fig. 12). The Ripley Island uplift defines a complex structural/erosional area that upwarped during the Late Ordovician/Early Silurian, coeval with the uplifting of the Cincinnati Arch. Across the Ripley Island uplift, the Lower Silurian Brassfield Formation was not deposited and the overlying Osgood Member of the Salamonie Dolomite thins (Foerste, 1891, 1904), resting unconformably upon Ordovician rocks (Nicholl and Rexroad, 1968). Conglomerates consisting of pebbles of Ordovician rocks in Brassfield matrix are found to the southeast of the structure (Foerste, 1891, 1904, 1919). This relationship demonstrates a southeastern direction of erosional transport from the Ripley Island uplift. Phosphatized hardgrounds (Laferriere et al., 1986) suggests an extended period of nondeposition related to the feature.

The Ripley Island uplift is located near the junction of the northwest margin of the English Graben and the perpendicularly oriented Flatrock subbasin (Hendricks, 1996). The absence of Sebree Trough sediments across Ripley Island indicates that the feature may have originated during the Champlainian and continued to uplift through the Cincinnatian. By Early Silurian time, the Ripley Island uplift may have evolved into an arch that effectively isolated the northwestern Flatrock subbasin from the area overlying the English Graben. Unusual, endemic echinoderm faunas are present in the Osgood Member (Paul, 1971) and the Laurel Member (Miller, 1892; Springer, 1926; Paul, 1971; Frest, 1975a, b) of the Salamonie Dolomite on or adjacent to the Ripley Island uplift. Moreover, the overlying Waldron Shale is abundantly fossiliferous in the Ripley Island uplift area. In areas to the north and south, the Waldron is essentially unfossiliferous (Paul, 1971). Possibly, the Ripley Island uplift exerted some

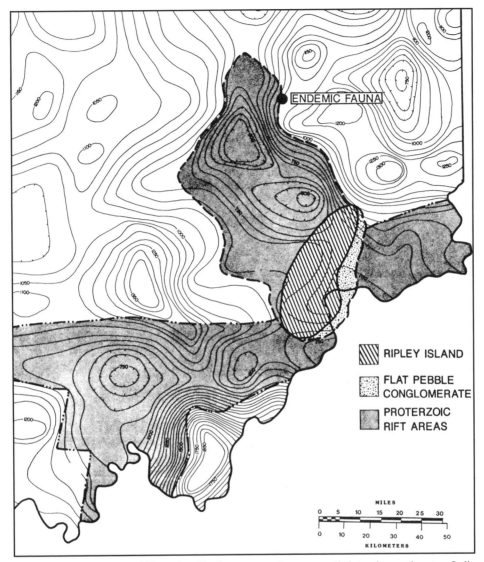

Figure 12. Ripley Island and Flatrock subbasin versus total aeromagnetic intensity, southeastern Indiana. The Ripley Island uplift defines a complex structural/ erosional area that upwarped during the Late Ordovician/Early Silurian, coeval with the uplifting of the Cincinnati Arch. The Ripley Island uplift may have been formed by remobilization of the English Graben rift margins during Ordovician and Silurian tectonism. Across the Ripley Island uplift, the Lower Silurian Brassfield Formation was not deposited and the overlying Osgood Member of the Salamonie Dolomite thins. Conglomerates consisting of pebbles of Ordovician rocks in Brassfield matrix are found to the southeast of the structure. By Early Silurian time, the Ripley Island uplift may have evolved into an arch that effectively isolated the northwestern Flatrock subbasin from the area overlying the English Graben.

degree of environmental control that resulted in the development of the unusual, endemic echinoderm faunas in the Salamonie Dolomite and the abundant faunas in the Waldron.

The position of the Sebree Trough and the Ripley Island uplift in southeastern Indiana indicates that the English Graben rift margins may have been remobilized during Ordovician and Silurian tectonism. Taconic deformation appears to have resulted in the development of a silled depocenter adjacent to and within the rift segment, in which the Sebree Trough sediments accumulated. Subsequent positive isostatic compensation of rift sedi-

ments may have in turn led to the development of the Ripley Island positive feature.

Stratigraphic and structural evidence within the Paleozoic section supports the conclusion that shallow structures occurring along English Graben margins reflect deep rift related features. Isopachous thinning of the Devonian aged New Albany Shale (Hasenmueller and Woodard, 1981) is correlative to position above postulated rift margin fault blocks. Regional syndepositional remobilization of basement faults during New Albany time is evident from growth-style deposition of the shale in the

Rough Creek Graben (Hester, 1983). Mississippian Valmeyerian Series strata exhibits isopachous thins (Keller and Becker, 1980) above the postulated basement blocks, and may reflect an additional episode of remobilization flexure during the mild tectonism of the Mississippian. Additional data supporting remobilization tectonism are the existence of shallow faults that cut post-Devonian Paleozoic strata upon the margins of these isopachous anomalies, and the development of associated rectilinear drainage patterns (Greb, 1989) and the erosional Mount Vernon Uplands (Droste and Keller, 1989a) upon the Absarokan unconformity surface.

Lithologic data

Four wells within or adjacent to the English Graben/LAS have penetrated pre–Mount Simon strata that suggest the existence of a Keweenawan-aged rift system. Two of the wells penetrated arkosic sediments similar to the Middle Run Formation at the confluence of the East Continent Rift Basin and the English Graben as follows: (1) Well 1, Ashland Oil #1 Collins (Switzerland County, Indiana), and (2) Well 2, Ashland Oil #1 Eichler (Switzerland County, Indiana)

After encountering a normal Paleozoic section, these wells penetrated an anomalous clastic sequence beneath the unconformable contact of the Proterozoic rocks. Current data is insufficient to determine whether these pre-Paleozoic arkoses overlie the Middle Run Formation, or are a facies equivalent. Similarity with strata found in Ohio and Kentucky indicates that the clastics may be Precambrian (Keweenawan-aged) sediments (Cincinnati Arch Consortium, 1992), possibly equivalent to the Oronto Group of the Lake Superior district (Dickas et al., 1992b).

The lithic arenites of the Middle Run Formation are commonly red/gray, fine- to medium-grained sandstones with minor red siltstones and shales. The sandstone is supported by framework grains of felsic volcanic rock fragments, with minor sedimentary and rare metamorphic (?) clasts. The lithic arenites are interpreted to have been deposited in an arid continental setting, in alluvial and fluvial depositional settings (Harris, 1992).

The remaining two wells penetrated mafic strata similar to the igneous assemblage noted in the East Continent Rift Basin as follows: (1) Well 3, DuPont #1 Wadd/Fee (Jefferson County, Kentucky), and (2) Well 4, Farm Bureau #1 Brown (Lawrence County, Indiana).

The DuPont #1 Wadd/Fee penetrated the eastern flank of the LAS; beneath a normal Paleozoic section, the well cored a fractured, medium grained gabbroic complex at a depth of 1,815 m (5,954 ft). Geochemical and thin-section analyses (Manuel Bass, 1990, written communication; Cincinnati Arch Consortium, 1992; John Green, 1992, written communication) indicate a compositional similarity to continental-style flood basalts found in the Midcontinent Rift System and in the East Continent Rift Basin. The occurrence of medium grained gabbro immediately beneath the Mount Simon demonstrates that significant unroofing of the mafic complex occurred prior to the deposition of Paleozoic rocks.

To the north, the Farm Bureau #1 Brown was drilled in Lawrence County, Indiana, on the downthrown side of the Mount Carmel Fault in the Leesville field. Drilled upon a major positive magnetic anomaly, the well penetrated a sequence of flood basalt flows beneath the Mount Simon Sandstone (Manuel Bass, 1992, personal communication) before reaching total depth. It was noted that "the similarity in texture, mineralogy, and chemical composition of the Precambrian basalts and diabases from Michigan … and the basalt in Lawrence County, Indiana, suggests that all are of similar age. The basalt from Lawrence County possibly … corresponds in age to the Keweenawan lava flows of the Keweenawan Peninsula, Michigan" (Greenburg and Vitaliano, 1962, p. 488). Recent geochemical investigations (Wilson, 1989; Drahovzal et al., 1992; Cincinnati Arch Consortium, 1992; Walker and Misra, 1992) support this conclusion. Additional evidence bracketing the age of the LAS is interpreted from transcurrent offset of the mafic complex along the wrench systems that modify the adjacent English Graben. This would imply that the LAS was emplaced prior to transcurrent offset, assumed to be Keweenawan or Grenvillian in age.

Mineralogical, textural, and compositional characteristics of the gabbro cored in the Du Pont #1 Wadd/Fee well suggest that the LAS may have been emplaced as a feeder dike system or associated fault-bounded magma chamber. This volcanism may have been associated with the English Graben margin faults. A possible mechanism for mafic emplacement may be the generation of a dilational strain field in the margin-fault footwall at the base of the crust (Ellis and King, 1991). During extension, deep strain allows for the propagation of a magma-driven dike into shallower strata. The feeder dike would provide a source for the Proterozoic flood basalts noted in the English Graben area and would provide the geologic link between the compositional and structural aspects of the LAS.

The occurrence of flood basalt flows in Lawrence County and the possible existence of an eroded feeder dike system or eroded magma chamber in Harrison County suggests that the Proterozoic sequence of southeastern Indiana contains continental volcanics similar to those found in the East Continent Rift Basin of Ohio and in the Midcontinent Rift System of the Lake Superior region. The occurrence of an associated Proterozoic clastic assemblage supports the conclusion that the English Graben was a depocenter for rift sediments prior to the deposition of the Paleozoic sequence.

Hydrologic data

In southeastern Indiana, the porous Siluro-Devonian carbonates are an important aquifer that provides fresh water in a large portion of the state. Wells drilled in search of potable water have uncovered an unusual anomaly in the Siluro-Devonian carbonates, interpreted to be associated with the position of the English Graben.

The freshwater recharge area for the Siluro-Devonian carbonates occurs along the outcrop upon the Cincinnati Arch.

Massive quantities of fresh water, probably derived from glacial meltwater during the Pleistocene, entered the vugular carbonates and flushed downdip. Eventually, an equilibrium state was reached with the saline pore fluids (greater than 10,000 ppm TDS) present in the structurally downwarped Illinois Basin. The dense brines of the Illinois Basin remained unflushed in the deeper basinal areas, while the more buoyant fresh water occupied the structurally positive regions.

An anomalous plume of saltwater was discovered, extending updip through surrounding freshwater wells to onlap the flanks of the Cincinnati Arch (Rupp and Pennington, 1987; Fig. 13). Seismic and geopotential evidence demonstrates that the anomalous tongue of saltwater closely corresponds to the position of the English Graben depocenter.

These data suggest that, during the Pleistocene, the Siluro-Devonian carbonates overlying the English Graben may have been structurally lower than the surrounding area. Fresh water flushed the surrounding nonrifted areas and preserved the brines in the structurally lower carbonates overlying the English Graben.

Surficial data

Additional evidence of rift remobilization can be found adjacent to the city of Louisville, Kentucky, at the Falls of the Ohio River. As the name implies, a series of falls and rapids developed in the Middle Devonian aged Jeffersonville Limestone riverbed create the single portage of the Ohio River. Louisville became established at this portage point and, subsequent to the construction of a bypass canal, continued to grow into a small city. The riverbed now lies exposed to the north of McAlpine Dam in the recently created falls of the Ohio State Park.

Once covered by Pleistocene gravels, the falls were created by the unearthing of a structurally complex carbonate ridge by this youthful stretch of the Ohio River. Field observations at the falls indicate an unusual, differentially expanded joint pattern, steep-sided crevices oriented oblique to the joints, and en-echelon fractures in the riverbed. The en-echelon fractures are oriented subparallel to the large open crevices, and may be related to stress relief expansion related to crevice formation.

Potential field analysis indicates that the Falls of the Ohio may be located upon a major wrench fault, and upon the northern flank of the LAS. Remobilization of the proposed basement wrench system is suggested by the occurrence of at least six earthquake epicenters in the vicinity of Falls of the Ohio (Braile et al., 1984; Hendricks et al., 1994). The anomalous structural features noted at Falls of the Ohio suggest that considerable tectonic activity has occurred. It is suggested that these deformation events may be genetically related to the remobilization tectonism noted along the margin of the English Graben.

SUMMARY

The interpretation of a wide variety of geological and geophysical data supports the conclusion that a structurally com-

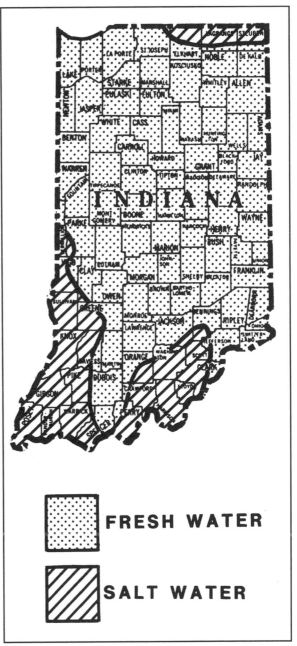

Figure 13. Siluro-Devonian salinity map, Indiana. The Cincinnati Arch provides the freshwater charge area for the Siluro-Devonian carbonates along the eastern shelf of the Illinois Basin. Saline pore fluids (>10,000 ppm TDS) are present in the deeper portions of the Illinois Basin. The dense brines of the Illinois Basin remained unflushed in the deeper basinal areas, while the more buoyant fresh water occupied the structurally positive regions. An anomalous plume of saltwater extends updip through surrounding freshwater wells to onlap the flanks of the Cincinnati Arch. Seismic and geopotential evidence demonstrates that the anomalous tongue of saltwater closely corresponds to the position of the English Graben depocenter (Rupp and Pennington, 1987).

plex deep depocenter is present in the English Graben area. Seismic and potential field data is interpreted to indicate the existence of a thick section of strata in an abruptly bounded, transversely offset rift segment. Depth to crystalline basement estimates suggests a strata thickness in excess of 6,100 m. The majority of this sedimentary section occurs as conspicuously layered seismic reflectors beneath the Paleozoic strata. Earthquake epicenter data indicate that modern fault movement may be strike slip, responding to the contemporary east-west horizontal maximum principle stress. Local earthquake epicenters appear to occur upon the transtensive wrench systems that offset the graben structure.

The rift fault boundary systems appear to have remobilized during later orogenic events, and are currently expressed by shallow fault systems that modify Paleozoic strata. Hydrocarbon productive reservoirs occur coincident with the bounding fault systems, and appear to be related to rift-margin remobilization deformation.

Within or adjacent to the English Graben/LAS, four drill holes that penetrate the Proterozoic indicate lithologic similarities to the Keweenawan-aged strata found in the East Continent Rift Basin of Ohio and Kentucky, and the Midcontinent Rift System strata found in the Lake Superior region. Anomalous hydrological and surficial data appear to have a high correlation to position within the English Graben.

The data presented in this study indicate that the eastern flank of the Illinois Basin is considerably more complex than previously described. A century of data acquisition in the area has discovered a significant number of geological and geophysical anomalies that show a high degree of correlation. The interpretation in this paper of the existence of a deep Proterozoic depocenter, deformed by wrench faults and subsequent remobilization events, assembles the data into an internally consistent regional framework. This report does not provide a complete resolution to the anomalous nature of the eastern margin of the Illinois Basin. Renewed evaluation of the area, perhaps fueled by the recognition of increased complexity and potential resources, will hopefully provide a more comprehensive understanding of the English Graben.

REFERENCES CITED

Ashley, G. H., 1898, Note on fault structure in Indiana: Indiana Academy of Science Proceedings for 1897, p. 244–250.

Ault, C. H., and Sullivan, D., 1982, Faulting in southwestern Indiana: United States Nuclear Regulatory Commission, NUREG/CR-2908, 50 p.

Bergstrom, S. M., and Mitchell, C. E.,1990, Trans-Pacific graptolite faunal relations; the biostratigraphic position of the base of the Cincinnatian series (Upper Ordovician) in the standard Australian graptolite zone succession: Journal of Paleontology, v. 64, no. 6, p. 992–997.

Black, D. F. B., 1986, Basement faulting in Kentucky, *in* Aldrich, M. J., and Laughlin, A. W., eds., Proceedings of the Sixth International Conference on Basement Tectonics, 1985, Santa Fe, New Mexico: Salt Lake City, Utah, International Basement Tectonics Association, p. 125–139.

Braile, L. W., Keller, G. R., Hinze, W. J., and Lidiak, E. G., 1984, An ancient rift complex and its relation to contemporary seismicity in the New Madrid seismic zone: Tectonics, v. 1, no. 2, p. 225–237.

Bryan, B. K., 1975, The Greenwood Anomaly of the Cumberland Plateau: A test application of GRAVMAP, a system of computer programs for reducing and automatic contouring land gravity data [M.S. thesis]: Knoxville, University of Tennessee, 118 p.

Cincinnati Arch Consortium, 1992, The geology and geophysics of the East Continent Rift Basin: Ohio Department of Natural Resources Division of Geological Survey, Kentucky Geological Survey, Indiana Geological Survey, Open File Report OFR 92-4, 130 p.

Dickas, A. B., Kholodkov, Y. I., Zarchikov, G. I., Chernikov, B. A., and Stark, T. J., 1992a, Intercontinental rift systems of the United States of America and associated mineral resources: North Caucusus Scientific Center of Higher Institutions Journal, v. 1 and 2, p. 48–60 (in Russian).

Dickas, A. B., Mudrey, M. G., Ojakangas, R. W., and Shrake, D. L., 1992b, A possible southeastern extension of the Midcontinent rift system located in Ohio: Tectonics, v. 11, no. 6, p. 1406–1414.

Drahovzal, J. A., Harris, D. C., Wickstrom, L. H., Walker, D., Baranoski, M. T., Keith, B., and Furer, L. C., 1992, The East Continent Rift Basin: A new discovery: Kentucky Geological Survey Special Publication 18, Series XI, 25 p.

Droste, J. B., and Keller, S. J., 1989a, Development of the Mississippian-Pennsylvanian unconformity in Indiana: Indiana Geological Survey Occasional Paper 55, 11 p.

Droste, J. B., and Keller, S. J., 1989b, Lithostratigraphy and petroleum of the Stevensport Group (Mississippian) in Indiana [abs.]: American Association of Petroleum Geologists Bulletin, v. 73, no. 8, p. 1029.

Ellis, M., and King, G., 1991, Structural control of flank volcanism in continental rifts: Science, v. 254, p. 839–842.

Esarey, R. E., 1923, The relationship of the Mt. Carmel and Heltonville Faults to the Dennison Anticline [M.A. thesis]: Bloomington, Indiana University, 25 p.

Fisher, J. A., 1981, Fault patterns in southeastern Michigan [M.S. thesis]: East Lansing, Michigan State University, 80 p.

Foerste, A. F., 1891, A report on the geology of the Middle and Upper Silurian rocks of Clark, Jefferson, Ripley, Jennings and southern Decatur Counties, Indiana: Report of Indiana Department of Geology Natural Resources, v. 21, p. 213–288.

Foerste, A. F., 1904, The Ordovician-Silurian contact in the Ripley Island area of southern Indiana, with notes on the age of the Cincinnati geanticline: American Journal of Science, Series 4, v. 18, p. 321–342.

Foerste, A. F., 1919, Echinodermata of the Brassfield (Silurian) Formation of Ohio: Denison University Bulletin of Science Laboratory, v. 19.3, p. 3–31.

Frest, T. J., 1975a, Caryocrinitidae (Echinodermata: Rhombifera) of the Laurel limestone of southeastern Indiana: Fieldiana: Geology, v. 30, p. 81–106.

Frest, T. J., 1975b, Marsupiocrinidae of the Laurel limestone of southeastern Indiana: Geology Magazine, v. 112, p. 565–574.

Genik, G. J., 1993, Petroleum geology of Cretaceous-Tertiary rift basins in Niger, Chad, and Central African Republic: American Association of Petroleum Geologists Bulletin, v. 77, no. 8, p. 1405–1435.

Greb, S. F., 1989, A sub-rectangular paleovalley system, Caseyville Formation, Eastern Interior Basin, western Kentucky: Southeastern Geology, v. 30, no. 1, p. 59–75.

Greenburg, S. S., and Vitaliano, C. J., 1962, Basalt from a deep well in Lawrence County, Indiana: Journal of Geology, v. 70, p. 482–488.

Harris, D. C., 1992, Petrology of the Middle Run Formation (Precambrian), East Continent Rift basin, western Ohio and north-central Kentucky: Geological Society of America Annual Meeting, Abstracts with Programs, v. 24, no. 7, p. A231.

Hassenmueller, N. R., and Woodard, G. S., 1981, Studies of the New Albany Shale (Devonian and Mississippian) and equivalent strata in Indiana: U.S. Department of Energy, DE-AC21-76MCO5204, 99 p.

Henderson, J. R., and Meuschke, J. L., compilers, 1951, Total intensity aeromagnetic maps: U.S. Geological Survey (in cooperation with Indiana Department of Conservation, Division of Geology), Geophysical Map GP Series, scale 1:24,000.

Henderson, J. R., and Zietz, I., 1958, Interpretation of an aeromagnetic survey of Indiana: Indiana Geological Survey Professional Paper 316-B, 37 p.

Hendricks, R. T., 1996, Stratigraphy and sedimentology of the Silurian (Wenlockian) Laurel Member of the Salomonie Dolomite in Kentucky and adjacent southeastern Indiana [M.S. thesis]: University of Kentucky, Lexington, in press.

Hendricks, R. T., Ettensohn, F. R., Stark, T. J., and Greb, S. F., 1994, Geology of the Devonian strata of the Falls of the Ohio area, Kentucky-Indiana: stratigraphy, sedimentology, paleontology, structure, and diagenesis: Annual Field Conference of the Geological Society of Kentucky Guidebook, Kentucky Geological Survey, Sept. 10–11, 1993, 28 p.

Hopkins, T. C., and Siebenthal, C. E., 1897, The Bedford oolitic limestone of Indiana: Indiana Department of Geology and Natural Resources, Annual Report 21, p. 289–427.

Howe, J. R., 1985, Tectonics, sedimentation, and hydrocarbon potential of the Reelfoot Aulacogen [M.S. thesis]: Norman, University of Oklahoma, 109 p.

Jillson, W. R., 1922, The conservation of natural gas in Kentucky: J. D. Morton and Company, 152 p.

Johnson, R. W., Haygood, C., Hildenbrand, T. G., Hinze, W. J., and Kunselman, P. M., 1980, Aeromagnetic map of east-central Midcontinent of the United States: U.S. Nuclear Regulatory Commission, NUREG/CR-1662, 12 p.

Keith, B. D., 1985, Facies, diagenesis, and the upper contact of the Trenton Limestone of northern Indiana, *in* Cercone, R. C., and Budai, J. M., eds., Ordovician and Silurian rocks of the Michigan Basin and its margins: Michigan Basin Geological Society Special Paper 4, p. 15–32.

Keith, B. D., 1989, Regional facies of the Upper Ordovician Series of eastern North America, *in* Keith, B. D., ed., The Trenton Group (Upper Ordovician Series) of eastern North America—Deposition, diagenesis, and petroleum: American Association of Petroleum Geologists Studies in Geology 29, p. 1–16.

Keller, G. R., Bland, A. E., and Greenberg, J. K., 1982, Evidence for a major Precambrian tectonic event (rifting?) in the eastern Midcontinent region, United States: Tectonics, v. 1, p. 213–223.

Keller, G. R., Russel, D. R., Hinze, W. J., Reed, J. E., and Geraci, P. J., 1980, Bouguer gravity anomaly map of the east-central Midcontinent of the United States: U.S. Nuclear Regulatory Commission, NUREG/CR-1663, 12 p.

Keller, S. J., and Becker, L. E., 1980, Subsurface stratigraphy and oil fields in the Salem Limestone and associated rocks in Indiana: Indiana Geological Survey Occasional Paper 30, 63 p.

Laferriere, A. P., Hattin, D. E., Foell, C. J., and Abdulkareem, T. F., 1986, The Ordovician-Silurian unconformity in southwestern Indiana: Indiana Geological Survey Occasional Paper 53, 12 p.

Lucius, J. E., and Von Frese, R. R. B., 1988, Aeromagnetic and gravity anomaly constraints on the crustal geology of Ohio: Geological Society of America Bulletin, v. 100, no. 1, p. 104–116.

McPhee, J. P., 1983, Regional gravity analysis of the Anna, Ohio, seismogenic region [M.S. thesis]: West Lafayette, Indiana, Purdue University, 100 p.

Mescher, P. K., 1980, Structural evolution of southeastern Michigan–Middle Ordovician to Middle Silurian [M.S. thesis]: East Lansing, Michigan State University, 75 p.

Miller, S. A., 1892, Palaeontology: Annual Report of the Geological Survey of Indiana, v. 71, p. 611–705.

Mitchell, C. E., Bergstrom, S. M., and Schumaker, G. A., 1991, The Sebree Trough project 5. Development and possible tectonic origin of a Middle to Late Ordovician intracratonic basin: Geological Society of America North Central Section, Abstracts with Programs, v. 23, no. 3, p. 49.

Nicholl, R. S., and Rexroad, C. B., 1968, Stratigraphy and conodont paleontology of the Salamonie Dolomite and the Lee Creek Member of the Brassfield Limestone (Silurian) in southeast Indiana and adjacent Ken-

tucky: Indiana Geological Survey Bulletin 40, 73 p.

Paul, R. C., 1971, Revision of the Holocystytes (Diploporita) of North America: Fieldiana, v. 24, Publication 1135, 166 p.

Pratt, T., Culotta, R., Hauser, E., Nelson, D., Brown, L., Kaufman, S., Oliver, J., and Hinze, W., 1989, Major Proterozoic basement features of the eastern Midcontinent of North America revealed by recent COCORP profiling: Geology, v. 17, p. 505–509.

Pratt, T., Hauser, E. C., and Nelson, K. D., 1992, Widespread buried Precambrian layered sequences in the U.S. Midcontinent: Evidence for large Proterozoic depositional basins: American Association of Petroleum Geologists Bulletin, v. 76, no. 9, p. 1384–1401.

Rosedahl, B. R., Kaczmarick, K., and Kilembe, E., 1992, The Tanganyika, Malawi, Ruzwa, and Turkana rift zones of East Africa: An inter-comparison of rift architectures, structural styles, and stratigraphies: International Basement Tectonics Conference, 10th, Keynote Address, Aug. 7, 1992, Duluth, Minnesota, p. 147–152.

Rudman, A. J., Summerson, C. H., and Hinze, W. J., 1965, Geology of basement in midwestern United States: American Association of Petroleum Geologists Bulletin, v. 49, p. 894–904.

Rupp, J. A., and Pennington, D., 1987, Determination of the 10,000 mg/l TDS surface within the bedrock aquifers of Indiana: Proceedings, Indiana Academy of Science, v. 97, p. 383–389.

Schwalb, H., 1980, Hydrocarbon entrapment along a Middle Ordovician disconformity, *in* Luther, M. K., ed., Proceedings, Kentucky Oil & Gas Association, Technical Sessions, Annual Meetings, 36th and 37th, 1972 and 1973: Kentucky Geological Survey Special Publication 2, 71 p.

Shrake, D. L., Wolfe, P. J., Richard, B. H., Swinford, E. M., Wickstrom, L. H., Potter, P. E., and Sitler, G. W., 1990, Lithologic and geophysical description of a continuously cored hole in Warren County, Ohio, including description of the Middle Run Formation (Precambrian?) and a seismic profile across the core site: Ohio Department of Natural Resources, Division of the Geological Survey, Information Circular 56, 11 p.

Soderberg, R. K., and Keller, G.R., 1981, Geophysical evidence for deep basin in western Kentucky: American Association of Petroleum Geologists Bulletin, v. 65, p. 226–234.

Springer, F., 1926, American Silurian crinoids: Smithsonian Institute Publication 2871, 239 p.

Stark, T. J., 1990, Tectonic controls on Silurian reef growth in southwestern Indiana [abs.]: Abstracts of the Technical Sessions, Kentucky Oil and Gas Association, 54th Annual Meeting, Louisville, Kentucky, 1990, p. 15.

Street, R., Zehulin, A., and Harris, J., 1991, The Meade County, Kentucky, earthquakes of January and March 1990: Seismological Research Letters, v. 62, no. 2, p. 105–111.

Tanner, G., 1985, The Mt. Carmel Fault and associated features in south-central Indiana [M.S. thesis]: Bloomington, Indiana University, 66 p.

Walker, D., and Misra, K. C., 1992, Tectonic significance of basalts of the Middle Run Formation (Upper Proterozoic) of the East Continent Rift basin, Indiana and Kentucky: Geological Society of America Annual Meeting, Abstracts with Programs, v. 24, no. 7, p. A330.

Wickstrom, L. H., and Gray, J. D., 1989, Geology of the Trenton Limestone in northwestern Ohio, *in* Keith, B. D., ed., The Trenton Group (Upper Ordovician Series) of eastern North America—Deposition, diagenesis, and petroleum: American Association of Petroleum Geologists Studies in Geology 29, p. 159–172.

Wickstrom, L. H., Gray, J. D., and Stieglitz, R. D., 1992, Stratigraphy, structure, and production history of the Trenton Limestone (Ordovician) and adjacent strata in northwestern Ohio: Ohio Department of Natural Resources, Division of the Geological Survey Report of Investigations 143, 78 p.

Wilson, M., 1989, Igneous Petrogenesis: London, Unwin Hyman Publishing, 466 p.

Zoback, M. L., and Zoback, M. D., 1980, State of stress in the conterminous United States: Journal of Geophysical Research, v. 85, p. 6113–6156.

MANUSCRIPT ACCEPTED BY THE SOCIETY JANUARY 16, 1996

Geological Society of America
Special Paper 312
1997

Role of the Reelfoot Rift/Rough Creek Graben in the evolution of the Illinois Basin

Dennis R. Kolata and W. John Nelson
Illinois State Geological Survey, 615 East Peabody, Champaign, Illinois 61820

ABSTRACT

The southern end of the Illinois Basin is underlain by the Reelfoot Rift and a dog-leg extension in western Kentucky known as the Rough Creek Graben. Seismic reflection profiles indicate that the rift system consists of several half grabens linked by transfer zones. In northeastern Arkansas and western Tennessee, the rift is characterized by opposing-polarity half grabens separated by a central uplift known as the Blytheville Anticline. In southern Illinois and western Kentucky, where the rift curves eastward into the Rough Creek Graben, a shallow half graben occurs at the southern edge of the rift. Eastward, across an apparent transfer zone, the polarity reverses and a deep half graben is developed along the north side of the Rough Creek Graben.

The origin and evolution of the Illinois Basin were controlled by geodynamic processes within the rift system. The rift formed concurrently with the breakup of a supercontinent during late Precambrian to Early Cambrian time. Lithospheric extension within the rift system resulted in block faulting and rapid subsidence. By Late Cambrian, the tectonic setting changed from a rift basin to a broad cratonic embayment centered over the rift. During the remainder of the Paleozoic Era, the proto–Illinois Basin was a broad trough extending from the continental margin in east-central Arkansas northeastward through Illinois, Indiana, and western Kentucky. The depocenters for all major Paleozoic sequences in the basin were situated over the rift system. Widespread structural deformation in the proto–Illinois Basin began during Late Mississippian time concurrently with the initial accretion of continents forming the Pangea supercontinent. Collision during the Alleghenian and Ouachita orogenies induced compression in the continental interior, reactivating faults within the rift system and evidently increasing subsidence rates within the proto–Illinois Basin. The compressional episode was followed by a post–Early Permian episode of extension coinciding with breakup of Pangea. Later extension again reactivated faults within and adjacent to the rift system. Post-Pennsylvanian, pre-Late Cretaceous uplift of the Pascola Arch structurally closed the southern end of the Illinois Basin, creating the present basin geometry.

Present-day horizontal compression is reactivating ancient faults within the Reelfoot Rift, causing frequent earthquake activity including the intense and destructive earthquakes near New Madrid, Missouri, during the winter of 1811 to 1812.

INTRODUCTION

The Illinois Basin covers an area of approximately 285,000 km² in parts of Illinois, Indiana, and Kentucky (Fig. 1). It contains about 500,000 km³ of Cambrian through Pennsylvanian sedimentary rocks (Fig. 2) that have a maximum thickness of more than 7,600 m (Goetz et al., 1992). Beneath the southern end of the basin is a late Precambrian to Early Cambrian rift that is dogleg shaped in map view. The northeast-trending segment is named the Reelfoot Rift; the east-trending segment is called the Rough Creek Graben (Figs. 1 and 3). The rift system strongly influenced the origin and evolution of the

Kolata, D. R., and Nelson, W. J., 1997, Role of the Reelfoot Rift/Rough Creek Graben in the evolution of the Illinois Basin, *in* Ojakangas, R. W., Dickas, A. B., and Green, J. C., eds., Middle Proterozoic to Cambrian Rifting, Central North America: Boulder, Colorado, Geological Society of America Special Paper 312.

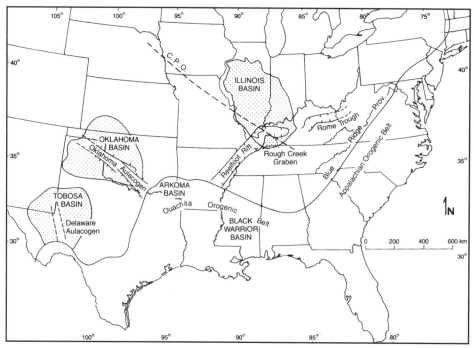

Figure 1. Midcontinent United States showing aulacogens, rifts, orogenic belts, and cratonic basins (modified from Buschbach and Kolata, 1991). Dashed line marked by C. P. O. is the northern margin of the Central Plains orogen.

Illinois Basin (Braile et al., 1986; Kolata and Nelson, 1991). In addition, the rift controlled development of the Mississippi Embayment during Mesozoic-Cenozoic time, and today, maximum horizontal compression is reactivating ancient faults within the rift, causing earthquakes.

Interest in the rift is high because of its hydrocarbon potential (Howe and Thompson, 1984; Bertagne and Leising, 1991; Goetz et al., 1992) and because of its earthquake potential (McKeown, 1982; Andrews et al., 1985; Crone et al., 1985). The goal of this paper is to provide an overview of the rift system and its role in the evolution of the Illinois Basin. Our primary focus, and the basis of our conclusions, is on a review of the most up-to-date structural and stratigraphic analyses in the basin.

EVOLUTION OF THE REELFOOT RIFT/ ROUGH CREEK GRABEN

Pre-rift history

The Reelfoot Rift/Rough Creek Graben and the Illinois Basin all lie within the Eastern Granite-Rhyolite Province, an anorogenic basement complex that extends from eastern Iowa, Missouri, and Arkansas through Illinois and Indiana into Ohio (Zietz et al., 1966; Bickford et al., 1986). Granitic rocks in this region formed at approximately 1,500 to 1,420 Ma (Bickford et al., 1986). Seismic reflection and potential field data suggest that a major Proterozoic crustal boundary lies within or beneath the granitic rocks in the southern part of the Illinois Basin (Heigold

and Kolata, 1993). The boundary lies along the projected extension of the northern margin of the Early Proterozoic Central Plains orogen and apparently marks the convergent margin of this orogen (Fig. 1). The dogleg bend of the rift system lies close to the projected front of the Central Plains orogen. Hence, the eastern segment of the rift system likely exploited a preexisting zone of weakness (Heigold and Kolata, 1993).

AGE OF RIFTING

The presence of Middle Cambrian shale in the upper part of the Rough Creek Graben succession, as seen in the Texas Gas Exploration No. 1 Shain well in Grayson County, Kentucky (Fig. 4; Schwalb, 1982), indicates that rifting was well underway by this time. No known Middle Cambrian strata occur outside the rift system within the immediate region. Furthermore, the Rome Trough, believed to be genetically related to the Rough Creek Graben (Goetz et al., 1992), contains Lower and Middle Cambrian sedimentary rocks, a succession that is not known to occur elsewhere on the craton. The Reelfoot Rift/Rough Creek Graben is overstepped by Late Cambrian siliciclastic and carbonate rocks.

The Reelfoot Rift/Rough Creek Graben is one of several failed rifts or aulacogens that formed along the eastern and southern margins of the North American craton during breakup of a supercontinent beginning during late Precambrian time (Braile et al., 1986; Thomas, 1989; Kolata and Nelson, 1991). Diachronous rifting along the Appalachian-Ouachita margin

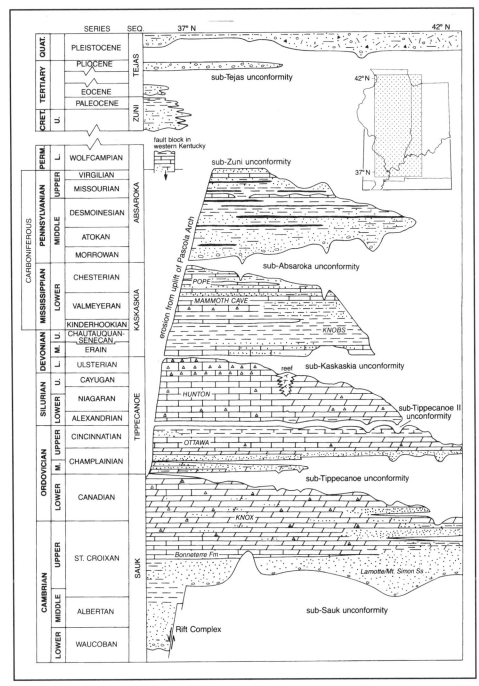

Figure 2. Stratigraphic cross section of the Illinois Basin representing a wide swath extending from near 37 to 42°N (modified from Buschbach and Kolata, 1991). Groups, supergroups, and some formations are labeled within the graphic column. Sequences, series, and systems are also shown.

took place beginning approximately at 730 Ma in the Appalachians and continued to about 525 Ma along the southern margin of North America (Rankin et al., 1989; Thomas, 1989). Igneous rocks associated with the Southern Oklahoma Aulacogen (Fig. 1) range in age from 570 to 525 Ma (Gilbert, 1983; Coffman et al., 1986; Perry, 1989). More recently, the southern Oklahoma structure and associated igneous rocks have been interpreted to be related to a transform fault (Alabama-Oklahoma transform) that propagated into the continent from the so-called Ouachita rift (Thomas, 1991). By either interpretation, formation of the Reelfoot Rift/Rough Creek Graben can be assumed to have commenced during the long period of diachronous rifting, sometime between latest Precambrian to Early Cambrian.

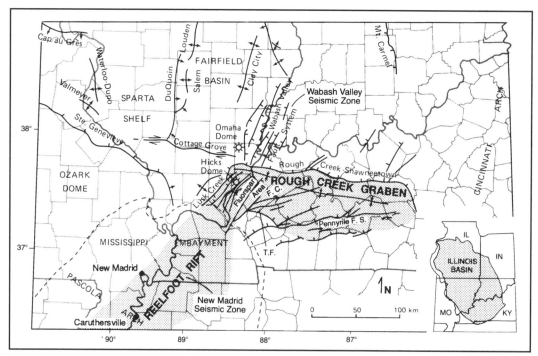

Figure 3. Major structural features (modified from Nelson, 1991) in the southern part of the Illinois Basin (inset). The rift complex, including the Reelfoot Rift and Rough Creek Graben, is shaded. Dashed line is the northern margin of the Mississippi Embayment of the Gulf Coastal Plain. The states include Illinois (IL), Indiana (IN), Kentucky (KY), and Missouri (MO).

Geometry and stratigraphy of the rift system

The Reelfoot Rift extends from east-central Arkansas into southernmost Illinois, lying beneath the Mississippi Embayment of the Gulf Coastal Plain (Fig. 4). The southern end of the rift was deformed and truncated during the Ouachita orogeny. The northern end is contiguous with the Rough Creek Graben. The rift is known from seismic reflection data (Howe and Thompson, 1984; Howe, 1985; Bertagne and Leising, 1991; Nelson and Zhang, 1991; Goetz et al., 1992; Potter et al., 1992) and refraction profiles (Ginzburg et al., 1983; Mooney et al., 1983) as well as gravity and magnetic data (Cordell, 1977; Hildenbrand et al., 1977, 1982; Keller et al., 1983; Kane et al., 1981; Braile et al., 1982; Hildenbrand, 1985).

Reelfoot Rift

Seismic reflection profiles show that the Reelfoot Rift/Rough Creek Graben consists of half grabens linked by transfer zones. Total sedimentary fill in the deeper part of the half graben, on the southeast side of the Reelfoot Rift, is approximately 8 km thick, with about 5 km being synrift deposits (Nelson and Zhang, 1991). Most synrift faults within the Reelfoot Rift trend northeasterly, parallel or subparallel to the regional strike of the rift (Howe and Thompson, 1984; Howe, 1985). COCORP seismic lines from northeastern Arkansas and western Tennessee indicate that the rift consists of opposite-facing half grabens separated by a complexly

faulted axial uplift known as the Blytheville Anticline (Fig. 4; Nelson and Zhang, 1991). Pre-Knox and Knox Group strata (Sauk Sequence) are upwarped along this northeast trending anticline, truncated, and overlain by slightly deformed Late Cretaceous and early Tertiary strata (Zuni and Tejas Sequences) of the Mississippi Embayment (Fig. 5A). The Blytheville Anticline appears to be structurally continuous with major faults exposed north of the Mississippi Embayment, particularly the Lusk Creek fault zone (LCFZ) in southern Illinois. We believe that the LCFZ formed initially during development of the Reelfoot Rift/Rough Creek Graben (Kolata and Nelson, 1991) and was reactivated during the late Paleozoic, with reverse faulting followed by normal faulting (Nelson et al., 1991; Weibel et al., 1993). The New Burnside and McCormick Anticlines, situated immediately west of and approximately parallel to the Lusk Creek Fault Zone, appear to have formed during the same episode of northwest-southeast horizontal compression. The northeast orientation and apparent structural continuity of the Blytheville Anticline with the faults and anticlines in southern Illinois suggests that the Blytheville Anticline was deformed under the same stress field. This style of intracratonic deformation induced by continental collision resembles the deformation extending deep into the Asian continent resulting from Cenozoic collision with India as discussed by Molnar and Tapponier (1975). The axial fault zone of the Blytheville Anticline is seismically active, as evidenced by spatial correlation with the northeast-trending linear belt of earthquake epicenters

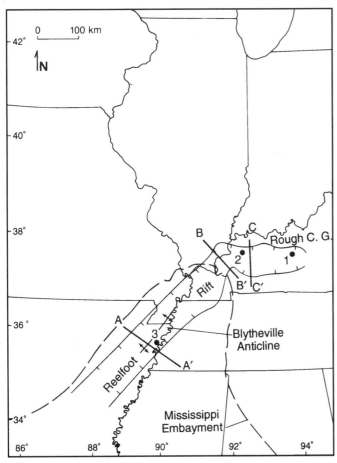

Figure 4. Rift system showing location of cross sections (see Fig. 5) and cited drill holes: 1, Texas Gas Transmission No. 1 Shain, Grayson County, Kentucky, 10-L-36; 2, Exxon No. 1 Duncan, Webster, County, Kentucky 5-M-22; and 3, Dow Chemical No. 1 Wilson, Mississippi County, Arkansas, 14-12N-9E. Boundary of the Mississippi Embayment also shown.

in the New Madrid Seismic Zone (compare Figs. 4 and 10 [in a following section]), (Crone et al., 1985).

Relatively few wells have been drilled into the stratigraphic succession in the Reelfoot Rift, and only the Dow No. 1 Lee Wilson well (Fig. 4) in Mississippi County, Arkansas, has penetrated unequivocal crystalline basement. Consequently, stratigraphy of the rift is poorly understood. Several different interpretations of the rift succession exist in the literature. In a study of drill cuttings and wireline logs from significant deep tests, Houseknecht (1989) proposed that the Reelfoot Rift contains strata older than any on the craton of the Midcontinent, a concept suggested by Schwalb (1982). According to Houseknecht (1989), the rift succession consists of a basal arkose (age unknown), an overlying carbonate formation, and an upper shale interpreted to be a facies, at least in part, of the Lamotte or Mount Simon Sandstones and possibly the basal part of the Bonneterre Formation (Fig. 2). The shale in turn is overlain by post-rift basin carbonates of the Bonneterre Formation and Knox Group.

Working with drill-hole data from southeastern Missouri, Palmer (1989) interpreted the rift succession to consist of the Lamotte Sandstone (mostly arkose) of undetermined age at the base. The Lamotte is believed to be overlain by argillaceous carbonates (mostly limestone) and shales interpreted as a basinal facies of the relatively pure, dolomitic rocks of the Bonneterre Formation occurring immediately north and west of the rift system. The basinal facies is interpreted by Palmer (1989) to be Late Cambrian (Dresbachian Stage), a younger age than that implied by Houseknecht (1989). Seismic reflection data (Howe and Thompson, 1984; Howe, 1985) appear to show abrupt thickening of shale in the rift, supporting the idea that Bonneterre carbonate rocks on the platform are in facies with a basinal shale. Palynomorphs from the basinal facies suggest a late Middle Cambrian to Late Cambrian age (Wood and Stephenson, 1989), however, additional biostratigraphic studies are needed to more accurately assess the age of the stratigraphic succession in the Reelfoot Rift.

Rough Creek Graben

A high-quality proprietary oil industry seismic reflection profile crossing the Illinois-Kentucky Fluorspar District shows that at the junction of the Reelfoot Rift and Rough Creek Graben (Fig. 4), the rift is a half graben with the major fault on the southeast (Potter et al., 1992). The southeastern margin of the rift appears to coincide with the Tabb Fault System (Fig. 5B) where the sedimentary section is approximately 6 km thick. Synrift deposits are interpreted from seismic reflection data to be about 2 km thick (Fig. 5B). No well has penetrated the rift succession in this region. Based on the Dow No. 1 Lee Wilson well in the Reelfoot Rift, the basal part of the sedimentary section in the graben probably consists of arkose overlain by shale and/or carbonate rocks. The Mount Simon Sandstone does not occur south of the Rough Creek Graben, so it either pinches out or grades to shale or carbonates in the rift. Rift-bounding faults in this region have been reactivated several times, particularly at the end of the Paleozoic (Nelson, 1991).

Several high-quality seismic profiles transecting the graben east of the fluorspar district show an abrupt reversal in graben polarity in the Rough Creek Graben (Bertagne and Leising, 1991). The reversal apparently occurs near the junction of the Reelfoot Rift and the Rough Creek Graben. East of the fluorspar district the major displacements switch from the south (Tabb Fault System) to the north (Rough Creek–Shawneetown Fault System—RCSFS) side of the graben (Fig. 5C). The RCSFS underwent late Paleozoic to early Mesozoic reverse, possibly strike-slip, and normal displacements (Bertagne and Leising, 1991; Nelson, 1991; Goetz et al., 1992). Sedimentary fill within the graben thickens northward apparently reaching a maximum of more than 8 km in McLean County, western Kentucky (Goetz et al., 1992). At inferred basement level, throw across the graben-bounding fault in this area is approximately 4 km, down to the south. The southern boundary of the graben corresponds to the

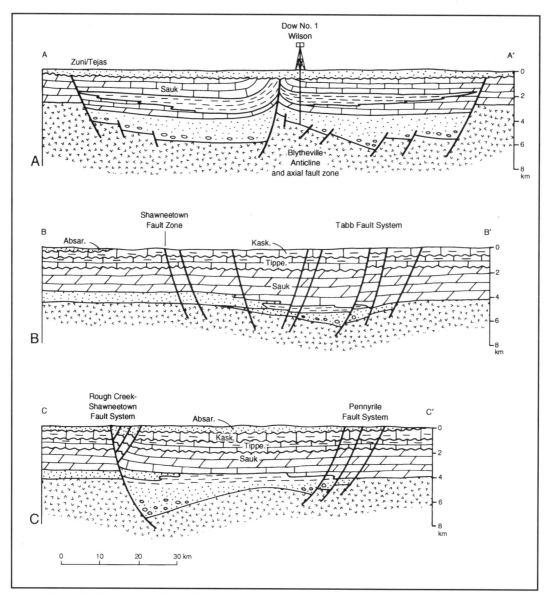

Figure 5. Cross sections in the rift system (see Fig. 4 for locations). Section A, interpretation based on seismic data from Howe and Thompson (1984) and Nelson and Zhang (1991); shows opposing half grabens with complexly faulted axial uplift known as the Blytheville Anticline; unconformity separates the Sauk from the Zuni and Tejas Sequences. Section B, based on proprietary industry seismic reflection data, shows small half graben at intersection of the Reelfoot Rift and Rough Creek Graben; southern end is bounded by the Tabb fault system. Section C, based on interpretation of seismic lines published by Bertagne and Leising (1991), shows half graben developed on north side of the Rough Creek Graben; the Rough Creek–Shawneetown fault system underwent reverse faulting during the late Paleozoic. Unconformity bound Sauk, Tippecanoe, Kaskaskia, Absaroka Sequences also shown.

surface trace of the Pennyrile Fault System (PFS) and an unnamed east-trending fault zone just south of the PFS (Fig. 3). Major displacements in the PFS are down to the north. These faults have undergone repeated normal displacement from Cambrian through post-Pennsylvanian time (Lumm et al., 1991).

The synrift stratigraphy within the Rough Creek Graben appears to be similar to that in the Reelfoot Rift. Two deep

wells in western Kentucky penetrated the graben sedimentary fill: (1) the Texas Gas Exploration No. 1 Shain well in Grayson County, and (2) the Exxon No. 1 Duncan well in Webster County (Fig. 4). Both wells encountered over 600 m of shale and shaly limestone and dolomite below relatively pure dolomite of the Knox Group. The shaly succession has not been observed outside of the rift system. Late Middle Cambrian trilobites were recovered from the shaly section in the Grayson County well (Schwalb, 1982), thus establishing the age of synrift rocks at one point in the graben. These shaly rocks are in the same stratigraphic position as the basinal shale observed in the Reelfoot Rift, and are approximately the same age. Synrift carbonate rocks in the Reelfoot Rift and Rough Creek Graben are relatively rich in limestone. Dolomite is more common at the margins of the rift and in post-rift strata that overlie the rift.

EVOLUTION OF THE ILLINOIS BASIN

The geometry of the Illinois Basin represents the cumulative effects of structural deformation that has occurred since late Precambrian–Early Cambrian time. Consequently, the shape and extent of the basin have undergone many changes since its origin. The present configuration developed during post–late Pennsylvanian to Late Cretaceous deformation. Prior to this time, the geometry of the proto–Illinois Basin was influenced by relatively rapid subsidence in the region of the rift system. The following discussion focuses on the evolution of the southern part of the basin in the vicinity of the rift system.

Late Cambrian through Middle Ordovician subsidence

By Late Cambrian time, most faulting within the rift system had ceased, except locally along rift-bounding faults. The tectonic setting changed from a rift basin to a broad, slowly subsiding cratonic embayment (Fig. 6). Concurrently, a worldwide rise in sea level flooded the Midcontinent and widespread siliciclastics of the Mount Simon, Lamotte, and Reagan Sandstones were deposited. Regional downwarping of the rift system and surrounding craton continued throughout most of the Paleozoic, forming a broad southward-plunging trough extending toward the cratonic margin in east-central Arkansas. The proto–Illinois Basin, thus, developed as a cratonic embayment centered over the rift system. Open-marine circulation dominated, and depositional and diagenetic facies aligned with tectonically defined bathymetric zones (Sloss, 1979). During Late Cambrian through Early Ordovician time (about 30 m.y.), approximately 2,500 m of primarily carbonate rocks (Knox Group, Fig. 2) were deposited in a depocenter situated over, but extending beyond, the rift system. This was the time of the highest rate of post-rifting subsidence (85 m/m.y. average; J. D. Treworgy, oral communication, 1992) in the proto–Illinois Basin. Regression of seas from the North American craton during the Early Ordovician resulted in subaerial exposure and formation of a widespread major unconformity. The region of the Reelfoot Rift/Rough Creek Graben, however, apparently continued to be a depocenter for marine sediments. Conodont zonation (Norby et al., 1986) suggests the region contains a continuous stratigraphic record between the Sauk and Tippecanoe Sequences. The relatively rapid rate of subsidence during this time has been attributed to thermal subsidence that accom-

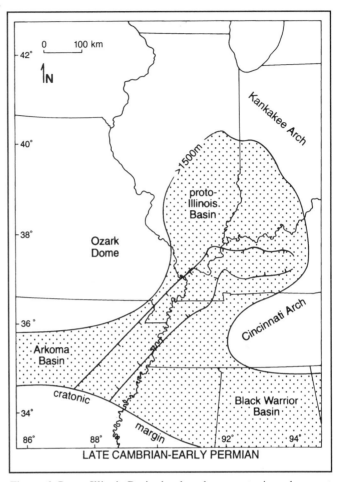

Figure 6. Proto–Illinois Basin developed as a cratonic embayment centered over the rift system. The boundary of the basin is arbitrarily delineated; shaded area indicates where Paleozoic rocks are in excess of 1,500 m (modified from Kolata and Nelson, 1991).

panied and followed rifting (Heidlauf et al., 1986; Klein and Hsui, 1987; Treworgy et al., 1989; Kolata and Nelson, 1991).

Middle Ordovician to mid-Mississippian subsidence

From Middle Ordovician to mid-Mississippian time, subsidence rates in the southern part of the basin were relatively slow, on average 15 m/m.y. (J. D. Treworgy, oral communication, 1992). As during deposition of the Sauk Sequence, depocenters for the Tippecanoe and Kaskaskia Sequences were situated over the rift system (Kolata, 1991). The presence of a positive gravity anomaly (Cordell, 1977) and associated high velocity in the lower crust (7.3 km/s) beneath the Reelfoot Rift (Mooney et al., 1983; Ginzburg et al., 1983) suggests that tectonic subsidence was driven by an isostatically uncompensated mass beneath the rift (Braile et al., 1986; Kolata and Nelson, 1991). Accordingly, when the rift formed, crustal extension was accompanied by a rise of the asthenosphere. Molten mantle material intruded the weakened lower crust along the incipient rift. When the magma solidified, it formed an isostatically uncompensated excess mass in the lower crust (DeRito et al., 1983). Downwarp of the crust was centered over the rift system but extended into regions beyond the rift due to flexural rigidity of the lithosphere. During Late Ordovician to mid-Mississippian time, the mass was largely supported by the strength of the lithosphere resulting in relatively slow subsidence rates.

Late Mississippian through Early Permian episode of compression

Widespread deformation swept through the proto-Illinois Basin beginning in latest Mississippian time, continuing at least into Early Permian. The timing of deformation corresponds with the Ouachita and Alleghenian orogenies. The resulting northerly and northwesterly directed convergence stresses probably were associated with the assembly of Pangea.

The Rough Creek–Shawneetown Fault System (RCSFS), which originated as a south-dipping listric normal fault during late Precambrian to Early Cambrian time, underwent reverse movement with the south side raised. Maximum vertical uplift was at least 1,100 m near the abrupt bend in the fault system in southern Illinois (Nelson and Lumm, 1987). Detached thrust-folds developed to the west of the RCSFS where the fault system curves abruptly to the southwest joining the Lusk Creek fault zone in southern Illinois. Convergent or transpressive displacement along the axial fault zone in the Reelfoot Rift apparently created the Blytheville Anticline as a positive flower structure (Fig. 7). The southern and southeastern boundary faults of the rift system evidently underwent little movement, acting as hinge zones for the larger displacements to the north and northwest. Right-lateral strike-slip displacement occurred along the Cottage Grove fault system (CGFS).

Igneous intrusions developed near the intersection of the Reelfoot Rift and Rough Creek Graben. The northwest-trending Tolu Arch, which terminates at Hicks Dome, apparently formed by the intrusion of ultramafic magma into the lower crust. Hicks Dome and associated diatremes formed by explosive release of mantle-derived gases associated with alkali igneous activity (Bradbury and Baxter, 1992; Potter et al., 1992). Magma entered tension fractures longitudinal to Tolu Arch and radial to Hicks Dome and also in the Cottage Grove fault system farther north. These are largely northwest-trending dikes consisting of lamprophyre and mica peridotite (Nelson and Lumm, 1987). K-Ar and Rb-Sr dates on the intrusives indicate an Early Permian age of 265 ± 15 Ma (Zartman et al., 1967) and $^{40}Ar/^{39}Ar$ methods yield an age of 272 ± 1 Ma for the Grant intrusive breccia near Hicks Dome (Snee and Hayes, 1992).

Tectonic subsidence rates within the basin increased, on average 35 m/m.y. (J. D. Treworgy, oral communication, 1992), beginning in mid- to late Mississippian time concurrently with the formation of clastic wedges that prograded cratonward from orogenic sources along the eastern and southern margins of North America. Compression associated with the orogenic

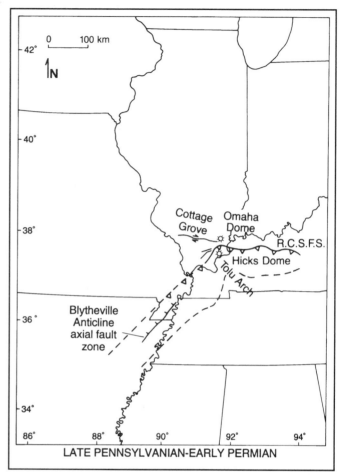

Figure 7. Late Pennsylvanian through Early Permian compressional structures in the southern part of the Illinois Basin (modified from Kolata and Nelson, 1991). R.C.S.F.S. = Rough Creek–Shawneetown Fault System.

activity apparently reactivated ancient structural features hundreds of kilometers within the craton. Increased subsidence rates in the Illinois Basin also appear to be linked to convergence at the plate margins.

DeRito et al. (1983) noted that interior cratonic basins commonly occur over ancient rift zones that are characterized by a positive linear Bouguer gravity anomaly caused by an excess mass emplaced in the lower crust during rifting. They relate discrete periods of increased basin subsidence rates to large-scale compression, the driving force for subsidence being the isostatically uncompensated excess mass. Accordingly, compressive stress causes a decrease in the effective viscosity of the lithosphere, allowing the excess mass to sink for a time at an increased rate. The positive Bouguer gravity anomaly beneath the Reelfoot Rift has been attributed to an excess mass that caused the proto-Illinois Basin to subside at an accelerated rate during the late Paleozoic episode of convergence (Klein and Hsui, 1987; Treworgy et al., 1989; Kolata and Nelson, 1991).

Post–Early Permian episode of extension

The southern Midcontinent United States, including the proto–Illinois Basin, underwent an episode of extension sometime after the Early Permian. A 200-m.y. hiatus in the stratigraphic record in the southern part of the basin, Early Permian to Late Cretaceous, prevents precise determination of when extension commenced (Fig. 2). We suspect, however, that extension coincided with the breakup of Pangea beginning during Late Triassic to Early Jurassic time. The earlier north-northwest to south-southeast contraction was succeeded by a regime of northwest-southeast extension (Kolata and Nelson, 1991).

Extension produced northeast-trending normal faults in the Fluorspar Area Fault Complex (FAFC) and the Wabash Valley fault system (Fig. 8). Some faults in the FAFC displace the northwest-trending Early Permian dikes, indicating that these faults are younger than the dikes (Oesterling, 1952; Hook, 1974). During the episode of extension, the crustal block bounded on the north by the RCSFS dropped back 1,000 m to near its approximate pre-Pennsylvanian position. Also the Lusk Creek fault zone underwent normal displacement, dropping the southeast block below its pre-Pennsylvanian position and producing a graben in southern Illinois. Dragging and wedging along the fault zone left slices of older rocks high within the fault zone. Reactivation of ancient faults bounding the Rough Creek Graben also led to formation of the east to northeast-striking normal faults of the Pennyrile fault zone (Nelson and Lumm, 1987; Nelson, 1991; Nelson et al., 1991; Weibel et al., 1993).

The presence of Early Permian cyclic marine rocks preserved in a small graben within the RCSFS in Union County, Kentucky (Kehn et al., 1982), suggests that much of the region was covered by Permian strata, removed by erosion prior to the Late Cretaceous.

Uplift of the Pascola Arch

Sometime between Late Pennsylvanian to Late Cretaceous time the Pascola Arch was uplifted, structurally closing the proto–Illinois Basin on the southwest side and creating the present basin geometry. The crest of the arch is situated in northwestern Tennessee and southeastern Missouri, where Cambrian strata are overlain unconformably by Late Cretaceous sedimentary rocks (Grohskopf, 1955). Broad, concentric belts of Ordovician, Silurian, and Devonian rocks dip regionally away from the Cambrian inlier (Fig. 9). The arch mapped by Grohskopf (1955) extends from the Mississippi River northwestward to the Ozark Dome. His maps and cross sections imply that the arch extends eastward into western Kentucky and northwestern Tennessee and is a regionally extensive structure. This fact has been demonstrated on sub-Cretaceous geologic maps by Schwalb (1982, p. 41) and Thomas (1989, plate 6) and other regional bedrock maps (King and Beikman, 1974).

Reconstructing the stratigraphic section, Marcher and Stearns (1962) estimated that 2,400 m to 4,000 m of Paleozoic

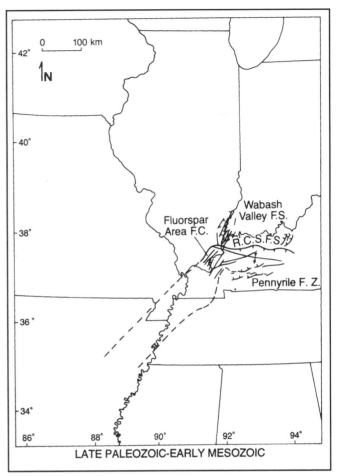

Figure 8. Mesozoic extensional structures in the southern part of the Illinois Basin (Kolata and Nelson, 1991). R.C.S.F.S. = Rough Creek–Shawneetown fault system; F.C. = Fault Complex; F.S. = Fault System; F.Z. = Fault Zone.

strata were eroded from the crest of the Pascola Arch. The crest is now covered by approximately 900 m of Late Cretaceous and early Tertiary siliciclastics of the Mississippi Embayment (Schwalb, 1982).

The age and mechanism for uplift of the Pascola Arch are not well understood, in part because of the 200 m.y. hiatus in the stratigraphic record, Late Pennsylvanian to Late Cretaceous. The uplift probably occurred after Early Permian. Preservation of the Early Permian carbonates and siliciclastics along the RCSFS suggests an open-marine environment, rather than a shallow, restricted terrestrial environment that would be expected on the flanks of an arch undergoing uplift. Late Jurassic and Early Cretaceous mixed siliciclastics, including conglomerates, in the subsurface of northeast Louisiana and southeast Arkansas (Braunstein et al., 1988) may be sedimentary evidence for uplift and shedding of sediment from the arch during this time. An Early Cretaceous age for uplift of the arch was proposed by Marcher and Stearns (1962). That would require the uplift and subsequent removal of as much as 4,000 m of strata, with erosion to sea level, all within Cretaceous time.

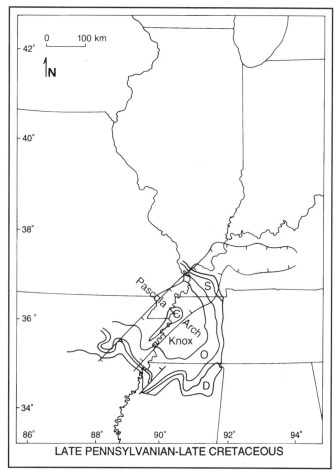

Figure 9. Sub-Cretaceous geologic map showing erosional pattern of Paleozoic rocks on the Pascola Arch (modified from Schwalb, 1982). Arch was uplifted during post–Late Pennsylvanian to pre–Late Cretaceous time. Є = Cambrian; O = Ordovician; S = Silurian; D = Devonian.

Post–Late Cretaceous history of the Mississippi Embayment

By Late Cretaceous time, the Pascola Arch was truncated by erosion. Downwarping of Paleozoic rocks within the rift began during the Late Cretaceous. By the end of Cretaceous time, a major marine invasion occurred, accompanied by deposition of sands, silts, and clays in the Mississippi Embayment of the Gulf Coastal Plain. The embayment, lying over the Reelfoot Rift, continued to subside as repeated transgressive-regressive cycles persisted into middle Eocene time, depositing thick sequences of marine and nonmarine sediments (Crone and Brockman, 1982). By the end of the Eocene, subsidence slowed or ceased within the northern part of the embayment. During Pliocene or early Pleistocene time, the area was blanketed by coarse fluvial gravels. Sedimentary fill within the embayment thickens southward from an erosional featheredge in southern Illinois to as much as 1,200 m thick in northwestern Mississippi.

One of the most significant post-Paleozoic events in the

Illinois Basin area has been a change from extension to horizontal contraction. The Midcontinent region presently is experiencing northeast-southwest-oriented maximum horizontal compression caused by westward drift of the North American plate away from the mid-Atlantic Ridge and accompanying subduction and transform motion along the Pacific coast (Irving, 1977; Zoback, 1992). Maximum horizontal compression appears to be reactivating ancient structures within the Reelfoot Rift including the axial fault zone of the Blytheville Anticline (Crone et al., 1985). Earthquake epicenters are concentrated along faults within the New Madrid Seismic Zone (Fig. 10). Focal-plane mechanisms for earthquakes along the Blytheville Anticline show primarily right-lateral strike-slip motion (Herrmann and Canas, 1978; Stauder, 1982). Earthquakes centered here in the winter of 1811 and 1812 were among the most intense in the recorded history of the United States.

CONCLUSIONS

The tectonic history of the Reelfoot Rift and Rough Creek Graben had a significant influence on the evolution of the Illinois Basin. Geodynamic processes within the rift system, greatly influenced by plate tectonic interactions from late Precambrian to the present, have controlled basin geometry, basin subsidence, sedimentation rates, depocenters, facies relations, structural styles, and localized intrusive activity during reactivation of rift structures. Present-day maximum horizontal compression within the North American plate is reactivating ancient faults locally within the rift causing contemporary earthquake activity.

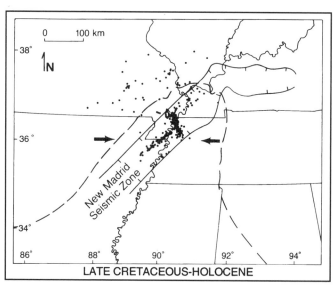

Figure 10. Present-day maximum horizontal compression within the Illinois Basin region is oriented east-northeast to west-southwest (Sbar and Sykes, 1973; Zoback and Zoback, 1980; Zoback, 1992). The compression is reactivating ancient faults within the rift system, causing earthquakes in the New Madrid Seismic Zone. Black dots are earthquake epicenters detected from July 1974 to June 1978 (Stauder, 1982).

ACKNOWLEDGMENTS

We thank Richard W. Ojakangas and an anonymous reviewer for the helpful comments and suggestions on an early draft of this report. We also thank Sandra K. Stecyk for preparing the figures.

REFERENCES CITED

Andrews, M. C., Mooney, W. D., and Meyer, R. P., 1985, The relocation of microearthquakes in the northern Mississippi Embayment: Journal of Geophysical Research, v. 90, p. 10223–10236.

Bertagne, A. J., and Leising, T. C., 1991, Interpretation of seismic data from the Rough Creek Graben of western Kentucky and southern Illinois, *in* Leighton, M. W., Kolata, D. R., Oltz, D. F., and Eidel, J. J., eds., Interior cratonic basin: American Association of Petroleum Geologists Memoir 51, p. 199–208.

Bickford, M. E., Van Schmus, W. R., and Zietz, I., 1986, Proterozoic history of the Midcontinent region of North America: Geology, v. 14, p. 492–496.

Bradbury, J. C., and Baxter, J. W., 1992, Intrusive breccias at Hicks Dome, Hardin County, Illinois: Illinois State Geological Survey Circular 550, 23 p.

Braile, L. W., Keller, G. R., Hinze, W. J., and Lidiak, E. G., 1982, An ancient rift complex and its relation to contemporary seismicity in the New Madrid seismic zone: Tectonics, v. 1, p. 225–237.

Braile, L. W., Hinze, W. J., Keller, G. R., Lidiak, E. G., and Sexton, J. L., 1986, Tectonic development of the New Madrid Rift complex, Mississippi Embayment, North America: Tectonophysics, v. 131, no. 1–2, p. 1–21.

Braunstein, J., Huddlestun, P., and Biel, R., 1988, Gulf Coast Region, *in* Lindberg, F. A., ed., Correlation of stratigraphic units of North America: American Association of Petroleum Geologists, COSUNA Chart Series.

Buschbach, T. C., and Kolata, D. R., 1991, Regional setting of the Illinois Basin, *in* Leighton, M. W., Kolata, D. R., Oltz, D. F., and Eidel, J. J., eds., Interior cratonic basin: American Association of Petroleum Geologists Memoir 51, p. 29–55.

Coffman, J. D., Gilbert, M. C., and McConnell, D. A., 1986, An interpretation of the crustal structure of the Southern Oklahoma Aulacogen satisfying gravity data: Oklahoma Geological Survey Guidebook 23, p. 1–10.

Cordell, L., 1977, Regional positive gravity anomaly over the Mississippi Embayment: Geophysical Research Letters, v. 4, p. 285–287.

Crone, A. J., and Brockman, S. R., 1982, Configuration and deformation of the Paleozoic bedrock surface in the New Madrid seismic zone, *in* McKeown, F. A., and Pakiser, L. C., eds., Investigations of the New Madrid, Missouri, earthquake region: U.S. Geological Survey Professional Paper 1236, p. 115–135.

Crone, A. J., McKeown, F. A., Harding, S. T., Hamilton, R. M., Russ, D. P., and Zoback, M. D., 1985, Structure of the New Madrid seismic source zone in southeastern Missouri and northeastern Arkansas: Geology, v. 13, p. 547–550.

DeRito, R. G., Cozzarrelli, F. A., and Hodge, D. S., 1983, Mechanism of subsidence of ancient cratonic rift basins: Tectonophysics, v. 94, p. 141–168.

Gilbert, M. C., 1983, Timing and chemistry of igneous events associated with the Southern Oklahoma Aulacogen: Tectonophysics, v. 94, p. 439–455.

Ginzburg, A., Mooney, W. D., Walter, A. W., Lutter, W. J., and Healy, J. H., 1983, Deep structure of northern Mississippi Embayment: American Association of Petroleum Geologists Bulletin, v. 67, p. 2031–2046.

Goetz, L. K., Tyler, J. G., Macarevich, R. L., Brewster, D., and Sonnad, J., 1992, Deep gas play probed along Rough Creek Graben in Kentucky part of southern Illinois Basin: Oil and Gas Journal, September 21, p. 97–101.

Grohskopf, J. G., 1955, Subsurface geology of the Mississippi Embayment of southeast Missouri: Missouri Geological Survey and Water Resources, 2nd ser., v. 37, 133 p.

Heidlauf, D. T., Hsui, A. T., and Klein, G. deV., 1986, Tectonic subsidence analysis of the Illinois Basin: Journal of Geology, v. 94, no. 6, p. 779–794.

Heigold, P. C., and Kolata, D. R., 1993, Proterozoic crustal boundary in the southern part of the Illinois Basin: Tectonophysics, v. 217, p. 307–319.

Herrmann, R. B., and Canas, J. A., 1978, Focal mechanism studies in the New Madrid seismic zone: Seismological Society of America Bulletin, v. 68, p. 1095–1102.

Hildenbrand, T. G., 1985, Rift structure of the northern Mississippi Embayment from the analysis of gravity and magnetic data: Journal of Geophysical Research, v. 90, no. B14, p. 12.

Hildenbrand, T. G., Kane, M. F., and Stauder, W., 1977, Magnetic and gravity anomalies in the northern Mississippi Embayment and their spatial relation to seismicity: U.S. Geological Survey, Map MF-914.

Hildenbrand, T. G., Kane, M. F., and Hendricks, J. D., 1982, Magnetic basement in the Upper Mississippi Embayment region—A preliminary report, *in* McKeown, F. A., and Pakiser, L. C., eds., Investigations of the New Madrid, Missouri, earthquake region: U.S. Geological Survey Professional Paper 1236, p. 39–53.

Hook, J. W., 1974, Structure of the fault systems in the Illinois-Kentucky fluorspar district: Kentucky Geological Survey Special Publication 22, p. 77–86.

Houseknecht, D. W., 1989, Earliest Paleozoic stratigraphy and facies, Reelfoot basin and adjacent craton, *in* Gregg, J. M., Palmer, J. R., and Kurtz, V. E., eds., Field guide to the Upper Cambrian of southeastern Missouri: stratigraphy, sedimentology, and economic geology: Rolla, Missouri, University of Missouri, Department of Geology and Geophysics, p. 25–42.

Howe, J. R., 1985, Tectonics, sedimentation, and hydrocarbon potential of the Reelfoot Aulacogen [M.S. thesis]: Norman, Oklahoma, University of Oklahoma, 109 p.

Howe, J. R., and Thompson, T. L., 1984, Tectonics, sedimentation and hydrocarbon potential of the Reelfoot Rift: Oil and Gas Journal, November 12, p. 174–190.

Irving, E., 1977, Drift of the major continental blocks since the Devonian: Nature, v. 270, p. 304–307.

Kane, M. F., Hildenbrand, T. G., and Hendricks, J. D., 1981, A model for the tectonic evolution of the Mississippi Embayment and its contemporary seismicity: Geology, v. 9, p. 563–567.

Keller, G. K., Lidiak, E. G., Hinze, W. J., and Braile, L. W., 1983, The role of rifting in the tectonic development of the Midcontinent, U.S.A.: Tectonophysics, v. 94, p. 391–412.

Kehn, T. M., Beard, J. G., and Williamson, A. D., 1982, Mauzy Formation, a new stratigraphic unit of Permian age in western Kentucky: U.S. Geological Survey Bulletin 1529-H, p. H73–H86.

King, P. B., and Beikman, H. M., 1974, Geologic map of the United States: U.S. Geological Survey, scale 1:2,500,000.

Klein, G. deV., and Hsui, A. T., 1987, Origin of cratonic basins: Geology, v. 15, p. 1094–1098.

Kolata, D. R., 1991, Overview of sequences, *in* Leighton, M. W., Kolata, D. R., Oltz, D. F., and Eidel, J. J., eds., Interior cratonic basins: American Association of Petroleum Geologists Memoir 51, p. 59–73.

Kolata, D. R., and Nelson, W. J., 1991, Tectonic history of the Illinois Basin, *in* Leighton, M. W., Kolata, D. R., Oltz, D. F., and Eidel, J. J., eds., Interior cratonic basin: American Association of Petroleum Geologists Memoir 51, p. 263–292.

Lumm, D. K., Nelson, W. J., and Greb, S. F., 1991, Structure and chronology of part of the Pennyrile fault system, western Kentucky: Southeastern Geology, v. 32, no. 1, p. 43–59.

Marcher, M. V., and Stearns, R. G., 1962, Tuscaloosa Formation in Tennessee: Geological Society of America Bulletin, v. 73, p. 1365–1386.

McKeown, F. A., 1982, Overview and discussion, *in* McKeown, F. A., and Pakisker, L. C., eds., Investigations of the New Madrid, Missouri, earthquake region: U.S. Geological Survey Professional Paper 1236, p. 1–14.

Molnar, P., and Tapponier, P., 1975, Cenozoic tectonics of Asia: Effects of a continental collision: Science, v. 189, p. 419–426.

Mooney, W. D., Andrews, M. C., Ginzburg, A., Peters, D. A., and Hamilton, R. M., 1983, Crustal structure of the northern Mississippi Embayment and a comparison with other continental rift zones: Tectonophysics, v. 94, p. 327–348.

Nelson, W. J., 1991, Structural styles of the Illinois Basin, *in* Leighton, M. W., Kolata, D. R., Oltz, D. F., and Eidel, J. J., eds., Interior cratonic basin: American Association of Petroleum Geologists Memoir 51, p. 209–261.

Nelson, W. J., and Lumm, D. K., 1987, Structural geology of southeastern Illinois and vicinity: Illinois State Geological Survey Circular 538, 70 p.

Nelson, K. D., and Zhang, J., 1991, A COCORP deep reflection profile across the buried Reelfoot Rift, south-central United States: Tectonophysics, v. 197, p. 271–293.

Nelson, W. J., and 11 others, 1991, Geology of the Eddyville, Stonefort, and Creal Springs Quadrangles, southern Illinois: Illinois State Geological Survey Bulletin 96, 85 p.

Norby, R. D., Sargent, M. L., and Ethington, R. L., 1986, Conodonts from the Everton Dolomite (Middle Ordovician) in southern Illinois: Geological Society of America, Abstracts with Programs, v. 18, p. 258.

Oesterling, W. A., 1952, Geologic and economic significance of the Hutson zinc mine, Salem, Kentucky—its relation to the Illinois-Kentucky fluorspar district: Economic Geology, v. 47, p. 316–338.

Palmer, J. R., 1989, Late Upper Cambrian shelf depositional facies and history, southern Missouri, *in* Gregg, J. M., Palmer, J. R., and Kurtz, J. E., eds., Field guide to the Upper Cambrian of southeastern Missouri: stratigraphy, sedimentology, and economic geology: Rolla, Missouri, University of Missouri, Department of Geology and Geophysics, p. 1–24.

Perry, W. J., Jr., 1989, Tectonic evolution of the Anadarko basin region, Oklahoma: U.S. Geological Survey Bulletin 1866A, p. A1–A19.

Potter, C. J., Goldhaber, M. B., Heigold, P. C., and Taylor, C. D., 1992, Regional seismic reflection line, southern Illinois Basin, provides new data on Cambrian rift geometry, Hick Dome genesis, and the Fluorspar Area Fault Complex: Geological Society of America Abstracts with Programs, v. 24, no. 7, p. A279.

Rankin, D. W., and 9 others, 1989, Pre-orogenic terranes, *in* Hatcher, R. D., Jr., Thomas, W. A., and Viele, G. V., eds., The Appalachian-Ouachita orogen in the United States: Geological Society of America, The Geology of North America, v. F-2, p. 7–100.

Sbar, M. L., and Sykes, L. R., 1973, Contemporary compressive stress and seismicity in eastern North America: an example of intraplate tectonics: Geological Society of America Bulletin, v. 84, p. 1861–1882.

Schwalb, H. R., 1982, Geologic-tectonic history of the area surrounding the northern end of the Mississippi Embayment: University of Missouri at Rolla Journal, no. 5, p. 31–42.

Sloss, L. L., 1979, Plate-tectonic implications of the Pennsylvanian System in the Illinois Basin, *in* Palmer, J. E., and Dutcher, R. R., eds., Depositional and structural history of the Pennsylvanian System of the Illinois Basin, Part 2, invited papers: Illinois State Geological Survey Guidebook Series 15a, p. 107–112.

Snee, L. W., and Hayes, T. S., 1992, $^{40}Ar/^{39}Ar$ Geochronology of intrusive rocks and Mississippi Valley–type mineralization and alteration from the Illinois-Kentucky Fluorspar District, *in* Goldhaber, M. B., and Eidel, J. J., eds., Mineral resources of the Illinois Basin in the context of basin evolution: U.S. Geological Survey Open-File Report 92-1, p. 59.

Stauder, W., 1982, Present-day seismicity and identification of active faults in the New Madrid seismic zone, *in* McKeown, F. A., and Pakiser, L. C., eds., Investigations of the New Madrid, Missouri, earthquake region: U.S. Geological Survey Professional Paper 1236, p. 21–30.

Thomas, W. A., 1989, The Appalachian-Ouachita orogen beneath the Gulf Coastal Plain between the outcrops in the Appalachian and Ouachita Mountains, *in* Hatcher, R. D., Jr., Thomas, W. A., and Viele, G. V., eds., The Appalachian-Ouachita orogen in the United States: Geological Society of America, The Geology of North America, v. F-2, p. 537–561.

Thomas, W. A., 1991, The Appalachian-Ouachita rifted margin of southeastern North America: Geological Society of America Bulletin, v. 103, p. 415–431.

Treworgy, J. D., Sargent, M. L., and Kolata, D. R., 1989, Tectonic subsidence history of the Illinois Basin [abs.]: American Association of Petroleum Geologists Bulletin, v. 73, no. 8, p. 1040.

Weibel, C. P., Nelson, W. J., Oliver, L. B., and Esling, S. P., 1993, Geology of the Waltersburg Quadrangle, Pope County, Illinois: Illinois State Geological Survey Bulletin 98, 41 p.

Wood, G. D., and Stephenson, J. T., 1989, Cambrian palynomorphs from the warm-water provincial realm, Bonneterre and Davis Formations of Missouri and Arkansas (Reelfoot Rift area): Biostratigraphy, paleoecology, and thermal maturity, *in* Gregg, J. M., Palmer, J. R., and Kurtz, J. E., eds., Field guide to the Upper Cambrian of southeastern Missouri: Stratigraphy, sedimentology, and economic geology: Rolla, Missouri, University of Missouri, Department of Geology and Geophysics, p. 84-103.

Zartman, R. E., Brock, M. R., Heyl, A. V., and Thomas, H. H., 1967, K-Ar and Rb-Sr ages of some alkalic intrusive rocks from central and eastern United states: American Journal of Science, v. 265, no. 10, p. 848–870.

Zietz, I., King, E. R., Geddes, W., and Lidiak, E. G., 1966, Crustal study of a continental strip from the Atlantic Ocean to the Rocky Mountains: Geological Society of America Bulletin, v. 77, p. 1427–1448.

Zoback, M. L., 1992, State of stress in the Earth's crust: Geological Society of America Abstracts with Programs, v. 24, no. 7, p. A40.

Zoback, M. L., and Zoback, M., 1980, State of stress in the conterminous United States: Journal of Geophysical Research, v. 85, no. 11, p. 6113–6156.

MANUSCRIPT ACCEPTED BY THE SOCIETY JANUARY 16, 1996

Geological Society of America
Special Paper 312
1997

Intrusive style of A-type sheet granites in a rift environment: The Southern Oklahoma Aulacogen

John P. Hogan and M. Charles Gilbert
School of Geology and Geophysics, University of Oklahoma, Norman, Oklahoma 73019-0628

ABSTRACT

A-type granites of the Wichita Granite Group are an integral part of voluminous magmatism that accompanied formation of the Southern Oklahoma Aulacogen during Pre-Cambrian to Cambrian crustal rifting of the Laurentian Supercontinent. This extensional environment produced sheet granites of distinctive characteristics. They were emplaced at a very shallow level in the crust on a gabbroic substrate beneath a growing rhyolitic volcanic pile. These granites appear to have intruded first by passively lifting the overburden and then by active diking and stoping along their lateral margins. This resulted in intrusions characterized by sheet-forms of high aspect ratios ($\approx$40 to 100) that are distinct from shapes considered more typical for epizonal plutons (e.g., ring complexes or laccoliths). The level of emplacement and intrusive style of these A-type sheet granites were dominantly controlled by variation in local stress regimes and mechanical properties of the crust. To the ascending magmas, the unconformity between rhyolite and underlying units acted as a *crack stopper* preventing further rise of buoyant magma. In addition, the intrusive style of these sheet granites in part also reflects physical properties of the magma and therefore magma chemistry. A-type Wichita granites crystallized from high temperature (>900 °C), relatively dry (initial H_2O < 2.75 wt.%) magmas with elevated fluorine contents. These compositional features apparently lowered the viscosity of these melts enabling them to intrude nonexplosively into the subvolcanic environment and spread out laterally to form relatively well mixed sheetlike magma bodies.

INTRODUCTION

A-type granites of the Wichita Granite Group are an integral component of voluminous bimodal igneous activity that accompanied formation of the Southern Oklahoma Aulacogen (SOA) during Cambrian extension of the Laurentian Supercontinent. These granites were consistently emplaced at the base of a penecontemporaneous voluminous volcanic pile that overlies an older layered gabbroic complex associated with earlier stages of rifting (Ham et al., 1964). They were intruded as a series of thin ($\approx$0.5 km) but laterally extensive ($\approx$25 to 55 km in length) horizontal *sheets*. The sheet form is clearly evident from mapping and geophysical studies and is in marked contrast to more typical intrusive styles previously documented for epizonal granites (i.e., ring complexes, laccoliths). Interestingly, A-type sheet granites have also been reported from the Bushveld Complex, South Africa (Kleeman and Twist, 1989) and from the Robertson River Igneous Suite in the eastern United States (Tollo et al., 1991), and thus this intrusive style may be more prevalent then previously recognized.

Major element and, to a large extent, trace element compositions of Wichita granites are indistinguishable from associated rhyolites (Weaver and Gilbert, 1986). Why then did some magmas stall beneath the volcanic pile to form granite rather than

Hogan, J. P., and Gilbert, M. C., 1997, Intrusive style of A-type sheet granites in a rift environment: The Southern Oklahoma Aulacogen, *in* Ojakangas, R. W., Dickas, A. B., and Green, J. C., Middle Proterozoic to Cambrian Rifting, Central North America: Boulder, Colorado, Geological Society of America Special Paper 312.

rising to the surface to form rhyolite? What enabled such high-silica magmas, normally thought of as being extremely viscous, to spread out horizontally over large distances and crystallize as sheet granites rather than as ring complexes or laccoliths? The purpose of this paper is (1) to document the unusual intrusive style of these A-type sheet granites, and (2) to assess the relative role that mechanical and compositional effects may have had in determining the final granite morphology.

REGIONAL GEOLOGY

The SOA (Hoffman et al., 1974) is one of several aulacogens that formed in response to breakup of the Laurentian Supercontinent during Late Proterozoic to Cambrian time (Fig. 1). Subsequent to rifting, large interior basins developed over some of these aulacogens and in Oklahoma, igneous rocks of the rifting event were buried by up to 4 to 5 km of sediments (Johnson et al., 1988; Gilbert, 1992). However, during the late Paleozoic, igneous rocks of the SOA were uplifted as large fault-bounded blocks. These blocks define a portion of the Amarillo–Wichita Mountains–Criner (Arbuckle) structural axis that trends N60° to 70°W across the southern Midcontinent. The absence of significant Mesozoic and Cenozoic deformation in the Midcontinent ensured preservation of the rift assemblage (McConnell and Gilbert, 1990). Subsequently, erosion has exposed a cross section through volcanic and plutonic rocks that floored this rift. Thus, in

North America the SOA is one of the best-preserved and -exposed examples of igneous activity associated with an ancient rifting event.

Igneous activity associated with the SOA appears to be bimodal both in composition and in time (Table 1). Field evidence for coeval mafic and felsic magmatism is present only locally (Vidrine and Fernandez, 1986; Hogan, unpublished mapping). Layered mafic rocks of the Raggedy Mountains Gabbro Group represent the earliest igneous activity associated with rifting. On the basis of field and petrographic data this group is subdivided into two distinct suites, the older Glen Mountains Layered Complex (GMLC) and younger Roosevelt Gabbro Group (RGG) (Powell et al., 1980). Four whole rocks from the GMLC yielded a Rb-Sr isochron age of 577 ± 165 Ma and a three-point mineral whole rock Nd-Sm isochron age of 528 ± 29 Ma (Lambert et al., 1988). U-Pb studies of zircon from a member of the RGG have been interpreted to yield a crystallization age of 552 ± 7 Ma (Bowring and Hoppe, 1982). Layering in RGG is subparallel but discordant to layering in GMLC indicating an episode of faulting separated these two intrusive events (Gilbert, 1983; McConnell and Gilbert, 1990). A series of basalt spilites, the Navajoe Mountain Group, has been tentatively correlated with the gabbros (Ham et al., 1964). However, information about these basalts is scant, as they are only known from subsurface well cuttings, and their relationship to the rifting event remains enigmatic.

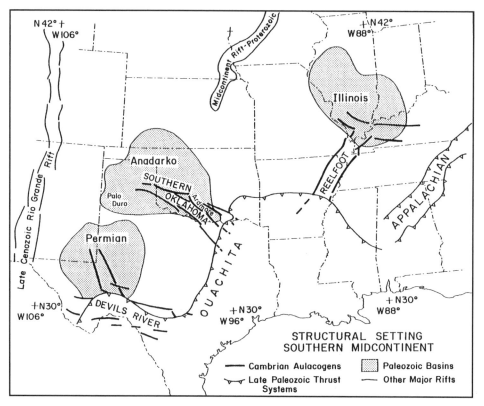

Figure 1. Aulacogens formed during Late Proterozoic to early Paleozoic breakup of the Laurentian Supercontinent. The Wichita Mountains, southwest Oklahoma, crop out where the letters OKLA are positioned (modified from McConnell and Gilbert, 1990).

TABLE 1. IGNEOUS STRATIGRAPHY OF THE WICHITA MOUNTAINS*

Radiometric Age (Ma)	Group	Constituent Units	Generalized Lithology
	Late diabase dikes		Fine-grained diabase of various ages.
	Rhyolite dikes		Porphyritic siliceous dikes with extremely fine-grained recrystallized goundmass.
514 ± 10(?)	Cold Springs Breccia		Basaltic enclaves intimately mixed in a highly variable matrix of quartz monzodiorite to granodiorite.
525 ± 25	Wichita Granite Group	Coarse-grained sheet granites	Coarse-grained (>1 cm) hypidomorphic leuco-alkali feldspar granites with minor amphibole biotite, magnetite, titanite, zircon, apatite, and fluorite.
525 ± 25	Wichita Granite Group	Fine-grained sheet granites	Fine-grained (<1 cm) porphyritic-xenomorphic, variably granophyric, leuco-alkali feldspar granite with amphibole, biotite, ± plagioclase, magnetite, titanite, zircon, apatite, and fluorite.
525 ± 25	Carlton Rhyolite Group		Rhyolitic lavas interbedded with minor tuffs and agglomerates.
552 ± 7(?)	Raggedy Mountains Gabbro Group	Roosevelt Gabbros	Medium- to fine-grained hornblende-biotite gabbro with ± olivine, ± opx.
577 ± 165	Raggedy Mountains Gabbro Group	Glen Mountains Layered Complex	Layered mafic complex consisting of anorthositic gabbro, anorthosite, troctolite, and olivine gabbro.
(?)	Navajoe Mountain Basalt-Spilite Group		Basalts, variably altered, only known from subsurface.

*See text for further discussion and sources of data.

Igneous activity associated with later stages of rifting appears to have been dominated by voluminous high-silica rhyolites and granites (Table 1). The Carlton Rhyolite Group consists of a sequence of thick subaerial flows, minor ignimbrites, nonmarine air-fall tuffs, and rare basalt flows deposited on an angular unconformity (Ham et al., 1964). An age of 525 ± 25 Ma was established for Carlton Rhyolite by U-Pb dating of zircon (Tilton et al., 1962). Ham et al. (1964) argue that nonmarine Carlton Rhyolite Group rested unconformably on marine basalts and spillites (Navajoe Mountains Group). Locally this erosional event was extensive enough to expose underlying gabbro. Thus, shortly after crystallization of the Raggedy Mountains Gabbro Group and prior to extrusion of most of the Carlton Rhyolite, 2 to 3 km of overburden were eroded (Ham et al., 1964; Powell and Phelps, 1977). Erosion was likely accentuated by normal faulting associated with uplift and extension of the brittle upper crust during rifting (McConnell and Gilbert, 1990).

A-type sheet granites of the Wichita Granite Group are intrusive into the base of Carlton Rhyolite. The granites occur as a series of thin (≈0.5 km) but laterally extensive (as much as 55 km in length) sheets. The sheet form is evident both from mapping and from geophysical data. The granites are high SiO_2 (≈71.0 to 77.6 wt.%) ± amphibole ± biotite ± magnetite alkali-feldspar leucogranites and are definitive examples of A-type granites intruded in a crustal rifting environment (Myers et al., 1981; Gilbert et al., 1990; Hogan et al., 1992). They are predominantly hypersolvus, metaluminous, have elevated high field strength element abundances (e.g., Zr > 500 ppm), contain magmatic fluorite, and have trace element abundances with a "within plate" signature (Weaver and Gilbert, 1986; Hogan et al., 1992). A U-Pb age of 525 ± 25 Ma on zircon from Quanah Granite permits a close temporal association of granite and rhyolite magmas (Tilton et al., 1962). The rhyolites and granites are inferred to be comagmatic based on similarity in age and in major element composition (Ham et al., 1964; Gilbert, 1982).

Rhyolitic and basaltic dikes represent the final stages of igneous activity associated with rifting. In the Wichita Mountains, rhyolite dikes are relatively rare and are known only to crosscut members of the Wichita Granite Group. Their presence demonstrates that felsic liquids continued to rise after emplacement and cooling of granite, but more importantly they signify that tectonism, resulting in uplift and erosion, continued during younger magmatic events (Gilbert and Myers, 1986). Basaltic dikes are considerably more numerous than rhyolite dikes and some crosscut all igneous units (Gilbert and Hughes, 1986). In the southeastern portion of the SOA (Arbuckle Mountains) more extensive erosion has presumably stripped away Cambrian granite and rhyolite from areas where they had been emplaced on Proterozoic basement. Here, porphyritic microgranite or "rhyolite-" as well as basaltic feeder-dikes crosscut exposed basement concordant to the structural grain of the rift (Taylor, 1915; Denison, 1982). These dikes indicate that throughout the aulacogen, rift-related faults and fractures are likely to have acted as conduits for transport of magma to high crustal levels.

FINE- AND COARSE-GRAINED GRANITE SHEETS

Introduction

Individual granite sheets are distinguished by differences in grain size, texture, mineralogy, and chemistry (Hogan and Gilbert, 1991). Granite sheets can also be separated into distinct classes based on chemical differences (Myers et al., 1981). In this paper, granite sheets are grouped into two broad textural categories: "fine-grained" (<1 cm) and "coarse-grained" (>1 cm) granite sheets.

Mount Scott, Cache, Lugert, and Headquarters Granites are typical examples of "fine-grained" granite sheets. They are medium (1 to 5 mm) to fine grained (<1 mm), equigranular to inequigranular porphyritic, and variably granophyric (Fig. 2). Alkali-feldspar and quartz phenocrysts, less than half a centimeter in size and commonly smaller, are set in a granophyric microgranitic groundmass. Although the modal abundance of granophyre varies significantly throughout individual sheets, there does not appear to be a corresponding change in major element chemistry (Myers et al., 1981).

Coarse-grained granites are less abundant in the Wichita Mountains and are represented by Reformatory and Quanah Granite sheets (Fig. 2). Reformatory Granite exhibits the coarsest grain size and is inequigranular seriate and nongranophyric. Alkali-feldspar crystals (1 to 2 cm) commonly exhibit macrozonation. The main facies of Quanah Granite is equigranular and non-granophyric. Alkali-feldspar crystals are approximately 1 cm in size.

Nearly all the granite sheets exhibit, to varying extent, finergrained border facies along their margins. For example, the majority of Mount Scott Granite (facies A) is a medium-grained

(2 to 4 mm) porphyritic amphibole ± biotite granite, with gray ovoid feldspar phenocrysts set in a variably granophyric groundmass. Facies A can be traced for over 55 km in length (Merritt, 1965; Myers et al., 1981). Facies B of Mount Scott Granite is an equigranular granophyric microgranite that crops out sporadically and appears to underlie facies A. Facies B may either be an earlier intrusion or a quenched margin of Mount Scott Granite sheet (Myers et al., 1981). The presence of "quenched" margins suggest these granites crystallized from high temperature crystal-poor melts.

Contact relationships

Introduction. At one time sheet granites of the Wichita Mountains were entirely capped by rhyolite. In subsurface, sheet granites as thick as 180 m were consistently encountered at the

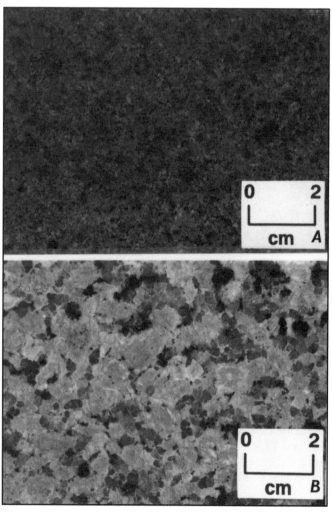

Figure 2. Polished slabs displaying textures typical of Wichita sheet granite: A, "fine-grained" (Mount Scott Sample SQ, Comanche County, Cooperton 7.5-minute Quadrangle; SW SE SE 4-3N-15W); and B, "coarse-grained" (Reformatory Sample JH-9-90 Greer County, Granite 7.5-minute Quadrangle; NE SW NE 22-6N-21W).

base of Carlton Rhyolite (Ham et al., 1964). At the surface, erosion has stripped away the majority of Carlton Rhyolite, and extensive rhyolite outcrops are found only on the northern (e.g., Bally Mountain) and southern (e.g., Fort Sill) flanks of the SOA (Fig. 3). Rhyolite xenoliths, as much as ≈50 m in length, can be locally abundant along lateral edges of granite sheets and are interpreted as stoped country rock (Schoonover, 1948; Green, 1952; Ham et al., 1964; Gilbert et al., 1990; Hogan and Gilbert, 1992). In between observed margins of the sheets, granite crops out on individual hills and small mountains isolated from one another by onlapping Permian sediments. Because of restricted exposure and scant geochronological data, spatial relationships of granite sheets are not entirely established. However, several critical exposures, combined with subsurface drilling, reveal consistent intrusive relationships between finer- and coarser-grained granite sheets and country rock. These intrusive relationships are discussed below.

Eastern Wichita Mountains. Contact relationships of Mount Scott Granite demonstrate the relatively thin, flat-lying and horizontal form of these sheet granites. Along the northeast slope of the eastern Wichita Mountains, the basal contact of Mount Scott Granite with underlying gabbro can be followed for tens of kilometers (Fig. 4). This contact dips gently southwest, and 5 km to the southwest, gabbro reappears in an erosional window through the sheet granite. This basal contact is characterized by a considerably more mafic Mount Scott Granite and the presence of hybrid rock types, some of which were described by Huang (1955). South of Mount Scott, in an erosional window, the Mount Scott Granite sheet can be seen to cut up section, and its floor (facies B) rests stratigraphically on Carlton Rhyolite (Schoonover, 1948; Gilbert, 1982). Here, facies A appears to gradually grade into facies B: the granite becomes finer grained (from ≈3 mm to ≈1 mm) and more granophyric towards the rhyolite-granite contact, but the color index remains

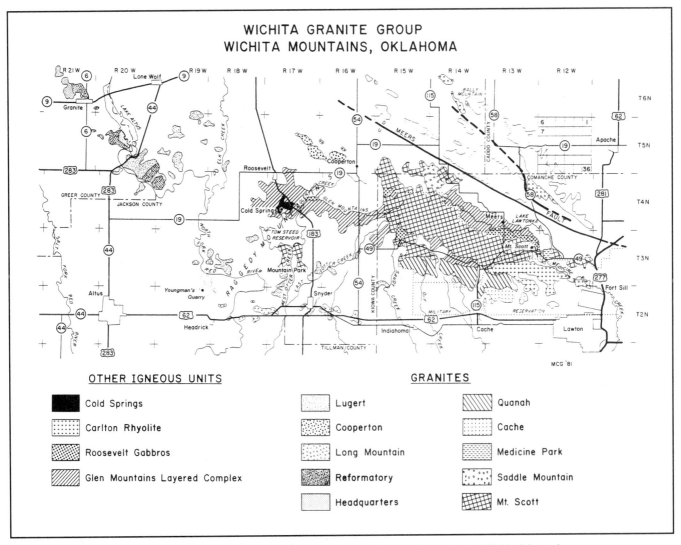

Figure 3. Generalized geologic map of major igneous units cropping out in the Wichita Mountains, southwestern Oklahoma.

Figure 4. View (looking southwest) of a portion of the northeastern slope of the Wichita Mountains. The basal contact of the Mount Scott sheet granite with the underlying gabbroic substrate is well defined, in a regional sense, by tree line. The flat-topped mountain to the east is Mount Scott, rising ≈343 m from the flat terrain, and the mountain to the west with the gently southeast dipping slope is Mount Sheridan.

described: rhyolite surrounded by Cache Granite has been treated as xenoliths (e.g., Schoonover, 1948; Green, 1952; Ham et al., 1964), but are more likely erosional windows (Gilbert, unpublished mapping).

Southwestern Wichita Mountains. In the southwestern portion of the Wichita Mountains, contact relationships between fine-grained granite and presumed Carlton Rhyolite are well displayed at Youngman Quarry in Jackson County, Oklahoma (Fig. 5). Here, on a quarry face approximately 30 to 40 m high, salmon red, medium- to fine-grained, enclave-rich, Lugert Granite intrudes a pink, very fine grained, highly granophyric igneous rock of problematical origin. This rock occurs as stoped blocks in Lugert Granite, and may represent an earlier quenched marginal facies of Lugert Granite. This unit also has been previously correlated with Long Mountain Granite (Myers et al., 1981). However, in thin section, this rock exhibits micrographic clusters with radial habits and a quasispherulitic appearance; a texture indicative of volcanic origin. Gilbert et al., (1990) suggested that these stoped blocks may have originally been glassy Carlton Rhyolite that was devitrified and recrystallized upon intrusion of Lugert Granite. If this is true, then the contact breccia in Youngman Quarry is likely to represent the southwest lateral edge of the Lugert Granite sheet.

Northwestern Wichita Mountains. In the northwestern Wichitas, exposed gabbro is rare to absent, and the floor of sheet granites is generally not exposed. However, presence of locally large gabbro masses, treated as xenoliths by Scull (1947) and Polk (1948), suggests that here also granite sheets rest on a gabbroic substrate and that this regional contact is not far below the present surface. This interpretation is supported by gravity data that indicate the presence of a large mafic mass close to the surface (Lyons and O'Hara, 1982; Zietz, 1982). Although the majority of the granite sheets appear to be floored by gabbro,

constant (Schoonover, 1948). Towards its northern border, in northerly dipping sections, Mount Scott Granite can be followed from areas with plutonic characteristics gradually into a spherulitic porphyry granite (i.e., Saddle Mountain Granite) with volcanic characteristics. Myers et al. (1981) interpret this unit as a portion of Mount Scott magma that breached the roof of the sheet. This is consistent with a subvolcanic emplacement of a relatively thin sheetlike magma chamber.

Coarse-grained Quanah Granite exhibits a quenched marginal facies along the contact with fine-grained Mount Scott and fine-grained Cache Granite sheets (Green, 1952; Gilbert, 1982). The contact between Quanah and Mount Scott Granites is an igneous breccia: angular blocks of Mount Scott Granite are surrounded by Quanah Granite (Gilbert, 1982). In contrast, apophyses of Quanah penetrating into Cache or xenoliths of Cache in Quanah Granite are rare along the Quanah-Cache contact. The Cache Granite–Carlton Rhyolite contact has not been well

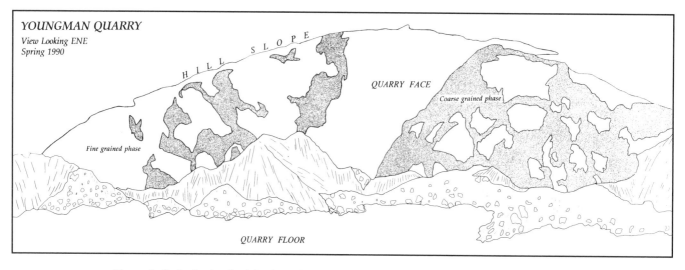

Figure 5. Geologic sketch of the face of Youngman Quarry (NE NE SW-1-2N-19W) showing the spatial relationship of Lugert sheet granite (patterned) and stoped blocks interpreted as recrystallized Carlton Rhyolite (unpatterned). Infiltrating contacts are characteristic of lateral margins of sheet granites. The quarry face is approximately 30 to 40 m high at its greatest dimension.

subsurface drilling indicates granite sheets may also be floored by either Navajoe Mountain basalts or Tillman metasediments (Ham et al., 1964).

Northwest of Granite, Oklahoma, contact relationships between rhyolite, fine-, and coarse-grained granite sheets are well exposed. Locally, coarse-grained Reformatory Granite becomes crowded with abundant xenoliths of fine-grained Headquarters Granite in some areas and Carlton Rhyolite in other areas (Hogan and Gilbert, 1992). A myriad of Reformatory Granite dikes infiltrate the Headquarters Granite sheet and its bounding rhyolite. Reformatory Granite and xenoliths are so intimately mixed that this zone had been mapped as a single unit of contact breccia (Taylor, 1915; Scull, 1947; Merritt, 1958). The contact between normal Reformatory Granite (where xenoliths of Headquarters Granite or Carlton Rhyolite are extremely rare) and contact breccia is sharp (Fig. 6). This contact locally strikes 0 to 030° and dips ≈45°SE. As this contact is approached, Reformatory Granite becomes finer grained, from an average of 10 to 20 mm down to 5 to 10 mm, and mineralogic layering becomes more evident (Scull, 1947; Hogan, 1992, unpublished mapping). The layering is defined by an increase in amphibole, biotite, magnetite, and accessory minerals with an attitude (≈357°, 45°E) that approximately parallels the contact.

Intrusion of Reformatory Granite complicated, but did not entirely obscure, intrusive relationships between fine-grained Headquarters Granite and Carlton Rhyolite. Xenoliths of rhyolite consistently occur along the eastern side of the contact breccia whereas xenoliths of Headquarters Granite occur on the western half. Definitive contact relationships between Head-

Figure 6. View (looking north) of the contact between "normal" Reformatory Granite, which comprises the relatively smooth sloping hill to the east, and the Reformatory contact breccia to the west (Granite 7.5-minute Quadrangle; SW NE NE 27-6N-21W). Here the contact breccia consists of stoped blocks of Carlton Rhyolite (rough or rubbley blocks) in Reformatory Granite (smoother appearing surfaces). Mineralogic layering (not to be confused with shadows from the guyed-wires) can be seen in the normal Reformatory Granite (i.e., just above the old washing machine in the lower right-hand corner of the photograph).

quarters Granite and Carlton Rhyolite in the xenoliths have not been observed. However, Headquarters Granite exhibits a sharp transition from equigranular granite to porphyritic granite and gradually becomes sparsely porphyritic and extremely fine grained and variably granophyric as the cryptic contact with rhyolite is approached. Similarly, rhyolite xenoliths in proximity to the cryptic contact are sparsely porphyritic with alkali feldspar as the sole phenocryst phase. With increasing distance from the contact, phenocrysts of alkali feldspar, then quartz become more abundant in porphyritic rhyolite xenoliths. The contact between the fine-grained granite sheet and its rhyolite cover does not appear to be marked by diking or stoping.

Conditions of emplacement

Pressure. Values for intensive parameters during crystallization of Wichita granites can be estimated from stratigraphic and geochemical data. Crystallization ages and field relationships for Carlton Rhyolite and Wichita granites suggest emplacement of sheet granites occurred shortly after build up of the rhyolite pile began. Ham et al. (1964) reported a maximum measured thickness of ≈1.4 km for Carlton Rhyolite. This overburden thickness corresponds to a lithostatic load of ≈500 bars for granite magmas crystallizing at the base of a developing rhyolite pile and probably represents a median value. An unknown amount of rhyolite may have been removed by erosion requiring higher pressures, but some granites may have been emplaced under less cover requiring lower pressures.

Comparison of normative compositions of granitic rocks with experimentally determined liquidus boundary curves, minima, and eutectics in the primary system SiO_2-$NaAlSi_3O_8$-$KAlSi_3O_8$ (Q-Ab-Or) can suggest partial melting and crystallization conditions. Such comparisons must be viewed with caution because (1) granite compositions do not necessarily represent melt compositions; and (2) the position of boundary curves, minima, and eutectic compositions are displaced by the presence of additional components (e.g., anorthite, fluorine). However, low anorthite contents (Mount Scott An_{4-6}, but most other granites are less than An_2) and leucocratic compositions (Σ of normative Qz, Ab, Or commonly is >95%) of Wichita granites and Carlton Rhyolite facilitate rather direct comparisons of their chemical compositions with experimentally determined phase relationships in this system.

With these concerns in mind, Figure 7 is a plot of normative compositions of Carlton Rhyolite (including Saddle Mountain Granite), fine-grained granites (Mount Scott, Headquarters, and Cache Granites), and coarse-grained granite (Reformatory and Quanah Granites) projected onto the Qz-Ab-Or ternary. Samples of Carlton Rhyolite and fine- and coarse-grained granites define elongated, overlapping fields, of approximately constant Or composition. These fields can be viewed as extending from the 5.0 kb $aH_2O = 0.40$ liquidus minimum to beyond the 0.5 kb $aH_2O = 1.0$ minimum. The trend of these fields, from high pressure H_2O-undersaturated minima to low pressure H_2O-saturated minima, may in part reflect feldspar accumulation. Evidence for

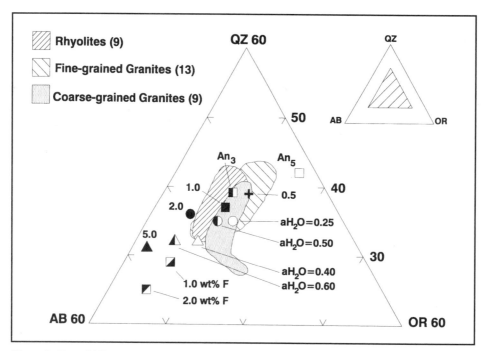

Figure 7. Cross-Iddings-Parsons-Washington (CIPW) normative compositions of minima, eutectics, and whole-rock compositions of fine- and coarse-grained Wichita granites and Carlton Rhyolite in the system SiO_2-$NaAlSi_3O_8$-$KAlSi_3O_8$-H_2O. Minimum at 0.5 kb aH_2O = 1.0 shown as a cross. Shown in squares are positions for 1 kb minima at aH_2O = 1 (solid square), with addition of 1.0 and 2.0 wt.% F (half-filled lower and upper corners respectively) or addition of normative $CaAl_2Si_2O_8$ (An_3 half-filled square, An_5 empty square). Shown in circles are positions for 2.0-kb minima at aH_2O = 1 (solid circle), aH_2O = 0.5 (half-filled circle), aH_2O = 0.25 (empty circle). Shown in triangles are positions for 5.0 kb eutectic at aH_2O = 1 (solid triangle), and the 5.0 kb minima at aH_2O = 0.6 (half-filled triangle), aH_2O = 0.40 (empty triangle). Sources of data for experimentally determined minima and eutectic are from Tuttle and Bowen (1958), James and Hamilton (1969), Manning (1981), and Holtz et al. (1992). Whole rock analyses are from Myers et al. (1981), Gilbert and Myers (1986), Gilbert et al. (1990), and Gilbert and Hogan (unpublished data).

high-level fractionation is present in the Reformatory Granite sheet (Hogan and Gilbert, 1992). Samples of differentiated Reformatory Granite plot closer to lower pressure minima. However, samples of undifferentiated Mount Scott Granite cluster tightly around the 2.0 kb aH_2O = 0.50 minima (which also approximately coincides with the 4.0-kb dry liquidus boundary curve of Steiner et al., 1975). McConnell and Gilbert (1990) interpreted the position of the Mount Scott Granite composition on the 4.0-kb dry boundary curve of Steiner et al. (1975) to be the pressure at which the bulk Mount Scott Granite magma formed by partial melting (a depth of ≈12 to 15 km).

Variation in normative Qz-Ab-Or compositions may reflect partial melting under relatively dry conditions at depth followed by subsequent evolution to saturated conditions under very low pressures at the subvolcanic level of emplacement. This scenario is consistent with estimates of ≤2.75 wt.% and ≤2.0 wt.% for initial H_2O contents of Mount Scott and Quanah Granites, respectively (Gilbert et al., 1990), and the localized presence of miarolitic cavities in the Wichita granites. In contrast, Holtz et al. (1992) interpreted a similar trend for A-type sheet granites from the Bushveld complex to represent protracted evolution of

relatively "dry" magmas with decreasing pressure at a low and constant aH_2O. Alternatively, Manning (1981) demonstrated that 1.0 kb H_2O-saturated minima are displaced, at relatively constant Or composition, towards the Ab-Or join with increasing fluorine content of the melt. The presence of magmatic fluorite and of high fluorine contents of mafic silicates from Wichita granites suggest these magmas did crystallize under the influence of fluorine. Thus, in addition to other effects, the fluorine contents of these melts may have influenced the final composition of these granites.

Temperature. Retrieval of crystallization temperatures from mineral equilibria has been difficult due to subsolidus reequilibration. However, Zr contents of granitic rocks can be used as a chemical geothermometer (Harrison and Watson, 1983). Crystallization temperatures calculated from Zr concentrations of rhyolite, and fine-, and coarse-grained granites range from approximately 800 to 950 °C (Table 2). This geothermometer is sensitive to the abundance of restite zircon and zircon fractionation. In many Wichita granites, zircon crystals with distinct cores are rare. The scarcity of cores and limited evidence of concordancy from the U-Pb age of zircons from the Quanah Granite

TABLE 2. Zr GEOTHERMOMETRY OF SELECTED FELSIC IGNEOUS ROCKS, SOUTHERN OKLAHOMA AULACOGEN*

Lithodemic Unit	Samples	M[†]	Zr (ppm)	Tzr[§] (°C)
RHYOLITES				
Carlton Rhyolite	7	1.3 ± 0.08	680 ± 44	942 ± 12
Saddle Mountain	3	1.4 ± 0.09	584 ± 9	914 ± 7
FINE-GRAINED GRANITE SHEETS				
Mount Scott	6	1.4 ± 0.06	507 ± 33	893 ± 11
Cache	1	1.3	281	850
Lugert	15	1.4 ± 0.15	384 ± 147	864 ± 32
Headquarters	6	1.3 ± 0.04	188 ± 25	806 ± 12
COARSE-GRAINED GRANITE SHEETS				
Quanah	2	1.3 ± 0.04	494 ± 70	904 ± 18
Reformatory	11	1.4 ± 0.08	462 ± 188	881 ± 32

*Sources of data: Weaver and Gilbert, 1986; Gilbert, Myers, and Hogan, unpublished data.
[†]M = (Na + K + 2Ca)/(Si • Al) calculated from whole rock analyses.
[§]Tzr calculated from Watson and Harrison (1983).

(Tilton et al., 1962) suggest the majority of zircons in these granites are melt precipitated. However, variation in Zr abundance (commonly several tens of ppm) in these granites can be attributed to zircon fractionation. Mechanical segregation of zircon by gravitational separation may explain lower calculated temperatures for more differentiated granites (Hogan et al., 1993). In contrast, rare zircon crystals are observed in thin sections of aphanitic rhyolites, suggesting Zr abundances of these rocks have not been altered by fractionation. Significantly higher temperatures calculated for rhyolites (≈950 °C) may more closely approximate initial liquidus temperatures for felsic melts of the SOA. Intrusion or eruption of these melts at very high temperatures is consistent with 1.0-kb experiments of Clemens et al. (1986) that indicate liquidus temperatures for A-type granites may exceed 975 °C at low water contents (Gilbert et al., 1990).

A MODEL FOR EMPLACEMENT OF RIFT-RELATED A-TYPE SHEET GRANITES

Felsic igneous activity associated with formation of the SOA was dominated by high temperature, relatively dry, A-type magmas. Throughout the Wichita Mountains, field relationships demonstrate that this magmatism progressed through a consistent sequence of events. Initially, felsic magmas erupted onto the floor of the rift to form an extensive volcanic pile (e.g., Carlton Rhyolite). However, at some point in time, a portion of felsic magma failed to breach the surface and ponded beneath the volcanic pile. Here, felsic magma spread out laterally, over large distances, to first form fine-grained sheet granites (e.g., Mount Scott, Headquarters) and then subsequently to form coarse-grained sheet granites (e.g., Reformatory, Quanah). The brittle style of intrusion, sharp contacts, and progressive coars-

ening of crystalline products suggest each event represents a distinct magmatic pulse in the rift history, although the temporal extent of hiatuses separating pulses of magmatic activity may not be great.

The rise of magma through the crust is greatly facilitated by fractures (Turcotte, 1982; Hutton, 1992). Within the SOA, magma transport toward the surface was likely to have occurred along linear systems of faults or fractures related to the rift. Weertman (1980) demonstrated that crustal fractures filled with a critical volume of liquid, less dense than the surrounding rock, will self propagate until they reach the Earth's surface. With this in mind, why did felsic magma pond beneath the volcanic pile and crystallize as Wichita granite rather than continuing to spill out onto the rift floor as Carlton Rhyolite?

Final emplacement level is known to be a function of temperature and volatile content, specifically H_2O, of the magma (Cann, 1970; Hyndman, 1981). Low-temperature, "wet" granitic magmas are unlikely to migrate far from their source before intersecting the wet granite solidus and freezing in place. In contrast, high-temperature, relatively "dry" granitic magmas have the potential to intrude to shallow levels in the crust before completely crystallizing. The common occurrence of sparsely porphyritic fine-grained marginal facies indicates many Wichita granites intruded as crystal-poor melts at or near their liquidus temperatures. Even at the low pressures of emplacement, vapor saturation and loss of volatiles from these melts did not occur until after considerable crystallization (Gilbert et al., 1990).

Intrinsically low H_2O contents of felsic magmas from the SOA may explain the absence of recognized collapsed calderas or ring complexes typical of hypabyssal A-type granitic intrusions in many other anorogenic environments (e.g., Oslo Rift, Norway). Ring complexes form when the energy released during rapid volume expansion of the magma, as a result of vapor saturation, exceeds the shear strength of the country rock and forms conical fractures above the intrusion that can be exploited by rising magma (Anderson, 1936; Billings, 1945; Chapman, 1976; Ike, 1983). Magma with higher H_2O contents may undergo more vigorous retrograde boiling, which triggers violent eruptions resulting in evacuation of magma from the underlying chamber to produce a collapsed caldera (Smith et al., 1961). The high temperatures and low H_2O contents of both Carlton Rhyolite and Wichita granites make it unlikely that magmas yielding granite rather than rhyolite did so because they quenched during vapor saturation at a higher confining pressure.

Emplacement of Wichita granites at or near the base of Carlton Rhyolite more likely reflects a mechanical anisotropy in the crustal column. Localization of hypabyssal intrusions along horizontal planes of weakness (e.g., joints, faults, foliation, bedding, unconformities) has long been recognized (Russell, 1896; Chamberlain and Link, 1927; Mudge, 1968). In the Wichita Mountains, the unconformity separating Carlton Rhyolite from underlying gabbro or metasedimentary rocks represents a crustal anisotropy that may have been further enhanced by the presence of a saprolitic soil horizon developed on the underlying gabbro substrate (Gilbert and Myers, 1986). After crystallization of Carlton Rhyolite, many of the rift-related faults may not have extended upward into the volcanic pile but could have terminated at the unconformity. This unconformity would have acted as a "crack-stopper," a surface that will terminate the upward motion of self-propagating liquid-filled cracks (Weertman, 1980). Thus, the upward migration of felsic magmas along rift-related faults was terminated at the unconformity. Here, rising magma spread out laterally beneath the Carlton Rhyolite to form sheetlike granite bodies. Accordingly, the shift from extrusive to intrusive igneous activity in the SOA may be the interplay between a crustal anisotropy and crystallization temperature and not due to temperature alone or to some compositional contrast (e.g., H_2O content) between magmas that formed rhyolites and those that formed granites. Felsic magmas did continue to reach the surface, possibly along rift-related faults that postdated formation of the major volcanic pile and sheet granites, as indicated by rare "rhyolite-dikes" observed crosscutting granite.

Sheetlike magma chambers can readily propagate, at very low positive magma pressures, by splitting the host rock along a preexisting discontinuity and passively lifting the roof (Chamberlain and Link, 1927; Anderson, 1951; Pollard, 1973; Johnson and Pollard, 1973; Pollard and Johnson, 1973; Weertman, 1980; Dixon and Simpson, 1987). As the magma chamber expands, tensional fractures develop in the surrounding host rock as a result of stress gradients produced by the intrusion (Pollard and Johnson, 1973). At the intrusion terminus, tensional fractures open toward the magma chamber and are commonly occupied by peripheral dikes (Jackson and Pollard, 1988). In contrast, tensional fractures that form over the central portion of intrusions close toward the magma chamber. This spatial variation in stress may explain the dichotomy in intrusive style exhibited by sheet granites: the common occurrence of stoped blocks, extensive diking, and contact breccias along lateral margins of sheet granites and an apparent absence of such features in central portions of these intrusions.

Studies of high-level intrusions suggest that sheet forms represent a limited initial stage of growth prior to formation of a laccolith. Laccoliths are concordant intrusions, approximately circular in plan, with flat floors, that have domed up the country rocks over their centers during emplacement. Since the studies of G. K. Gilbert (1877) and C. B. Hunt (1953), the formation of laccoliths has received considerable attention (Johnson and Pollard, 1973; Pollard and Johnson, 1973; Dixon and Simpson, 1987; Jackson and Pollard, 1988). The initial stage of laccolith formation begins as a thin concordant sill of circular plan. After reaching a critical radius, the central portion of the sill thickens by passively lifting the overburden and the intrusion takes on the characteristic domed laccolith form. The point at which lateral growth of the intrusion rapidly diminishes and the sill begins to dramatically thicken into a laccolith can be shown to take place at a radius of $2T_E$, where T_E is the effective mechanical thickness of the overburden (Pollard and Johnson, 1973).

In the Wichita Mountains, maximum measured thickness for Carlton Rhyolite suggest sheet granites crystallized at a depth of approximately 1.5 km. The presence of slip surfaces (e.g., bedding-plane faults, etc.) can reduce T_E to be less than the total overburden thickness. However, if Carlton Rhyolite is treated as a single massive unit, then, theoretically, granitic magmas could have spread laterally to a maximum diameter of 6 km before undergoing significant thickening to form laccoliths. This estimate is in marked contrast with lateral dimensions of the largest individual, homogeneous, sheet granite (Mount Scott) whose extent is at least 55 km in length and 17 km in width, while probably not much exceeding a maximum thickness of 0.5 km (Fig. 8).

If the laccolith model is applicable to sheet granites, the discrepancy in lateral extent of these intrusions implies sheet granites may form from coalescing magmas fed by multiple point feeders or by fissures. Sheet granites from the Bushveld Complex, South Africa (Kleemann and Twist, 1989), the Kinnaird pluton, Canada (Halwas and Simony, 1992), and the Robertson River Igneous Suite, United States (Tollo et al., 1991), are composite intrusions, presumably reflecting multiple-feeder sources.

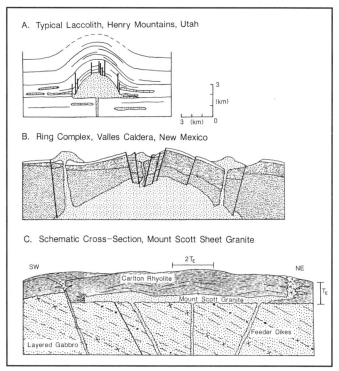

Figure 8. Comparison of shapes typical for: A. laccoliths, Henry Mountains, Utah (from Jackson and Pollard, 1988), B. ring complexes, Valles Caldera, New Mexico (from Smith et al., 1961), and C. sheet-granite intrusions, Mount Scott sheet granite, Oklahoma. Cross-sections are at the same scale, no vertical exaggeration. The high aspect ratio of the Mount Scott sheet granite, even along its width, is readily apparent in comparison to these other epizonal intrusions (length of the Mount Scott sheet granite is at least 55 km). Mount Scott magma rose along feeder dikes that exploited rift-related faults which predated formation of the penecontemporaneous rhyolite pile.

Subtle, internal contacts have been recognized in several Wichita sheet granites. However, the lateral extent of these contacts and whether they represent separate magmatic pulses or internal differentiation of a single batch of magma remains to be determined. On the other hand, a more likely scenario is that the absence of laccoliths in the SOA reflects the geometry of a feeder-dike system rather than point sources. Results of both physical and numerical experiments for laccolith formation are valid for radial intrusions, circular in plan, fed by a central pipe with a circular cross section (e.g., Pollard and Johnson, 1973; Dixon and Simpson, 1987). In contrast, hypabyssal magma chambers fed from an elongate dike will be elliptical in plan with the long axis parallel to the trend of the dike (Pollard and Johnson, 1973). Magma will spread out evenly in both directions above feeder dikes intersecting normal to the "crack stopping" surface and only to one side of dikes that intersect at an acute angle (Pollard, 1973). En echelon dike feeder systems could readily yield wider sheets in some regions than a simple $2T_E$ calculation would indicate. Thus, Wichita granite intrusions could have maintained sheet-granite geometries without thickening into laccoliths if magma chambers were fed from multiple feeder dikes exploiting a system of rift-related faults and fractures, structures that would have been common in this rift environment.

CONCLUSIONS

Emplacement of A-type sheet granites in the SOA was strongly determined by mechanical properties of the country rock. These magmas were principally emplaced along the same stratigraphic horizon: an erosional surface developed on rocks of the Raggedy Mountain Gabbro Group and Tillman metasediments. This surface acted as a "crack-stopper," a horizontal plane of weakness that prevented further rise of low-density liquids along self-propagating crustal fractures. With emplacement of each additional batch of magma, this stratigraphic horizon was buried by a greater thickness of overburden. Thus, felsic magmatism progressed through a distinct cycle, beginning with eruption of voluminous rhyolites, followed by emplacement of fine-grained, laterally extensive granite sheets, then by intrusion of less abundant, coarse-grained granite sheets. Coarsening of crystals may reflect slower cooling rates for younger magmas undergoing crystallization in more insulated surroundings, possibly still warm from previous intrusions (Gilbert and Myers, 1986). Magma transport along large-scale linear fractures or faults related to rifting enabled magma chambers to maintain high-aspect ratios and sheetlike forms rather than doming into laccoliths. The dichotomy in intrusive style observed within individual sheet granites (i.e., passive uplifting of the roof over central portions of intrusions to abundant diking and stoping along lateral margins) can be attributed to variation in local stress regimes surrounding the intrusion. Therefore, the shift from extrusive to intrusive igneous activity in the SOA is predominantly the result of a crustal anistropy rather than a fundamental physical (e.g., temperature) or compositional difference

(e.g., H_2O contents) between magmas that formed rhyolites and those that formed granites.

At the same time, emplacement characteristics were also influenced by physical properties of the magma and therefore, in part, by magma composition. Intrinsically high liquidus temperatures and low H_2O contents enabled these magmas to intrude to very shallow levels without undergoing appreciable crystallization or violent exsolution of volatiles in an explosive eruption. The presence of magmatic fluorite and relatively high fluorine contents of mafic silicates suggest fluorine was an important volatile component in addition to H_2O. Fluorine, like H_2O, dramatically lowers solidus temperatures and viscosities of granitic magmas (Manning, 1981; Dingwell et al., 1985). At a temperature of 1,000 °C, the addition of 1.0 wt.% F will lower the viscosity of a dry, high SiO_2 granitic melt one order of magnitude lower than an equivalent fluorine-free melt (Dingwell et al., 1985). However unlike water, fluorine has a high melt/fluid partition coefficient (Hards, 1978) and at low confining pressures should be retained in the melt. Decreased viscosities and solidus temperatures may have enhanced fluid flow of these magmas into extensive sheet forms.

ACKNOWLEDGMENTS

This investigation was funded in part by NSF grant #EAR-9316144 to J. P. Hogan and M. C. Gilbert and by Eberly Chair Funds available to M. C. Gilbert. This manuscript benefited from thoughtful reviews by R. E. Denison, R. P. Tollo, and M. B. Wolf.

REFERENCES CITED

Anderson, E. M., 1936, The dynamics of the formation of cone-sheets, ringdykes, and cauldron-subsidences: Proceedings, Royal Society of Edinburgh, v. 56, p. 128–157.

Anderson, E. M., 1951, The dynamics of faulting and dyke formation with applications to Britain: Edinburgh and London, Oliver and Boyd, 206 p.

Billings, M. P., 1945, Mechanics of igneous intrusion in New Hampshire: American Journal of Science, v. 243, p. 40–68.

Bowring, S. A. and Hoppe, W. J., 1982, U-Pb ages from the Mount Sheridan gabbro, Wichita Mountains: *in* Gilbert, M. C., and Donovan, R. N., eds., Geology of the eastern Wichita Mountains, southwestern Oklahoma: Oklahoma Geological Survey Guidebook 21, p. 54–59.

Cann, J. R., 1970, Upward movement of granitic magma: Geological Magazine, v. 107, p. 335–340.

Chamberlin, R. T., and Link, T. A., 1927, The theory of laterally spreading batholiths: Journal of Geology, v. 35, no. 4, p. 319–357.

Chapman, C. A., 1976, Structural evolution of the White Mountains magma series: Geological Society of America Memoir 146, p. 281–300.

Clemens, J. D., Holloway, J. R., and White, A. J. R., 1986, Origin of A-type granite: Experimental constraints: American Mineralogist, v. 71, p. 317–324.

Denison, R. E., 1982, Geologic cross section from the Arbuckle Mountains to the Muenster Arch, southern Oklahoma and Texas: Geological Society of America, Map and Chart Series MC-28R, 8 p., scale 1:250,000.

Dingwell, D. B., Scarfe, C. M., and Cronin, D. J., 1985, The effect of fluorine on viscosities in the system Na_2O-Al_2O_3-SiO_2: implications for phonolites, trachytes and rhyolites: American Mineralogist, v. 70, p. 80–87.

Dixon, J. M., and Simpson, D. G., 1987, Centrifuge modelling of laccolith intrusion: Journal of Structural Geology, v. 9, p. 87–103.

Gilbert, G. K., 1877, Report on the Geology of the Henry Mountains: U.S. Geographical and Geological Survey, Rocky Mountains Region, 160 p.

Gilbert, M. C., 1982, Geologic setting of the eastern Wichita Mountains, with a brief discussion of unresolved problems, *in* Gilbert, M. C., and Donovan, R. N., eds., Geology of the eastern Wichita Mountains, southwestern Oklahoma: Oklahoma Geological Survey Guidebook 21, p. 1–30.

Gilbert, M. C., 1983, Timing and chemistry of igneous events associated with the Southern Oklahoma Aulacogen: Tectonophysics, v. 94, p. 439–455.

Gilbert, M. C., 1992, Speculations on the origin of the Anadarko basin, *in* Mason, R., ed., Proceedings, International Conference on Basement Tectonics, 7th, Kingston Ontario: Kluwer Academic, p. 195–208.

Gilbert, M. C., and Hughes, S. S., 1986, Partial chemical characterization of Cambrian basaltic liquids of the Southern Oklahoma Aulacogen, *in* Gilbert, M. C., ed., Petrology of the Cambrian Wichita Mountains Igneous Suite: Oklahoma Geologic Survey Guidebook 23, p. 73–79.

Gilbert, M. C., and Myers, J. D., 1986, Overview of the Wichita Granite Group, *in* Gilbert, M. C., ed., Petrology of the Cambrian Wichita Mountains Igneous Suite: Oklahoma Geologic Survey Guidebook 23, p. 107–116.

Gilbert, M. C., Hogan, J. P., and Myers, J. D., 1990, Geology, geochemistry and structure of low-pressure sheet granites, Wichita Mountains, Oklahoma, *in* Guidebook for Field Trip no. 5: Geological Society of America, Annual meeting, Dallas Geological Society, 65 p.

Green, J. H., 1952, Igneous rocks of the Craterville area, Wichita Mountains, Oklahoma [M.S. thesis]: University of Oklahoma, 63 p.

Halwas, D., and Simony, P., 1992, The Kinnaird pluton: A multiple granitic sheet: Geological Association of Canada and Mineralogical Association of Canada, Joint Meeting, v. 17, p. A45.

Ham, W. E., Denison, R. E., and Merritt, C. A., 1964, Basement rocks and structural evolution of southern Oklahoma: Oklahoma Geological Survey Bulletin, v. 95, 302 p.

Hards, N., 1978, Distribution of elements between the fluid phase and silicate melt phase of granites and nepheline syenites: NERC Progress in Experimental Petrology IV, p. 88–90.

Harrison, T. M., and Watson, E. B., 1983, Kinetics of zircon dissolution and zirconium diffusion in granitic melts of variable water contents: Contributions to Mineralogy and Petrology, v. 84, p. 66–72.

Hoffman, P., Dewey, J. F., and Burke, K., 1974, Aulacogens and their genetic relationship to geosynclines, with a Proterozoic example from Great Slave Lake, Canada, *in* Dott, R. H., Jr., and Shaver, R. H., eds., Modern and ancient geosynclinal sedimentation: Society of Economic Paleontologists and Mineralogists Special Publication 19, p. 38–55.

Hogan, J. P., and Gilbert, M. C., 1991, Contrasting crystallization styles of sheet granites from the Wichita Magmatic Province: Implications for source region heterogeneity: Geological Society of America Abstracts with Programs, v. 23, p. 33.

Hogan, J. P., and Gilbert, M. C., 1992, Reversely-zoned A-type sheet granites of the Southern Oklahoma Aulacogen, U.S.A.: Eos (Transactions, American Geophysical Union), v. 73, p. 354–355.

Hogan, J. P., Gilbert, M. C., Weaver, B. L., and Myers, J. D., 1992, High level A-type sheet granites of the Wichita Mountains Igneous Province, Oklahoma, U.S.A., *in* Brown, P. E., and Chappell, B. W., eds., Second Hutton Symposium, The Origin of Granites and Related Rocks: Royal Society of Edinburgh, Transactions, v. 83, p. 491–492.

Hogan, J. P., Dewers, T., and Gilbert, M. C., 1993, Minor/accessory mineral segregations in the Reformatory granite: Geological Society of America Abstracts with Programs, v. 25, no. 1, p. 14.

Holtz, F., Pichavant, M., Barbey, P., and Johannes, W., 1992, Effects of H_2O on liquidus phase relations in the haplogranite system at 2 and 5 kbar: American Mineralogist, v. 77, p. 1223–1241.

Huang, W. T., 1955, Occurrences of leucogranogabbro and associated igneous rocks in the Wichita Mountains, Oklahoma: American Journal of Science, v. 253, p. 341–357.

Hunt, C. B., 1953, Geology and geography of the Henry Mountains region, Utah: U.S. Geological Survey Professional Paper 228, 234 p.

Hutton, D. H. W., 1992, Granite sheeted complexes: evidence for the dyking ascent mechanism, *in* Brown, P. E., and Chappell, B. W., eds., Second Hutton Symposium, The Origin of Granites and Related Rocks: Royal Society of Edinburgh, Transactions, v. 83, p. 377–382.

Hyndman, D. W., 1981, Controls on source and depth of emplacement of granitic magma: Geology, v. 9, p. 244–249.

Ike, E. C., 1983, The structural evolution of the Tibchi ring-complex: a case study of the Nigerian Younger Granite Province: Journal of the Geological Society of London, v. 140, p. 781–788.

Jackson, M. D., and Pollard, D. D., 1988, The laccolith-stock controversy: New results from the Henry Mountains, Utah: Geological Society of America, v. 100, p. 117–139.

James, R. S., and Hamilton, D. L., 1969, Phase relations in the system $NaAlSi_3O_8-KAlSi_3O_8-CaAl_2Si_2O_8-SiO_2$ at 1 kilobar water pressure: Contributions to Mineralogy and Petrology, v. 21, p. 111–141.

Johnson, A. M., and Pollard, D. D., 1973, Mechanics of growth of some laccolith intrusions in the Henry Mountains, Utah, I: Tectonophysics, v. 18, p. 261–309.

Johnson, K. S., Amsden, T. W., Denison, R. E., Dutton, S. P., Goldstein, A. G., Rascoe, B., Jr., Sutherland, P. K., and Thompson, D. M., 1988, Southern Midcontinent region, *in* Sloss, L. L., ed., Sedimentary cover—North American craton; U.S.: Boulder, Colorado, Geological Society of America, DNAG, v. D-2, p. 307–359.

Kleemann, G. J., and Twist, D., 1989, The compositionally-zoned sheetlike granite pluton of the Bushveld Complex: Evidence bearing on the nature of A-type magmatism: Journal of Petrology, v. 30, p. 1383–1414.

Lambert, D. D., Unruh, D. M., and Gilbert, M. C., 1988, Rb-Sr and Sm-Nd isotopic study of the Glen Mountains layered complex: Initiation of rifting within the Southern Oklahoma Aulacogen: Geology, v. 16, p. 13–17.

Lyons, P. L., and O'Hara, N. W., 1982, Gravity anomaly map of the United States: Tulsa, Oklahoma, Society of Exploration Geophysicists, scale 1:2,500,000.

Manning, D. A. C., 1981, The effect of fluorine on liquidus phase relationships in the system Qz-Ab-Or with excess water: Contributions to Mineralogy and Petrology, v. 76, p. 206–215.

McConnell, D. A., and Gilbert, M. C., 1990, Cambrian extensional tectonics and magmatism within the Southern Oklahoma Aulacogen: Tectonophysics, v. 174, p. 147–157.

Merritt, C. A., 1958, Igneous geology of the Lake Altus area, Oklahoma: Oklahoma Geological Survey Bulletin 76, 70 p.

Merritt, C. A., 1965, Mt. Scott Granite, Wichita Mountains, Oklahoma: Oklahoma Geology Notes, v. 25, p. 263–272.

Mudge, M. R., 1968, Depth control of some concordant intrusions: Geological Society of America, v. 79, p. 315–332.

Myers, J. D., Gilbert, M. C., and Loiselle, M. C., 1981, Geochemistry of the Cambrian Wichita Granite Group and revisions of its lithostratigraphy: Oklahoma Geology Notes, v. 41, p. 172–195.

Polk, T., 1948, A study of the igneous rocks of the Devils Canyon Mountain Group, Wichita Mountains, Oklahoma [M.S. thesis]: University of Oklahoma, 87 p.

Pollard, D. D., 1973, Derivation and evaluation of a mechanical model for sheet intrusions: Tectonophysics, v. 19, p. 233–269.

Pollard, D. D., and Johnson, A. M., 1973, Mechanics of growth of some laccolith intrusions in the Henry Mountains, Utah, II: Tectonophysics, v. 18, p. 311–354.

Powell, B. N., and Phelps, D. W., 1977, Igneous cumulates of the Wichita Province and their tectonic implications: Geology, v. 5, p. 52–56.

Powell, B. N., Gilbert, M. C., and Fischer, J. F., 1980, Lithostratigraphic classification of basement rocks of the Wichita Province, Oklahoma: Geological Society of America Bulletin, Part I, Summary: v. 91, p. 509–514; Part II, v. 91, p. 1875–1994.

Russell, I. C., 1896, On the nature of igneous intrusions: Journal of Geology, v. 4, no. 2, p. 177–194.

Schoonover, F. E., 1948, The igneous rocks of the Fort Sill Reservation, Oklahoma [M.S. thesis]: University of Oklahoma, 123 p.

Scull, B., 1947, A further study of the igneous rocks in the Granite-Lugert area, Oklahoma [M.S. thesis]: University of Oklahoma, 65 p.

Smith, R. L., Bailey, R. A., and Ross, C. S., 1961, Structural evolution of the Valles Caldera, New Mexico, and its bearing on the emplacement of ring dikes, article 340, *in* Short papers in the geologic and hydrologic sciences: U.S. Geological Professional Paper 424-D, p. D145–D149.

Steiner, J. C., Jahns, R. H., and Luth, W. C., 1975, Crystallization of alkali feldspar and quartz in the haplogranite system $NaAlSi_3O_8$-$KAlSi_3O_8$-SiO_2-H_2O at 4 kb: Geological Society of America Bulletin, v. 86, p. 83–98.

Taylor, C. H., 1915, Granites of Oklahoma: Oklahoma Geological Survey Bulletin 20, 108 p.

Tilton, G. R., Wetherill, G. W., and Davis, G. L., 1962, Mineral ages from the Wichita and Arbuckle Mountains, Oklahoma, and the St. Francis Mountains, Missouri: Journal of Geophysical Research, v. 67, p. 4011-4019.

Tollo, R. P., Lowe, K. T., Arav, S., and Gray, K. J., 1991, Geology of the Robertson River Igneous Suite, Blue Ridge Province, Virginia, *in* Schultz, A., and Compton-Gooding, E., eds., Geologic evolution of the eastern United States. Virginia Museum of Natural History, Guidebook no. 2, p. 229–262.

Turcotte, D. L., 1982, Magma migration: Annual reviews of Earth and Planetary Sciences, v. 10, p. 397–408.

Tuttle, O. F., and Bowen, N. L., 1958, The origin of granite in light of experimental studies in the system $NaAlSi_3O_8$-$KAlSi_3O_8$-SiO_2-H_2O: Geological Society of America Memoir 74, 153 p.

Vidrine, D. M., and Fernandez, L. A., 1986, Geochemistry and petrology of the Cold Springs Igneous Breccia, Wichita Mountains, Oklahoma, *in* Gilbert, M. C., ed., Petrology of the Cambrian Wichita Mountains Igneous Suite: Oklahoma Geological Survey Guidebook 23, p. 86–106.

Watson, E. B., and Harrison, T. M., 1983, Zircon saturation revisited: Temperature and composition effects in a variety of crustal magma types: Earth and Planetary Science Letters, v. 64, p. 295–304.

Weaver, B.L., and Gilbert, M. C., 1986, Reconnaissance geochemistry of silicic igneous rocks of the Wichita Mountains: The Wichita Granitic Group and the Carlton Rhyolite Group, *in* M. C., Gilbert, ed., Petrology of the Cambrian Wichita Mountains Igneous Suite: Oklahoma Geological Survey Guidebook 23, p. 117–125.

Weertman, J., 1980, The stopping of a rising, liquid-filled crack in the Earth's crust by a freely slipping horizontal joint: Journal of Geophysical Research, v. 85, no. B2, p. 967–976.

Zietz, I., compiler, 1982, Composite magnetic anomaly map of the United States. Part A Conterminous United States: United States Geological Survey, GP954A, 2 sheets, scale 1:2,500,000.

REVISED MANUSCRIPT RECEIVED FEBRUARY 13, 1994
MANUSCRIPT ACCEPTED BY THE SOCIETY JANUARY 16, 1996

Index

[Italic page numbers indicate major references]

Typeset and printed in U.S.A. by Johnson Printing, Boulder, Colorado